FAO FORESTRY PAPER 17

economic analysis of forestry projects

prepared by
hans m. gregersen
and
arnoldo h. contreras

FOOD AND AGRICULTURE ORGANIZATION OF THE UNITED NATIONS
Rome 1979

First printing 1979

Second printing 1981

The designations employed and the presentation of material in this publication do not imply the expression of any opinion whatsoever on the part of the Food and Agriculture Organization of the United Nations concerning the legal status of any country, territory, city or area or of its authorities, or concerning the delimitation of its frontiers or boundaries.

M-30

ISBN 92-5-100827-2

FOREWORD

This publication has been produced as part of a continuing FAO programme to help transfer skills in analysis and planning to those responsible for these activities in the forestry sector in developing countries and, thereby, to assist them in strengthening their capability to identify, prepare and implement viable and useful forestry development projects and programmes.

At an early stage in this programme, which is conducted mainly through the medium of seminars and workshops, it became clear that to be effective there would need to be a publication directed to the particular needs of forestry planners. Though the literature on cost benefit analysis is already large and still expanding, most of it is still concerned with developing methodological approaches at a rather theoretical level. Few of the existing works give much practical guidance to the analyst and planner in the field in grappling with the detailed problems of identification, measurement and valuation that constitute project planning at their level. While some guides do seek to provide help of this nature, they tend to be oriented towards sectors other than forestry.

The present publication has been developed to meet this need in forestry. As the authors explain in their preface, it is intended to be a practical document - one that does provide guidance on how to solve the pragmatic problems encountered in actual practice, but which does so in a way which helps the user understand the conceptual reasons for doing what is suggested.

It has been developed over a period of several years, the early part of which was devoted to an extensive and in-depth analysis of actual experience gained in applying economic and financial analysis to projects in the forest-based sector. In order to make this experience available as fully as possible, several case studies have been developed based upon the more important and representative of these projects. These are being published concurrently with the present publication in a companion volume.1/ In addition, a second companion volume is being prepared which reviews, more broadly, the range of methods and approaches to analysis and decision-making available for use in forestry, in order to indicate what tools are available to supplement or complement economic analysis.2/

This publication has been prepared within the Planning and Investment Studies Unit of the Forestry Department of FAO, under the direction of the head of that Unit, J.E.M. Arnold. It has been written by H.M. Gregersen, Professor of Forestry and Agricultural and Applied Economics at the University of Minnesota, who spent a year as consultant to FAO for this purpose, and by A.H. Contreras of FAO. In addition we have been fortunate to benefit from the advice and help of many people both within FAO and from outside the Organization, among whom I wish to acknowledge A.M. Eid, M. Gane, A. Grayson, I.I. Holland, T. Houghtaling, J. MacArthur, S. McGaughey, J. Spears, R. Steele and P.A. Wardle. I would wish in particular to thank W.A. Ward and J. Price Gittinger for their contribution in suggesting not only relevant technical material, but also appropriate means for presenting such material in a way that is more clearly understood and usable by professionals involved with forestry project analyses.

1/ Economic Analysis of Forestry Projects: Case Studies, FAO Forestry Paper No. 17, Sup. 1, FAO, Rome, 1979.

2/ Economic Analysis of Forestry Projects: Readings, FAO Forestry Paper No. 17, Sup. 2, FAO, Rome, forthcoming.

I should also like to take this opportunity to record the particular contribution of the Swedish International Development Authority (SIDA). This publication and the companion case studies have been made possible through special budgetary contributions to FAO from SIDA for this purpose. This is but the most recent instance of SIDA's long and generous support and encouragement to FAO's training activities related to forestry and forest industries planning in developing countries.

In concluding, I wish to stress that, at least at present, there is no one accepted right way to carry out economic analyses of projects. Much remains to the judgement of the analyst, the planner and the decision-maker. This publication, therefore, does not pretend to present the final solution; nor does it represent FAO's view of what that solution should be. What we do hope it provides is a tool, which I think will be a powerful tool, to help those involved with forestry projects in exercising their judgement, and in moving towards a greater degree of unanimity about an acceptable methodology for project analysis. We welcome comments on this volume and the two companion volumes mentioned above.

M.A. Flores Rodas
Assistant Director-General
Forestry Department

TABLE OF CONTENTS

LIST OF TABLES

Table | Page

LIST OF FIGURES

LIST OF ABBREVIATIONS

a	year
ACE	allowable cut effect
ARR	alternative rate of return
B/C	benefit to cost ratio
BE	breakeven
CIF	cost, insurance and freight
CRI	consumption rate of interest
EAFP	economic analysis of forestry projects
ERR	economic rate of return
FE	foreign exchange
FOB	free on board
FRR	financial rate of return
ha	hectare
IRR	kcal (page 11-16 footnote)
kcal	kilocalorie
kg	kilogram
km	kilometre
l	litre
LC	local currency
m^3	cubic metre
$m^3(r)$	cubic metre of roundwood
MRR	marginal rate of return
NPW	net present worth
OC	opportunity cost
OER	official exchange rate
PV	present value
SER	shadow exchange rate
SEV	soil (or land) expectation value
w.t.p.	willingness to pay

PREFACE

Economic Analysis of Forestry Projects (henceforth referred to as EAFP) provides guidance and guidelines for those in public forestry agencies who are responsible for planning and appraising forestry projects.

For the forester with a background in economics and experience in applying it, or for the professional economist, some of the material covered will probably add little to what he already knows, although it may serve as a useful review. For the forester with no background in economic analysis and no experience with project appraisals, EAFP will probably not provide sufficient guidance for carrying out actual project analyses, unless it is used in conjunction with a training programme where concepts can be more fully explained. EAFP is written for those between these two extremes - those with some training in economics and/or practical experience in applying economics in the field and those who find themselves in positions where they have responsibilities for appraising projects and providing information to decision-makers on the economic values associated with alternative forestry project opportunities.

Since EAFP is written primarily for those working in forestry, it is assumed that forestry concepts and terminology are understood by the reader. For this reason technical forestry topics such as mensuration, silviculture, and engineering are not discussed. EAFP is meant to complement other works in the forestry field. Forestry involves some conditions not found in most other sectors. For example, the long time period involved in growing trees relative to production of most other goods and services makes the time factor and uncertainty in project analysis take on a greater significance. The present guide emphasizes these topics and others that are particularly relevent to forestry projects.

The purpose is to provide a practical, workable approach to economic analysis of forestry projects. Unfortunately, what can be and has been done in practice tends to differ from what ideally and theoretically should be done to make a complete economic analysis. The overall approach presented broadly reflects what is being done in actual project analyses. Some of the suggested elements have not been used in practice yet, at least not in forestry project studies.

EAFP does not contain a rigorous methodology which has to be followed step by step in order to come out with a "good" economic analysis of a project. While in some chapters specific guidelines are suggested for a particular approach to a problem, this has been done for the sake of clarification and not to suggest that the analyst follow them exactly in each appraisal. A good analysis depends greatly on the circumstances surrounding a project and the analyst's judgement based on a thorough technical understanding of the potential approaches to various problems encountered in an analysis. EAFP contributes to the latter condition. The analyst will have to decide how well the various guidelines are applicable under the particular circumstances surrounding the project being analysed.

It is well recognized that decisions on projects generally are based on a number of different criteria and considerations, some economic and financial, but others which relate entirely to social, political, organizational, and environmental considerations.

The relative weight given to any particular criterion or objective will vary with the decision-making situation. However, in most instances consideration is given to economic factors and in many cases such factors influence heavily the decisions concerning public forestry projects. Thus, economic analysis of projects deserves special attention, and it is with this in mind that EAFP has been written.

One further point needs to be made at this stage. Economic analysis should be an important component of decision-making for all forestry activities, whether or not they are described and presented as projects. The concept of a project, as an identificable and separable set of inputs and outputs and the activities which transform inputs into outputs, is useful as a basis for investment decisions. However, in practice only relatively large and new activities usually get formally defined as projects. Much of what is undertaken in forestry is in the nature of incremental additions to continuing activities, often made up of small component parts which are repeated from year to year, which are not analyzed in any depth each time a decision is made to undertake them. Yet analytically they are no different from the larger "projects", and it is equally important to know about their economic impact and efficiency.

A full-scale economic analysis is unlikely to be either warranted or needed for each activity on each occasion that it is considered. For incremental investments in an ongoing programme rough guidelines or rules of thumb can often be developed which will suffice to indicate the economic validity of a particular action in a particular situation. However, a detailed economic analysis will be needed in order to develop the guidelines if the latter are to be useful. EAFP, therefore, is intended to help facilitate the wider application of economic analysis not just to those activities designated as "projects", but to all forestry activity planning.

Chapter 1

INTRODUCTION

1.1 THE NATURE OF THE PROBLEM

Forestry and forest industry activities are basic to most countries. In some areas the focus is on conserving forests and building up new forest resources to meet future requirements for forest products and to protect land and environments from destruction. In other areas, the emphasis is on increased utilization of abundant forest resources to gain needed products and perhaps export revenues which can be used to purchase other goods and services. Regardless of the situation, governments are developing a greater interest in better utilization of forests, both for the production of goods and for various environmental and protection services which forests can provide. With this increasing interest in the forest has come a greater awareness that the potentials are limited in terms of satisfying all demands on the forest at one time. Increased utilization of existing forests and establishment of new forests require new investments of a nation's other limited resources - land, managerial and technical skills, and capital. There are a number of competing uses for these other resources, both within the forestry sector and in other sectors in the economy. Thus governments have to develop some means of choosing between alternative uses for the same limited resources.

The usual approach involves setting up some criteria for choice and then appraising or evaluating each proposed alternative use of resources in terms of these criteria. To aid in this process, a whole body of concepts and techniques has evolved under the heading of "project planning, evaluation and analysis", where the term "project" refers to a particular use of resources that is to be evaluated. A project involves inputs (costs) and outputs (benefits) and the measures and activities which transform the inputs into outputs. The project becomes something concrete which can be identified with specific purposes. The process of project planning involves identifying alternative means for achieving a given purpose or objective, narrowing down the alternatives to the one which seems best to meet this purpose, detailed design of that chosen alternative and then appraisal of the chosen alternative in terms of the relevant criteria for acceptance which relate to how the project affects society in terms of its objectives.

A major function of project planning, from start to finish, is to generate information on alternatives being considered in order to ascertain what their effects will be. This function is called analysis. Most project planning exercises involve a number of different types of analyses to provide different types of information on the effects of the project in terms of various specific objectives. One major objective associated with public projects is to increase the aggregate of goods and services available for society from the use of the nation's limited resources. This is the "economic efficiency" objective, and the analysis which looks at a project in terms of this objective is commonly called an "economic efficiency analysis" or an "economic analysis". This is the subject of EAFP. Specifically, EAFP attempts to explain how an economic analysis of a forestry project is carried out and how the results are used in the various stages of the project planning process.

EAFP stops at the point where a finished appraisal of a given project is completed. How the results of final appraisals are used in decision-making - i.e., how the result of an economic analysis is weighted and considered by decision-makers in relation to all the other relevant objectives and constraints associated with project choice - is a separate matter that deserves separate treatment. Decision-making is a highly complex subject which cannot easily be covered in one chapter. Thus, all EAFP does is to point out what an economic appraisal should include in order to be considered complete, relevant and useful to decision-makers. To repeat, EAFP is primarily concerned with how economics is used in developing and appraising a particular forestry project, from the time the initial idea is put forward until a final project design is presented to decision-makers.

1.2 THE NATURE OF FORESTRY PROJECTS

Projects in the forest-based sector vary widely in nature, scope and size from a small fuelwood plantation to a large integrated forestry and forest industry complex including wood production, harvesting, transport, processing and marketing, and from an industry project to a national park or a watershed protection project.

The differences between forestry projects and other types of projects are more a matter of degree than uniqueness. However, there are some particular features that tend to characterize forestry projects. They are:

(a) a long production period or period between the time an initial investment is made and output results (e.g., in the case of some plantation projects);

(b) the tree is both the production unit and the product (both the "factory" and the output);

(c) related to (a) and (b) is the characteristic of one-way flexibility in production. A tree can be left to grow, but once it is harvested, the "factory" is also destroyed and it can take a long time to build it back to a given level. Thus, there is great flexibility in terms of when to harvest, but little flexibility in terms of building up inventories;

(d) in contrast with production processes involving machinery and engineering controls, biological production processes tend to be characterized by a great deal of heterogeneity within any given system. Such systems are characterized by great variation in output. Uncertainty can be substantial;

(e) any given forest tends to have multiple uses e.g., timber production, wildlife protection, watershed and soil protection, aesthetics and recreation. Thus the problems of joint products, conflicting objectives and trade-offs between uses take on particular importance in forest project analyses.

1.3 OBJECTIVES, CONSTRAINTS AND PURPOSES ASSOCIATED WITH FORESTRY PROJECTS

The objectives associated with forestry projects will vary with the type of project, the nature and purpose of the institution which will carry out the project, the point of view adopted, and of course, the broader political, cultural and economic environment within which the project will function. All pulp and paper projects are not associated with the same objectives; this also applies to national park or watershed projects.

For the private entrepreneur or corporation, a forestry project is generally undertaken to make a profit or increase the chances for making future profits, or to reduce the danger that expected profits will not be forthcoming. By definition, most private projects are associated mainly with some aspect of the profit motive.

The public sector is interested in undertaking forestry projects, or supporting them in the private sector, for a variety of reasons associated with basic societal objectives. These can be many and varied and seldom is a public forestry project associated with only one of them. Common objectives are to:

- improve economic efficiency, i.e., increase the aggregate benefits (goods and services available for society) derived from the use of the nation's limited resources;
- improve conditions for the poorer members of society (or decrease the gap between rich and poor, e.g., through employment creation);
- increase social, political and economic stability (e.g., through improving the balance of payments situation, providing public services, or following sustained yield policies, etc.);
- improve environmental conditions and land use;
- generate increased revenues for government which can be used for various social needs.

Most public forestry projects involve a combination of these objectives and some may include all of them. Similarly, public support for private projects is generally based on the belief that such projects will contribute to a number of social objectives. How such multiple objectives are handled in practice in project planning and decision-making varies with the situation. Most commonly, some of them are put in the form of constraints on the project. A plantation project may be contemplated to provide fuelwood for consumption and to protect steep hillsides from erosion. Increased consumption may be taken as the main objective, with the environmental improvement objective expressed as a constraint on the way in which the project will be undertaken (e.g., related to management and harvesting options which are acceptable in terms of the protection objectives). Ultimately, public administrators or decision-makers have to provide the basis for weighing different objectives and setting constraints. [1]

---

[1] One constraint which is of particular interest for foresters is that associated with a sustained yield policy for public forests. This constraint is in turn related to the basic objectives of economic and social (community) stability. It may conflict with the economic efficiency objective, and decision-makers often have a difficult task in reconciling the two. One relationship between a sustained yield policy and economic efficiency is discussed further in Section 3.3.3.3.

These concepts might seem foreign to foresters used to thinking in physical terms of growing trees, producing wood, protecting soil, etc. However, there is a direct link between these types of physical activities and the objectives mentioned. Forestry activities are not financed and carried out for the benefit of the forest. They are carried out to satisfy human wants, whether these be more housing or paper, greater enjoyment from looking at better forests, or protection of soils and watersheds to prevent decreases in production of food, fibre or water or deterioration of water quality. The stated objectives reflect that forestry investments are ultimately made to increase human satisfaction.

Thus, when the purpose of a particular forestry project is to grow 200 hectares (ha) of trees for fuelwood, this is one step removed from the objective of increasing the satisfaction of human wants, though it may be perfectly consistent with it. A purpose such as growing trees for fuel provides no guidance in terms of getting at a solution to the basic problem of resource allocation or economic efficiency. Thus, in terms of the economic efficiency analysis, the analyst must also identify how the increased wood will affect society and its wants. In other words, the basic question asked by the economic analyst is whether or not this use of resources (in producing fuelwood) increases aggregate benefits to the nation (the value of goods and services available for consumption) more than some alternative use of the resources involved.

In reality, the answer cannot be divorced from the other objectives which society has, nor from the various constraints which influence decisions in a country at any given time. Thus, the process of developing, analysing and evaluating project alternatives becomes something much more complex and diffuse than merely looking at projects in terms of economic efficiency. The term project planning is used to describe this broader process.

1.4 PROJECT PLANNING [1]

A main function of project planning in the public forestry sector is to identify and design forestry projects that are workable, effective and consistent in moving the nation toward its various objectives.

In a more practical sense this function involves finding the technically feasible solution to a given situation which gives an acceptable economic return to society (i.e., is economically efficient), is adapted to the institutional and managerial situation of the country, and which can be financed with the resources available. Thus, most public forestry projects generally end up being compromises that move the nation ahead in terms of one or more dominant objective(s), while avoiding conflict with all other objectives (i.e., it meets certain constraints associated with these other objectives). The compromise solution is reached through a process of give and take,

---

[1] This section provides only a very brief overall view of project planning in relation to the subject of EAFP. The reader desiring a more detailed view is referred to FAO, 1974.

during which the project idea is looked at broadly from each point of view, or in terms of each objective and constraint, conflicts in views are identified, alternatives are examined and adjustments are made until finally all points of view (objectives and constraints) are satisfied at least at some minimum acceptable level, where "acceptable levels" are defined implicitly or explicitly by decision-makers and policies/laws. As the give and take process progresses, alternatives are narrowed down and the amount of detail and the level of analysis required for the project increase. Finally, a detailed design for one well-defined alternative and a plan for implementing that alternative evolve. A decision is then made whether or not to undertake the project, based on a comparison of its desirability with the desirability of other project opportunities requiring the same limited resources.

The project planning process is sometimes described as an orderly, sequential series of steps that are separable and well defined. While such a neat view is appealing, it is misleading. In reality project planning is a flexible continuing process of successive approximations and refinements as different points of view and objectives are considered, argued, and reconsidered continuously from the point in time when the initial project idea and a range of alternatives are identified, until efforts focus on one alternative, and a final decision is made on whether or not to implement that alternative. Experience shows that even after implementation, the process of give and take and change continues. There are very few projects that are implemented in exactly the way they were envisaged when the decision to go ahead with them was made. Flexibility, adjustment and readjustment characterize most real world project planning exercises. There is no one well-defined way to plan a project.

While it is unrealistic to view the overall process of project planning as a neat set of sequential steps characterized by orderliness, it is necessary to recognize the need for and existence of orderliness in the analytical efforts which provide the background information on which decisions are made as a project evolves. Indeed, the firm belief that a systematic orderly approach to analysis is possible and desirable is an underlying reason for EAFP.

1.5 USE OF ANALYSIS IN PROJECT PLANNING

The term "analysis" as used here means an examination of a project to distinguish its component parts and the relationship of those parts to the whole. During the entire process of project planning, from the time an initial project idea is identified, various types of analyses are being carried out in order better to understand the project from different points of view and to help guide the process of moving toward the best means of achieving objectives.

The types of analyses carried out vary in orientation and scope depending on the objectives and constraints associated with the project and the stage in the planning process. Most projects involve a number of objectives and constraints and, therefore, a number of different types of analyses. For example, if a pulp and paper project is being considered by the public sector, it is likely being analyzed from a technical point of view (related to the constraint that it must be technically feasible), from a budget point of view (related to the constraint that it must be consistent with existing and expected institutional conditions, availability of resources, etc.), from an environmental point of view (related to an environmental improvement objective or environmental maintenance constraint), and from an economic efficiency point of view (related to the objective of increasing the benefits which will flow from the use of the nation's limited resources). There also may be analyses associated with many other aspects of the proposed project, e.g., impacts associated with local development, balance of payments, employment, markets, etc.

Ideally, one integrated analysis would be developed to deal with all these factors and the various project objectives at the same time. In reality, all of these factors cannot be considered in one analysis, partly because some will involve different units of measurement than others, partly because the various objectives with which the analyses are associated are not complementary and there is no practical, realistic way of combining or weighting the various objectives in quantitative terms, and partly because different analyses are required at different times in the planning process. Generally different analyses (related to the different objectives and constraints) are carried out independently or separately by specialists, or by one or a few general foresters in the case of smaller less complex forest-based projects or activities. At best, these specialists are interacting throughout the process of planning and developing the project. More often, some analyses follow others and interaction takes place after initial results have been obtained.

Economic analysis is just one of the inputs into this process. Its importance depends on the importance given to the economic efficiency objective and the way in which this type of analysis is introduced or used in the planning process.

1.6 ORGANIZATION OF EAFP

At an early stage in the project planning process, when alternatives for achieving a given purpose are being identified, a number of alternatives will be analysed in a partial fashion to throw light on which alternatives should be considered further and which should be rejected, i.e., how the project should be designed. The analysis may only be concerned with alternative costs, or whether a specific technology is preferable to another, etc. At a later stage, when interest focuses on one specific alternative design for achieving the project purpose, the required level of analysis is generally more comprehensive in scope. The type and level of analysis thus varies with the intended use of the results, but the basic steps and techniques are the same. With this in mind, EAFP has been divided into two main parts.

Part I deals with the steps and techniques involved in an economic analysis. Chapter 2 presents an overall view of the process together with some comments on the relationships between economic analysis and other important types of project analyses which are generally associated with the economic analysis. Chapter 3 presents a discussion on how to define project context and the scope of the analysis. Chapter 4 discusses identification of costs and benefits. In Chapter 5 the basic principles of valuation are discussed. Chapter 6 discusses valuation of costs and benefits when market prices are considered acceptable measures of value for an economic analysis. Chapters 7 and 8 discuss valuation of outputs and inputs when market prices are not acceptable, or when inputs and outputs are involved for which no market prices exist (i.e., inputs and outputs not traded in the market). Chapter 9 presents a discussion of measures of project worth, or how costs and benefits can be compared in an economic efficiency context. Recognizing that most projects involve uncertainty, Chapter 10 provides practical guidelines for considering uncertainty in an economic analysis.

The use of economic analysis in project planning varies by the stage in the planning process, and Part II explores in some detail the uses in (a) design and preparation, and (b) final appraisal.

It is in the early stages of project planning that economic analysis can have its greatest impact. "If economic analysis is to make a maximum contribution to the attempt to ensure that scarce resources are used to best advantage for the country, it should be used from the earliest phases of this process of successive sifting and narrowing down of options that are open to the country."[1] While an economic analysis at the final appraisal stage provides useful information for decision-making, by that time the main choices related to technology, size, location, and scope have already been made. Most developments in the forestry sector in most countries take place in an incremental fashion through gradual changes in orientations, policies regarding conservation, forest exploitation, etc. and corresponding gradual modifications in the ways in which forestry activities are carried out. Economic analysis, by providing information which can be used in project identification and design, can help shape such gradual changes. If economic analysis does not enter the project planning process until a well-defined project with a detailed design is presented for final appraisal and decision, then its only contribution will be to help shape the decision as to whether or not to undertake the project. At this stage it is generally too late to have any influence on all the incremental decisions concerning project dimensions which, when added up, could amount to a significant impact on development and increased efficiency in the use of the nation's resources.

Chapter 11 concentrates on the use of economic analysis in project identification and design, i.e., in helping to shape decisions concerning project scope (components to include and exclude), project size (in relation to economies of size associated with various activities and requirements for project output), project technology (in relation to factor scarcity and availability), and project location (in relation to regional needs and opportunities).

Chapter 12 explores the need for and usefulness of economic analysis in the final appraisal stage. It includes a description of the overall economic appraisal process for projects and gives one example from an actual project.

---

[1] Squire and van der Tak, 1975.

PART I

APPROACHES TO ECONOMIC ANALYSIS

Chapter 2

FINANCIAL AND ECONOMIC ANALYSES

AN OVERALL VIEW

2.1 INTRODUCTION

Resources are controlled by many different entities - individuals, private or public corporations, government agencies etc. Each of these allocates the resources it controls on the basis of how the returns from a project contribute to its objectives.

Private entities generally have an objective related to monetary profits, or the difference between what they have to spend or give up and what they expect to receive in money receipts from a project. The term "commercial profitability" is used to describe the relationship between outflow and inflow of funds for goods and services.

The term "financial analysis" is used to describe the type of analysis that develops an estimate of commercial profitability for a project. A financial analysis is carried out from the point of view of specific entities involved in a project. It considers the monetary returns expected by such entities from investment of their funds (resources) in a project. A financial analysis also provides information on when funds will be required (outflows) and when receipts (inflows) can be expected. This latter type of information is essential for budget planning. As such, financial analyses are also relevant for public projects.

An 'economic efficiency' analysis is in a sense merely an extension of the financial analysis concept, where the entity from whose point of view the analysis is being carried out now becomes society as an undifferentiated whole rather than a specific entity (or entities) within the society. As such, the economic efficiency analysis is also concerned with "profitability", but in this case it is the profitability from society's point of view, which is related to the return to society as a whole which can be obtained with a given use of its limited resources. This is called "economic profitability" to distinguish it from commercial profitability. Economic profitability relates directly to the economic efficiency objective defined in the previous chapter. The relationship can be characterized as follows:

Resources have value to society because there are not enough of them at a given time to satisfy all society's wants. Society desires to allocate its limited resources so that they make the greatest possible contribution to satisfying its wants for goods and services. This desire is expressed through what was earlier called the economic efficiency objective. If the existing allocation of resources can be improved, in the sense that more goods and services can be produced with given resources, then there is an improvement in economic efficiency. A use of resources (a project) which improves economic efficiency is an economically profitable project.

Just as the concept of economic profitability parallels the concept of commercial profitability, so the economic efficiency analysis parallels the financial analysis in terms of procedure. However, they differ in terms of what is included as costs and benefits and how costs and benefits are valued. In the financial analysis, benefits are defined in terms of actual monetary returns to a specific entity (or group of entities) in society from whose point of view the analysis is being carried out. These returns result from the sale or rental of goods and services in a market, and thus returns are measured in terms of market prices. Costs in the financial analysis are represented by outflows of money from the entity (ies), mainly paid out for goods and services purchased in the market. In the economic analysis, on the other hand, the concern is with what society gives up and what society gains from a project. Costs are thus defined in terms of value of opportunities foregone by society because resources are used in the project rather than in their best alternative use. Thus, costs in an economic analysis are referred to as "opportunity costs". Project benefits are defined in terms of increases in goods and services available to society as a whole due to the project. As discussed later in this chapter, these two different concepts of costs and benefits (or the differences between commercial and economic profitability) give rise to some specific differences in the ways in which costs and benefits are identified and valued in the two types of analyses.

Both economic and financial analyses are needed for public projects or private or mixed projects for which the public sector is considering support. The economic analysis is needed to provide information on whether or not the project would provide an economically efficient use of the resources available to society. The financial analysis is needed to provide information on actual amounts and timing of inflows and outflows of funds needed to undertake the project. As mentioned, this latter information is essential for budget planning and control.

Further, for a mixed public-private project or a private project being considered for public support (special permissions, subsidies, etc.), the results of financial analyses undertaken from the viewpoints of the different entities involved in the project also provide useful information on whether or not these different entities will have the incentive and the funds to undertake a project which is economically efficient.

This point goes back to the comment made earlier that specific entities control resources and make decisions on whether they want to commit the resources they control to a given project. A project can be extremely attractive from a national economic efficiency point of view, but if it is not also financially attractive to all private entities which have to commit resources to it, then it will not be undertaken as planned. A financially unattractive project can be made financially attractive if the government (the public) provides subsidies (incentives). Whether or not such subsidies are considered justifiable in a social economic context depends directly on their required magnitude in relation to the economic surplus associated with the project (economic benefits minus economic costs, appropriately adjusted to take time into account). Similarly, analyses which show that a project appears to be more attractive financially than economically may provide some indication of the desirability to tax the financial entity(ies) involved.

Since both financial and economic analyses have much in common in terms of information requirements and procedure, they are generally carried out together. The steps in a financial analysis are more straightforward to carry out and clearer in concept. Therefore they generally precede the comparable steps in the economic efficiency analysis. In practice, a step in the financial analysis is completed and then the results of this step are used as a starting point for the parallel step in the economic analysis. This is also the approach considered in EAFP. Actually there is no necessary reason why the economic analysis should follow the financial analysis steps; it is more a matter of convenience. In cases where no financial analysis has been carried out, the economic analysis procedure is exactly the same as described in EAFP.

2.2 FINANCIAL AND ECONOMIC ANALYSES: SIMILARITIES AND DIFFERENCES

A financial analysis carried out to estimate expected commercial profitability for a project involves four major steps. First, inputs bought in the market are identified in terms of when they are needed (purchased or rented). Similarly, outputs traded in the market are identified in terms of when they are sold. This information results in a "physical flow" table. Second, market prices for the inputs and outputs are estimated for the times at which inputs will be bought and outputs sold. This information is entered into "unit value" tables. The third step involves combining the information from the two previous steps into a "cash flow" table which shows the value of total inputs and outputs at the times that such values (outflows and inflows of money) accrue to the entity from whose point of view the analysis is being undertaken. To complete the cash flow table, certain financial transactions which involve transfers of control over resources (but no use of real resources) are added to the table. These include such items as taxes and loan repayments (outflows) and subsidies and loan proceeds (inflow) plus a number of other expenses or receipts, depending on the project and the purpose for the analysis. Finally, the inflows and outflows of funds are totalled by years in which they occur to arrive at a net cash inflow (outflow) line. The fourth step then involves using these net value figures by years to derive some measure(s) of commercial profitability. [1]

The economic efficiency analysis involves the same basic four steps, and each of the first three steps can start with the results of the comparable step in the financial analysis, if certain key differences are kept in mind. The differences and the adjustments needed to develop the economic analysis are outlined in Table 2.1. The following paragraphs summarize each step, and the remaining chapters in Part I discuss the steps in detail.

2.2.1 Physical input and output tables - identifying inputs and outputs

The physical input and output tables for the financial analysis include those inputs which have to be purchased or which are owned by the entity and have an opportunity cost and those outputs which are sold by the entity(ies) from whose point of view the analysis is being carried out.

2.2.1.1 Adding indirect effects

To arrive at the appropriate physical flow tables for the economic analysis certain additional inputs and outputs may have to be added to reflect the fact that the analysis is considering all effects of the project (Item 1 in Table 2.1). Some of the relevant economic effects of the project may not have been included in the financial accounts because they occurred outside the market (i.e. they are not directly traded in a market) and do not

---

[1] Some examples of this process are provided by the case studies in Appendix A (from FAO, 1979). See also Chapter 12.

directly affect the project cash flow. These are commonly referred to as "indirect effects" or "externalities", i.e., effects external to the project from a financial point of view since they involve no direct monetary inflows or outflows. (They are also often referred to as "spillover effects", "non-market effects"). This concept can be illustrated with examples.

If a pulp and paper project creates pollution downstream from the mill (lowers the quality of water available to downstream users) and the entity establishing the mill does not have to pay to clean up or avoid that pollution, then this is not considered as a cost in the financial analysis undertaken from that entity's point of view. However, in the economic analysis, this is a relevant cost, since it represents a reduction in the availability of clean (or cleaner) water to members of society due to the project. In an economic analysis, a cost is any reduction in the availability of resources or goods and services (quantity or quality)due to the project, regardless of who is affected by the reduction. Similarly, if a project produces a pleasing environment (which people value), then this is a benefit in the economic analysis, but not in the financial analysis, unless consumers pay for it to the entity carrying out the project. Benefits in an economic analysis are represented by increases in goods and services available to society due to the project regardless of who actually receives them and who pays for them.

In cases where a number of indirect effects are identified, or where it is difficult to quantify and/or value them, the analyst may want to develop a separate table which lists these indirect effects for the years in which they are expected to be relevant. Even if they cannot readily be quantified and valued, they should be mentioned explicitly in the analysis.

In developing the economic accounts from the financial ones, the analyst should look carefully at all the supporting infrastructure needed to make the project work. Sometimes, for a private project, such infrastructure costs are not included in the financial analysis if government has promised to provide supporting facilities - roads, power generation, water, housing, etc. with public funds. Thus, if the financial analysis has been carried out only from the point of view of a private entity involved in the project, such inputs are indirect and have been excluded, since they will not involve outflows or loss of funds from the entity. However, if they are needed due to the project (i.e. they would not have been provided in the absence of the project), then they represent a use of resources and are relevant in terms of the economic analysis undertaken from society's point of view.

Table 2.1

RELATIONSHIP BETWEEN STEPS IN A FINANCIAL AND AN ECONOMIC ANALYSIS

Financial Analysis	Economic Analysis	Comments
	1. Develop physical flow tables (inputs and outputs)	
Direct inputs provided by the financial entity and outputs for which the entity is paid are included.	In addition to direct inputs and outputs, indirect effects are included, i.e., effects which are not included in the financial analysis since they are not directly traded in a market. These are effects on others in society.	See Chapter 4
	2. Develop unit value tables	
Market prices are used. For inputs and outputs which occur in the future, future market prices are estimated.	Consumer willingness to pay (w.t.p.) is used as the basic measure of value. In cases where market prices adequately reflect w.t.p., such prices are used. In other cases, "shadow prices" are estimated to provide the best measure of w.t.p.	See Chapters 5 - 8
	3. Develop cash flow/economic value flow tables	
Inputs and outputs are multiplied by market prices to arrive at total costs and returns which are then entered in the cash flow table. Transfer payments (taxes, subsidies, loan transactions, etc.) are added to the cash flow table.	Inputs and outputs are multiplied by unit economic values to arrive at total economic costs and benefits which are then entered in a total value flow table. Transfer payments are not treated separately, but included as part of economic costs or benefits as appropriate.	See Chapter 9
	4. Calculate measures of project worth	
Using cash flow table, calculate chosen measures of project worth or commercial profitability. Test results for uncertainty by varying values of key parameters in a sensitivity analysis.	Calculate chosen measures of economic efficiency or economic worth, using the information in the total value flow table. Test results for uncertainty by varying values of key relationships/parameters in a sensitivity analysis.	See Chapters 9 and 10

2.2.1.2 Using the "with and without" concept

Any effect of a project should be identified and measured on the basis of the difference in a given situation with and without the project. This "with and without" concept is basic to project analysis. It is important to keep in mind that the situation as it exists today would likely not remain the same in the absence of the project. Thus, the "before" project situation should not be taken to be the same as the "without" project situation in identifying project effects. Changes would likely take place without the project and these need to be estimated. An example will illustrate this point.

In a soil conservation project to restore fertility to a moderately eroded piece of land, the benefit is sometimes estimated as the difference between production with the present level of moderate fertility and production which will be achieved with the improved fertility associated with the project. However, assume that if the conservation project were not introduced, the situation without the project would eventually deteriorate to one of total loss of production, due to the cumulative nature of the erosion process. The correct benefit measure in this case would include the difference between no production and the level achieved with the project. (The timing of the deterioration process without the project would have to be considered in deriving the output quantities to enter into the physical flow table.) It would not be the difference between the present moderate level of production and full production. If the analyst ignored the "with and without" concept, he would understate the benefits due to the project.

In applying the 'with and without' concept to economic costs (or "opportunity costs"), particular care has to be taken to identify properly the best actual opportunity foregone, i.e., the best alternative use of an input that would actually have taken place in the absence of the project, taking into account the various institutional (social and political) constraints or policies that are expected to exist.

> "The technical opportunities that cannot be made use of, given social constraints, are not real opportunities, and the identification of costs as maximum benefits sacrificed must be based on real feasibility... The starting point of all project evaluation is to ask the question: If we did not choose the project, what difference would it make? And the assessment of the differences that would result depends on a clear identification of political and social constraints that limit economic opportunities." 1/

When substantial change is not expected without the project during the period of the project, the analyst may be justified in saving time and money by assuming the "without" situation to hold constant over time. However, there are many forestry projects - which generally involve long time periods - where ignoring potential changes over time without the project will involve some major under - or over-estimation of costs and benefits due to the project.

The above provides a summary of the main differences between the physical flow accounts for the financial and economic analyses. Details are provided in Chapter 4, which deals with identification of costs and benefits in an economic analysis.

1/ UNIDO. 1972, p. 53.

2.2.2 Unit value tables - valuing inputs and outputs

The next step in both the financial and economic analyses is to develop unit value tables for the inputs and outputs with due consideration given to trends in prices and forecasts or projections of future prices. 1/ (Item 2 in Table 2.1) When inputs and outputs have been identified for the economic analysis, a large overlap will likely remain with the financial analysis, i.e., most of the input and output items included in the financial analysis will also be represented by similar ones in the economic accounts. However, there can be differences in the values attached to such common inputs and outputs and these have to be considered in deriving the economic unit value tables.

Unit values used in the financial analysis are market prices. In the economic analysis, outputs are valued on the basis of consumers' willingness to pay (w.t.p.) for them. Market prices may or may not adequately reflect w.t.p. Similarly, inputs in the economic analysis are valued on the basis of consumers' w.t.p. for the benefits (goods and services) foregone by using resources in the project rather than in their best alternative use, i.e., their "opportunity costs" or maximum benefits foregone. Market prices for inputs may or may not provide an adequate measure of opportunity cost in a given project environment. Chapter 5 provides some guidelines on how to determine whether or not to use a market price for an input or output in the economic analysis. It is not an easy task nor one that is amenable to precise rules. Much depends on what information is available and what it will cost in time and money to obtain additional information on which to base a revaluation of an input or output.

If it is decided that a market price provides an adequate reflection of economic value for an input or output, then it also can be entered in the economic unit value table. Chapter 6 provides some guidelines for proper estimation of market prices.

2.2.2.1 Shadow prices

If it is decided that a market price does not provide an adequate reflection of economic value, then a more appropriate value (related to the definitions above) has to be derived. This process of revaluation is called "shadow pricing" and the resulting values are called "shadow prices". Detailed guidelines for shadow pricing inputs and outputs are presented in Chapters 7 and 8, together with the appropriate concepts of value on which the guidelines are based.

Indirect effects, or those costs and benefits entered in the economic analysis which were not included in the financial analysis will obviously have to be shadow priced, since by the previous definition (see Section 2.2.1) they are not directly traded in a market. In some cases, market prices can be used as shadow prices for indirect effects.

1/ See Chapter 6 for discussion of forecasting future values.

2.2.2.2 Income distribution and economic value measures

A major assumption underlying this approach to valuation of benefits and costs for the economic analysis is that the willingness to pay one dollar (or monetary unit) by one individual is as valuable to society as the willingness of any other individual to pay one dollar for the same good or service or a different good or service. In contrast with the financial analysis, the economic efficiency analysis does not distinguish between who loses and who specifically gains consumption benefits due to the project. The analysis is essentially neutral in terms of the distribution of benefits and costs among members of society.

The implicit assumption is that the existing distribution of income (and therefore purchasing power) is correct from society's point of view. Increasingly, this assumption is being questioned and projects are being designed with the explicit objective of redistributing income (consumption opportunities) from the richer to the poorer members of society. This objective can be considered in the economic analysis by attaching higher weights to benefits received and costs incurred by the poorer members of society.

Recently, attempts have been made to combine efficiency and redistribution considerations into one integrated system of social economic analysis.[1] While such systems are conceptually sound, they are not at a stage where they can be applied realistically in practice in most cases, mainly due to the lack of generally acceptable income weights for different groups in society. EAFP follows the practice of analysing a project in economic efficiency terms first (i.e., assuming equal weights). A redistribution analysis can be undertaken separately as warranted by the circumstances. Henceforth, the term "economic analysis" is used as meaning the same as an "economic efficiency analysis", the separate analysis of income redistribution as an "income redistribution analysis", and the combined analysis as a "social economic analysis".

To summarize, derivation of unit values for the economic analysis involves a two stage process. First, for inputs and outputs traded in a market, a judgement is made on the adequacy of market prices used in the financial analysis as measures of economic value. This is discussed in Chapter 5. If they are judged to be adequate, they are entered in a table showing unit economic values. Chapter 6 provides a discussion on use of market prices. If they are judged to be inadequate, then they are treated in the second stage just like indirect effects for which no market prices exist. This second stage involves a judgement on whether or not an acceptable shadow price can be developed. If the judgement is negative, then it is better to treat the effect in a qualitative or physical quantitative fashion, by making explicit mention of the effect in the economic analysis report. The analyst should not try to develop some spurious value measure which will merely serve to confuse and mislead decision-makers. If the judgement is that a shadow price can be developed, then the analyst proceeds to do so and the resulting values are entered in the unit value table. This second stage in valuation is detailed in Chapters 7 and 8.

1/ Cf. IDB 1977, UNIDO 1972, Little and Mirrlees 1974, Squire and van der Tak 1975.

2.2.3 Total value flow tables

Once market prices have been estimated in the financial analysis, they are multiplied by the quantities of inputs and outputs from the physical flow table(s) and entered in a "cash flow" table in the financial analysis (Item 3 in Table 2.1). The cash flow table provides a picture of the inflows and outflows of cash expected for a given project alternative by years or other time intervals (see Chapter 9).

In order to arrive at a picture of the total financial cash flow from a particular entity's point of view, it is also necessary to add any direct subsidies received or loan proceeds as receipts in the table and any direct taxes, other government payments, and loan repayments as costs or expenditures in the cash flow table. These are all called "transfer payments", since they involve transfers of control over resources but do not involve any direct changes in the use of real resources in the project as defined.

Since the economic analysis is concerned only with real resource flows (and real output flows), transfer payments should not be shown separately in the total value flow tables for the economic analysis. This point is explained further in Chapter 9, together with some examples and guidelines for treatment of transfer payments.

To summarize, in moving from the financial cash flow table (either for a commercial profitability analysis or an analysis of return on a particular entity's equity capital) to an economic value flow table for a project, adjustments for direct transfer payments associated with the project have to be made in the economic value flow table, since they do not represent any change in real resource use or project output for a given project alternative. There are some exceptions in cases where taxes, royalty payments, loans, etc., involve foreign exchange or payments outside the country. These are discussed further in Chapter 9. They are relevant for forestry projects in some cases.

2.2.4 Economic profitability or efficiency

Once inputs and outputs have been properly identified and valued for the economic analysis, there is no further advantage to be gained from using the results of the financial analysis in completing the economic analysis. As indicated in Item 4 of Table 2.1, the financial analysis involves calculation of one or more measures of commercial profitability. Parallel, but completely independent calculations are made to determine the economic efficiency or economic profitability of a project.

A given project (use of resources) is considered efficient in economic terms if:

- its benefits are equal to or greater than its costs;
- benefits are at least equal to costs for each separable component of the project; [1]
- there is no known lower cost means actually available (given existing constraints) to achieve the same project effects (or benefits).

---

[1] Separable in the sense that a project can exist (technically) with or without the component. For example, a plantation project can exist with or without a fertilizer component. Thus, the fertilizer component is separable in terms of the above definition (see Chapter 3).

In the case of all three conditions, appropriate adjustments have to be made to take timing of costs and benefits into account (see Chapter 9). If any of the three conditions or criteria are not met, then the project alternative does not represent an economically efficient use of resources, and it should be rejected or revised in terms of the economic efficiency objective. As mentioned a number of other objectives will likely influence the final decision on a project, but these are not considered in the economic efficiency analysis.

The rationale for the first condition is the easiest to grasp. If the costs for a project exceed the benefits that will flow from it, then this means that society will have a net loss in value of goods and services available for consumption if the project is undertaken.

The rationale for the second condition concerning separable components of a given project can be illustrated with an example. Assume a project designed to produce sawnwood and plywood (the two have been included in the same project for administrative or other reasons). It may be that one of these separable components - say sawnwood production - has costs higher than benefits if considered alone. The plywood component returns may be high enough to carry this loss and still make the overall project acceptable. However if the sawnwood component were removed from the project, then the total net return would increase and the benefits to society would be greater per unit of resources committed. If the components are not analysed separately, then this information will never come to light and the second condition for economic efficiency cannot be tested.

Commercial profitability of a project also could be improved if unprofitable separable components were eliminated from the project. Although the overall project might be able to carry the unprofitable component, this would certainly not make financial sense, if the component could be eliminated without making the rest of the project unfeasible. Thus, since there is a parallel condition related to financial profitability, it is likely that separable components have also been treated separately in the financial analysis. In that case, the economic analysis and the identification of direct inputs and outputs by components can proceed using the results of the financial analysis. Otherwise, the analyst will have to go back to the basic project technical studies to determine the interrelationships between components and the separability of components in terms of their costs and benefits.

The third condition makes intuitive sense. If there is a known cheaper way of achieving a given purpose than the alternative being analysed, then it would make little sense not to use that cheaper way provided it achieves exactly the same benefits of effects. Thus, in comparing the costs of alternatives, adjustments may have to be made in the benefit rows of the value flow table as well as in the cost rows. For example, clearing land in the tropics for plantations may be accomplished at lower economic cost using heavy machinery rather than labour, but there can be some negative indirect effects (costs) associated with the heavy machinery in terms of environmental deterioration. These costs also have to be considered in the economic analysis.

How the economic analyst deals with the three conditions or criteria in an analysis depends on which of the followkng two situations is relevant:

The first situation or possibility is one where the project purpose is not now being met and no decision has been made at the time of analysis as to whether or not the project output should be produced and added to the supply available to society. In this case, the analyst has to look at the project in terms of all three conditions for economic efficiency. In other words, in this case the decision-maker not only is interested in knowing whether the proposed project is the lowest cost means of achieving the project purpose, but also whether benefits exceed costs by a great enough margin to make the project worth undertaking, i.e., whether it is worthwhile to add the project output to the goods and/or services available to society. This first situation is the one commonly discussed under the heading of "cost-benefit" analysis.

The second situation is one where the decision has already been made that the project purpose will be achieved or will continue to be achieved. For political or other reasons the project benefits are desired and will be produced, perhaps because they traditionally have been provided or because society (through its government) feels that the benefits should be provided (for example a minimum level of fuel for poor members of society who cannot afford to pay for it). Benefits with or without the project will be the same. Thus, in terms of the economic efficiency analysis, the main task is to concentrate on the third condition, or a comparison of the costs of alternative means of achieving the project purpose, but taking different indirect effects into account. To make this comparison the analyst uses what is called a "least cost analysis" or a "cost-effectiveness analysis", which simply means that costs of known feasible alternatives are compared to find the lowest cost means of achieving the project purpose of effects. The lowest cost alternative is the most economically efficient alternative.

In this second situation, the project could involve production of an output that will substitute for a good or service that is already being consumed (and produced by alternative means, either domestic production or imports). If the consumption from the existing source is expected to continue in the absence of the project, then the relevant task of the economist is to focus on a comparison of the opportunity costs of the existing source of supply with those associated with the proposed project source of supply. If the project costs are lower than those for any known, feasible alternative, then it should be accepted in terms of the economic efficiency objective.

The point to stress in terms of the two situations is that all projects should be subjected to least-cost analysis (consideration of the third condition for economic efficiency), while only some involve full scale cost-benefit analyses. The differences will become clearer later in Part I and in the discussion in Part II of the appropriate treatment of the two types of situations in project planning.

2.2.5 Dealing with uncertainty

One additional point needs to be mentioned in connection with estimating the economic efficiency of a project. This is the question of how uncertainty is treated in the analysis. So far costs and benefits and their appropriate measure of value have been defined and how they are used in determining efficiency. Little has been said about the empirical problems associated with identifying, valuing and comparing costs and benefits. As techniques and empirical questions are discussed in the remainder of Part I, it will become evident that a great deal of uncertainty surrounds most empirical analyses of economic efficiency. A major function of the economic analysis should be to explore the implications of uncertainty surrounding the values of project parameters for the measures of economic efficiency derived. This function is usually incorporated into what is called a "sensitivity analysis", or an analysis of the sensitivity of a chosen measure of project worth to changes in the assumptions concerning inputs and outputs and the values attached to them. The concepts and techniques are explored in detail in Chapter 10.

2.3 ECONOMIC ANALYSIS OF PROJECTS VERSUS POLICIES

Before getting into a detailed discussion of the four major steps in an economic efficiency analysis and particularly into the details on valuation, it is necessary to point out the difference between the interpretation used here for an economic efficiency analysis for a project which will exist in a given political and social environment, and an economic analysis of the policies which shape that environment (i.e., an analysis of the costs and benefits to the nation in efficiency terms associated with the existence of a policy). The two are considered here to be quite separate.

In an economic efficiency analysis for a project, costs and benefits are defined and valued in terms of the actual conditions which are expected to exist in the project environment. These conditions are influenced by government policies. Some of these policies are aimed at supporting objectives other than increasing economic efficiency. For this reason some of the policies can lead the economy away from maximum possible economic efficiency. If they were eliminated, the allocation of resources could be improved in terms of the economic efficiency objective. Some argue that only in a distortion-free environment (i.e., in the absence of all policies restricting economic efficiency) can the conditions be found which are adequate for identifying and valuing costs and benefits for an economic analysis.

While these arguments have merit, it is felt that the worthiness of a project should be estimated in terms of the difference it makes to society, given expected actual conditions and the actual available opportunities which will exist given such conditions. As mentioned in Section 2.2.2, a project that is technically feasible but cannot be implemented because of some policy restriction is not a real opportunity and should not be considered as being feasible. To do so might result in even worse distortions in resource allocation when all objectives and constraints are considered. Thus, the recommendation made here and the approach followed in EAFP is to take into account all policies which are expected to exist during the life of the project when calculating opportunity costs for inputs and valuing project outputs.

Under certain circumstances, it may also be desirable to examine the efficiency conditions that would exist without a given policy. This should be done separately. This type of analysis is called a "policy efficiency analysis". Such an analysis can be useful in two main ways:

- it allows an examination of the changes that would occur in the economic profitability associated with a given project if the policy were changed. Since the permanence of a given policy over the life of a project is not certain, this analysis is really aimed at exploring one area of project uncertainty by testing the sensitivity of the project to changes in policies;
- it can generate information on the general effect of a policy on resource allocation in a given total economic setting. (Most changes in policies

will have impacts reaching far beyond the confines of a specific project.) Such information provides the basis for assessing the overall desirability of a policy in terms of the national efficiency objective and provides background information for making the future policy decisions.

The main elements involved in an economic analysis and the relationship between economic and financial analyses have been discussed. In the following chapters details are given on how to carry out an economic analysis (i.e. techniques) and where and when to apply economic analysis (i.e., uses for economic analysis in project planning) but first there is a brief discussion on how the financial and economic analyses provide (or do not provide) information related to other objectives which are commonly of concern to decision-makers.

2.4 ECONOMIC AND FINANCIAL ANALYSES AND OTHER OBJECTIVES

As stated before, decisions are not based on financial and economic criteria alone. Decisions makers are concerned with effects of projects which are related to other objectives. Some of the main ones of concern include (a) income redistribution effects, (b) balance of payments effects, (c) employment effects and (d) environmental effects.

It was pointed out in Section 2.2.2.2 that the economic analysis tells the user nothing about income redistribution effects, although it is possible to weigh the costs and benefits used in the economic analysis to reflect income redistribution objectives. The financial analysis does tell something about the incidence of expenditures and receipts associated with a project, but weights are not attached to expenditures and receipts associated with different income groups. Thus, if there is an objective to redistribute income to the poorer members of society through projects, a separate analysis will have to be undertaken which appropriately weights costs and benefits.

With regard to the employment objective, shadow prices for labour used in the economic analysis should reflect conditions of unemployment and thus favour use of labour in cases where there is substantial unemployment. In addition, the physical flow tables provide an indication of the impact of the project in terms of numbers of persons employed.

In some situations balance of payments effects are of primary concern. The financial and economic analyses can be set up in such a way that costs and benefits (expenditures and receipts) are listed by foreign and domestic sources. A summary can then be prepared to indicate the net effect of the project in terms of foreign exchange or balance of payments. The shadow price used for foreign exchange is a means of directly incorporating the balance of payments objective in the economic analysis. Other approaches are also possible.[1]

The economic and financial analyses tell nothing directly about environmental impacts related to environmental improvement or maintenance objectives. A separate environmental impact analysis will have to be undertaken for this purpose. Similarly, there may be other social and political objectives which are relevant in particular cases. The impacts of a project in terms of such other objectives will also have to be analysed separately.

---

[1] See paper by McGaughey, in FAO, forthcoming.

Chapter 3

PROJECT CONTEXT

3.1 INTRODUCTION

As mentioned in Chapter 1, in the process of shaping a project idea into a well-defined project to meet an objective, a number of alternatives are likely to be considered and compared. Before alternatives can be analysed in an economic framework and compared, it is necessary to know in each case the context of the analysis and the dimensions of the alternatives being considered. First, the level of detail required, the relevant time frame for the analysis, and the constraints on the analysis, have to be defined in terms of the purpose for the analysis. Second, the scope of the project alternative in terms of its components (and their separability and interdependencies) has to be determined, at least on a preliminary basis.

The first consideration is needed in order to allocate the time and budget available for analysis to the various tasks involved and to the alternatives being considered; the second is needed so that inputs and outputs of separable project components can be properly identified and valued in order to evaluate alternatives in terms of the second condition for economic efficiency mentioned in Chapter 2, namely that each separable component should have benefits at least equal to costs.

3.2 LEVEL OF DETAIL AND RELEVANT TIME FRAME FOR THE ANALYSIS

An economic analysis takes time and costs money. In most cases both are limited for any given project. Thus, the time and funds available have to be allocated to the various tasks involved in the analysis. The appropriate allocation of time to input and output identification and valuation will depend on the nature of the alternative being analysed and the purpose for the analysis. If the analyst is involved in the early stages in project planning, where an idea exists and the task is to sift through a large number of alternative means for implementing the idea, then he will likely want to concentrate on looking at the alternatives in a very general (as opposed to detailed) fashion in order to eliminate the obviously unacceptable ones. As the planning process progresses, the number of alternatives will be narrowed down through successive eliminations and more time and effort will be devoted to detailed analysis of a few alternatives. Finally, once one alternative is settled on and detailed design of that alternative takes place, a comprehensive economic appraisal will be required. These various uses for economic analysis are discussed in more detail in Part II. For the present it is enough to point out that before any given economic analysis proceeds it is necessary to have in mind the level of detail required for the particular purpose and the level of detail possible, given time and funding limitations and existing data.

A time frame needs to be established for the project idea being considered, i.e., the analyst has to settle on a relevant project time span, or how far into the future effects should be considered. Similarly, the time interval to use in identifying effects needs to be established. In reality, effects take place on a continuous basis and may change constantly over the life of a project. It is obviously impractical to consider and identify effects on a daily basis, so what other basis should be used for identifying and listing inputs and outputs and values?

With regard to the appropriate project time span, a general recommendation is to consider a time period that is long enough to include all the major effects of the project that can be foreseen. For example, a project which involves growing trees on a fifty year rotation should have a defined time span of at least fifty years. Some complications can arise in terms of input and output identification if a project only involves a part of an on-going programme. These problems and how to handle them are discussed under the heading of "time-slice" projects in Section 3.3.3.1.

With regard to the appropriate time interval to consider in identifying inputs and outputs, the usual procedure - and the one followed here - is to consider one year intervals, i.e., to list inputs and outputs on a yearly basis. A year can be defined to begin on any date, e.g., January 1, June 1, etc. For forestry projects, a "year" is often taken to begin on the day of planting or some other major initial investment in the project (e.g., site preparation). However it does not matter when the year is defined to start, so long as the same date each year is assumed in the physical flow table (and in the subsequent value flow tables). Similarly, there is no convention regarding how a given input or output is allocated to a given year when it occurs sometime in between the chosen date for the beginning of a "year" and the beginning of the next year. A convenient procedure is to assign any input or output that occurs within six months of the beginning date of a given year to that year and any effect which occurs more than six months after that beginning date to the next year in the physical flow table. The main point is that once a rule is set up for assigning inputs and outputs to a particular year, it should be followed uniformly throughout the analysis.

It is only by convention that the period of one year is chosen for use in investment projects. If a very short project and/or a very high discount rate is being dealt with, then a shorter time interval (say 3 months or even one month) can be used. The procedure is exactly the same, although the data requirements and calculations usually become more cumbersome.

3.3 ANALYSING INTERDEPENDENCE AND SEPARABILITY OF PROJECT COMPONENTS

By its very nature, a project consists of interrelated components. Indeed, diverse activities or components are combined into one "project" because they are interrelated in some way. However often some of the components of a project can be defined separately in the sense that most of their costs and benefits are independent from the rest of the project and the components can be added to (eliminated from) the project without affecting its technical feasibility, although they may obviously affect its overall profitability or economic efficiency.

If such separable components can be identified, then inputs and outputs should be allocated to them and they should be analysed separately, since as explained in Chapter

2, each separable project component should have benefits at least equal to costs in order for the total project to be considered an economically efficient use of resources.

When looking at component separability there are two relevant questions. First, does it make sense in the context of the purpose for the analysis to separate components, and if so, which components? Second, can components reasonably be separated for analytical purposes, i.e., can the inputs required for each component meaningfully be separated from each other?

With regard to the first question, the answer depends very much on the viewpoint of the institution for which the analysis is being carried out. If the institution does not want to change a given project scope, or if it has already decided on the size of the project, then it makes little sense to waste valuable analytical time and effort on detailed separation of components. The answer also depends to some extent on the stage in the project planning process. At the early stages, when alternative combinations of components and project sizes are being explored, it makes sense to separate out components and to analyse them individually and in combinations. This is indeed one of the main functions of the project identification and preparation stages in the planning process, and one of the main uses for economic analysis at these stages (see Chapter 11). However once a project alternative has been shaped and designed in detail, it may make little sense to spend much time on detailed analysis of components that have already been analysed and accepted in the earlier stages of planning. This is the case, for example, in the final project appraisal stage. The question at this stage is whether or not in fact alternatives were looked at at an earlier stage. If not, then there may be some justification for separate analysis of components even at the final appraisal stage, if such is not ruled out by the relevant decision-makers or the institution undertaking the analysis.

For the present, let it be assumed that there is a need and desire to look at separable project components in economic efficiency terms. The second question then arises, namely, what are the considerations that are relevant in determining whether or not components can be reasonably separated for analysis?

There are at least four types of interrelationships between project components and between a project and other projects or activities which have to be considered in looking at the question of separability. These are:

- horizontal interrelationships, i.e., interrelationships between components at the same level in the production process (see Section 3.3.1);
- vertical interrelationships, i.e., interrelationships between project components at different levels in the production process, i.e., where the output from one is an input into the next (see Section 3.3.2);
- interrelationships through time, i.e., the problem of identifying costs and benefits in a "time-slice" project, or a project that only involves one time segment of an on-going activity or programme (Section 3.3.3.1);
- interrelationships between a given project and other activities which should be considered within the project scope if a meaningful economic analysis is to be carried out. This relates to the problems associated with identifying and valuing indirect effects. (Section 3.3.3.2).

3.3.1 Horizontal project components

Forestry projects may involve two types of horizontal project components. One type is found in projects that are designed to produce several different outputs, for example sawnwood and plywood, or joint products such as timber, watershed or soil protection, and wildlife habitat. The other type is related to project scale, i.e., where a number of relatively independent production units producing the same output(s) are combined for administrative or other reasons into one "project". Examples would be a community fuelwood plantation project that includes sub-units or components in a number of independent communities, or a smallholder farm forestry project that involves support for establishment of numerous small independent plantations on private farms in a given region. 1/

For both types — several outputs or several producers of the same output(s) — there will always be some inputs which are jointly required by all components. If nothing else, since they are encompassed in one project, they will have project administration inputs in common. But quite often they will also have other inputs in common, e.g., infrastructure, marketing services, etc.

A typical situation where separate analyses can be undertaken is where several parallel processing activities are included within the scope of the same project. For example, in a project designed to produce both plywood and sawnwood, the major input items can generally be assigned separately to the two activities (although they also will likely have some inputs in common, e.g., administration, some infrastructure, etc.). 2/

In many other types of forestry projects with joint outputs, there is little scope for separate analyses of components, since most of the inputs required to produce the outputs are common to all of them. For example, a plantation project may produce wood, provide soil protection and wildlife habitat. All three outputs ("multiple use" of the plantation) result from the same production system and inputs and are thus difficult, if not impossible, to separate from each other in terms of inputs.

In this latter case the cost of adding on one purpose or output could be analysed. For example, the extra cost of management and harvesting associated with improvement in the soil protection function of a plantation on a hillside aimed primarily at producing wood and wildlife habitat could be analysed. But this would not be the same as analysing the soil protection output as a separate horizontal component, since the additional costs required to obtain the soil protection would not be the same as the total costs for it if taken in isolation. This type of analysis of the cost of adding on an additional purpose to the main project purpose is relevant in some cases, as will be discussed in Part II.

1/ For examples see the Case Studies outlined in Appendix A.

2/ The question of allocating a fixed and limited wood supply (input) to alternative processing activities is a separate question and is treated in Chapter 11.

For a type of project that involves a number of relatively independent units producing the same outputs, (such as the fuelwood plantation or small-holder agro-forestry examples cited above), there is a different set of questions which is relevant in determining the value of looking at separable units. First and foremost is the question of data and information on which to base such separation. If, as is often the case, estimates of "average" or "typical" conditions are used for all the components because of lack of more detailed information, then separate analyses make little sense, since all components will have the same assumed conditions and thus the analysis of each will produce the same results. For example, for an agro-forestry project in the Philippines involving subsidization of several hundreds of smallholder farmers, the data base was such that the best the analyst could do, given limitations on time and funds, was to use estimates of "typical" input requirements and "typical" yields for the area in which the farmers were located. 1/ Information was not available on which to base a disaggregated analysis of the relative profitability of different types of farms or different sites. Thus, components were not separated out for separate analyses. Instead, an "average" farm was analysed and the results extrapolated to take into account all the expected participants in the project.

Even if more detailed data had been available, it would hardly have been worth the analyst's time and effort to analyse each potential participant separately. However, separate analyses might have been made for several broad productivity and/or location classes to provide some indication of the relative profitability of different groups within the total project scope. Such information would be useful for establishing priorities in cases where there were more potential participants than funds to support them.

Whether or not separation of such components makes sense, even if the information on which to base separation is available, depends on the nature of the particular project situation, the time and funds available for analysis, and the objectives and constraints faced by the relevant institutions involved in the project. It is seldom worthwhile to separate out all such components. But it is generally worthwhile to look at some major classes of components in these types of projects. Once the relevant separation has been determined (agreed upon) then the analyst can proceed to identify inputs and outputs by such categories, developing separate physical flow and unit value tables for each.

However in these cases or the ones previously cited which involved joint outputs, the analyst still faces the problem of allocating some inputs which are jointly required by several or all components. As mentioned, even in the clearest of cases, there will always be some joint inputs (costs).

Some argue that as long as all inputs cannot be separately assigned to specific components there is little justification for separate analyses of components. The argument is that arbitrary assignment of joint costs is artifical and may lead to wrong decisions. The question is really one of degree. In cases where joint costs are significant in relation to separable costs (say around 25 percent or more of the total costs) separate analyses of components might lead to problems.

1/ See Case Study No. 1, FAO, 1979.

In cases where the estimated joint costs are relatively insignificant, or where the analyst is asked or required to undertake separate analyses of components, the analyst has to resort to judgement in allocating joint costs in a systematic, albeit arbitrary fashion.

To summarize, for horizontally related project components, the analyst should explore the extent to which inputs and outputs (costs and benefits) can be separated in a meaningful way. If three-quarters or more of the costs required for a given component can be separated out, then it is probably worthwhile to analyse the component separately, using whatever information and judgements are available to allocate joint costs. If none of the project components appear to be reasonably separable in terms of their inputs, then inputs should merely be identified for the project as a whole.

The above relates to analysis of a given project which is already defined in scope. If the economic analysis is being used to help determine an appropriate project scope and content (i.e., in the early stages of project planning), then horizontal components and alternative combinations of components can be looked at in more detail. This is discussed in Chapter 11, under the uses of economic analysis in project design.

3.3.2 Vertical project components

Most forestry and forest industry projects also involve distinguishable vertical components or activities, where the output or result from one component is an input into another component in the project. For example, wood produced in plantations is an input into a processing activity, with both being part of a defined integrated forestry and forest industry project. The wood production and the processing are quite well-defined separate activities, if the wood has alternative uses or value other than in the project processing activities. (If it does not have other uses, then it cannot be analysed separately. See section 3.3.3.2).

The main point to keep in mind when dealing with vertically related components is the concept of one-way dependence. This concept can be illustrated with a parallel. If a column is being built out of bricks, the bricks on the top depend directly on the bricks below; building cannot start from the top down. Each successive brick placed on the column depends directly on all those below it. On the other hand, the top brick can be eliminated then the next one and so forth down the column without affecting the bricks below. Similarly with a project (if the bricks are considered as being vertically related project components), components at the bottom can be undertaken without undertaking those at the top, but a component at the top cannot be undertaken without also undertaking all those that lie below it. Thus, it makes sense to analyse lower components separately from those above, but it makes little sense to analyse a higher component separately without considering all those that lie below it and on which it depends.

For example, assume a plantation project for which fertilization is being considered. Applying the 'with and without' concept, the one-way dependence involved between the plantation and the fertilization can be seen. Without the plantation project, the fertilizer obviously would not be applied. Therefore, if it is applied with the project, the <u>total</u> costs and benefits involved are properly of concern in the analysis. The plantation project can be undertaken without the fertilization (it is independent of the fertilization), while the fertilization cannot be undertaken without the plantation (it is dependent on the plantation). Thus the two can be logically separated in terms of analysing the profitability of the plantation without the fertilization, but it would be meaningless to analyse fertilization without also considering the plantation in this particular case.

At the same time, it should be emphasized that it makes sense to analyse the incremental costs and benefits associated with adding on a component (adding a "brick" to the column). Thus, for vertical components, the analyst should attempt to separate out inputs and outputs, so an analysis can be made of whether or not adding the next higher component involves an addition to the present value of total net benefits of the project. To take again the example of a project that envisages possible application of fertilizer to a plantation, assume that the overall return of the project, including the fertilization component, is \$1 500 and the total cost is \$1 200, both adjusted to take timing into account. If the project is looked at as a whole, the net benefits would be \$300 and the project would be considered economically profitable. However, looking at the fertilizer component in terms of additional costs and benefits, the added value yield (benefit) due to the fertilizer is \$100, while the cost of fertilizer and its application is \$150. Therefore the fertilizer component involves a net cost of \$50 (i.e., \$150 minus \$100). Total net benefits would be \$50 higher, or \$350, if the fertilizer component were excluded. According to the second condition for economic efficiency, the project would not be considered economically efficient unless the fertilizer component was eliminated. Only by analysing the incremental costs and benefits involved can it be seen whether or not a dependent component should be included in the project.

Two points should be emphasized. First in this example, it was assumed that both establishment of the plantation and fertilization were being considered as components of a project proposal. If the plantation were already established, then the fertilization component would be considered as a separate project and only the incremental costs and benefits involved in fertilization would be analysed. The same conclusion as above would be reached, namely that the fertilization costs would exceed the benefits.

Second is the assumption that the wood produced in the plantation without fertilization would have an economic use. The assumption would likely be true in this case. However, in some cases this assumption might not hold; then non-separable components would have to be dealt with. For example, if the wood to be produced as part of an integrated project has no value other than in the particular processing activity being considered as a project component, then the wood growing separately from the processing activity cannot be meaningfully evaluated. (The two components are not separable.) Section 3.3.3.2 discusses this point further.

Most forestry projects involve a number of vertical components. Some can be meaningfully separated as discussed above; others cannot. For example, in a project involving land clearing and planting of trees, the clearing component and the planting component need not be separated, since the value of the wood output is dependent on both clearing and planting and it is not possible to derive a meaningful output value for the land clearing in isolation from what will be done on the land after it is cleared.

To summarize, inputs and outputs should be listed by separable vertical components so an analysis can be made of whether or not it makes economic sense to add successive components to the overall project (such as in the example of adding fertilizer to a plantation project).

From these comments on hortizontal and vertical components, it can be seen that for most projects there will be a number of intermediate physical flow tables needed and not just one. Thus, if a project has two horizontally separable components and three vertically separable components for each of the two, it could have six separate flow tables, or one for each of the two horizontal components with three vertical components separated within each horizontal component. A total flow table would also be prepared, once the separable components have been analysed.

3.3.3 Interdependencies with other projects

The above two types of relationships refer to interdependencies and separability of components within a given, defined project. Two additional types of relationships also have to be analysed in order properly to identify inputs and outputs. The first relates to interdependencies between the project and other projects over time, i.e., in the case where the project merely represents a part of an on-going activity or programme. This type of project is called a "time-slice" project. The second is the type of interdependency which exists when the output of a given defined project only has one use and there is no practical way of estimating the value of the benefits of the project other than as an input into that use. These two types of interdependencies and their implications for input and output identification are discussed below, together with a special case of interdependency found in forestry, namely, the case of the "allowable cut effect".

3.3.3.1 "Time-slice" projects and interdependencies over time

It is quite common to find projects which include only a given part of an ongoing programme. These are called "time-slice" projects. [1] Identification of costs and benefits in this type of project can be tricky, since care is needed to identify carry-over values from previous activities (projects) which should be entered as costs in the new project and residual values associated with the new project which should be entered as benefits at the end of the new project. This task involves, among other things, distinguishing between sunk and nonsunk or recoverable costs. A sunk cost is one which has already been committed and which cannot be recovered and thus should not enter into consideration in an analysis of appraisal of a project involving decisions about future expenditure or use of resources. With or without the project, the resources are committed in the case of sunk costs. Thus, they involve no change in the project. These types of values are treated as follows:

Initial carryover or "inherited" costs and treatment of sunk costs. In a time-slice project, i.e., an investment in continuation or expansion of an on-going operation, resources used in the present operation which will also be used in the continuation or time-slice project should be treated as follows (this guide relates to the general rule that analysis should be based on the difference "with and without" the project):

- if the resource would actually have been used in some other productive use in the absence of the proposed continuation project, then it must be included as an input in the economic appraisal of the continuation project and given some positive value;

---

[1] FAO, 1979. Case studies nos. 2, 4 and 5 deal with projects which represent plantation activities for a certain number of years of ongoing forest plantation programmes. (See also Appendix A.)

- if the resource that will be used in the continuation project has no use other than for the proposed continuation project, then it should be included as an input but valued at zero in the economic analysis.

Residual value at the end of a project.[1/] Most projects have capital assets (land, buildings, equipment, etc.) which have differing lives. If some capital asset has a life that is longer than the project period chosen, i.e., the asset has some other use at the end of the project, then the value in that other use should be entered as a "residual value" or benefit at the end of the project. The agrument is exactly the same as in the case of carryover or inherited costs, except residual values are entered as benefits instead of costs, since when the project is terminated, it releases resources (or goods and services) which can be used in producing other consumption goods and services.

Residual values are common in financial analyses, since most often a purchase cost of an asset is entered into the accounts at the time it is paid for, and this purchase cost takes into account the expected stream of benefits foregone during the entire life of the asset, not merely for the time during which the asset will be used in the project. Thus, when a land purchase cost is entered in the financial analysis, it theoretically takes into account the value of the alternative benefits which the land could produce forever, not merely during the project time span. Thus, if the land has a use beyond the time span of the project (as it normally does) then a residual value should be entered at the end of the project to take into account the fact that the land will be sold or put into some other use when it is released from the project.

In an economic analysis, the theoretically correct way to enter the opportunity cost of land is to enter each year an annual value foregone by using it in the project in that year. In this case, since only the opportunity cost of the land during the time in which it is used in the project is entered into the accounts, there is no residual value to account for in the economic analysis. The same goes for other capital assets, again, however only in a theoretical sense. In reality it is difficult to allow for annual values foregone or opporutnity costs for most capital assets. Thus, commonly they are entered at full value at the time they are first committed to the project and thus a residual value is relevant. In terms of input identification, this means that an asset is entered once in the analysis as a cost in the year in which it is first committed to the project and then it is entered at the end of the project as a benefit and assigned a residual value which reflects the initial real cost for it plus the value of any improvements resulting from the project which have raised its real opportunity cost.

Residual value should not reflect any real value increase that would have taken place without the project. At the same time, if the real opportunity cost of an asset is increasing over the life of the project, then this should be reflected as a cost to the project in the unit value tables (see Chapter 8).

1/ Residual value is often referred to as "salvage" value. However, in the case of land, it seems awkward to refer to value of land at the end of a project as "salvage" value. Thus the more general term, "residual" value is used.

Time-slice projects can involve some serious problems in terms of identification and valuation of inputs and outputs. Such complications can be avoided by combining all directly interdependent activities (time-slices) and appraising them as one project. (The time-slice component of relevance can at the same time be analysed separately in terms of its incremental costs and benefits, such as discussed earlier).

There are often time-slice projects that involve expenditure of funds for a few years of an on-going programme, with benefits occuring many years after the "project" is finished in an administrative sense. The question arises as to how to handle such projects. The answer is clear. All costs up to the point at which the output from the project occurs must be included in the economic analysis and the outputs must also be included, even if they occur a number of years after the administrative life of the project has terminated. In other words, the economic analysis deals with a project as including all the interrelated costs and benefits associated with achieving a given purpose or output.

3.3.3.2 Vertical interdependencies between separate projects

In some cases, meaningful decisions about one project cannot be made separately from decisions regarding other projects. Thus, they need to be combined as components of one project. Specifically, the output of one project cannot be valued properly if it only has one use and that is as an input into one other specific project or activity. This case relates closely to that discussed in Section 3.3.2, except here a "project" has been proposed which is in fact not separable from certain other activities. In other words, in defining a project all elements (or components) needed to make the project feasible have to be included. This problem can be illustrated with a simple example.

A country is contemplating establishment of a pulp and paper mill to produce for the local market. There is no current pulp and paper production in the country. All consumption is based on imported paper. As a start, the country planners propose establishment of a pulpwood plantation project. The pulp and paper mill will come "later". This plantation project has to be analysed. A problem then arises since decisions on the plantation project can be made only in terms of decisions concerning the size and type of pulp and paper mill that will be constructed (and when it will be constructed and come on stream to consume the pulpwood output). Further, since there is no market for pulpwood in the country, there is no practical way to value the pulpwood output from the project.

The best way to get around this problem would be to take one step back and redefine the "project" to include both the plantation activities and the pulp and paper processing activities. If this were done, then the dimensions of the plantation component could be better defined in the context of the intended use for the wood output, and the wood could be treated as an input into the processing activities rather than as a project output that is difficult to value as such. The output of the project in this case would be paper.

If the analyst runs up against this type of situation, the best he can do is to suggest that the separate projects be combined into one, or if that is impossible, then merely look at the cost side of the wood production. Of course, if there is an alternative use for the wood from the plantations, then a measure of value could be derived on the basis of the willingness to pay for the wood in that other use. However, in many cases, particularly in developing countries where totally new activities are being introduced, such alternative uses do not exist.

Since this problem really relates centrally to the problem of output valuation, it is discussed further in Chapter 7. Here the subject is raised as a point to watch in defining the scope of a project.

There are other interrelationships between various activities which are relevant in defining the best project scope to meet a given objective or purpose. For example, several activities which have initially been defined as independent projects may be complementary in one of several ways. It may be that to take full advantage of such complementarities these activities should be combined into one project. For example, if the residues from a sawmilling project could be used in particleboard production, then consideration should be given to designing a project that includes both. These types of questions and others related to project identification and design are discussed in Part II.

3.3.3.3 The special case of the "allowable cut effect" (ACE)

Increasingly, foresters are being introduced to the concept of the ACE and its potential for raising rates of return from plantation projects. The basic concept is that if a country has a non-declining even flow sustained yield policy and has a lot of old growth or mature forest that is not adding any appreciable net increment, then by establishing a plantation, the allowable cut of the old growth can immediately be increased under the assumption that the plantation volume will become available to meet the even-flow constraint in the future. The value of the increased volume of old growth harvested immediately is then attributed to the plantation project as a benefit. Since this benefit occurs immediately, rather than in the future when the plantation wood is merchantable, it tends to increase the present value of the net benefits of the project.

Whether or not this is an appropriate approach to benefit identification depends directly on the assumptions made with regard to policies. If it is assumed that the even-flow sustained yield policy will remain in effect, then the allowable cut effect would appear to be appropriate. This follows from application of the "with and without" concept. Without the plantation project, the additional wood would not be harvested now due to the even flow sustained yield policy constraint. If the allowable cut is an actual constraint (i.e., if there is demand at prevailing prices for more wood than is allowed each year), then with the project the additional old growth will be harvested. Thus, due to the project (and how it relates to policy) the additional wood is made available to society now and this is identified as the benefit due to the project. (Of course, in this case the actual wood output from the plantation in the future is not considered as a benefit due to the project).

A commonly heard argument is the following: Since the wood could be obtained by merely changing the policy, how can the benefits be attributed to the project? The answer goes back to the basic assumptions underlying the measures of value used in the type of economic analysis discussed in EAFP. (See Sections 2.2.1.2 and 2.3). Opportunity costs as defined here relate to opportunities that are actually feasible, given the expected political and social environment which is expected to exist. Any policy could be changed. But the relevant question is: Will it be changed? If the sustained yield even flow policy is expected to remain in effect, then the ACE is a legitimate approach to benefit valuation, and the appropriate output to identify and enter into the physical flow table is the volume of old growth timber that will be harvested immediately due to the project. In a sense the ACE becomes a way of modifying or circumventing the sustained yield even flow policy impacts.

Another criticism of the allowable cut effect is the fact that certain present wood supply is being substituted for uncertain future wood supply, i.e., what happens if the plantation burns sometime in the future, or if for some other reason all or part of the new plantation wood does not actually become available when planned? These questions have to be considered by decision makers in each case. Application of the allowable cut effect in project analyses is in fact a matter of policy choice, and one which is quite separate from the decision regarding sustained yield policies. If a government has decided to use the allowable cut effect and if the conditions are such that it matters (i.e., if there is a large enough volume of old growth timber to which the ACE can be applied), then it legitimately can be used in economic analyses of projects.

In applying the ACE in identifying outputs, care is needed to analyse the assumption that the quality and use for the plantation grown wood will be the same as the quality and use for the old growth timber which is attributed to the project as an "output". This becomes a judgmental factor. For example, if a fast-growing, low density species is planted and then under the ACE a long fibre, dense species with high use value for structural products is harvested, it becomes highly questionable whether this higher valued present harvest should be attributed to a plantation project that will involve production of wood with a different use and use value. It is because of these types of questions that ACE depends on government policy concerning its use. Some countries use it; others do not. The project analyst generally follows accepted practice in his country, although he can at the same time try to argue for changes in the practice.

The sustained yield/even flow policy and the associated ACE policy are two classic examples of policies that are not designed with maximum economic efficiency in mind. They are, therefore, prime candidates for a policy efficiency analysis, which was discussed in Section 2.3. However regardless of what such an analysis might indicate in terms of the cost in economic efficiency terms associated with policies, they should be considered as given and taken into account when identifying and valuing inputs and outputs in an economic analysis of a project, if they are expected to remain in effect over the life of a given project.

Chapter 4

INPUT AND OUTPUT IDENTIFICATION

4.1 INTRODUCTION

After the context of a project alternative has been defined and the scope of the analysis determined, the next step is to identify the effects (or the inputs and outputs) associated with the project. In the economic analysis, any effect which results in an increase in desired goods and services available for society is a "positive" effect (output) and any effect which results in a reduction of goods and services available is a "negative" effect (input). Increases or decreases can relate to either or both quantity and quality of goods and services. The theoretical goal at this stage is to identify all the effects of the project on society. In practice, it is only possible to identify some of them due to lack of available information and lack of time and funds to generate additional information.

For the purposes of identification, a distinction is made between direct inputs and outputs and indirect effects. This is done more for convenience than for any conceptual or theoretical reasons. The terms are defined in relation to the financial analysis and the physical flow tables derived for use in estimating commercial profitability. In this context, direct inputs and outputs are those which enter into the financial analysis (i.e., are directly traded for money in a market) and indirect effects are all those other (often non-market) effects which are not considered in the financial analysis). 1/

A point to note is that a given effect may be direct or indirect, depending on whether or not it is traded directly in the market in a particular project situation and environment. For example, in one case, fuelwood may be traded in the market, while in another case it is produced and distributed "free" using some quota or other allocation mechanism. In the latter case, it would not have entered into financial accounts as a revenue (receipt). In the former it would have been considered in a financial analysis.

Similarly on the input side, a given input can be direct or indirect in the context of the definitions, depending on whether or not it is paid for by the entity for which the financial analysis is carried out. For example, if the government provides and pays for certain roads required for a private plantation project, then the cost of such would not enter the financial analysis for the private entity for which the analysis is being done. It would still be an input into the project from the economic point of view and should be identified as such. If the private project built the road, even though it was fully paid for (subsidized) by the government, then it would have appeared in the financial analysis. (See Chapter 9 where treatment of subsidies in the economic analysis is discussed.)

1/ Indirect effects are often referred to as "externalities" or "spillover" effects.

It does not matter whether an effect is labelled as direct or indirect. The distinction is made for convenience and to remind the analyst that he has to look beyond the financial analysis for effects associated with a project.

With this in mind, the identification procedure suggested here, and discussed in the remainder of the chapter, is as follows:

First, using the physical flow table(s) developed for the financial analysis and/or the various technical studies available for the project, identify direct inputs and outputs. To the extent that separable project components have been identified, divide up the direct inputs and outputs by components. These can be listed in separate physical flow tables for components and added together at a later, summary stage in the analysis (see Section 4.2).

Second, identify the indirect effects due to the project. List these by separable components if possible, as indirect positive effects, if they add to the aggregate quantity/quality of goods and services available for consumption, or as indirect negative effects if they involve reductions in the quantity/quality of goods and services available. Such efforts can again be associated with both quality and quantity changes (see Section 4.3).

In identifying both direct and indirect effects, it is important to distinguish them on the basis of what the resulting information will be used for in succeeding stages in the analysis. Thus, they should be divided and distinguished in categories which make sense from the point of view of valuation and in terms of the types of sensitivity tests which will be included in the analysis. Generally, project activities should not be listed as "inputs", since values will normally be attached to the inputs required to carry out the activities and not the activities themselves. 1/ For example, it is not enough to identify "land clearing" as an input in a plantation project analysis. Rather, "land clearing" can be a heading in the physical flow table, but under it should be listed requirements for various types of labour and supervision, machinery, tools, etc. Similarly, if at all possible, structures that will be constructed as part of the project should be broken down by the component inputs required to build them, and roads should be broken down by labour, machinery, and various materials required instead of just listed as "roads". If this is not done, it becomes difficult at later stages to develop proper values, since it is the inputs which are required to build the roads which are shadow priced or valued.

4.2 IDENTIFYING DIRECT INPUTS AND OUTPUTS

The direct inputs and outputs are generally the most important in terms of total project costs and benefits and are central to the economic as well as the financial analyses of a project. In most analyses of forestry projects, they are the only effects which have been given explicit consideration in terms of monetary values.

Most of the direct inputs and outputs which are relevant for the financial analysis are also relevant for the economic analysis. Commonly, the identification of such effects is done at the same time for both analyses.

1/ Summary tables may present costs by activities, but these summaries can only be derived by estimating the inputs actually required to implement them.

4.2.1 Direct Inputs

The main source of information on direct inputs will be the engineering and other technical studies available at the time of the economic analysis. The various input categories for the project and its separable components are defined and the relevant quantities are then entered in physical flow tables by each category and for the year(s) in which they are needed. The listing of inputs is done in a form that will facilitate valuation at a later stage. The types of main input categories which are relevant for most projects are shown in Table 4.1. The table provides only a convenient checklist which will have to be expanded both in breadth and detail for particular cases. [1]

Input categories shown in Table 4.1 can be listed in a number of different ways by subcategories related to (a) phases of the project, (b) activities or components within phases, and (c) by foreign and domestic sources for each phase and activity. There may be three major phases:

- project planning (preinvestment phase);
- investment phase (construction i.e., fixed investment and preproduction capital costs);
- production phase.

Activities within each phase will differ with the project being analysed. Production activities (or components) will often include raw material production, processing activities, storage, sales and distribution. In many types of forestry projects it makes little sense to separate the investment phase from the production phase for the economic analysis. It is often preferable to treat the two together and distinguish activities such as site preparation, planting, crop maintenance and management inputs during the growing period and harvest and transport. The only general rule for establishing appropriate categories is that the analyst classify inputs in a way that makes sense in terms of the objective of the analysis, i.e., the derivation of the total value flow table and the measures of project worth. Some examples for specific projects are given later.

If balance of payments effects are of particular concern to decision-makers, then all inputs can be listed separately by domestic and foreign sources.

The amount of detail required for the tables depends on the stage in the planning process. During initial phases, when project identification, preparation and design is the main focus, the analyst may start with very general, rough estimates which can be used to make initial comparisons between alternative technologies, scales, locations, etc. As attention focuses on one alternative design, the detail required increases. When the alternative has been designed and prepared, the analyst may wish merely to summarize inputs by categories and activities or components with headings such as shown in Table 4.1. The final appraisal document should not contain excessive details. Rather, reference can be made to the supporting studies, so the decision-maker can find details if so desired. He should not be forced to wade through them to put the logic of the project and its appraisal clearly in perspective.

---

[1] cf. FAO 1973, UNIDO 1978, OECD 1968.

Table 4.1

CATEGORIES OF DIRECT INPUTS[1/]

Inputs Category	Comments
1. Manpower	A distinction should be made between unskilled and skilled labour, staff, consultants, etc.
2. Land	Land can be further broken down into categories to reflect different uses and values.
3. Equipment	Working tables will be needed with detailed listings of equipment required and timing of such requirements. In the final tables, some major subcategories can be used as derived from the detailed tables. Replacement requirements have to be included.
4. Raw materials	Such items as utilities (energy, fuels, etc.), wood raw material, if purchased, chemicals and other purchased inputs, and water can be listed separately. 2/
5. Structures and civil works	If structures and civil works (housing, roads, other facilities such as dock and harbour services) are purchased or rented outright, then they would appear as separate inputs. However, if the project itself involves construction of such works, then they should not be listed as inputs as such. Rather, the component labour, land, equipment and raw material requirements for constructing them are listed.

Note: See text for further discussion of how these inputs should be listed by subcategories related to (a) phases of the project, (b) activities or separable components, and (c) foreign and domestic sources.

1/ As mentioned in the text, depending on the situation some of the listed inputs may be indirect instead of direct, e.g., in the case of infrastructure such as roads, community facilities, etc. It all depends on whether or not they are directly paid for by the project entity for which the financial analysis is being done.

2/ If raw materials, such as wood, are produced as part of the project itself, then the component input requirements are listed rather than the raw materials such as roundwood. See text.

Table 4.2

POSSIBLE DIRECT OUTPUTS FROM FORESTRY PROJECTS

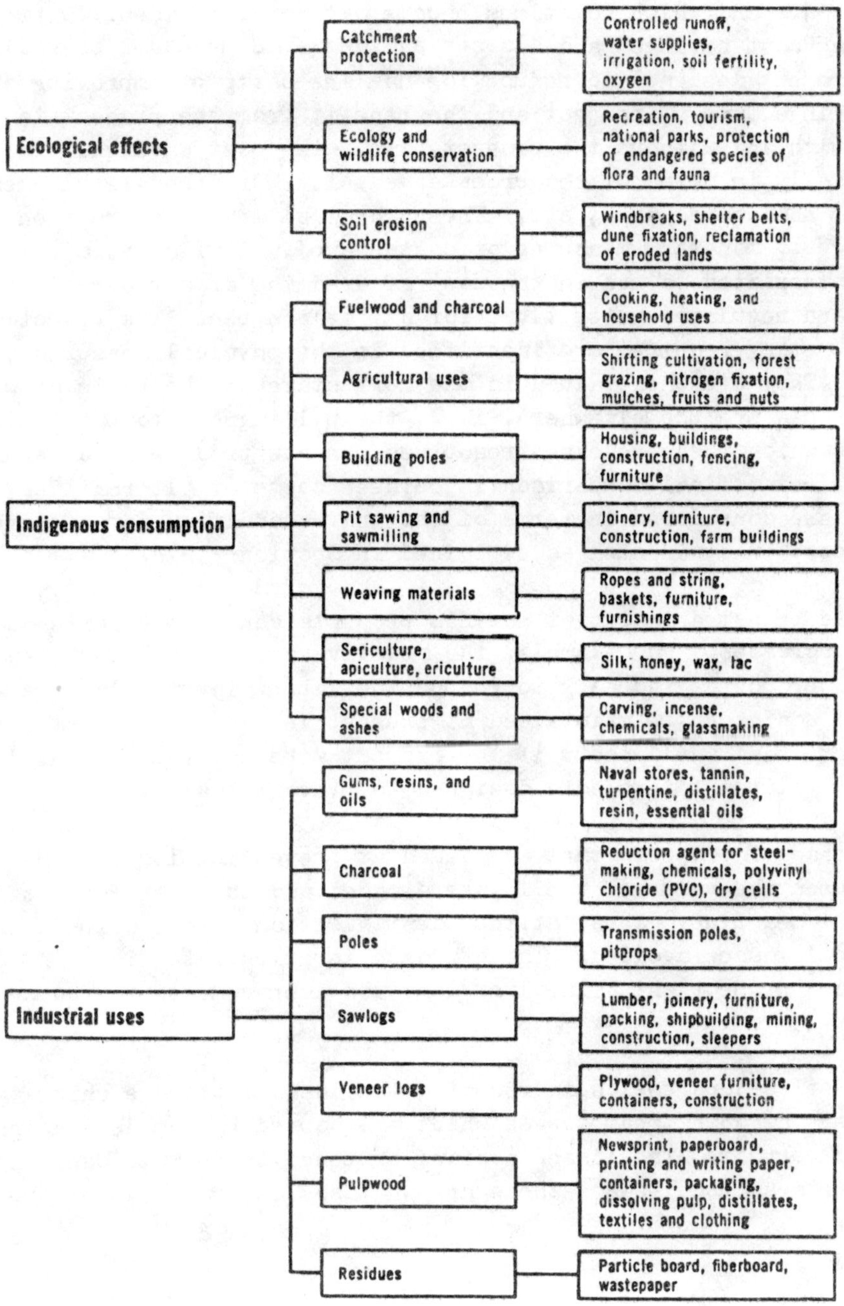

Source: World Bank, Feb. 1978.

4.2.2 Direct Outputs

Direct outputs can also be derived from the basic technical studies and from market studies which are a basic element for projects involving direct outputs. Some main categories of direct output types which can be associated with forestry projects are shown in Table 4.2. Depending on the project circumstances, some of these may be indirect instead of direct outputs, particularly in the case of "ecological effects".

There are two types of potentially direct project outputs which are not shown in Table 4.2 and which sometimes become difficult to identify properly. These can be labelled as "cost savings" and "losses avoided". Some examples will illustrate them. Assume a project designed to reduce log hauling costs by improving a logging road. This is a cost savings type of project and the benefit from the project is the difference in hauling costs with and without the project, i.e., the cost savings. The "output" can be specified initially in terms of resources saved, i.e., reduced requirements for trucks, maintenance labour and spare parts, etc. These physical measures are then transformed at the valuation stage to monetary measures of costs saved. Similarly, a watershed protection project may be contemplated to reduce the cost of dredging of a reservoir that provides flood protection and regulates water flows for dry season use. The reductions in dredging equipment, labour, etc., required are identified as the physical measures of "output" or resources saved. (They are then valued in the next stage on the basis of what these released resources can produce elsewhere, i.e., the willingness to pay for the additional goods and services which these released resources can now produce in alternative uses). In both cases, the relevant final comparison is between costs of alternatives, i.e., "cost savings" projects are considered in terms of the third condition for efficiency or by applying a least cost analysis such as explained in Section 2.2.4.

It should be noted that cost savings projects can also be oriented toward preventing future cost increases. For example, the relative price for labour may be increasing and a project could be proposed gradually to reduce the labour input into a particular activity so that total unit costs can be maintained at present levels or at least prevented from increasing at a rate that would occur if the project were not undertaken. This type of project is closely related to projects designed to prevent losses.

In the case of projects that are aimed at preventing losses, the relevant comparison is between the value of the losses avoided and the costs of avoiding the losses through the project measures. Thus, at the identification stage, outputs are identified in terms of physical losses avoided. The approach is illustrated in a FAO document for a watershed protection project which involves land use improvements to reduce siltation in a reservoir. 1/

Reduced siltation results in reducing the loss of storage capacity, which in turn results in reducing the downstream losses which are caused by the decreasing water availability from the reservoir. The losses avoided or benefits in this case are identified in terms of such downstream uses (since these are what society values, not the capacity of the reservoir itself).

1/ Example no. 2 in Gregersen and Brooks paper in FAO, forthcoming.

Similarly, forest protection projects are aimed at reducing the risk of loss due to fire, insects, disease, etc. In these cases the probability of loss without the project and the reduced probability of loss with the project have to be estimated. The difference is the "output" or benefit due to the project. This task is appropriately done by the technical experts. Once such information is available, the task of the economist is to take the appropriate estimates of physical losses avoided and attempt to value them in a time context. Since the estimates of physical losses avoided will be subject to probabilities so will be the values of these losses avoided. At the input and output identification stage, there are no particularly unique problems involved, although analyses involving probabilities are always more complicated to carry out (and require more data) than those involving the assumption of certainty.1/

Finally, there is the situation mentioned earlier where a project involves both losses avoided and production (output) increases over present levels. For example, assume a situation where an area of hill land is deteriorating due to erosion taking away the productive top soil. It has been estimated that the production from the land will decrease over a 20-year period from level A to zero (point B) in figure 4.1. Now a project is proposed to build up production to level C in ten years. The appropriate measure of output is area ACDE, plus the loss avoided, or area AEB. If only the production increase over present level were included, it would understate the output or benefits of the project.

If production is expected to continue at level C beyond the 20 year life of the project, then the benefits or output of the land beyond that period should also be included in the project calculations net of any additional costs required to maintain production at that level. In other words, at the end of the project period, there is a residual value (such as explained in Chapter 3) that can be attributed to the project. It can be seen that application of the 'with and without' concept is critical to proper benefit identification in these cases.

Table 4.3 provides an example of a physical flow table for a forestry project, showing how direct inputs and outputs are organized and how inputs are listed in the year(s) in which they are used and outputs by the year(s) in which they occur. An additional example is provided in Chapter 12.

1/ The assumed probabilities can be tested in the sensitivity analysis (see Section 2.2.5 and Chapter 10).

Figure 4.1

SOIL PROTECTION BENEFITS

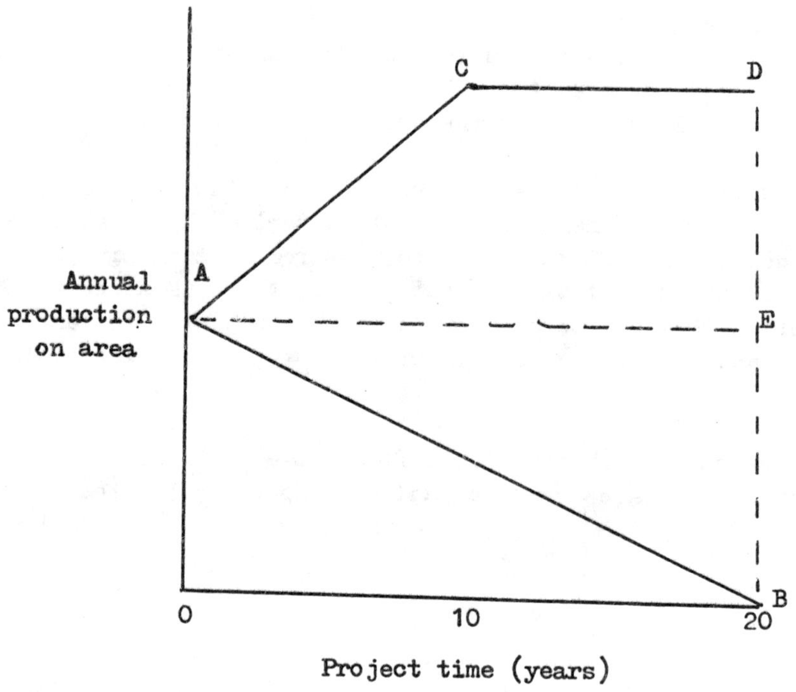

Table 4.3

TIMING AND MAGNITUDES OF PHYSICAL INPUTS AND OUTPUTS

for Assumed "Average" 10 ha Farm 1/

Item	Units	Years 0	1	2	3	4	5	6	7	8	9	10	11	12	13	14	15
Inputs	2/																
Land preparation lining, digging and planting	m.d.	74	74	74	74												
labour	m.d.	38	38	38	38												
seedlings	no.	1200	1200	1200	1200												
Replanting																	
labour	m.d.	16	16	16	16												
seedlings 3/	no.	300	300	300	300												
Fertilization																	
labour	m.d.	25	25	25	25												
fertilizer	kg.	4	4	4	4												
Weeding	m.d.	68	68	68	68				34	34	34	34	34	34	34	34	34
Singling	m.d.									25	25	25	25	25	25	25	25
Output																	
Pulpwood stumpage m^3(r)									184.1	205.8	205.8	227.0	227.0	247.8	247.8	268.2	205.8

1/ From Case Study No. 1. See FAO 1979.
2/ man days
3/ Assumed that 25 percent would have to be replanted on the average.

4.3 IDENTIFYING INDIRECT EFFECTS

An indirect effect was defined earlier as any change in the quantity or quality of goods and services available to society due to the project which does not enter into the accounts for the financial analysis, since it is not directly bought or sold in a market by the financial entity for which the financial analysis was done.

A first point to note about indirect effects is that many of them cannot be meaningfully valued in monetary terms. However they should still be identified in quantitative physical terms, if possible, and otherwise at least specified in descriptive terms. Regardless of whether or not they have an identifiable monetary value, they may be important in the broader context of decision-making, where many considerations other than monetary values are important.

A second point is that whenever an indirect positive effect is identified, the analyst should be careful to search for any corresponding indirect negative effect (cost) required to bring about the positive one. It is only the net indirect effect that can be attributed to the project. The following discussion will illustrate this point.

4.3.1 Indirect positive effects

The following are the main indirect positive effects of concern in forestry projects:

- soil and watershed protection and wildlife and recreation habitat improvements which are not directly traded in a market and thus are not accounted for in the financial analysis;
- benefits accruing to society due to the fact that the project has trained persons to be more productive or has demonstrated the viability of some activity which is then undertaken by entities outside the project boundaries;
- cost savings which result in output expansion and increased use of excess capacity outside the project, but due to the project activities.

Some examples of each type follow -

Soil and watershed protection and wildlife recreation habitat improvements. Many projects involving establishment and/or management of forests for wood production also produce certain indirect effects in the form of improvements in soil or watershed protection "services" from the land (forest) and, possibly, improvements in wildlife habitats and recreation opportunities. In rare instances, these services are paid for directly to the project and thus enter the financial analysis as direct outputs. (See Section 4.2.2). However in most cases they are not directly priced in a market.

Quantification of such indirect effects depends on the availability of input/ output information which describes the changes in output that will take place with a given forestry activity. In the absence of such information, there is little that the economic analyst can do to quantify them. He can describe them to the extent possible and point out that they will likely result.

There are some studies which have been carried out for specific regions which link various forestry activities to watershed protection changes and further link these changes to consumption changes downstream. [1] The transferability of such specific results to a broad range of project situations may be possible. The best that can be done is to rely on the judgments and figures provided by the technical experts. If such effects have been identified in quantative terms, they enter the analysis in exactly the same way as any other quantified input or output.

Training and demonstration effects. A project may involve training of labour to increase its productivity. The training expenses are likely to be direct inputs into the project; however, the indirect effects due to the training are not accounted for in the financial analysis, since the project financial entity does not collect the increased revenues made possible by use of this better trained labour in other projects when the project is terminated or the labour leaves the project for other employment. It is very difficult to quantify this benefit and particularly to value it. Thus, it is generally included in the analysis in a descriptive fashion, for example, "100 labourers will be trained to operate power saws and this will increase their productivity in future years". The training expenditure also results in benefits in the form of increased output per unit of input in the project itself. These should be accounted for in the direct output measures for the project.

Similarly, in many forestry project situations, there can be significant demonstration effects. For example, a public project may involve support for establishing fuelwood plantations in selected communities. Once surrounding communities see the benefits to be derived from such plantations, they may on their own undertake to establish such plantations to meet their increasing requirements for fuel and/or to reduce the increases they are experiencing in fuel costs. The net benefits resulting from this type of demonstration effect can appropriately be attributed to the project being analysed (even though the additional plantations resulting due to the demonstration effect are totally outside the project scope). The 'with and without' concept can be applied to see which net benefits would not have been expected to result without the project. They can legitimately be attributed to the project. It should be emphasized though that it is only the net benefits that can be attributed to the project. If the additional outputs are to be attributed to the project, then care should be taken to attribute as inputs the resources and goods and services needed to bring about the additional output.

Cost savings and increased use of excess capacity in other sectors. If a forestry project results in production of lower cost wood than previously (i.e., more efficient wood production) there may be an increase in the use of wood in existing idle processing capacity outside the project boundaries. (The increases will be due to the fact that the price of the final product can be lowered since costs are lowered; demand for such products will increase because of the lower price, and therefore processing can increase to meet this demand.) The indirect benefits in this case will be the increased output resulting outside the project less the costs (the inputs) required to bring about

---

[1] FAO, forthcoming. Paper by Gregersen and Brooks.

this new production.

Similarly, a road project designed to reduce the cost of delivered wood (i.e., increase efficiency in wood delivery) may have indirect effects beyond the project. Such improved roads may be used by farmers who can lower their effective costs of delivery, thereby lowering farm product prices, which can result in increased demand and expansion of production (i.e., goods available for society to consume). Such increases can be attributed to the project in question (the road project) net of any increases in costs (use of resources) required to bring about these production increases. The appropriateness of attributing these net benefits can again be ascertained by applying the 'with and without' concept.

This type of indirect positive effect should be distinguished from what is generally called a "multiplier effect", i.e., a short run increase in income generated outside the project when surplus capacity in an economy is activated by additional rounds of spending resulting from investment in the project. Forest recreation projects are often justified in terms of the additional expenditures which will occur in the communities adjacent to the recreation project. From a national point of view, such "benefits" need to be questioned. In most cases they are merely transfer payments in the sense that the expenditures would occur elsewhere in the absence of the project. Again, application of the 'with and without' concept is critical in identifying true net indirect positive effects associated with such additional expenditures. They generally can be justified only in cases where the funds available for the project could only be used for the project being analysed and not for any other project in the economy. This would be the case for tied grants and loans which could not be used for anything other than the project in question. In this case, it still only is the net effect which should be included, i.e., there may be additional non-tied expenditures - outside the project boundaries - which are required to achieve the benefits or indirect positive effects in question.

4.3.2 Indirect negative effects

There are also certain indirect negative effects which may be associated with forestry projects. The main categories are:

- pollution or negative environmental effects not accounted for by direct costs to the financial entities involved;
- increases in costs outside the project boundaries which influence production (cause decreases) elsewhere in the economy;
- infrastructure costs not included as direct costs, but required for the project.

Pollution and negative environmental effects. The common example given is a pulp mill that pollutes water that is put back into rivers, thereby reducing the quality of water downstream and consumption benefits of downstream water users. Similarly, such a project may reduce air quality. Often a measure can be derived of the amount of pollutants which the mill discharges into a river or lake. In some instances such increased pollution levels can be associated with losses in consumption benefits (e.g., loss of fish catch, increased health problems, etc.). Quite often, however, this type of indirect negative effect is merely described in the project document without making any attempt to value it since the necessary data on input-output relationships are not available. Increasingly, pollution is being internalized in projects through the use of effluent charges or requirements for pollution control equipment or project water purification expenses. In these cases, such effects enter the project analysis as direct effects or inputs, since they involve financial costs and enter the financial analysis.

A similar situation exists with most other types of indirect negative effects involving deterioration of the environment, for example, soil deterioration and watershed benefits foregone due to a project that involves manipulation of vegetation upstream. Considerable research and study has been devoted to watershed problems and potentials for improving watersheds through forestry activities. There are some estimates of quantitative relationships available which may be usefully transferred from one situation to another. The judgment on transferability should be made by the technical personnel familiar with watershed management and the project.

Cost increases affecting production of non-project output. In some instances, a forestry project may result in the prices for certain inputs being increased. Such increases will affect other producers who have to curtail production. Reduction of their production releases resources, some of which may not be usable in producing other goods and services. If there is a net loss in the value of goods and services available to society due to such a project effect on prices, then this is legitimately attributed to the project as an indirect negative effect. Such a net loss would result if some of the resources released had no alternative uses and thus remained idle when the project-caused cost increases put them out of work. For example, if the project demand for imported machinery results in the price for such machinery increasing to the point where certain other activites cannot afford it and they have to shut down, then they release labour and other resources that may not be able to find alternative employment. The reduced output value from the activities that shut down, less the new value produced by those released resources which find alternative uses, would be a measure of the indirect cost of the project being analysed.

Infrastructure costs. As mentioned earlier, it is common that some of the infrastructure - roads, community facilities, power generation and communication facilities - which have to be produced for the project are not paid for directly by the financial entity for which the financial analysis is being undertaken. In such cases, the costs associated with such infrastructure have to be included in the economic analysis as indirect negative effects of the project, to the extent that their provision involves use of resources that could have been used in the absence of the project to produce other goods and services valued by society. Both capital and operating costs associated with such infrastructure have to be considered. At the same time if certain infrastructure items will be used outside the project boundaries, then allowance should be made for such use as an indirect positive effect. Again, the "with and without" test is applied to infrastructure.

If infrastructure is produced and operated directly by the project entity or entities for which the financial analysis is undertaken, then it should have been included in the financial analysis, even if it is entirely subsidized by the government or some other entity not considered in the financial analysis. The exception is if the financial analysis netted the project expenditure against the subsidy. In this case the cost to the financial project entity would not appear in the financial accounts. In such cases, the cost should still be included in the economic analysis. The cost of the infrastructure is real. It is impossible to generalize on how such subsidies and infrastructure expenditures are handled in the financial analysis. In each case the analyst preparing the economic analysis has to look at the project's financial accounts and make sure that costs to society are included and subsidies are appropriately treated as "transfer payments" as suggested in Chapter 9.

Common categories of infrastructure which the analyst should examine critically include those shown in Table 4.4.

4.3.3 Additional points: indirect effects

What is to be done in terms of identification of indirect effects ? There is no one best way to proceed, since there are few ready and available sources of information on most of such effects. Success in identifying indirect effects depends a great deal on experience and knowledge of relevant interrelationships based on study of other projects and technical literature. Interaction between various technical experts is essential, since identification of most indirect effects depends on information related to technical relationships.

Given some general ideas on potential indirect effects of given types of activities the analyst can proceed to estimate whether any given type will be relevant for the particular project he is analysing. If he decides that it is likely to be relevant, then he can discuss with technical experts the likely physical magnitudes of the effects (both positive and negative) and list these in a separate table (or tables). Where it does not appear possible to estimate magnitudes (quantities involved) the analyst should still develop a statement describing the nature of the effect expected in as specific terms as possible.

Some indirect effects will be accounted for in the economic analysis through shadow pricing of direct inputs and outputs and will, therefore, not appear as separate cost or benefit items (see Chapter 5). For example, if water used in a pulp mill is shadow priced to reflect its true opportunity cost, then this shadow price (cost) should incorporate the value of opportunities for using clean water downstream that are foregone due to the project polluting downstream water. Since identification and valuation are closely interrelated, in practice the two steps are often carried out simultaneously, i.e., a given effect is identified and then valued at the same time. The distinction between identification and valuation in EAFP is made for clarity of exposition and to emphasize the point that even though a given effect cannot be valued in monetary terms, it should still be identified and specified as explicitly as possible.

Table 4.4

INFRASTRUCTURE CATEGORIES CHECKLIST FOR ECONOMIC ANALYSIS

Rail (track and rolling stock)
Road (highways and vehicles)
Port
Shipping
Logging facilities (vehicles, equipment, roads)
Power (generation, distribution)
Telephone
Freshwater supply
Stormwater drainage
Sewerage (drains and treatment)
Housing
Education (schools)
Health (hospitals)
Government Agencies (post office, tax department, justice, etc.)
Churches
Recreation facilities (sporting and cultural)
Commercial facilities (shops, banks, hotels, etc.)

Source: R.G. Steele, FAO internal working paper (restricted), 1979.

4.4 LOCATION RELATED INPUTS AND OUTPUTS (EFFECTS)

As mentioned earlier, inputs and outputs or effects associated with a project should be identified in such a way that the process of valuation is facilitated. Since many inputs and outputs will be valued directly or indirectly on the basis of (market) prices that are established in locations other than those where projects produce outputs or use inputs, it is important to pay special attention to the handling, marketing and transport functions and properly to identify the inputs used in these functions due to the project or saved by producing an output in the project rather than importing it or producing it somewhere else in the domestic economy. This category of effects relates closely to infrastructure inputs discussed in Section 4.3.2.

As in the case of infrastructure (and other inputs and outputs), location related effects can be identified as direct inputs and outputs or as indirect effects depending on the nature of the project and the financial analysis being carried out. The important point is that they be included in the analysis and not that they are classified correctly as direct or indirect.

Location related effects which need to be considered can be divided into general ones, i.e., relevant for all types of projects and specific ones, i.e., specific to certain types of projects which involve substitutions (as explained below). In both cases, they only arise when (a) the value measure (price) which is to be used for a direct project input is established in a market, or at a point which is different from the point of use of the input in the project, and (b) the value measure (price) to be used in valuing a project output is established in a market or a location **that is different from the point of use of the output and/or different from the point of production of the output. Thus, this type of effect is one that can only be identified properly in the context of the valuation system which will be used. This emphasizes the point made earlier that in practice identification and valuation often have to be carried out simultaneously for some types of project effects.**

4.4.1 General effects

(a) For all direct project outputs the analyst needs to identify inputs required to handle project outputs and move them to their intended point(s) of consumption (or export) at which their values are determined. For example, in the case of an export output, it will be valued on the basis of its export price, generally determined at the port of export. (This will be discussed in the following chapters). In this case, the inputs - handling and transport - associated with getting the output from the project point of production to the port in which the export price is determined should be included as inputs in the project accounts (the physical flow table and, later, the value flow table).

(b) In the case of all direct inputs used in a project, the additional inputs required to handle and to move such direct project inputs (resources, goods or services) from their point(s) of origin (or the location(s) at which their prices are determined) to the point(s) of use in the project need to be included in the project accounts. For example, in the case of imported inputs, which will be valued on the basis of an import price established at the port of import, the handling and transport inputs from that port to the point(s) of use in the project need to be included.

4.4.2 Specific effects

In addition to these two general considerations (which should be considered for all inputs and outputs) there are two special cases where a project can result in positive effects (cost or resource savings) which must be considered and identified where relevant.

(a) In the case of a project output which substitutes for an import or a domestically produced output, the project will often result in a savings of handling and transport inputs which would have been incurred in the absence of the project. These inputs are saved because the good or service being substituted by the project output will not have to be handled and transported from its point of origin (e.g., port of import) to the market(s) or point(s) of consumption in which its local market price (or w.t.p. for it) is determined. For example, in the case of import substitutes which will be consumed in market A, it will no longer be necessary to handle and transport the import from the port of import to market A. The resources saved due to the substitution are a positive effect of the project, if they have productive uses elsewhere in the economy. This will be determined in the valuation stages. At the identification stage such resources saved due to the project should always be included. (Of course, the effects described under (a) in Section 4.4.1 would also be included).

An example will illustrate this point. Assume that an import price of a good at the point of import (converted to local currency equivalent) will be used to value the output of a project that will substitute for the import. The local currency equivalent at port of import (point X in Figure 4.2) is P100. Adding on marketing costs (transport and handling, etc.) to the point of consumption (point Y in Figure 4.2) a local price of P140 is arrived at (which is here assumed to equal the w.t.p. for the output at the point of consumption). It can be seen that in additon to saving the local currency equivalent of the import price, the additional handling and transport costs of P40 from the port of import (point X) to the market or consumption point (Y) is also saved. This can be legitimately attributed to the project as a separate positive effect (resource savings) in addition to the direct import cost savings which will be used to value the direct project output. Of course, by producing the output at project location X, the transport and handling costs of P30 between point Z and the market (point Y) are also incurred. This additional requirement for transport and handling services is taken care of under (a) in Section 4.4.1 as an additional cost (or input requirement) due to the project.

(b) In the case of a project which uses as an input a local resource or locally produced good or service which would have been exported in the absence of the project, use of the input in the project will result in savings in additional resources which would have been required to handle and move the particular project input in question from its point of origin to the port of export to point(s) of use in which its price is determined. These savings of additional inputs are legitimately identified as positive indirect effects due to the project in the stage being discussed here.

Figure 4.2

IDENTIFYING LOCATION EFFECTS: IMPORT SUBSTITUTES

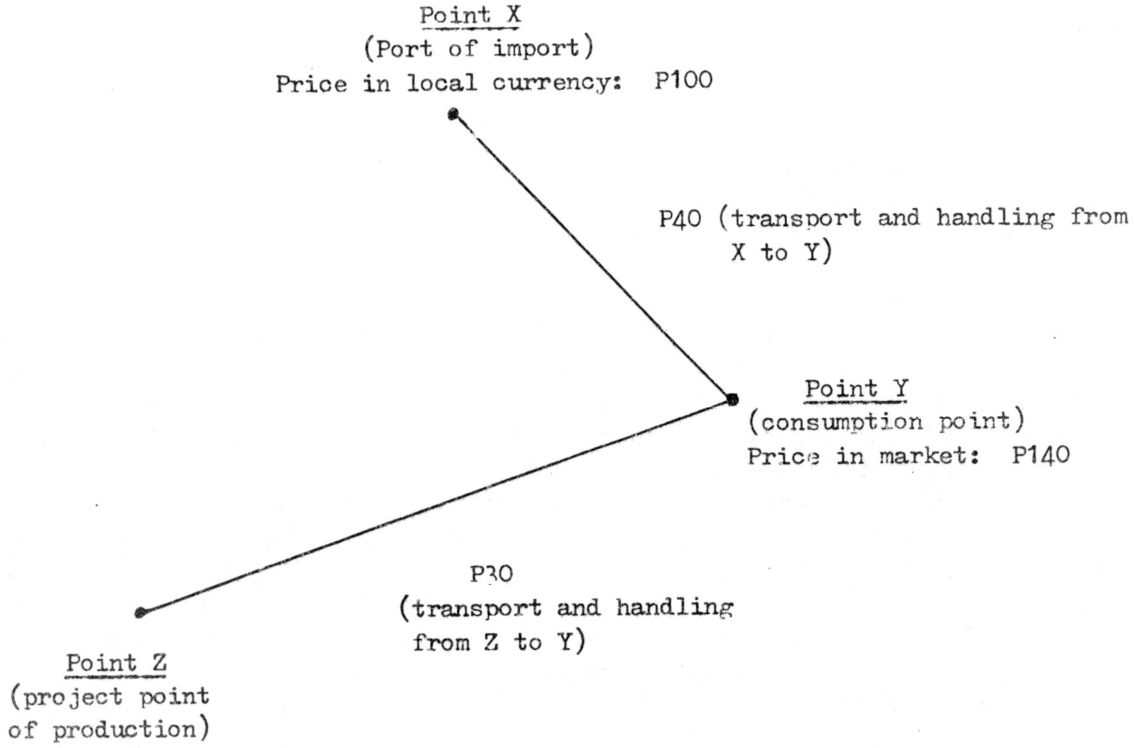

An example will illustrate this point. Assume conditions as shown in Figure 4.3. The local currency equivalent of the export price which would have been received for the input if it were not used in the project is P200 at the port of export (point M in Figure 4.3). 1/ This value is used as the basis for valuing the opportunity cost of the input being used in the project. However, by using the input in the project rather than exporting it, the P50 worth of transport and handling resources which would have been required to get the resource, good or service in question from its point of origin (point N in Figure 4.3) to the port of export (the point at which the P200 is determined) is saved. This can legitimately be attributed to the project as an indirect positive effect. Of course, the additional cost of P30 required to get the input from its point of origin (point N) to the project point of use would also be included as additional direct inputs due to the project. (This follows from application of the 'with and without' concept and is taken into account under (b) in Section 4.4.1.

It should be noted that the P50 of resources saved by not having to move the input (resource, good or service) from its point of origin to the port of export from which it would have been exported could also have been netted out of the P200 to arrive at the net opportunity cost associated with using the input in the project rather than exporting it. Both approaches, treating the transport and handling resources (valued at P50) as separate project effects, or netting them out of the P200 - would give exactly the same result. Thus, the question is really which of the two approaches provides the best information for decision makers or causes the least confusion. It is felt that the former approach causes the least confusion and the least chance for making errors in arriving at the final picture of direct and indirect costs and benefits associated with a project. The recommended approach results in a more systematic process of identification and valuation of all project effects.

In both cases (a) and (b) in Section 4.4.2 the need to include the effects mentioned arises from the nature of the measures of value which are commonly used and the fact that such measures are determined in locations which are different from either the project location or the point(s) of consumption (or export in the case of (b)). In all cases, whether or not the identified additional inputs or indirect positive effects will be assigned a positive or zero value depends on whether or not the additional inputs used or saved have any alternative productive uses (i.e., opportunity costs). This is determined in the valuation stage.

In the chapters that follow, which deal with valuing inputs and outputs, it will be assumed that location related project effects have been explicitly recognized and identified in this earlier stage in the analysis and thus will be valued independently of values assigned to direct project inputs and outputs. This approach has the advantage of clearly pointing out handling and transport inputs and not confusing decision-makers by netting out transport and handling costs from established prices used to value direct project outputs or inputs.

1/ Derivation of the local currency equivalent will be discussed in Chapters 5 and 8.

Figure 4.3

IDENTIFYING LOCATION EFFECTS: PROJECT INPUTS WHICH WOULD HAVE BEEN EXPORTED

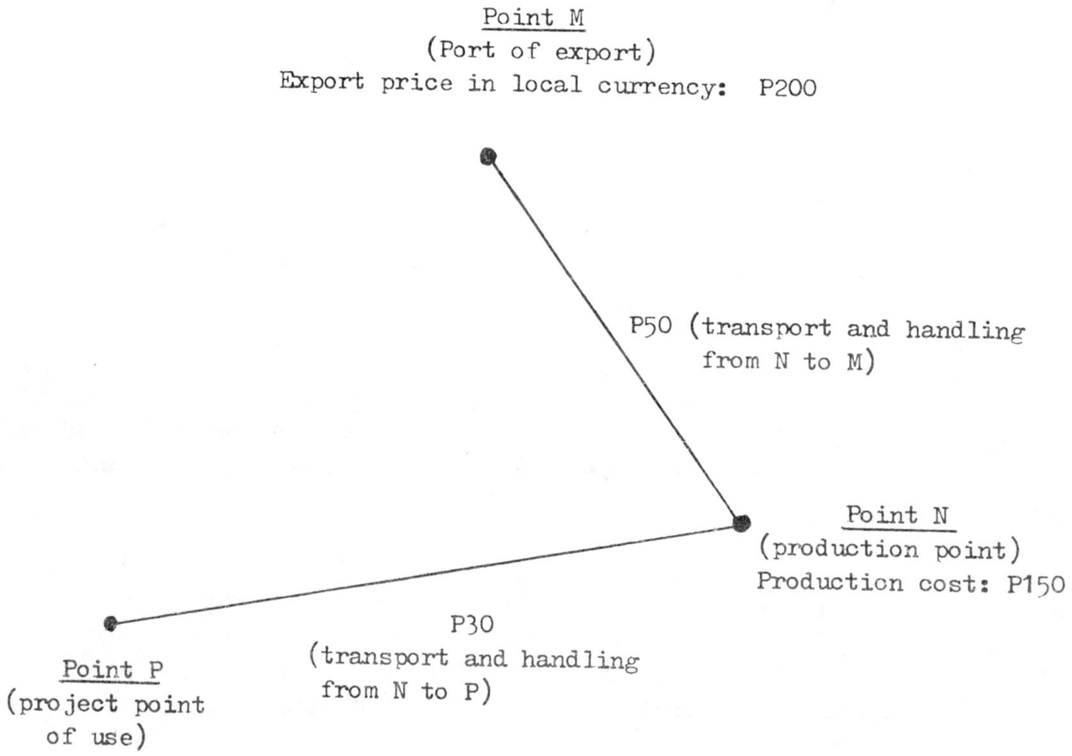

Chapter 5

VALUING INPUTS AND OUTPUTS

5.1 INTRODUCTION - THE APPROACH

Once inputs and outputs have been identified, the next step is to develop values for them. Chapters 5-8 are concerned with the process of determining appropriate values to use in an economic analysis.

In a financial analysis, the valuation process is fairly straightforward and market prices are used for all inputs and outputs. Non-market effects (externalities or indirect effects) are not valued in the financial analysis, since they do not enter into the physical or cash flow tables of the financial entity.

The valuation process is more complex for an economic analysis, first, because some market prices for direct inputs and outputs may not be appropriate measures of economic value as defined in Chapter 2; and, second, because indirect effects are considered to the extent possible.

Market prices should be used in the economic analysis to the extent that they reflect economic values. Therefore, the first logical step is to separate out all those inputs and outputs which are associated with market prices and then to determine whether existing market prices provide appropriate measures of economic value for those inputs and outputs. The remainder of this chapter discusses this step. In those cases where market prices are acceptable, the next step is to project them into the future (since a project involves future periods) and to make adjustments in prices for market location relative to the project and for inflation. These points are covered in Chapter 6. In the case of market priced inputs and outputs for which market prices do not provide acceptable measures of economic value, shadow prices have to be developed. Also, an attempt has to be made to derive shadow prices for the indirect effects associated with the project. Shadow pricing is discussed in Chapter 7 (for outputs or benefits) and in Chapter 8 (for inputs or costs).

Before the adequacy of existing market prices as measures of economic value can be usefully discussed, it is necessary to have clearly in mind the meaning of the terms "market prices" and "economic values". These are defined in the following section.

5.2 MARKET PRICES AND ECONOMIC VALUES - SOME DEFINITIONS

A market price is the amount of money which a buyer (consumer) has to pay at a given time in a given market for a good or service, or the amount of money which the seller of a good or service receives in the market. A market price is determined by the interaction of (a) consumers' willingness to pay for a good or service (demand), (b) suppliers' costs and willingness to sell it (supply), and (c) policies which constrain the free interaction of supply and demand. Regardless of how policies, market conditions, and other considerations affect the final magnitude of the market price, the basic point is that it becomes a fact once a transaction has taken place.

"World market prices" are particularly useful in an economic analysis as a basis for valuing inputs and outputs that are traded in international markets. There are two main types, namely, export prices (FOB prices) and import prices (CIF prices).

Export or FOB price. The term "FOB" means "free on board" and includes all costs to get goods on board the ship in the harbour of the exporting country, that is, project gate price, local marketing and transport costs, local port charges and export tariffs and subsidies.

Import or CIF price. The term "CIF" means "cost, insurance, and freight" included. It is defined as the price of the good delivered on the importing country's dock and includes the cost of the good at point of export (i.e., FOB price) freight charges to point of import, insurance charges and in some cases the cost of unloading from ship to pier at the port of the importing country. It excludes import duties and subsidies, port charges at point of entry (e.g., taxes, handling other than unloading, storage and agent fees), and local marketing and transport costs.

As indicated in these definitions, there is a direct relationship for a given good between FOB price in an exporting country and the CIF price in the importing country. Sometimes the analyst is dealing with a situation where a project output is intended for export, but it is not yet being exported from the project country. The analyst must then determine the most likely port of import for the project output, find the CIF price at that port and work backwards to estimate a FOB price in the project country. At the same time, there is no necessary direct relationship between the CIF price (import price) and the FOB price (export price) for a given product in a given country. Further, in a competitive world market situation, it is never possible for the FOB price in a given country to be higher than the CIF price for the same good in that same country, if both are adjusted to a common border point in the country. It is possible for the FOB price at a port on one side of the country to be higher than the CIF price at a point of entry on the other side of the country. But when adjusted to a common location and in a competitive situation, FOB can never be higher than CIF.

The basic measure of economic value adopted here is consumers' w.t.p. for goods and services, given existing policies which affect w.t.p. In the case of inputs or costs, the term "opportunity cost" (OC) is often used. As discussed in Chapter 2, the cost of using an input in the project being analysed is the value foregone by not being able to use it in its next best alternative use, i.e., its OC. However the value foregone is measured in terms of consumers' w.t.p. for the goods and/or services foregone. Thus, both in the case of benefits (outputs) and in the case of costs (inputs) w.t.p. is used as the basis for valuation in the economic analysis.

While this provides an adequate conceptual definition of economic value, it is necessary to be more specific when it comes to applying the concept in practice and deciding on exactly what measure of w.t.p. or OC, must be estimated when valuing different types of inputs and outputs. For this purpose, five categories of outputs and five categories of inputs can be defined, each of which is associated with a different measure of w.t.p. or OC. These categories are discussed below, first for outputs and then for inputs.

5.2.1 Appropriate economic value measures for different types of outputs

For analytical purposes, three types of effects which project outputs can have can be distinguished (See Figure 5.1). First, a project output can increase the total supply of a good or service available to society. Two categories of outputs related to this effect can be distinguished for the purpose of valuation. The first is consumer goods or services and the second is intermediate or producer goods or services (i.e., project outputs which will serve as inputs in other production processes which produce consumer goods). In the former case, the appropriate measure of value is the consumers' w.t.p. for the output of the project itself. In the latter case, the appropriate value measure is producers' w.t.p. for the project output, which in turn is based on consumers' w.t.p. for the other goods and services which will be produced with the output from the project being analysed. (See I and II in Figure 5.1)

As a second effect, a project can increase the availability of foreign exchange to the economy. There are two categories of outputs which fit in here. The first is exports and the second is import substitutes. The value of these types of outputs (of the project) are measured in terms of the local w.t.p. for the goods and services which can be purchased with the foreign exchange earned (in the case of exports), or the foreign exchange saved (in the case of import substitutes). Since it is necessary to measure economic values in terms of local consumers' w.t.p. for goods and services expressed in local currency, the foreign currency earned or saved has to be converted to local currency, and government policies which make local w.t.p. differ from what the country actually has to pay for imported goods and services in terms of foreign currency have to be taken into account. Unadjusted CIF and FOB values (converted to local currency) will not provide adequate measures of economic values (nor of local market prices) in cases where a government imposes tariffs or provides subsidies for exports and imports. (See Categories III and IV in Figure 5.1).

The third effect which a project output can have in terms of contribution to real national income occurs when the project output substitutes for other domestic supply, thereby releasing resources from these other domestic supply sources for use elsewhere in the economy. (See Category V in Figure 5.1). The relevant measure of economic value of the benefits due to the project in this case is the opportunity cost of the released resources, which is based on the w.t.p. for the goods and services which will be produced with the released resources.

Section 5.5 discusses how to determine the appropriateness of market prices and the factors which are likely to cause a discrepancy between the local market price and the economic value for these five categories of outputs.

Figure 5.1

OUTPUTS: MEASURES OF ECONOMIC VALUE [1/]

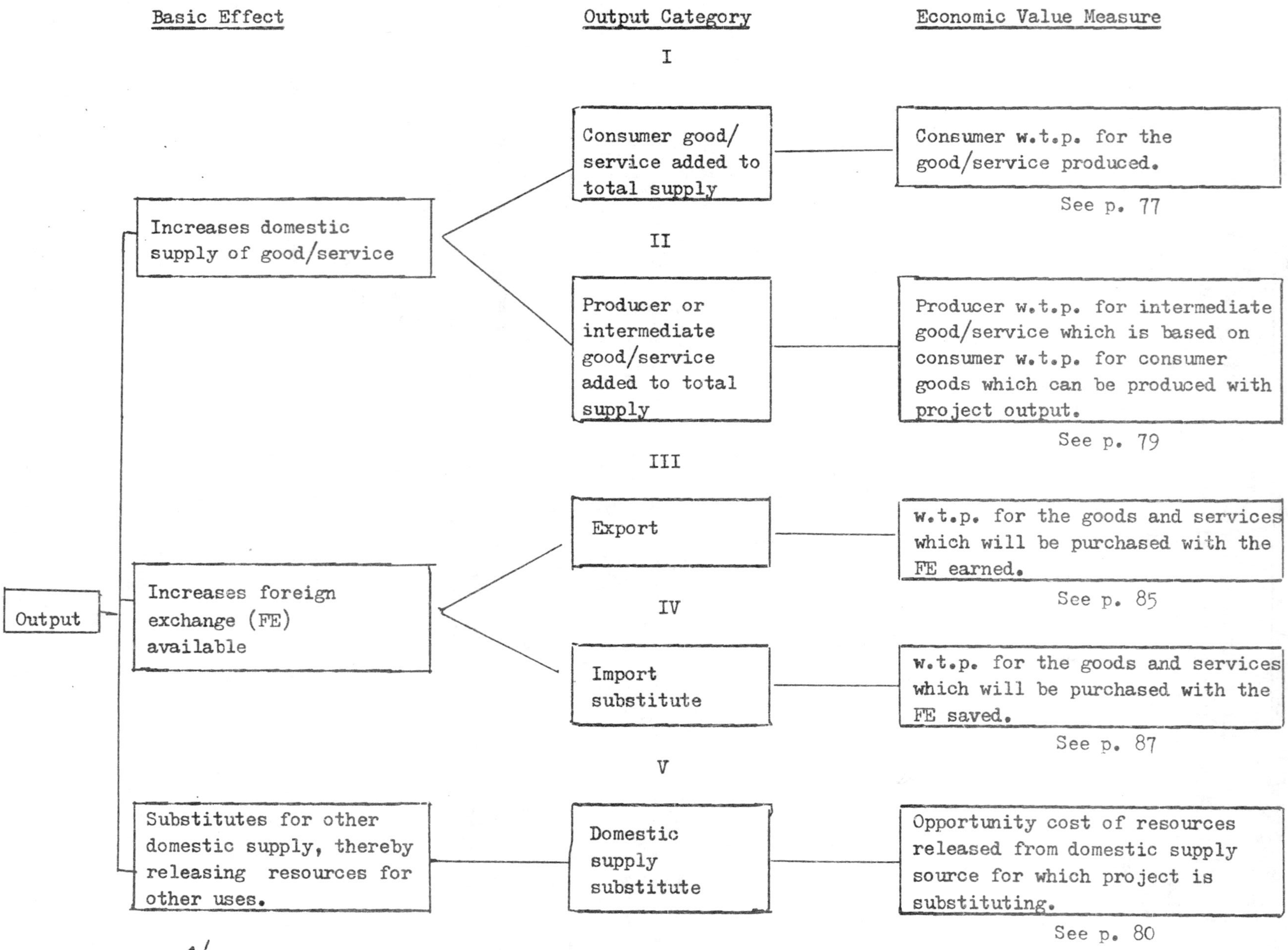

1/ The decision tree approach to presenting valuation rules was suggested by W.A. Ward. See Ward (1978).

5.2.2 Appropriate economic value measures for different types of inputs.

Two types of effects can be associated with using inputs in a project. The appropriate measures of opportunity cost or economic value are directly related to these two effects. First, an input into a project can result in a reduction of foreign exchange available for the rest of the economy. There are two categories of inputs which have this effect, namely, imported inputs when no quota exists and locally produced inputs which would have been exported in the absence of the project (see A and B in Figure 5.2). In the case of imported inputs, the appropriate measure of opportunity cost is based on the w.t.p. for the goods and services which would have been bought with the foreign currency or foreign exchange which will be spent on the imported inputs for the project. This category also includes the case where the project uses a locally produced input in short supply and forces a previous user of the input to import the input. In either case - direct use of imported input or forcing someone else to import it - the effect is the same and so is the value measure that is appropriate.

For locally produced inputs which would have been exported if they were not used in the project, the appropriate measure of opportunity cost is based on the w.t.p. for the goods and services which would have been purchased with the foreign exchange which would have been earned from exporting the inputs if they were not used in the project.

The second effect associated with inputs is a reduction in domestic real resources or inputs available to the rest of the economy when an input is diverted from other domestic use to the project. Here three categories of inputs can be distinguished (see C, D and E in Figure 5.2). The first is the locally produced input which does not reduce exports or induce new imports. The second is the imported input when a quota exists, i.e., a quantitative restriction exists on imports of the input. The third includes local resources, primarily land and labour, which are not "produced" as such.

In the case of the first category, a further distinction can be made for valuation purposes between the situation where the project induces additional domestic production of the input and the situation where the project reduces the availability of the input to the rest of the economy. In the former case, the appropriate measure of economic value is the opportunity cost of the resources used to produce the input. In the latter case, it is the opportunity cost of the input itself, i.e., the value foregone by using it in the project rather than in its next best alternative use. (Note that if other users now import the input, then it fits into Category A.)

In the case of imported inputs when a quota exists, the reasoning is that there is no additional outflow of foreign currency, since the total amount of the input imported into the country remains the same (at the level of the quota). Thus, the relevant opportunity cost is the value foregone from shifting the imported input from some other domestic use to the project. Naturally, if the quota is not being met, i.e., imports of the input without the project are below the level of the quota, then the quota is not effective and from an analytical point of view the input is reclassified as an imported input with no quota (see item A in Figure 5.2).

In the case of local resource, the appropriate measure of value is simply the opportunity cost of the resource or the value foregone by using the resource in the project rather than in its best alternative use.

Figure 5.2 1/

INPUTS: MEASURES OF ECONOMIC VALUE

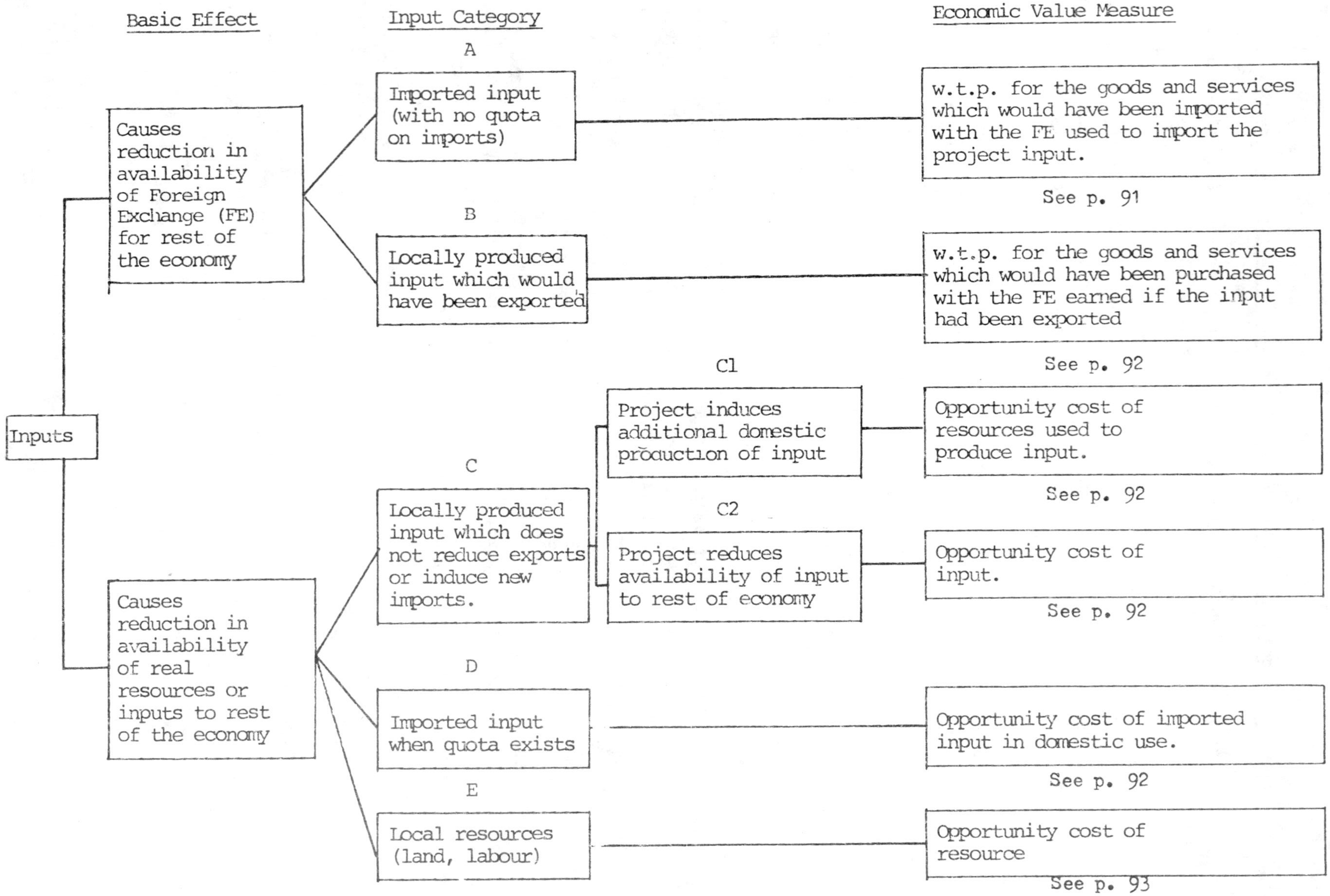

1/ The decision tree approach to presenting valuation rules was suggested by W.A. Ward. See Ward (1978).

Means for deriving these measures of economic value for inputs are discussed in Chapter 8 for cases where market prices are not considered as appropriate approximations of economic values. The following section discusses some of the factors to be considered in determining the appropriateness of market prices.

5.3 DETERMINING ADEQUACY OF EXISTING MARKET PRICES AS MEASURES OF ECONOMIC VALUE

In practice most analysts tend to accept market prices as proxies for the measures of economic values and then search for reasons why they are not acceptable. This contrasts with the theoretician's approach, which starts with the assumption that all inputs and outputs should be shadow priced, even though some shadow prices and market prices may end up being similar. Since the latter, more conceptually correct approach will involve considerably more time and funds. In practice, the analyst will generally have to use the first approach and only attempt to shadow price major inputs and outputs for which the market price is not considered to be an "adequate" measure of economic value. It should be up to the analyst to show convincing evidence that the magnitude and importance of the difference between a market price and the w.t.p. is great enough to justify the extra effort involved in shadow pricing an input or output. Quite apart from the additional time and funds required to develop shadow prices, there is the danger that inappropriate shadow prices will lead to decisions that, taken in the context of the actual workings of the economy, will be worse for the country than if market prices had been used (which take into account the influences of policies, customs, attitudes, and non-economic objectives which actually direct the economy).

"Adequacy" or "acceptability" of a given market price as a measure of w.t.p. or OC thus is a relative concept which depends on the situation. For any given situation, acceptability of a market price for an input or output depends on (a) the importance of the input or output in the overall project, (b) the estimated degree of discrepancy between market price and w.t.p. or OC for the input or output, and (c) the practicability of developing an acceptable shadow price (which relates centrally to the time and budget available for the economic analysis in each case, and the purpose for the analysis). Each of these three factors is discussed in the following sections.

5.4 ESTIMATING THE IMPORTANCE OF INPUTS OR OUTPUTS

Most market priced forestry project outputs will be important in terms of their values relative to total project benefits. Thus, most direct outputs are potential candidates for shadow pricing.

With regard to inputs, there will likely be many items which are relatively insignificant in terms of total costs measured in market price terms. Every project involves purchases of a myriad of small items - office supplies, hand tools, etc. Such items generally need not be shadow priced. However, one word of caution is needed here. A project may involve a number of different inputs that individually are unimportant in terms of total cost, but when they are added together, they may have a significant influence on total costs. While it may not be worth the time and effort to shadow price each item individually, it is possible through a sensitivity analysis to test the effect on project outcome of an increase in some or all of the values of these inputs combined. (See Chapter 10).

Most forestry projects also involve major input items such as land, labour, heavy machinery, processing equipment, etc. Such items may or may not be considered for shadow pricing, depending on time and funds available and the conditions which influence the market prices which exist for them.

As a rough rule of thumb, if an input valued in market price terms represents 5 percent or more of the total present value of the cost of the project, then it is a logical candidate for shadow pricing. Whether or not it is actually worth shadow pricing the input depends on the magnitude of the estimated difference between its market price and its economic value (as discussed in the next section). For example, if an item that represents 5 percent of total cost in present value terms has an estimated shadow price that is 80 percent below the market price, then the effect of shadow pricing will be to reduce the total cost by 4 percent (80 percent of 5 percent) for the economic analysis. This could be significant in terms of a project's economic profitability. On the other hand, if such an item has an estimated difference between market and shadow prices of only 10 percent, then the difference in terms of total cost would only be one-half of one percent, which would not be as significant in terms of economic profitability.

Some types of items often listed as inputs can present problems. For example, projects generally include various physical structures, roads, etc. If these have been listed as "inputs" they are likely to be major items in terms of total costs. As pointed out in the previous chapter, such items should be broken down into their component inputs of labour, equipment, various types of materials, etc. In this case, the judgement relates to whether the component inputs are important enough to merit the extra effort involved in developing shadow prices.

5.5 IDENTIFYING DISCREPANCIES BETWEEN EXISTING LOCAL MARKET PRICES AND ECONOMIC VALUES

To determine whether or not a shadow price should be developed for a given input or output which is considered important, an estimate is needed of the nature and direction of any likely discrepancy that might exist between its existing market price and its economic value.

On the surface the definitions of economic value and market price given in Section 5.2 may seem similar. In many cases the two measures will coincide. However, in some cases, what a consumer has to pay for a good or service in the market (the local market price) may not be a reflection of what he actually is willing to pay for it (its economic value). He may very well be willing to pay more than he actually has to pay, but because of various policies - e.g., a price ceiling - he does not have to pay an amount equal to his w.t.p., which was defined as the economic value of the good or service. Similarly, a producer (the project) may have to pay in the market an amount for an input which is higher or lower than the value which that input could produce in the best alternative activity, i.e., its OC. For example, a minimum wage law may be in existence which sets a wage that is higher than the OC of labour. The producer has to pay the minimum wage and this amount would be used in a financial analysis. However, in the economic analysis the lower OC of labour would be used.

5.5.1 Discrepancies caused by government policies

In the above examples, it is existing government policies which cause market prices to be different from economic values as defined here. However, in other cases where government policies affect market prices in the project environment there will be no discrepancy between market prices and economic values. For example, a government imposed tariff or tax on imported goods will affect the price of the good in the local market and the quantity of the good imported relative to what the price and quantity would have been in the absence of the tariff. However, in terms of the economic value definition used here, the resulting local market price with the tariff would still provide a reasonable approximation of consumers' w.t.p. for the good or service at the margin in the existing market. If the tariff were expected to remain in effect during the period of the project, then the local market price would be used in the economic analysis as well as in the financial analysis.

It can be seen that just because a government policy influences a local market price (and the quantity sold in the market), there is no reason to assume that the existing local market price for a particular good or service will differ from its economic value as defined above. It all depends on whether the policy restricts movement of the market price in a given situation, and on the type or category of input or output which is being considered. Thus, in order to develop a more systematic approach to identifying discrepancies between market prices and economic values each of these two considerations needs to be looked at.

In general, any policy which allows the free movement of prices will not cause a discrepancy to exist between market price and economic value as defined here. Such policies may have a significant effect on the size of the market, by changing supply and/or demand. However, so long as the policy does not prevent the changed demand and supply from coming into balance, then the local market price is likely to reflect w.t.p. for outputs and the OC of inputs. For example, a high tax on consumption of good X will restrict effective demand, but if the tax is not accompanied by a minimum price or price ceiling, supply and demand will come into balance at a price which reflects the "marginal" consumer's w.t.p. for good X, even though the quantity consumed will be less and the price will likely be higher than if the tax did not exist. The term "marginal consumer" is used to refer to the consumer who is just willing to pay the resulting market price, but no more. He is marginal in the sense that if the price were increased slightly, he would drop out of the market, i.e., not purchase the good or service in question. As a matter of practicality, most project analyses are undertaken with the simplifying assumption that output will be used or purchased by "marginal" consumers.

Similarly, in the case of a subsidy for consumption (or production) of good Y, the price that results in the market will reflect w.t.p. of the marginal consumer in that market, although the quantity sold will be higher and the price likely lower than in the absence of the subsidy. If a subsidy is provided to consumers it means that such consumers will be willing to buy more of a good or service at any given price or will be willing to pay a higher price for a given quantity. Similarly, a tax on producers means that they will be willing to sell less at any given price or that they will want a higher price for any given quantity of the good or service.

Governments use a great number of policy tools to guide their economies toward what are considered to be the economic and social development goals of the country. Common ones include taxes (including tariffs), subsidies, minimum price and price ceiling laws, and quantative restrictions on market transactions.

Of those mentioned, the only ones which restrict local price movements (i.e., result in excess demand or supply) are minimum price and price ceiling laws. Even these policies will cause discrepancies between local market prices and economic values in the case of only some of the input and output categories discussed in Section 5.2 (and shown in Figures 5.1 and 5.2). The potential influence of each of these types of policies and how it is possible to identify in practice whether an existing policy is effective is examined further. This latter question is particularly important, since in many countries policies exist on the books which are, in fact, ineffective and, thus, for analytical purposes they can be disregarded. For example, a government may have placed a price ceiling on lumber for construction but, in fact, the market price for lumber is below the price ceiling. In this case, the price ceiling is ineffective in the sense that the price of lumber would be the same with or without the policy, or price ceiling. As in the case of identification of inputs and outputs, the "with and without" concept can be applied in identifying the effectiveness of policies.

Minimum price imposed by government - If a minimum price imposed by a government is effective for a given good or service, then supply would tend to exceed demand for that particular good or service at the established minimum price. If supply does not exceed demand at this price, then the minimum price is an ineffective policy tool, since the market price would settle at the same level with or without the policy (the minimum price).

In the case of an effective minimum price associated with an output - i.e., where there is excess supply of that good or service at the minimum price - it is unlikely that a project would be proposed to expand supply (or add to total supply) of that output. This follows from the fact that buyers can obtain as much as they want at the minimum price without the project, so any addition to total supply would likely remain unutilized so long as the policy remains in effect. Consumption would remain the same with or without the project and, therefore, the output of the project would be valued at zero. Of course, if the minimum price is expected to be reduced, then this would have to be taken into account.

The only type of project that would likely be proposed (or make sense) for an output for which an effective minimum price exists is one that would substitute for existing supply. In this case, as pointed out in Section 2.2.4, the appropriate measure of the value of benefits due to the project is related to the w.t.p. for the goods and services which would be produced with the resources released, i.e., the cost savings, and not the w.t.p. for the project output itself. [1] Thus, a minimum price on the output itself would not be of concern in the analysis of this type of project.

In the case of inputs, where a minimum price for a project input is effective, supply of that input would exceed demand. The existing local price in the market would reflect the producers' willingness to pay for the input as long as there is competition among the producers who are buying it. However in this surplus situation, if the project employs additional units of the input out of the surplus supply, the sacrifice in terms of alternative consumption benefits foregone will be low or nil. The economic cost of using

---

[1] This follows from application of the "with and without" test. Consumption of the product produced by the project would not change.

an input in the project in circumstances where the input would otherwise be unemployed is then zero or at least less than the existing local price for purposes of an economic analysis. As will be pointed out in Chapter 8, there may be reasons why the shadow price in this situation will be greater than zero, e.g., in the case of labour.

Evidence of an effective minimum price policy can be observed by looking at the supply situation for goods and services. Some indications of effective minimum prices are as follows:

- accumulation of stocks when a minimum price is imposed;
- excess capacity coupled with an increase in the minimum price;
- creation of a black market (or informal labour markets where labour is hired below the minimum wage set by government);
- producers abandoning the market because of lack of sales;
- the existence of unemployment coupled with an effectively enforced minimum wage law. The minimum wage (the "market" wage) is higher than the opportunity cost of labour which should be used in the economic analysis.

Price ceiling imposed by the government - In the case of an effective price ceiling imposed by government, buyers will not be able to buy as much as they would like to at the controlled price, i.e., there will be an excess demand at the existing market price. In this situation the market price will be lower than marginal buyer's actual w.t.p. for the input or output. If there were no control on prices, buyers would bid up the market price until the available supply equaled demand at some higher price. This price would then be equal to w.t.p. for the input or output given the supply situation.

One place to look for evidence of the existence of a price ceiling is the regulatory legislation affecting the marketing of an input or output. However, the mere existence of legislation does not guarantee that price ceiling legislation will be effective. It is necessary, therefore, to look for evidence easily observable in the market that consumption is, in fact, restricted by the government control. Manifestations of an effective prioc ceiling policy include the existence of queues, black markets, and various forms of rationing. These are all indicators that the marginal buyer is willing to pay more than the going price for the quantity offered in the market, i.e., that the controlled market price is lower than the w.t.p. at the margin.

The nature and direction of discrepancies between local market prices and economic measures of value caused by effective government controls on local prices (minimum prices and price ceilings) are summarized in Table 5.1 for each of the output and input categories shown in Figures 5.1 and 5.2. As indicated for several categories of inputs and outputs the policies can either not be effectively applied, or if they are effectively applied, they are not applicable in terms of the measures of economic value used for the particular category of input or output being considered.

Table 5.1

DISCREPANCIES BETWEEN LOCAL MARKET PRICES AND ECONOMIC VALUES CAUSED BY EFFECTIVE GOVERNMENT CONTROLS ON LOCAL PRICES 1/

		Controls on Local Prices	
		Minimum Prices	Price Ceilings
Outputs (see Figure 5.1)			
I.	Consumer good/service added to total supply	NA	LP < w.t.p.
II.	Producer or intermediate good/service added to total supply	NA	LP < w.t.p.
III.	Export	NA	NA
IV.	Import substitute	NA	NA
V.	Domestic supply substitute	NA	NA
Inputs (see Figure 5.2)			
A.	Imported input - no quota	LP > OC	NA
B.	Locally produced input which would have been exported	NA	NA
C1.	Locally produced non-exportable input when project induces additional supply	NA	NA
C2.	Locally produced non-exportable input when project reduces availability to rest of economy	LP > OC	LP < OC
D.	Imported input when quota exists	NA	LP < OC
E.	Local resources	LP > OC	LP < OC

1/ LP = local price; w.t.p. = willingness to pay; OC = opportunity cost; NA means "not applicable" in terms of the appropriate measure of economic value shown in Figure 5.1 (outputs) or 5.2 (inputs) or that the policy cannot be effective for the particular type of input or output being considered.

This discussion relates to discrepancies between local market prices and economic values which are due to the influence of government policies. Government policies, through the existence of an official fixed exchange rate (OER), can also cause discrepancies between the existing "market" price for foreign currency (the OER) and the real economic value of foreign currency in terms of what it actually can purchase in the local market in terms of domestic prices. Since several of the output categories (III and IV) in Figure 5.1 and input categories (A and B) in Figure 5.2 involve use of world market prices (CIF and FOB values) to derive economic values, such discrepancies need to be taken into account if they exist, and foreign currency related effects shadow priced. A "shadow exchange rate" (SER) is commonly used. Its derivation is discussed in Chapter 7.

In the case of an identified discrepancy between the OER and the real value of foreign currency in local terms, a general SER is derived and used in all calculations of economic values for inputs and outputs which involve earning, saving or using foreign exchange or foreign currencies. Thus, once the SER has been derived, it can be used for all the categories of inputs and outputs shown in Figures 5.1 and 5.2 which directly or indirectly involve foreign exchange (i.e., output categories III and IV and input categories A and B). Not only should a general SER be used to evaluate all such inputs and outputs for a specific project, but it should also be used in analysing all other projects in the country. Thus, estimation of a SER should be done at the national level and used systematically for all projects. This contrasts with the situation for many domestic (non-traded) inputs and outputs. For example, in the case of labour there will likely be shadow prices that are unique to given small areas within a country. If unemployment is high in one area (and mobility of labour is low) then a lower opportunity cost (shadow price) for labour would exist for projects in that area than would exist for labour in other areas with lower unemployment.

5.5.2 Discrepancies caused by other factors.

In addition to government policies, there are other conditons which may exist in the project's economic environment which can cause discrepancies between existing local market prices and economic values for some input and output categories.[1] The main ones needing consideration are:

- existence of monopoly or monopsony elements in the markets for project outputs or inputs;
- existence of speculation or status influences, particularly in the case of land prices.

---

[1] Note that existing market prices are emphasized. In all cases where existing market prices are used as a basis for economic values, a number of factors that will influence projected future prices and economic values have to be taken into account. This chapter attempts to identify discrepancies between existing market prices and economic values. Expected changes in conditions in the future that will affect market prices as well as economic values are discussed in Chapter 6. They should also be considered in deriving estimates of future market prices that will be used in the financial analysis.

5.5.2.1 Monopsony and monopoly

The existence of monopsony power and monopoly power is common in many countries. Government set prices are a form of monopoly exercised by the government. The distinction made here (in relation to the two categories of pōlicy influences discussed in the previous section) is that this discussion involves monopoly and monopsony power exercised by private (non-governmental) individuals and/or groups, e.g., corporations or unions. The relationship between the existence of either of these two forms of market control and government policies is often difficult to ascertain, i.e., whether in fact a government is encouraging either or both of them, or it merely permits or condones them in the economy. In some cases, a government is against the existence of monopoly and monopsony in the private sector, but it does not have the political power to do away with them.

The point here, in terms of valuation, is that if monopoly and/or monopsony exist in the project environment and are expected to persist during the period of the project, then their effect should be taken into account when looking at discrepancies between market prices and economic values. For example, if a strong union exists and is expected to persist during the period of the project, then the discrepancy between union set wages and actual opportunity cost of labour should be taken into account in the same way that the discrepancy caused by a government set minimum wage needs to be considered.

Monopsony power - If one or a few purchasers acting together can alter market prices by modifying their buying policies, then a "monopsony" condition exists.[1] A buyer enjoying some degree of monopsony power will change the market price of the input or output to his advantage. If his power is effective, then the price that he will pay in the market for the input or output will be somewhat lower than what he actually would be prepared to pay if he did not have the monopsony power. The market price will reflect only partially his real willingness to pay.

If there is monopsony power associated with the product (good or service) being produced by the project, or monopsony power associated with the inputs required for the project being analysed, the same kinds of discrepancies can develop as in the case of price ceiling policies. Thus, the direction of such discrepancies for different categories of inputs and outputs can be identified in column 2 of Table 5.1.

Evidence of the existence of monopsony forces can be obtained in some cases by examining sales records. If one buyer dominates the market, then there is good reason to suspect that that buyer is influencing market prices in his favour. In the case of collusive agreements among several buyers, market influence is much more difficult to identify and to measure. Since these types of arrangements are forbidden by law in most countries, they tend to be made in secret or through informal, tacit understanding among buyers. Generally, when this type of influence is suspected to exist in the market, it can best be considered in a sensitivity analysis, since it is extremely difficult to adjust the market price for this type of effect.

1/ Strictly speaking, when there are a few purchasers rather than just one controlling a market, economists talk about "oligopsony". For convenience, the term "monopsony" is used loosely to refer to both situations.

Monopoly power - If one or a few sellers in a market have the power to influence prices by altering their selling policies, a monopoly condition exists.[1] Whether this monopoly condition will cause a discrepancy between market prices and economic values depends on the type of output or input being valued.

In general discrepancies caused by the existence of monopoly power in the markets for project outputs or inputs will be the same as those shown in column 1 of Table 5.1 for government minimum price policies. In other words, monopolists will in a sense have the power to set "minimum" prices. On the output side, if a monopolist sets a higher price than previously existed in the market, it will likely influence the volume of sales since less will be consumed when the price is increased. However, consumers will adjust their purchases until the marginal consumer (or last buyer of the good or service) is just paying what he is willing to pay. If the price were raised slightly, the marginal consumer at the initial price would drop out of the market. Thus, the price set by the monopolist will reflect w.t.p. at the margin, which is the concept of relevance in terms of valuing outputs in a practical project analysis.

On the input side, discrepancies can arise between the local market price and the opportunity cost associated with an input if a monopolist sets the price. For example, in the case of an imported input when no quota on imports exist, if the monopolist controls the local price of imports, he can set it above the opportunity cost of foreign exchange used to import the input. Similarly, in the case of local resources, say labour, a monopolist (e.g., a union) can set the minimum wage level at a level above the opportunity cost of labour.

5.5.2.2 Existence of speculation or status influences on market prices

Prices for land are often set in the market on the basis of speculation concerning future values of land and/or on the basis of status associated with owning land. Such influences can cause a divergence between market prices for land and the value of land in terms of its alternative productive uses (i.e., contribution to real national income). The valuation approach used in EAFP (see Chapter 8) eliminates the need to be concerned with such divergencies. It is suggested that land should always be shadow priced on the basis of its alternative productive uses over the period of the project and not on the basis of land sale prices.

5.5.3 Comments on identifying discrepancies

It is clear that very often the analyst will be able to identify some of the discrepancies discussed and to estimate the direction of the divergence between market prices and economic values, i.e., whether economic values will be higher or lower than existing market prices. However, it is quite a different matter to be able to measure the magnitude of such differences. Even so, the qualitative knowledge obtained provides a very useful guide for focusing attention in those areas in which the discrepancy although not quantifiable, is judged to have a potentially important impact on project worth. Thus, this type of analysis helps to identify areas of uncertainty to be treated later in the sensitivity analysis.

1/ The term "monopoly power" is used to refer to oligopoly (several-sellers) as well as the traditonal monopoly (one seller) situation.

There is one other point which should be mentioned. This relates to "second-round" effects, or discrepancies which exist between market prices and economic values one or more steps removed from the market prices for direct inputs or outputs associated with a project. The discussion of discrepancies has concentrated on only the direct or first-round discrepancies which can be identified. It is quite possible that the analyst will have information on which to judge whether there are discrepancies further down the line which should be taken into account. For example, it may be determined that there are no apparent policies or other factors which are directly affecting the price of locally produced tractors to cause a discrepancy between the local market price for a tractor and its economic value in terms of the project. However, it is known that there is a discrepancy between the local market price and the economic value for the steel used in producing the tractor. Ideally, an attempt should be made to shadow price the tractor taking into account the shadow price for steel. While in rare instances it may be possible to make such corrections, in general, from a practical point of view it will not be possible to trace through all the effects of every input that enters into the production of the inputs used in a project.

Normally, the analyist will have to be content to deal with the obvious discrepancies directly associated with the prices for project inputs and outputs. Obviously, if the analyst has the time and funds to carry the shadow pricing exercise to its logical extreme and to take into account all such second-round effects then he can proceed to do so using exactly the same concepts and techniques explained here for treating the first-round effects.

5.6 EASE WITH WHICH ACCEPTABLE SHADOW PRICES CAN BE DEVELOPED

The final factor which needs to be considered in deciding whether to use the market price for an input or output or whether it should be shadow priced relates to the ease with which an acceptable shadow price can be developed. In nearly all project analyses, the analyst is faced with a time and a budget constraint. He will not have the time to spend on shadow pricing every input or output item which is important and for which a discrepancy is expected between market price and economic value. For many inputs or outputs which are difficult to shadow price, the choice will have to be made between using a rough "guesstimate" of an appropriate shadow price (that at least covers some of the estimated discrepancies between market price and economic value) or using the market price, even though it is recognized to be less than a perfect measure of economic value. (In the latter case, the discrepancy is explicitly acknowledged in the analysis report and alternative values can be tested in the sensitivity analysis.) The choice between these two alternatives will have to be made on the basis of the circumstances surrounding the analysis (its purpose) and the judgement of the analyst as to just how critical the value of the particular item is in terms of the selected measure of economic efficiency.

Finally, for most forestry projects the analyst will encounter indirect effects (externalities or non-market priced effects) for which it is difficult, if not impossible, to develop acceptable shadow prices (e.g., in valuing scenic beauty, increases in self-reliance, reduction of drudgery, etc.). In such cases, the best the analyst can do is to describe the effects in physical and/or qualitative terms and suggest how they are likely to affect the project outcome and its impact on society.

Chapters 7 and 8 discuss in more detail some of the practical considerations which influence the decision on whether or not to attempt to shadow price an input or output.

Chapter 6

USE OF MARKET PRICES: GENERAL CONSIDERATIONS

6.1 INTRODUCTION

For most inputs and outputs, market prices are likely to be used as direct measures of economic value or indirectly in deriving shadow prices (as will be discussed in Chapters 7 and 8). There are some general considerations which need to be kept in mind when market prices are used as a basis for economic values.

Observed market prices for inputs and outputs reflect present and past conditions of supply and demand. Values used in project analyses involve consideration of future supply and demand conditions. Thus, to arrive at future value estimates for effects or inputs and outputs based on observed market prices, some adjustments in observed prices and trends in such prices may have to be made. First, it is recommended that economic analyses be carried out using price or value estimates net of the effects of inflation. This means that adjustments will have to be made to take out the likely effects of inflation on future price levels. Second, even after the expected effects of general price inflation have been eliminated, there may be changes in relative prices for certain inputs and outputs, i.e., changes in prices of specific inputs and outputs relative to general price changes due to inflation. Third, the project being analysed may itself have an influence on future relative prices, and thus on economic values. These three considerations are the subject of this chapter. They apply whether market prices are used directly or indirectly in deriving economic value measures.

The assumption stated in Section 4.4 namely that all location related effects are explicitly identified and recognized as independent project effects is stressed. This is mentioned since some manuals or guides to project analysis treat location effects in the valuation stage, by adjusting market prices for the location related effects. For example, instead of valuing an export output on the basis of its FOB value alone, the costs of transportation from the project (often called the project "gate") are subtracted from the FOB value of the output appropriately converted to local currency to arrive at what is called an "export parity price". [1]

While either approach - that suggested in Chapter 4 or the above approach which involves adjustments at the valuation stage - should produce the same result if done properly, the suggestion in Chapter 4, that location effects be considered separately in the analysis is followed here.

---

[1] The reader interested in this approach is referred to Ward 1976.

6.2 ESTIMATING FUTURE PRICES

An existing or past market price is a "fact". It represents what a good or service actually is or was traded for in a given market. Existing market prices can be used directly in valuing inputs and outputs that occur today. When deciding on how to adjust existing prices so they can be used to value inputs and outputs that occur sometime in the future, the analyst has to consider two factors. The first is general price inflation and the second includes influences that are likely to affect particular prices in the future, or prices of specific inputs and outputs relative to prices for other inputs and outputs.

Inflation relates to general price increases which affect all goods and services. Inflation reflects a decline in the real value of money. In addition, there may be relative price changes for some goods or services. Some prices may be expected to change in value more or less than the general level of inflation and, therefore, change relative to other prices. In developing future value estimates for project appraisal purposes, it is important that the analyst be aware of the distinction between general inflation and relative price changes and make appropriate adjustments for both in estimating future values.

6.2.1 Treatment of Inflation

The general approach recommended is to work with prices net of inflation, but to include any relative price changes which are expected. Thus, if prices of all inputs and outputs for the project are expected to increase at the same rate (i.e., at the rate of general inflation), then the analyst can merely use existing prices as a measure of future prices (keeping in mind that actual money prices will increase with general inflation). If certain prices are expected to increase or decrease relative to others, (i.e., faster or slower than the rate of general inflation) then adjustments can be made in such prices according to the assumptions accepted regarding the rate and direction of relative price changes. The result will be an estimate of the expected prices in the future, taking into account relative price changes expected, but excluding the effects of inflation.

The advantages of using price estimates net of inflation, i.e., relative prices, are (a) the analyst does not have to try to estimate general price inflation over the life of the project, which is always difficult to ascertain and justify, (b) the results can be understood more easily, and (c) the analyst will be able to show more clearly the assumptions used in the analysis concerning relative price changes.[1]

As an empirical point, it should be stressed that relative price changes tend to be more pronounced in situations of high inflation, as investors search for means to hedge against inflation. For example, high inflation tends to encourage investment in land and other real assets that increase in value at a rate greater than or equal to the rate of general inflation. Bank accounts and certain other fixed return investments, on the other hand, have a tendency not to keep up with inflation because the fixed return becomes less valuable in the future as inflation increases. Thus, relative prices of land and certain other assets may be bid up relative to other prices in periods of inflation as demand for such assets increases.

---

[1] For a more detailed discussion on the tratment of inflation and some examples of application, see H. Gregersen, 1975.

6.2.2 Estimating Relative Price Changes

It is quite common that some prices are increasing (or decreasing) relative to others. For example, in many countries, the price of stumpage is increasing relative to other prices, i.e., it is increasing at a rate faster than the rate of general price inflation. In cases where relative price changes are expected, the question arises as to how the analyst should attempt to estimate or forecast such changes. Forecasting is an area of specialization in and of itself and can be quite complicated to carry out in practice if it is to be done properly. Thus, no attempt is made to cover the techniques and approaches in EAFP.1/

In most circumstances the analyst dealing with forestry projects will not find himself in a position to use sophisticated forecasting techniques to estimate future prices; he will have to rely on simple approximation techniques.

If acceptable data on past prices are available, then the simplest approach is to plot prices over time on a graph. If a trend is evident, then the resulting trend line can be extended into the future. This can be done with regression analysis or simply by extending visually the historical price line on the graph into the future. In working with historical data, there are a number of ways of smoothing out variations that occur from year to year and adjusting for inflation to arrive at a long-term trend estimate which excludes cyclical influences and the effects of inflation.2/

In this type of trend projection, it is assumed that certain forces (other than the project itself) have affected prices in the past. It might not be possible to identify or define these forces with precision, nor how they are interlinked in the market. However, their combined effect on relative market price changes over time can be observed. For practical purposes, the analyst assumes that these forces will persist into the future and that the same trend can be projected over the period of the project. This simple type of projection technique is most often used in practice.

For some forest products, records on domestic prices over time may be limited. In fact this is the usual situation in countries where the contribution of the forest sector is modest and where statistical services are not well developed. In these cases, the analyst can do little more than try to obtain opinions of knowledgeable people and look at trends in relative prices in other countries and adapt these to his needs. Alternative price assumptions can be introduced in the sensitivity analysis of project results.

In other cases, accurate records exist covering extended periods, and clear trends are readily perceived. This might be the case, for example, for wages, or for some internationally-traded goods, where records can be obtained from international agencies or from the exporting/importing countries' statistical services. 3/

1/ The reader can find various techniques discussed in IUFRO 1971 and in Chisholm 1971, as well as in references cited in these two documents.

2/ See Chisholm, op cit. Some hand calculators are now available which will make forecasts based on various assumed relationships.

3/ cf. United Nations and FAO 1977.

Some approximations of future price movements are relatively easy to imagine without having much data. For example, if a given region's forests are being rapidly depleted, and population density and the development of general economic activity show a clear trend to increase, the analyst has sufficient reasons to expect growing scarcity of forest products and rising prices if it is assumed that the trend will persist. Thus he can pick some reasonable rate of increase in prices and test others in the sensitivity analysis.

On the input side, records usually exist on prices of imported goods (in the project country or in neighbouring importing country customs files, or in the files of importers). Price trends can be derived from such records and projected into the future.

6.2.3 The "Big Project" effect

Naturally, a new project being analysed has not influenced the way in which prices have changed in the past relative to other prices or the rate of general price inflation. However the proposed project may be large enough in relation to input or output markets to be able to influence prices in the future. Thus, a pulp and paper project's output might add significantly to supply and result in a decrease in future prices. Or the project requirement for given inputs may be large enough to push up the price of these inputs. If information is available on which to base an estimate of how the project is likely to affect future prices, it should be taken into account. It may only be possible to state the direction of the expected influence. Even though the magnitude of the effect cannot be estimated, the analyst should still include information on the expected direction of change, so various potential prices can be tested quantitatively in the sensitivity analysis.

Treatment of project influences on future input or output prices - or what is often called the "big project" effect - can also be considered as a form of shadow pricing. It does not matter whether such adjustments are dealt with under the heading of "shadow pricing" or "use of market prices", so long as they are considered to the extent possible.

Chapter 7

SHADOW PRICING OUTPUTS

7.1 INTRODUCTION

This chapter deals with an approach to shadow pricing outputs for an economic analysis when existing market prices are considered inadequate direct measures of economic value.

As shown in Figure 5.1, there are five basic categories of project outputs which can be distinguished for purposes of valuation. These are:

- consumer goods or services which add to total domestic supply available;
- producer or intermediate goods or services which add to total domestic supply available;
- output substituting for existing domestic supply;
- exports;
- import substitutes.

The eventual effect in all cases is an increase in the goods and services available for domestic final consumption. However, the appropriate approaches to shadow pricing such increases depend on the category of output being considered and the nature of the links from immediate or direct project output to the increase in availability of domestic consumption goods and/or services. In the case of the first two categories, the relevant measure of value is the w.t.p. for the output of the project. For the third category, the relevant measure of value is based on opportunity cost of the resources released. The last two categories of outputs involve earning or saving foreign exchange. Thus, the relevant measure of benefits is based on what the foreign exchange earned or saved can buy for domestic consumers in terms of local prices, i.e., w.t.p. for imported goods in local price terms. The remainder of this chapter discusses appropriate approaches to deriving these measures of value.

7.2 CONSUMER GOODS AND SERVICES THAT ADD TO TOTAL DOMESTIC SUPPLY

This category of output is often considered to be the most difficult type to value for an economic analysis when the local market price is rejected as a measure value. Fortunately, most forestry project outputs are not final consumer goods that are added to total supply. If they are, then it is frequently found that their existing market prices provide a reasonable approximation of economic value or w.t.p. The main exception is a market priced good or service for which a price ceiling has been set(see below). As mentioned in Chapter 5, in cases where a minimum price has been set which creates a discrepancy between market price and w.t.p., it is unlikely that a project will be proposed to add to total supply. This follows from the fact that an effective minimum price is associated with excess supply, so a project would not likely be proposed that would merely increase that excess supply.

The appropriate measure of value for this first category of output is consumers' w.t.p. for the increased output. If the existing market price is judged to be inappropriate as a measure of w.t.p., then the analyst has to try to estimate an approximate schedule of w.t.p. for the output. The usual way is to conduct a survey among prospective consumers.

Two points should be kept in mind concerning consumer surveys. First, in many cases and particularly in those situations where the project affects persons outside the market economy, potential consumers often will not understand monetary values well enough to provide an accurate monetary measure of their w.t.p. for the potential output, particularly considering that the expressed w.t.p. must reflect ability to pay to be meaningful. In other words, if a community family earns a cash income of $50 per year and says it is "willing to pay" $60 per year for, say, fuelwood, this is a meaningless result in terms of an economic efficiency analysis.

Second, experience indicates that w.t.p. surveys sometimes produce biased values, even for consumers within the market economy. For example, even if a family could well afford to pay what it says it is willing to pay, it may not actually do so if the good becomes available. Along the same lines, questions related to how much people would consume at a given price if the output were available sometimes elicit quantity estimates that are different from the quantities that people actually are willing to purchase at that given price. However, despite these potential shortcomings such surveys may be the only, and therefore best way to get some idea of local w.t.p. Thus, they can be a useful tool.

In some cases a forestry project output will add to total supply of a group of goods which have the same end use (i.e., relate to the same consumption objective). The goods themselves may be different, but the use is the same; thus they should be considered together. For example, fuelwood and coal may be used interchangeably for fuel by local villagers. A fuelwood project may add to the total supply of fuel available. It may be substituted for coal by local villagers, but the released coal in turn will be added to the supply available for and used by urban and industrial fuel users. If there is no market price established for fuelwood which is acceptable as a measure of economic value for the additional project output, then the market price for coal may provide an acceptable measure of value when appropriately converted to some common measure of fuel/energy value (e.g., calorific value). This would be the case if the market price adequately reflects w.t.p. for coal.

In this case a first reaction might be to value the fuelwood as a substitute for coal. In fact, while the fuelwood is being substituted for coal by villagers, the coal will be used elsewhere, i.e., the total supply of "fuel" available has been increased. Thus, the appropriate measure of value is the w.t.p. for the additional fuel indirectly made available to society by the project. Section 7.4 will discuss this distinction in more detail. The point is that w.t.p. is based on use or consumption value and there may be several seemingly diverse products which have the same use value. For the purposes of the economic analysis, they are considered together when defining "supply available" and determining whether a project output adds to or substitutes for existing supply.

In the case of a project which would add to total supply of a consumer good or service for which a price ceiling exists, a situation of excess demand may be encountered, i.e., at the prevailing maximum allowable market price, consumers are willing to buy more than suppliers are willing to sell. As indicated in Chapter 5, evidence of such a situation (an effective price ceiling) includes the existence of queues, black markets, etc. The black market price can provide an upper limit on the actual w.t.p. for the good, but should generally not be used as a proxy for w.t.p., particularly if the black market is fairly small relative to the total market. Rather, some value in between the administered price (the ceiling price) and the black market price could be used. The best approach in this case is probably to test a number of value assumptions in the sensitivity analysis. If the project produces an acceptable measure of economic profitability using the administered price, then there is less need to consider higher prices (such as the black market price) since they would merely serve to make the project even more profitable (or to increase the measure of economic efficiency).

A final comment relates to the suggestion sometimes made that world market prices can be used as proxy measures of economic value for this category of outputs. Based on the valuation system adopted in EAFP, if for policy or other reasons a market priced good or service could not, or would not be imported in the absence of the project, then its world market price (CIF value) should not be used as a measure of value for it. Similarly, if a good could have been exported, but is produced by the project for domestic consumption, then the export price should not be used as the basis of value for local w.t.p. In this latter case, it can be said that the decision-makers who decided that the good will be consumed domestically instead of being exported must consider the local consumption value to be at least as great as the export value to the nation. Thus, the export value provides an estimate of the minimum value of the output from the viewpoint of decision-makers. However actual w.t.p. by local consumers may be quite different from the decision-makers' interpretation of the minimum value of the product and it is this local w.t.p. which is relevant.

7.3 INTERMEDIATE GOODS WHICH ADD TO TOTAL DOMESTIC SUPPLY

Many forestry project outputs fall in this category. The appropriate measure of value should be based on the relative contribution of the project output to the value of the final consumer goods or services which will be produced with the project output, when such value is measured in terms of consumers' w.t.p. for those final goods or services. For example, lumber produced by a project should be valued on the basis of its contribution to the value of final consumer goods - housing, etc. - which will be produced with the lumber. In practice it is exceedingly difficult and time consuming to develop such a value measure, and this difficulty has led to the common practice of valuing such project outputs on the basis of producers' or "converters' " w.t.p. for them, where the "converters" are those who will take the project output and convert it into final products for consumption.

If there is a local market for such intermediate goods and it is competitive enough to make the price an acceptable reflection of w.t.p. for the output, then there are no problems of shadow pricing. However, if there are administered prices associated with the market or there is evidence of monopsony power on the part of those buying the output or monopoly power on the part of the sellers of the final products to be produced with the project output, then problems arise, for the market price no longer can be taken as an acceptable measure of economic value.

Similarly, in other cases there will be no established market for the project output (i.e., the final processing activities which will use the project output have not yet been established). Thus, there will be no established market price. The best approach in such cases is to evaluate the proposed project as part of a larger integrated project which would include everything up through the final production of consumers goods. For example, if the initial project were defined as one to produce pulpwood for a proposed pulp and paper mill, and there is no other market for pulpwood, then the pulpwood output could be considered as an intermediate input (cost) in an overall project (pulpwood, pulp and paper project) and the pulpwood could be valued as an input on the basis of the opportunity costs involved in producing it (see Chapter 8).

The analyst also could attempt to survey converters' w.t.p. for the project output. Such surveys are fraught with various difficulties similar to those mentioned for consumer surveys. The problems are even more difficult if the project output will not be sold competitively, since converters or producers of the final goods are not likely to reveal their true w.t.p. if they realize that they will be the only ones buying the output. Yet, under the circumstances, this type of survey coupled with judgement on the part of the analyst may provide the best information possible.

Another common approach used in financial analyses of forestry projects is to calculate a "surplus value" for the intermediate output and then attribute that value to the project producing the intermediate output. The surplus value is derived by estimating the final product price and then subtracting all costs other than the value of the project output (which will be an input in production of the final product). The amount left after these subtractions is then divided between profit and the surplus value to be attributed to the project output. This approach can provide an approximation of what the producer of the final product could afford to pay for the project output and still make an acceptable return or profit. (In calculating the surplus value allowance should be made for a profit element, usually equal to the going rate of return on similar types of investments.) In the absence of other means for approximating values, and if it is not possible to combine the proposed project with the further processing stages so the total integrated operation is treated as a whole, then the surplus value approach can at least provide some order of magnitude estimate of value.

It should be emphasized that the process of calculating a surplus value can be extremely difficult and time-consuming and also is fraught with potential errors, if adequate information is not available on the economic value of the final product and all the intermediate costs down to the proposed intermediate output of the project being analysed. For example, in the pulpwood plantation project mentioned earlier it would be necessary to develop an estimate of the value of the final paper production, estimates of all the costs involved in producing it and an estimate of the normal profit which could be expected. If there is no paper production in the country then these estimates can only be derived by going through a complete economic analysis of the proposed pulp and paper project, in which case, an evaluation of both the pulp and paper project and the plantation project as an integrated whole might be done.

7.4 OUTPUT SUBSTITUTING FOR EXISTING DOMESTIC SUPPLY

In this case, total supply available remains the same. The project would substitute for another domestic source of supply, which, when curtailed, would release resources for other uses (production of other goods and services). It is consumers' w.t.p. for these other goods and services (which would not have been produced in the absence of the project) that is used as a measure of value of the project to society.

This approach is quite different to the one for an output which increases supply. In a project that adds to total supply, it is the w.t.p. for the additional output of the project itself that is relevant. In a project that involves substitution, the relevant comparison is between the opportunity costs of alternative sources of the same output, since with or without the project the total amount of the good or service would be the same.

In some cases tracing the relevant substitution impact of a project can be a difficult process. Assume, for example, a situation where a fuelwood project output would partly substitute for non-commercial fuels, such as animal dung and crop residues which would be used for fuel in the absence of the project. Assume further that if these alternative fuels were left on the land they would increase the value of agricultural crops because of their properties as soil builders and fertilizers. In this case the net value of the increase in crop output or the value of crop losses avoided (i.e., "with" and "without" leaving the dung and residues on the land) can provide a measure of the benefits of the project. The fuelwood is considered as an indirect substitute for fertilizer and soil builders, and its value is determined by the value of these resources released and now available for agricultural production.

This value is being used as a measure of w.t.p. in the sense that it is assumed that farmers would be willing to pay to the fuel gatherers an amount up to the value of the crop loss avoided, say $20/ha, if the gatherers would leave the dung and crop residues on the fields. In turn, if these gatherers were given this amount of money, they would be willing to pay up to this amount to buy fuelwood with the same calorific value as the dung and residues left on the ground. Both farmers and gatherers would be just as well off as before. But crop consumers would be $20 better off, assuming that this value of crop loss avoided is based on consumer's w.t.p. for the crop. Thus, this is the benefit. An example shown in Table 7.1 illustrates the approach.

As another common example, assume that a project is being proposed to establish fuelwood plantations for a local community. The output would substitute for fuelwood presently being collected by local community members from natural forests on surrounding hillsides.

At the present time (without the project) village families have to spend time gathering fuelwood from natural forests some distance from their homes. If there is alternative productive work available for these families, then they have to give up the income from such alternative work in order to get the fuelwood, and society gives up the value that the fuelwood gatherers could have produced by working in alternative employment. This income given up (or the benefits society gives up) provides an estimate of the value of fuelwood. For example, suppose that a given family takes two days a week to gather its weekly fuelwood requirements of 20 kg and that the family members involved in the gathering would have produced a total of $2.00 in alternative work (either producing food for home consumption or in the employ of someone else) if they did not have to gather the fuelwood. This $2.00 that they give up would provide an estimate of their w.t.p. for the fuelwood, or the value to society of the resources saved.

In order to use this approach the analyst has to accept the assumptions that:

- the value to the fuelwood gatherer of additional fuelwood (beyond 20 kg in the case of the example) is not worth the additional income he would have to give up by going out to collect more fuel. In other words, the value of an additional unit of fuelwood to the gatherer is just equal to the value of the income he would have to give up to collect it. If it were more, then he would go out and collect more fuel (and give up income). If it were worth less, then he would give up an additional unit of fuelwood and work more;

Table 7.1

DERIVATION OF SHADOW PRICE FOR PROJECT FUELWOOD SUBSTITUTING FOR CROP RESIDUES 1/

Basic Information :

- Crop residues removed per ha/year (a)	2 tons
- Corn crop value increase per ha/a if residues left on fields	\$20
- Heating value of 2 tons crop residues	376 000 kilocalories (kcal)
- Heating value, 1m^3 of project fuelwood	188 000 kcal

Calculation of fuelwood shadow price:

- Heating value of 1m^3 of project fuelwood	=	heating value of 1 ton of crop residues
- Corn crop value increase due to 1 ton of crop residues	=	$\frac{\$20}{2} = \$10$
- Value of 1m^3 of fuelwood	=	\$10

1/ Hypothetical example

- the value of each unit of fuelwood consumed is the same to the gatherer. In fact, this assumption is common to all the valuation approaches suggested. While the first units consumed are likely to have a higher value than the last, there generally is no practical way of taking this into account quantitatively. Thus, the assumption is made that all units will have equal value and that this will be equal to the estimated value of the last unit. The result in most cases is a tendency to understate the real value of (or w.t.p. for) the total output. (This issue - and a confirmation of the fact that it is almost impossible to deal with it in practical valuation problems - is amply discussed in the literature under the heading of "consumer surplus.")1/

To take another example, assume that the family members involved in the fuelwood gathering have no alternative productive uses for their time. Does this mean that the proposed plantation fuelwood should be valued at $0? So long as there is fuelwood available for families to collect elsewhere, then an appropriate measure of value for the plantation output may be close to $0 from an economic efficiency point of view.2/ It would not likely be zero since fuelwood collection may involve a higher food intake than complete idleness, i.e., the collectors must have a higher calorie intake for them to be able physically to carry out the arduous task of collecting the wood. If the family is willing to incur this additional "cost" then the value of the fuelwood is at least equal to this cost, i.e., it is above zero. Similarly there may be health and fatigue costs. However, these are difficult to measure and value. Normally, they are merely described qualitatively in project reports.

Even if the value of the alternative uses of fuelwood gatherers' time is zero, there may be some benefits associated with a fuelwood project that permits natural vegetation to remain on areas that should be protected to prevent erosion or to provide habitat for wildlife (food). To the extent that these benefits can be quantified and valued, they should be included. If they cannot be valued, they should at least be treated explicitly in qualitative or physical quantitative terms in the analysis document.

Finally, it may be that while the local families can currently go out and collect fuelwood, scarcity of wood is increasing (e.g., as indicated by increasing amounts of time required to collect fuelwood). If this is the case, then the analyst has to allow for this changing situation in his analysis (by applying the 'with and without' concept). If the families are likely to have increasing opportunity costs over time, then the analyst can value future project output on this basis. For example, the one day of fuelwood gathering per week required now may not carry any opportunity cost, but if the time required is expected to increase to three days, then the family's home food production may suffer and this could constitute a basis for attributing a positive benefit to a fuelwood plantation project that would avoid this loss of home food production.

1/ cf. United States Department of Agriculture, Economic Research Service, October 1977. It is stated, for example, that "dissatisfaction among economists about the usefulness of consumer's surplus has brought outright condemnation by Samuelson...who remarks: 'The subject is of historical and doctrinal interest, with a limited amount of appeal as a purely mathematical puzzle.'" (p. 117).

2/ It is emphasized that efficiency is not the only concern in the economic analysis. The project may have value on the basis that it reduces the drudgery and toil of people (i.e. reduces "costs"), which was labelled in Chapter 2 as a legitimate goal for a project.

Many projects may involve both substitution and additions to total supply of consumption. For example, there may be an increase in the consumption of fuel because a fuelwood project provides it at a lower financial cost to the consumer than the price of the present fuel (the one for which the fuelwood will substitute). In this case, the two components have to be separated - the substitution part and the increase in total supply part - and each has to be appropriately valued according to the guidelines above. Suppose in a case where fuelwood substitutes for coal that without the fuelwood plantation project one million calories per day is consumed now using coal and that with the project, because of a lower financial price put on plantation fuelwood, consumption increases to 1.2 million calories per day. Substitution of fuelwood for the 1 million calories could be valued on the basis of the opportunity cost for production of the coal for which the wood is substituting. However, the additional consumption - 200 000 calories per day - should be valued on the basis of the consumer's w.t.p. for the additional consumption as discussed in Section 7.2, since it is adding to total supply and not substituting for the domestic coal. It is only being consumed because the financial price is lower for the fuelwood. The w.t.p. for it (its economic value) is also likely to be lower.

As another example, assume a project designed to improve a forest road so that hauling/transport costs for logs delivered to mills can be reduced. Part of the benefits can be measured in terms of the costs saved for the volume of wood that normally would travel over the road. In other words, as in the typical substitution project, the new road releases resources which had been used in transporting wood and which now can be used in other activities (production of other goods and services). This part of the benefits is appropriately valued on the basis of the opportunity cost of the resources released. However, it is also possible that the improved road results in an expansion of wood output. The logic is as follows. With lower transport costs, total production costs decrease. Producers of the wood products, if dealing in a somewhat competitive market will tend to lower prices as their costs go down. With lower prices, consumers will be willing to purchase more. Thus, the project also has resulted in an expansion of consumption of the products being produced. The net increase in the value of the expanded consumption (i.e., net of additional costs) can be attributed to the project as a benefit. Since this part of the output adds to total supply, it has to be valued as suggested in Section 7.2 or 7.3, depending upon whether the expanded supply involves a direct consumer good or an intermediate producer good.

To summarise, for a category of output that will substitute for other domestic supply of the same product or another product with the same use value, the appropriate measure of value of the benefits due to the project is the opportunity cost of the resources released, or the value of what these resources would produce if they were released. If the resources released have no other use, then the value of the project output may be zero or close to it. On the other hand, if the resources released are otherwise fully employed in the economy, and they are traded in a fairly competitive market, then the prices of the resources released provide an adequate approximation of the value of the project output. In between there will be cases where some of the resources released will have alternative uses and some will not. The analyst's task is to identify the various inputs released and then to determine their alternative use values or their opportunity costs. Finally, if the proposed project output, say lumber, will substitute for other domestic lumber supply, and that other supply will now be exported, then the project output is treated as an export output for valuation purposes (see following section).

7.5 EXPORTS

In this case the relevant measure of value is the local w.t.p. for the goods and services which will be purchased with the foreign currency earned. The foreign currency earned is reflected in the FOB value for the exports. If there is a free market exchange rate and no tariffs or subsidies attached to goods or services which will be imported with the foreign currency earned through the project, then the FOB value expressed in foreign currency (say dollars) can be converted to local currency using the market exchange rate that is expected to exist at the time the project output is exported.

However in reality there will seldom be a situation where there is a free exchange rate and no tariffs or subsidies. This means that something other than the existing exchange rate has to be used to convert the FOB value to local w.t.p. terms. For this purpose a shadow exchange rate (SER) can be used.

7.5.1 The shadow exchange rate

Before suggesting guidelines for use of a SER, it is necessary to look at how it is derived by national planners. The SER is defined as the real purchasing power of a unit of foreign currency expressed in local market price terms. It measures the average difference between local prices including tariffs and subsidies and prices calculated using the existing exchange rate, i.e., the average level of price distortion caused by tariff barriers. In an economic analysis the analyst is interested in actual w.t.p. or opportunity cost in local price terms. Therefore, the influences of tariffs and subsidies have to be included in the estimates. Sometimes the SER is adjusted to reflect non-tariff barriers, e.g., import and export quotas and controls on buying and selling foreign exchange.

The SER is generally calculated to reflect the average price distortion in the economy, considering all imports and exports. Thus, for example, assume a country situation where the existing exchange rate is set at 10 units of local currency (LC) per unit of foreign currency (say $). The average level of import tariffs and export subsidies (treated as "negative" tariffs) is calculated to be 10 percent. Simplifying somewhat, it can be assumed that the local currency is actually overvalued by 10 percent by the existing exchange rate. While officially the local price of foreign currency, or the rate of exchange is LC10 per $1, in fact, when people go to buy foreign goods in the local market, they pay on the average 10 percent more (because of the import tariffs) or LC11 per dollar worth of imported foreign goods. The SER in this case is 11 to 1 in contrast with the existing rate of 10 to 1. Similarly in the case of exports. Assume a project that earns $100 by exporting lumber. In terms of local currency converted at the existing exchange rate of LC10 per $, the benefits of the project would be $100 x LC10 or LC1 000. In fact, given the average tariff distortion of 10 percent, goods and services can be bought with $100 that are worth $100 x LC11, or LC 1 100 in terms of local w.t.p. Thus, in terms of the economic analysis, the benefits of the project in terms of local w.t.p. in local prices should be LC1 100 rather than LC 1 000.

Some guidelines for forestry project planners follow. As mentioned in Chapter 5, the SER used in a country should be a general one that reflects the entire trade picture and the average tariff or trade barrier effect on trade, where the average is calculated as a weighted average of all tariffs and subsidies on trade (i.e., tariffs and subsidies weighted by the amount of the trade to which they apply). Thus, it should be calculated by national planners for use in all project analyses in the country.

If such a national SER is available, it is recommended that the analyst use it. If the analyst believes that he has a strong case for modifying the SER imposed by the Central Planning body, he can try to persuade this body to change it. Until it is changed, he should use the given SER. In any case, he can develop a test of sensitivity of the project to potential alterations in the SER.

If a generally accepted SER is not available in the country (developed by the national planning office or some other national planning body), then the existing exchange rate can be used. The analyst of forestry projects should generally not try to develop a SER of his own, since the task is quite complex, and if it is not done correctly, it could easily lead to distortions and to results which are not comparable with those for other projects. However, he should test the sensitivity of the project results to alternative rates considered to be closer to the actual w.t.p. than the existing rate of exchange.

7.5.2 Valuing exports using FOB values and the SER

As mentioned earlier, the gross amount of foreign currency earned by an export project is measured by the FOB price for the output times the volume of output. In other words, the FOB price becomes the unit value of the export output expressed in foreign currency. Since local w.t.p. for goods and services expressed in local currency is being used to measure economic value, the foreign currency has to be converted to w.t.p. for what the foreign currency can buy in terms of local prices expressed in local currency. This is done by multiplying the FOB value by the SER.

In deriving the FOB value the market to which output will be exported can be determined, and using the CIF price in that market the FOB value for the output at the port of export can be derived. Obviously, if a FOB value already exists at the port of export, that value can be used. If several possible markets are being considered, then the FOB values associated with each can be derived, and if they differ the highest can be picked under the assumption that exports would go to the most profitable market. If the output is intended for several specified markets and they result in different FOB values, then a weighted average FOB value can be used, basing the weights on the proportion of output that will go to each market.

Projects that indirectly result in increased exports can also be considered in this category. For example, assume that the project output of sawnwood will replace other locally-produced sawnwood in the local market and this other sawnwood will now be exported. In this case, the FOB value still provides the relevant basis for measuring benefits, since the project will result in an increase in the nation's exports which will permit expanded imports of other goods and services. Applying the 'with and without' test, the shadow priced foreign exchange value of the exports is the relevant measure of value for the benefits due to the project.

7.6 IMPORT SUBSTITUTES

If the project output will substitute for imports which actually would have taken place in the absence of the project, then the correct basis for valuing the output is the foreign exchange savings made possible by the project. The CIF price in foreign currency of the substituted imports is multiplied by the SER (discussed in section 7.5.1) to obtain the local w.t.p. value, just as in the case of exports.

The project output may also substitute for another completely different imported product which has the same use. In this case, the project output can be valued on the basis of the CIF price for the other product times the SER, when appropriate adjustments have been made to equate the use-value of the project output with that of the other product.

For example, assume that project fuelwood will substitute for imported kerosene. In this case, the CIF price for the imported kerosene for which the fuelwood will substitute can be used to derive the shadow price for the fuelwood, by converting fuelwood and the imported kerosene to a common basis, e.g., cost/kilocalorie. An example is given in Table 7.2.

7.7 AVOIDING SOME POTENTIAL OUTPUT VALUATION ERRORS

Several of the valuation approaches suggested are based on the assumption that a project's output will substitute for some other good or service. In using this approach the analyst should pay particular attention to the following questions:

- are the goods indeed substitutable technically and in terms of consumer preferences?
- if they are, will the assumed level of consumption (substitution) actually take place ?

With regard to the first question, the analyst can draw on technical information and perhaps carry out a survey of consumer willingness and ability to substitute the two products. For example, some cooking and heating systems might be able to burn kerosene but not be properly designed for wood. In this case, the analyst looking at a fuelwood project that would substitute wood for kerosene would either have to doubt substitutability or suggest to the project planner that the project also include a component for redesign or remodelling of cooking/heating systems, if such is not already included. At the same time he would also want to check very carefully the substitutability of wood for kerosene in terms of some common heating and/or cooking values. Such measures may be calories per unit volume or weight, or it may be in terms of less accurate measures such as "average" amounts needed to cook common foods or to provide heat in homes, etc.

The same types of considerations would be necessary in looking at the substitutability of lumber or plywood for other building materials, domestic paper production for imported paper, etc. For example, a domestic newsprint project is proposed, based on the use of mixed tropical hardwoods. The resulting newsprint would have different quality characteristics than the imported newsprint for which it would supposedly substitute. Is it valid to use the price of imported newsprint to value the domestic (project) output? That will depend on whether the project output would be acceptable as a direct substitute in terms of use. Or assume a project to produce exterior or marine plywood intended as a

Table 7.2

ESTIMATING PROJECT OUTPUT VALUE ON THE BASIS OF THE VALUE OF ANOTHER PRODUCT FOR WHICH IT WILL SUBSTITUTE

Project output:	Fuelwood
Substitute product:	Kerosene, now imported with an estimated CIF price of $.40/litre (1)
Calorific values:	Kerosene: 3 200 kcal/l Air-dry wood: 188 000 kcal/m^3
Inputed substitution for wood:	$\frac{(\$/m^3)}{188\ 000\ \text{kcal}} = \frac{\$.40}{3\ 200\ \text{kcal}}$

or

$$(\$/m^3) = \frac{(188\ 000 \times \$.40)}{3\ 200}$$

$$(\$/m^3) = \$23.50$$

(This value could be used for the fuelwood if it is actually going to substitute for imported kerosene. It would be converted to local currency equivalent using the SER.)

substitute for non-treated interior plywood that is being used for exterior uses at present. In this case, the price of the local interior plywood would not be an adequate measure of the value of the project output, since the life (use value) of the two products would be quite different. Thus, the replacement rate over time would be different. A consumer w.t.p. survey would have to establish whether consumers would be willing to pay more for the better use value of the marine plywood. Such a survey would have to establish price-quantity relationships. This type of survey would be needed in any case as part of the market study for the financial analysis, so the additional effort for the economic analysis would be slight.

The last point brings up the second question raised. Assume that it is found that the marine plywood would be substituted for the interior plywood that had been used in exterior uses. Would the same quantity be consumed? This would likely depend on the actual pricing policy adopted for the project output (i.e., a financial consideration). If it were to be sold at the same price as the interior plywood, it might be substituted in equal quantities. But if the price were to be higher (because cost would be higher), then volume would likely be lower. In this case, the analyst has to watch the assumption about quantities of project output that would actually be directly substituted for interior plywood. Similarly, if it is to be sold at a subsidized price below the price of the substitute, volume may increase.

Finally, it should be reemphasized (as was done in Chapter 6 dealing with market prices) that relative values often change over time, i.e., the value estimated for an output today may not be the relevant or appropriate value for some future period, even after taking out the influence of expected general price inflation. Thus, to the extent possible, the analyst should attempt to estimate what likely changes in output values will take place over time due to the same types of factors discussed for market prices in Chapter 6. It is often difficult to project values into the future. There is uncertainty and many unquantifiable variables involved. Often the best thing to do is to assume constant relative values over time and then test the sensitivity of project results to potential changes in values. This is discussed further in Chapter 10, which deals with the treatment of uncertainty.

Chapter 8

SHADOW PRICING INPUTS

8.1 INTRODUCTION

This chapter deals with approaches to shadow pricing project inputs. As shown in Figure 5.2, inputs can be classified into five main categories for the purposes of empirical estimation of shadow prices:

- inputs that are imported when no quota exists on imports
- locally produced inputs which would have been exported if not used in the project
- locally produced non-exportable inputs
- imported inputs when a quota on imports exists
- resources (land and labour)

Each of these categories is discussed separately in the following sections. As in the case of outputs there are several categories of inputs that involve foreign exchange effects. The SER as discussed in Section 7.5.1 is used to value such inputs. Specific uses of the SER are discussed below where they are needed.

8.2 IMPORTED INPUTS WHEN NO QUOTA EXISTS

Imported inputs not limited by any quota are valued on the basis of the local value of the foreign currency required to import them. This is measured in terms of the CIF value for the input times the SER. There are two exceptions to this approach:

- In some cases, inputs are financed by a grant which is tied to the project, i.e., a grant which only can be spent on importing the input for its exclusive use in the project. If this is the case, then there is no difference in total foreign exchange availability for other uses with or without the project. Therefore no alternative benefits are sacrificed by using foreign exchange in importing the input. The economic cost to the domestic economy of the input financed by a tied grant is equal to zero.

- When the input is financed with a tied loan, the economic cost does not materialize when the input is paid for (imported), since there is no alternative use permitted (no opportunity cost) for the foreign loan. The cost occurs at the time of repayment of the loan, when alternative imports could have been financed with the foreign exchange used up in paying the debt (principal plus interest).

8.3 EXPORTABLE LOCALLY PRODUCED INPUTS

If the input used by the project actually would have been exported in the absence of the project, then the value foregone by the economy by using the input in the project is represented by the reduction in the availability of foreign exchange. The domestic w.t.p. for the imported goods and services foregone is the correct measure of the economic cost of using the input in the project. The basis for this value is the FOB price of the input (the foreign currency earnings foregone) converted to local prices of imported goods/services using the SER.

8.4 NON-EXPORTABLE LOCALLY PRODUCED INPUTS

The appropriate value measure for a non-exportable locally produced input (i.e., an input for which local production cost is greater than FOB value or where prohibited by government policy) is related to whether or not use of the input in the project reduces total supply of the input available to the economy. (See input categories C1 and C2 in figure 5.2.).

- If the project's use of the input reduces total supply of the input available to the rest of the economy, then the relevant shadow price of the input is based on the net benefits which are sacrificed (i.e., opportunity cost) in using the input in the project rather than in the next-best alternative use.

- If the project use of the input induces additional local production of the input, then the relevant cost is measured in terms of the value of the resources used up in increasing the supply of the input, i.e., their opportunity costs.

Note that if use of the input in the project induces additional or new imports of the input for use elsewhere in the economy, then the input can be treated as an imported input for valuation purposes, i.e., the foreign currency cost (CIF value) becomes the relevant measure of economic value when converted to local prices using the SER. (This parallels the case of a local consumed project output which induces exports of the same product from other producers.)

8.5 IMPORTED INPUT FOR WHICH A QUOTA EXISTS

If there is an import quota affecting an imported input, its value should be measured in terms of the w.t.p. for its contribution to the value of alternative outputs that would have been produced with the input elsewhere in the economy if the project were not implemented. The reasoning is that the total amount of the input allowed by the quota would have been imported with or without the project and, therefore, in these circumstances there would be no net drain of foreign exchange induced by the project. If imports of the input are below the quota, then the quota is ineffective and, from the point of view of the analysis, it does not exist. Thus, the input's CIF value could be used as a basis for valuing the input (see Section 8.2).

8.6 RESOURCES: LABOUR

The objective in valuing labour is to arrive at a measure of the value of the benefits foregone by employing labour in the project rather than in its next best alternative use. If labour is hired away from other productive work and there is little unemployment in the project region, the value of the labour in the other work, or the market wage, provides an acceptable measure of opportunity cost for the economic analysis. This chapter discusses situations where these conditions do not hold, i.e., the market wage does not adequately reflect opportunity cost.

8.6.1 Unskilled labour

The main questions of interest in shadow pricing unskilled labour relate to the following situations:

(i) Labour hired in the project is from the pool of unemployed persons in the project region. The value of these unemployed workers is equal to the production foregone by putting them in the project. If they were producing food or materials at home for their own consumption, and they have to give this up when they work in the project, then the value of what they give up is an appropriate measure of opportunity cost. If they were producing nothing (which is an exceptional case), then a shadow price close to zero can be used. The cost will probably never reach zero since there is generally some cost involved in training, housing or otherwise taking care of unskilled labour that has been unemployed for some time. This cost has to be added in somewhere in the accounts as a cost.

(ii) Labour hired in the project is taken from other productive jobs, but there is unemployment in the project region (i.e., persons willing and able to work in paid jobs). In this case the assumption generally adopted is that even if the project merely hires workers away from other jobs, these other now vacant jobs will then absorb new workers from the pool of the unemployed. Thus, the project will result indirectly in a reduction in unemployment and the labour used in the project should thus be valued in the same way as for (i). Application of the 'with and without' test demonstrates the logic of this approach.

(iii) Labour is hired only part-time in the project. In the case of unskilled seasonal labour, it is generally desirable to take into account general periods of seasonal employment and unemployment. First, the analyst can determine by observation or from records the periods of general seasonal employment existing in the market area for the labour that will be used in the project. He can then compare these periods with the periods during which temporary employment is required by the project. To the extent that the two periods do not overlap, he can use the shadow wage for unemployed labour as derived above in valuing seasonally unskilled labour employed in the project, since, by definition, such labour is unemployed during the off-season. However, if the project's requirements overlap with the general period of seasonal employment (for crop harvest, planting, etc.), and if there is no general unemployment during the period of seasonal employment, then he has to attribute a shadow price for seasonal labour employed in the project equal to the actual wage paid for seasonal labour in the regional economy.

A case study from Korea indicates how this was done in the case of fuelwood plantation project. [1] Since the fuelwood labour requirements overlapped somewhat with the seasonal requirements for agriculture, an average shadow wage rate based on the full seasonal wage rate and the off-season income (monetary and in-kind) of unskilled village labour was used as the shadow wage rate. The weighting was based on the proportion of project employment which overlapped with the period of general seasonal full employment.

In handling these three types of situations, it is necessary to look at the nature of the market and distinguish between unemployment in an economic sense and unemployment in the sense that it appears that people are "doing nothing". In an economic analysis, it is unemployment in an economic sense that matters, and this is determined by both supply and demand. As an example, assume a situation as follows:

In the project region there are some 1 000 persons in the unskilled labour category presently employed. About 100 persons are "unemployed" in the sense that they are not working in paid jobs (i.e., receiving wages). The project will require 10 full-time unskilled workers. How should they be shadow priced? The answer depends partly upon what the apparently "unemployed" workers are willing to work for (i.e., what they give up by going to work in the project). It may very well be that they are producing at home for their own consumption. If they go to work in the project they may have to give up this production (consumption). If there is a competitive labour market (and no minimum wage set by government or unions), then there is no unemployment in an economic sense. Those who are not working feel that spending their time doing other things is worth at least as much as the minimum wage paid in the competitive market. Thus, this minimum wage would provide a reasonable measure of labour value (or opportunity cost for labour) at the margin.

In cases where unemployment exists due to some policy and/or regulation of minimum wages, a shadow wage rate based on alternative production foregone has to be ascertained, and this will likely be lower than the regulated minimum market wage. For example, if a government-set minimum wage is in effect, it may be possible to locate an "informal" competitive labour market in rural areas where the actual wage paid is below the set minimum. This would provide an approximation of the appropriate shadow wage rate. If no informal market can be located, then the analyst will have to rely on rough estimates of what the "unemployed" would give up in terms of other production if they were employed in the project. This information might be obtained from surveys of local community households.

Confusion sometimes arises if the unemployed who will now be employed in the project are receiving unemployment payments (benefits) while they are unemployed. This type of payment is a transfer payment, or a transfer of consumption from some members of

---

[1] Case Study No. 2, FAO 1979

society to others. While it is relevant in a financial analysis carried out from the government's point of view, it will not be relevant in the economic analysis, where the analyst is attempting to estimate the opportunity cost of labour, or the value of consumption foregone by employing labour in the project being analysed.

8.6.2 Professional and Skilled Employees

Professional and skilled employees are required by most projects. In many developing countries there is an acute shortage of this type of employee. It is also common in these countries that the government imposes wage and salary increase limits (salary ceilings). As in all cases where an effective maximum price is imposed, the result is that the willingness of employers to pay the skilled labour might be higher than the current salary level. Skilled persons may be fully employed, but they are being paid less than the producers are prepared to pay, i.e., their real opportunity cost. In such cases the analyst may wish to use a shadow wage or salary level above the market level.

If skilled or professional labour is unemployed in the economy, then it can be treated in exactly the same way as unemployed unskilled labour, i.e., valued on the basis of its opportunity cost without the project.

8.7 RESOURCES: LAND

The appropriate measure of value for land is the highest net return that actually would have been obtained from the land in the absence of the project. The analyst thus needs to estimate what the net return would be from the best actual alternative use. This he uses as the shadow price for land.

In estimating the opportunity cost of land, the analyst can use information obtained from interviews and data on land use in the project region, particularly as such relate to land availability and uses of land similar to the proposed project lands.

In valuing land, the analyst should guard against over-valuation of land cost due to:

- attributing to the land a net value from alternative use which will be obtained from some otherwise idle area if the project is implemented;
- ignoring the fact that in some cases an alternative use which would take place in the absence of the project would not continue over the entire project period;
- forgetting to subtract all costs (other than land) needed to obtain the gross benefits from the best alternative use (i.e., it is the net value foregone which is the relevant opportunity cost).

Each of these points is discussed in greater detail in the following paragraphs.

In many cases, there are no actual alternative uses for lands devoted to forestry projects. This may be because of the low quality of the land for other uses, but it also may be because there is no land pressure in the project area and abundant other lands exist to accommodate other potential uses. For example, suppose there are two large idle land areas, A and B. It is proposed to put area A into the forestry project. Cattle production in the project region is expanding. The analyst estimates that the project area could support a net return from grazing of $10/ha/a over the project life. So could area B. If idle area B will likely absorb the foreseeable demand for such grazing land over the life of the project, if the project is undertaken, then there is no cost to society by putting land area A into project use and using area B for the grazing expansion. Thus, the opportunity cost of putting the land into project use would be zero. On the other hand, if the foreseeable expansion of grazing would require more than area B - i.e., if demand for area A is anticipated over the project life - then some cost would have to be attributed to the land area A put into project use, since some net grazing value would be foregone. The timing of this opportunity cost would have to be adjusted to the time when A actually would be needed.

Another potential over-valuation error to avoid relates to the assumption that a piece of land considered for a forestry project will have an alternative use which will continue to be viable over the entire project period. An example will illustrate this type of situation. Consider the case of a tropical land area having poor soils. There may be an immediate alternative annual crop value that would have been obtained in the absence of the project. If such an alternative use would have taken place, then this is an appropriate value to consider for the period during which the use would take place. However, someone knowing little about tropical soils may suggest that the estimated initial annual net crop value foregone should be used as a cost during every year of the forestry project's life - say fifteen years in this example. In general, for most tropical soils and environmental conditions, it will not be possible to have continuous production of annual food crops on the same land without introducing drastic measures, including very heavy applications of fertilizer which would increase costs and reduce potential net returns (i.e., the opportunity cost). The cost of such fertilizer and other treatments could result in the net value of the crop (the opportunity cost) reaching zero after only a few years of initial production. To shadow price land correctly the analyst might develop a shadow pricing schedule for the land such as shown in the hypothetical figures in Table 8.1. Note that the calculations in Table 8.1 would give quite a different answer if it were merely assumed that the opportunity cost per year would be the same over the entire life of the project and equal to the opportunity cost in the first year.

A third potential over valuation error relates to what is included in the opportunity cost calculations. It is the net value foregone which is relevant as an opportunity cost, not the total value of the output foregone. Thus, in a particular situation, a plantation project may be taking land out of crop production where the total or gross value of the crop foregone is $100/ha/a. To get an appropriate shadow price for the land, the analyst would have to subtract all the costs (other than land) required to bring forth that $100 of gross value. It may be, because of a depressed price due to over-supply of the crop that the costs would be equal to the $100 of gross value, in which case the opportunity cost of the land would be zero in terms of the forestry project. Society would not be giving up any net consumption benefits, since the costs would equal the benefits and the net value foregone would be zero.

Table 8.1

SCHEDULE OF NET CROP VALUE FOREGONE FOR USE IN SHADOW PRICING LAND 1/

Year	Shadow price based on annual net food crop value foregone ($/ha)
0	$75
1	$75
2	$70 - productivity starting to decrease
3	$65
4	$50 - heavy fertilizer application
.	.
.	.
9	$0 - value of required fertilizer is equal to net crop value increment
10	$0 - all nutrients removed; soil has essentially become sterile and of no further use for annual crop production
.	.
.	
.	
n (end of project)	

1/ Hypothetical data

8.8 CHANGES IN SHADOW PRICES OVER TIME

As indicated in Chapter 6, when market prices are used as a basis for shadow pricing, the analyst should keep in mind that the opportunity costs associated with inputs may change over the life of the project. Such expected changes have to be taken into account.

For example, in the case of land, although there are no apparent alternative uses for the land at the time of appraisal of the project, such uses may easily develop during the project period. Thus, a land cost should be included for the appropriate period. In a typical forestry project, the period of time involved can be substantial - say twenty years or more. Thus, the analyst should be concerned with what developments in the region would likely take place in the future which would make the land valuable for other uses during the project period. For example, even slight shifts in agricultural prices can make previously idle land attractive for agricultural production or livestock grazing, i.e., move the opportunity cost from zero to some positive value. To the extent possible, following the "with and without" principle, the analyst should try to anticipate such future uses and value them so they can be entered as a cost for the project. Note, however, that this does not mean that all idle resources will have some productive use in the future. It is very possible that a shadow price of zero is appropriate. The point is that the analyst needs to consider the possibility that there will be an opportunity cost involved during the project period. In cases of great uncertainty, he may merely wish to test alternative assumptions in a sensitivity analysis (Chapter 10).

Similarly, in situations where there is some indication that the employment situation existing at the beginning of the project will not hold over the entire project period, e.g., unemployment is expected to decrease due to general improvement in economic conditions even without the project, the analyst may wish to make adjustments in the shadow wage rate for latter years of the project. Again, this remains a matter of judgement. If the situation is very uncertain, the analyst may merely wish to consider such possibilities in the sensitivity analysis.

Chapter 9

COMPARING COSTS AND BENEFITS

9.1 INTRODUCTION

Once inputs and outputs have been identified and quantities designated in the physical flow table and unit values have been estimated for inputs and outputs (or at least for those for which values can be estimated), the next step is to combine the information from the physical flow and unit value tables into a total "value flow" table.

The value flow table provides total cost and benefit information in a form needed for the calculation of measures of economic efficiency or worth. Development of the value flow tables and measures of economic worth for projects are the subjects of this chapter.

Section 9.2 looks at the derivation and nature of the total value flow table and discusses some differences between this table and the "cash flow" table derived for and used in the financial analysis. Section 9.3 discusses the question of how to treat time in an economic analysis. Section 9.4 looks at the most common measures of project worth which take time into account and discusses their differences and similarities. Section 9.5 discusses the relationships between the most common measures of economic worth of projects.

The discussion in this chapter applies to the situation where a value flow table is being prepared for an entire project (i.e., total costs and benefits are being analysed) as well as to the situation where a component of a larger project is being analysed, i.e., total component cost and benefit comparisons.

9.2 THE "VALUE FLOW" TABLE AND ITS RELATION TO THE "CASH FLOW" TABLE

Chapter 2 mentioned the "value flow" table which displays aggregate values (quantitites multiplied by unit-values). Table 9.1, used in an economic analysis of tree-farm plantations in the Philippines, 1/ illustrates the common format for a value flow table. There are three major row headings in a value flow table: benefits, costs, and net benefits (costs). Columns are arranged by years, starting with the initial year of the project, which is labelled "year 0", and ending with "year n", the last year of the project. Thus, the value flow table describes the pattern of project associated real costs and benefits over time, by years. 2/

1/ Case Study No. 1, FAO, 1979.

2/ See Chapter 3 for further discussion on time period designations and appropriate project period. In this chapter a time interval of one year is used. Other intervals can be used without changing the basic points of the discussion.

Table 9.1

VALUE FLOW TABLE: 10 ha PLANTATION, PHILIPPINE PROJECT

(value in constant pesos)

	YEARS															
	0	1	2	3	4	5	6	7	8	9	10	11	12	13	14	15
Benefits																
1. Thinning	-	-	-	-	-	-	-	-	-	-	-	-	-	-	-	-
2. Harvest	-	-	-	-	-	-	-	5523	6174	6174	6810	6810	7434	7434	8046	6174
Total								5523	6174	6174	6810	6810	7434	7434	8046	6174
Costs																
3. Land preparation	295	295	295	295	-	-	-	-	-	-	-	-	-	-	-	-
4. Purchase of seedlings	78	78	78	78	-	-	-	-	-	-	-	-	-	-	-	-
5. Lining/dig/plant	150	150	150	150	-	-	-	-	-	-	-	-	-	-	-	-
6. Replanting	65	65	65	65	-	-	-	-	-	-	-	-	-	-	-	-
7. Fertilizing	200	200	200	200	-	-	-	-	-	-	-	-	-	-	-	-
8. Weeding	275	275	275	275	-	-	-	137	137	137	137	137	137	137	137	137
9. Singling	-	-	-	-	-	-	-	-	50	50	50	50	50	50	50	50
10. Administrative	100	100	100	100	100	100	100	100	100	100	100	100	100	100	100	100
Total	1163	1163	1163	1163	100	100	100	237	287	287	287	287	287	287	287	287
Net Benefits (costs)	(1163)	(1163)	(1163)	(1163)	(100)	(100)	(100)	5286	5887	5887	6523	6523	7147	7147	7759	5887

The bottom row of the table shows for each year the difference between benefits and costs, or the net beneftis (costs) for that year. (If costs are greater than benefits for a given year, then the figure appears in brackets, a common way of expressing a negative figure or a net outflow.)

It is instructive to look at the main differences between the value flow table for the economic analysis and the cash flow table used in the financial analysis (as described in Chapter 2). Not only will such a comparison provide insights into the differences between the economic and financial analyses, but it will also be useful in cases where the analyst will adjust directly the cash flow table to derive the economic value flow table. Three types of adjustments need to be made; they involve:

- adding some costs and benefits that are not included in the cash flow table;
- revaluing some costs and benefits in the cash flow table, using shadow prices instead of market prices;
- removing transfer payments from the cash flow table and adjusting for differences in timing of economic and financial costs and economic benefits and financial returns.

The first two of these adjustments have already been discussed (the first in Chapter 4 and the second in Chapters 5, 7 and 8). The third adjustment - the treatment of timing problems and transfer payments which show up in the cash flow table - is discussed below.

The main types of transfer payments of interest are taxes, subsidies, and loan receipts and repayment of loans and interest. Total value flow tables should be adjusted so that taxes and loan costs are not subtracted from benefits (or treated as costs) and so that subsidies and loan receipts are not added to benefits or netted out of costs.

In the case of loans, Squire and van der Tak (1975) explain the adjustments needed as follows:

> ...the payment of interest by the project entity on a domestic loan merely transfers purchasing power from the project entity to the lender. The purchasing power of the interest payment does reflect control over resources, but its transfer does not use up real resources and to that extent is not an economic cost. Similarly, the loan itself and its repayment are financial transfers. The investment, however, or other expenditure that the loan finances involves real economic costs. The financial cost of the loan occurs when the loan is repaid, but the economic cost occurs when the loan is spent. The economic analysis does not, in general, need to concern itself with the financing of the investment: that is with the sources of funds and how they are repaid.

Similar arguments hold for taxes and subsidies, although one additional point needs clarification to avoid a common confusion. Chapter 5 argued that tariffs (taxes) and subsidies should be considered in deriving measures of local w.t.p., i.e., their effect on local prices should not be removed if they are expected to persist during the period of the project. Why is it now argued that taxes levied on the project and subsidies provided to the project should be removed (or not be considered) in the economic analysis?

The answer is that two different considerations are being dealt with. In the case of derivation of values to use for inputs into the project and outputs from the project, the interest is in measures which reflect local w.t.p. for these items in the existing markets. The effect on w.t.p. of transfer payments is relevant, given the definition of economic value used in EAFP.

On the other hand, in deriving the appropriate economic measure of project worth, the interest is only in real resource flows and real flows of consumer goods or services coming from the project, valued in terms of the opportunity cost and w.t.p. value measures discussed earlier. A tax on the project output value merely means that some of the control over the benefits due to the project are transferred from the project entity to the public sector (government). The real benefits (the increase in consumer goods and services due to the project) do not change because a financial entity pays a tax. To society, the tax is not a cost associated with the project. To the financial entity it is a cost. Similar considerations hold in the case of subsidies given to the project (i.e., where the government shares the money cost of the project). The real costs (the opportunity costs) of the resources used in the project remain the same with or without a subsidy, and these are the costs which are of interest in the economic efficiency analysis.

To summarize, taxes and subsidies do influence the w.t.p. for goods and services (and the size of the market and the local price which is established), but they do not alter the real costs of a project nor the real benefits produced by the project. The two considerations are quite separate.

Depreciation should not be included in the economic analysis (nor should it have been included in the cash flow table). Depreciation is merely an accounting item and represents an internal transfer of some of the money profit from one account to another, in order to provide for replacement of assets. In the economic analysis, it is the real cost of an input that is relevant and its cost is entered at the time it is used in the project.

Finally, it should be pointed out that if the value flow table for the economic analysis is derived directly from the cash flow table, the analyst has to be careful to adjust the timing of entries in the value flow table to take into account the fact that costs in the economic analysis occur at the time resources are actually used in the project and benefits occur when outputs are consumed.

In the financial analysis, costs occur when payments are made, and this may be at some time other than when resources (inputs) are actually used in the project. For example, a given input may be used in the project in year 5, but paid for in years 6 through 8 (on an instalment basis). In the cash flow table, the cash outflow would occur in years 6 through 8, while in the economic analysis, the value of the input should be entered in year 5.

Similarly in the case of outputs or benefits. In the cash flow table for the financial analysis, the cash inflows or returns are entered when they actually occur. A given output may be paid for (to the project financial entity) after (or before) it is actually used (consumed). Thus, the return may appear in the cash flow table in a year that is different from that in which the output actually becomes available. In the economic analysis, the benefit should always be entered in the year in which the output is consumed or used.

These considerations are only of concern when the total value flow table is derived directly from the cash flow table. If the total value flow table is derived from the physical input and output tables and the unit value tables, then financial transactions such as taxes and subsidies will not appear and will thus not be of concern in preparing the total value flow table.

9.3 THE NET VALUE FLOW AND THE "TIME VALUE" OF CONSUMPTION

The main focus of the value flow table is on the bottom line, or the net value flow. If all costs and benefits of a project occurred at the same point in time, then the analyst could merely add up costs, add up benefits, and compare them without further adjustment. However, costs and benefits of a project occur over the life of the project. Typically, the life of forestry projects can cover a substantial number of years.

Project costs and benefits which occur at different points in time (in different years) cannot be directly compared. That is because value is intimately associated with time. The "value" of costs and benefits depends on when these costs and benefits occur. Thus, $1 of benefits occurring ten years from now is not as "valuable" in today's terms as $1 of benefits occurring immediately. If $10 is spent today and $15 is received back tomorrow that may be acceptable. But if $10 is spent today and the $15 is not received back for 40 years, that would probably not be acceptable. The amounts are the same. The difference is time.

From Table 9.1 the P 6 523 of net benefits occurring in year 10 are not worth P 6 523 in present value terms, simply because 10 years elapse to get them. It would be preferable to have the P 6 523 to use or to invest today and get a great deal more than P6 523 back 10 years from now.

For any given year, all costs and benefits have the same relative time value in terms of the present since they occur at the same point in time. In terms of the previous discussion, there is no problem in summing costs and benefits for any given year (such as shown on the bottom line of Table 9.1). The problem is how to compare net benefits (costs) which occur in different years. Since time does have an influence on value as considered at any given time, the analyst will want to develop information that permits the decision-maker to compare the costs and benefits which occur at different times and to compare projects which have different cost and benefit streams over time.

More specifically, the question is: How can a value occurring in year n (some future year) be equated with a value occurring in year 0 (the present), i.e., how can the net benefit (cost) items occurring in the bottom line of the value flow table be compared?

The common approach is to apply an "adjustment" factor to future net costs/benefit values so they can be expressed in terms of values occurring today. The adjustment factor is derived from the accepted time value of money; it is commonly called the "discount rate". The adjustment process is called "discounting".[1]

---

[1] The discount rate is often called the "interest rate".

9.4 THE DISCOUNT RATE

Since EAFP uses consumer's willingness to pay for goods and services as the common yardstick for valuing both costs and benefits, the discount rate used to discount costs and benefits should be the "consumption rate of interest". This rate should measure the discount attached to having additional consumption next year rather than this year. The appropriate magnitude of this discount rate (or rate of interest) is determined by a number of factors, including society's preference for present consumption at the expense of more rapid growth (higher savings and investment now with higher consumption in the future). 1/

As it turns out in practice, just as in the case of the SER, the forestry project analyst will generally not have to concern himself with the derivation of an appropriate consumption rate of interest (or shadow discount rate) to use in his analysis. The rate used should be one that is in general use in the project country. Thus, the analyst should obtain the appropriate discount rate from a central planning unit (e.g., national planning office) or from his administrative agency. 2/

At the extreme, if there is no discount rate available from the central planning office at the time the analysis is being undertaken, the analyst can pick a rate such as 8 or 10 percent and use that in the main analysis, and then test the sensitivity of the worth of the project to alternative rates of discount. (As will be discussed later, one widely used measure of economic efficiency does not directly require determination of the appropriate discount rate in order to calculate the measure).

There is sometimes a tendency to argue for use of "lower" discount rates in forestry project analyses. The argument is that there are certain "non-quantifiable benefits" from such projects which justify the use of a discount rate that is lower than the one used to evaluate other projects in the general economy. 3/ This is not recommended. Instead, analysts should use the established or acceptable discount rate used for evaluation of other projects and then discuss in qualitative terms the "unique" conditions associated with their project that make it "different" from other projects. This forces analysts and project planners to be explicit about their assumptions, thus avoiding the possible hiding of the efficiency shortcomings of a project behind a lower than normal rate of discount.

1/ See Squire and van der Tak 1975, p. 27.

2/ This recommendation provides a convenient excuse for not getting into the problems involved in determining the appropriate rate of discount. Since there is no general agreement among economists or policy makers concerning the appropriate derivation of the discount rate to use for public projects, it would, in any case, be futile to try to resolve the problem in this type of guide. An excellent review of the agruments is provided in Mikesell 1977.

3/ The same argument is often used by planners in the water resources field.

9.5 MEASURES OF PROJECT WORTH CONSIDERING TIME VALUE

Several indexes or indicators of project worth which take the influence of time into account (i.e., involve discounting) are in common use. There is no single measure of a project's worth which is universally accepted, since all share the characteristic of providing only partial information on project performance. Different indicators are needed and used for different purposes. There are, however, two measures that are widely used in economic analyses. 1/ These are the net present worth (NPW), and the economic rate of return (ERR). Since both are derived from the same basic data, namely, the project's costs and benefits, the two measures are intimately interrelated. The analytical information they provide is, however, somewhat different because of the different ways in which they combine cost and benefit data.

This section discusses the process of discounting and then the two commonly used measures of project worth. Other, specialized indicators of project performance, e.g., related to employment, foreign exchange effects, etc., are discussed elsewhere. 2/

9.5.1 Discounting costs and benefits - deriving "present value" estimates

The process of adjusting a future value to the present is called discounting. The resulting "adusted" value is called "present value" (PV).

The basic formula for discounting is the following:

$$PV = FV_n \left(\frac{1}{(1+i)^n} \right)$$

where

PV = present value

FV_n = future value in year n

i = discount rate (expressed in decimal form)

n = number of years until future value occurs

$\left(\frac{1}{(1+i)^n} \right)$ is commonly called the "discount Multiplier"

Given a discount rate of 8 percent, the present value of a $100 payment occuring 2 years from now can be calculated as follows:

1/ A third measure which some institutions use is the benefit-to-cost ratio. It is not dealt with further in EAFP, although its derivation is considered in Appendix D.

2/ Papers by McGaughey and Schuster in FAO, forthcoming.

$$PV = \$100 \left(\frac{1}{(1.05)^2}\right)$$

$$PV = \$100 \left(\frac{1}{(1.1664)}\right) = \$100\ (.8573) = \$85.73$$

If the 8 percent discount rate represents the consumption rate of interest, then the result, PV = \$85.73, indicates that \$100 of consumption occurring 2 years from now is equivalent in present value terms to \$85.73 of consumption occurring today. Put another way, it can be said that society is indifferent between (a) consuming today goods and services valued at \$85.73 and (b) waiting two years and being able to consume \$100 worth of goods and services. In other words, \$10.43 more of goods and services would be required 2 years from now (or a total of \$100 worth) in order to forego \$85.73 of consumption at present.

In this discounting example the value of $\left(\frac{1}{(1.08)^2}\right)$ was calculated directly. There are tables prepared and widely available which give the value of the discount multiplier $(1/(1+i)^n)$ for a wide range of interest rates and years. Further, it can also be calculated with simple pocket calculators, if they have a constant or a y^x key. Thus, the analyst will have no problems deriving the value of the discount multiplier for any number of years. For example, using Table 9.2, $1/(1.08)^2$ is equal to 0.8573, and this value times \$100 gives the result of \$85.73 arrived at earlier.

The basic discounting formula and tables are all that is needed to derive a NPW or an ERR for a project. However, in some cases, other formulas - derived from the above basic formula - can provide useful shortcuts in carrying out calculations. For example, sometimes equal annual or periodic payments are associated with a project for a number of years during its life. In this case, there are formulas and tables which provide the present value of such payments without having to discount each of the annual or periodic amounts separately. Similarly, in some cases the analyst will want to find an annual equivalent of a given value occurring at some time, or to find the present value of an annual series of payments occurring every year. The most common of these formulas are shown in Appendix B.

9.5.2 Net Present Worth

Going back to the Philippines tree-farming project example and its value flow (Table 9.1), and using the basic discounting process described, a measure of the present value (PV) of all net benefits (costs) occurring in the various years of the project can be developed once an appropriate discount rate has been chosen. If a discount rate of 5 percent is used, the present value of each of the net future benefit (cost) entries is as shown in row 2 of Table 9.3. Adding these items up (taking into consideration whether they are positive or negative) the NPW for the project is P29 310.

What does this NPW of P29 310 indicate? It indicates that, given the assumptions concerning the opportunity costs of the resources used in the project and the w.t.p. for the project output, this project will return a net surplus of P29 310 of consumption benefits in present value terms taking into account the assumed consumption rate of

Table 9.2

DISCOUNTED SINGLE PAYMENT MULTIPLIER -

THE VALUE OF A ONE DOLLAR PAYMENT DISCOUNTED FOR N YEARS

	RATE OF INTEREST			
YEARS	.045	.050	.055	.060
1	.95694	.95238	.94787	.94340
2	.91573	.90703	.89845	.89000
3	.87630	.86384	.85161	.83962
4	.83856	.82270	.80722	.79209
5	.80245	.78353	.76513	.74726
6	.76790	.74622	.72525	.70496
7	.73483	.71068	.68744	.66506
8	.70319	.67684	.65160	.62741
9	.67290	.64461	.61763	.59190
10	.64393	.61391	.58543	.55839
11	.61620	.58468	.55491	.52679
12	.58966	.55684	.52598	.49697
13	.56427	.53032	.49856	.46884
14	.53997	.50507	.47257	.44230
15	.51672	.48102	.44793	.41727
16	.49447	.45811	.42458	.39365
17	.47318	.43630	.40245	.37136
18	.45280	.41552	.38147	.35034
19	.43330	.39573	.36158	.33051
20	.41464	.37689	.34273	.31180
21	.39679	.35894	.32486	.29416
22	.37970	.34185	.30793	.27751
23	.36335	.32557	.29187	.26180
24	.34770	.31007	.27666	.24698
25	.33273	.29530	.26223	.23300
26	.31840	.28124	.24856	.21981
27	.30469	.26785	.23560	.20737
28	.29157	.25509	.22332	.19563
29	.27902	.24295	.21168	.18456
30	.26700	.23138	.20064	.17411
31	.25550	.22036	.19018	.16425
32	.24450	.20987	.18027	.15496
33	.23397	.19987	.17087	.14619
34	.22390	.19035	.16196	.13791
35	.21425	.18129	.15352	.13011
36	.20503	.17266	.14552	.12274
37	.19620	.16444	.13793	.11579
38	.18775	.15661	.13074	.10924
39	.17967	.14915	.12392	.10306
40	.17193	.14205	.11746	.09722

Table 9.2 (continued)

YEARS	RATE OF INTEREST .070	.080	.090	.100
1	.93458	.92593	.91743	.90909
2	.87344	.85734	.84168	.82645
3	.81630	.79383	.77218	.75131
4	.76290	.73503	.70843	.68301
5	.71299	.68058	.64993	.62092
6	.66634	.63017	.59627	.56447
7	.62275	.58349	.54703	.51316
8	.58201	.54027	.50187	.46651
9	.54393	.50025	.46043	.42410
10	.50835	.46319	.42241	.38554
11	.47509	.42888	.38753	.35049
12	.44401	.39711	.35553	.31863
13	.41496	.36770	.32618	.28966
14	.38782	.34046	.29925	.26333
15	.36245	.31524	.27454	.23939
16	.33873	.29189	.25187	.21763
17	.31657	.27027	.23107	.19784
18	.29586	.25025	.21199	.17986
19	.27651	.23171	.19449	.16351
20	.25842	.21455	.17843	.14864
21	.24151	.19866	.16370	.13513
22	.22571	.18394	.15018	.12285
23	.21095	.17032	.13778	.11168
24	.19715	.15770	.12640	.10153
25	.18425	.14602	.11597	.09230
26	.17220	.13520	.10639	.08391
27	.16093	.12519	.09761	.07628
28	.15040	.11591	.08955	.06934
29	.14056	.10733	.08215	.06304
30	.13137	.09938	.07537	.05731
31	.12277	.09202	.06915	.05210
32	.11474	.08520	.06344	.04736
33	.10723	.07889	.05820	.04306
34	.10022	.07305	.05339	.03914
35	.09366	.06763	.04899	.03558
36	.08754	.06262	.04494	.03235
37	.08181	.05799	.04123	.02941
38	.07646	.05369	.03783	.02673
39	.07146	.04971	.03470	.02430
40	.06678	.04603	.03184	.02209

Table 9.3

NET PRESENT WORTH - PHILIPPINE PROJECT. (5 PERCENT DISCOUNT RATE; VALUE IN CONSTANT PESOS)

	YEARS															
	0	1	2	3	4	5	6	7	8	9	10	11	12	13	14	15
1. Net Benefits (cost) 1/	(1163)	(1163)	(1163)	(1163)	(100)	(100)	(100)	5286	5887	5887	6523	6523	7147	7147	7759	5887
2. Present value of Net Benefits (costs) 2/	(1163)	(1107)	(1055)	(1055)	(82)	(78)	(75)	3757	3784	3795	4004	3814	3980	3790	3919	2832
								NPW = 29310 3/								

1/ Last row of Table 9.1

2/ Item in row 1 divided by $(1.05)^n$ for years 1-15.

3/ The sum of items in row 2.

interest (discount rate) of 5 percent, or the relative weight which society places on present consumption versus investment and future consumption. By using the discount rate it has ensured that the NPW result is comparable with those obtained for other projects that would involve different cost and benefit streams over time, i.e., the effect of different time values associated with consumption gained or foregone at different times in the future have been eliminated.

In general, given the above, it can be said that in economic efficiency terms any project that provides a positive NPW is an efficient use of the resources involved, assuming that each separable component also has a NPW $\geq$ 0 and the project is the least cost means of achieving the particular benefits. (See Chapter 2 for review of the three conditions for economic efficiency.)

While a project meeting these conditions is economically efficient, it still may not be chosen for implementation. That depends on the total budget available and the NPW associated with other projects on which the budget could be spent (see Section 9.5.4).

A project for which the estimated NPW is negative is not economically acceptable. The negative NPW indicates that there are better uses for the resources involved in the project, i.e., given their opportunity costs and timing and the discount rate, they could be used elsewhere to produce more consumption benefits in present value terms.

9.5.3 The Economic Rate of Return

In the previous example of NPW calculation, the NPW was P29 310 when a 5 percent discount rate was used. The question could be asked: What rate of discount would have to be used to obtain a NPW of zero, i.e., what is the implied discount rate that would make the PV of project benefits equal the PV of project costs? That rate is called the internal economic rate of return, or the ERR. It is essentially a "breakeven" discount rate in the sense that the PV of benefits equals the PV of costs.

One of the most commonly used measures of project worth in a financial analysis is the internal financial rate of return (FRR). It is comparable to the ERR in terms of derivation, although it means something slightly different. The FRR shows the investor what the average earning power is associated with a given investment of his funds. More specifically, it is the average rate of return on the invested funds outstanding per period while they are invested in the project, or that rate of interest which makes the NPW (using market prices) equal to zero.

Thus, a FRR of 10 percent indicates to the investor that he will receive \$.10 back per year for each \$1 invested during the years in which the investment is left in the project. This is a useful measure for an investor, since it provides a clear means for comparing alternative uses of his funds. Say that his best use of funds, other than putting them in the project, is to put them in the bank at 6 percent interest per year. He compares the rate of return on the project (10 percent) with the rate of return from the bank (6 percent), which is called the alternative rate of return (ARR), or his opportunity cost of capital 1/, and he then knows that the project use will give a greater return than the best alternative use. 2/

1/ This concept of "opportunity cost" is analogous to the one used throughout EAFP.
2/ The FRR and the ARR should be calculated net of inflation, i.e., in real terms.

The ERR is similarly interpreted, except it shows the decision-maker what society can expect to receive back in consumption benefits for a given investment of its scarce resources. In other words, if the calculated ERR is 10 percent, this tells the decision-maker that the average annual return of consumption benefits on resources outstanding per period while they are invested in the project will be $10 for every $100 of resources invested and left in the project. The ERR will be compared with the consumption rate of interest to see if the project earns enough to make it worthwhile to invest (forego consumption now in favour of future consumption). Say that the relevant consumption rate of interest is 5 percent. This means that society wants to get at least a 5 percent rate of return on investment of its resources to make it worthwhile to forego present consumption in favour of investment and future consumption. If the ERR turns out to be 10 percent for a given project, this means that on the average society will get more than the minimum acceptable 5 percent back. Thus, the project is economically efficient in terms of its use of scarce inputs assuming that the other two conditions for economic efficiency are met.

The Philippine example is used to show how the ERR is calculated. The undiscounted net benefit (cost) items for each year are shown in row 1 of Table 9.4. By discounting these by 32 percent the PV figures as shown in row 2 are obtained. If these values are totalled, the NPW is zero which by the definition occurs when the economic rate of return is used to discount all net benefits(costs). Thus, 32 percent is the ERR.

The calculation to find the ERR or the interest rate which makes NPW equal to zero has to be by trial and error.1/ Since the NPW is positive at 5 percent (Table 9.2) the ERR must be greater than 5 percent. 2/ By using a simple braketing approach, the mechanics of which are shown in Appendix C, the estimated ERR is obtained.

What does the ERR of 32 percent indicate in the Philippine example? It represents the "yield" of the resources used in the project over the project period. It means that $1 invested in the project will generate $0.32 per year for every year that the $1 remains committed to the project. It also indicates that this return is greater than the assumed consumption rate of interest of 5 percent, which measures the trade-off between consumption in a given year t_0 and consumption delayed until the following year, t_1. Society should be interested in leaving its resources in a project such as this rather than consuming them now because it will receive more back in the future than is needed to satisfy its perceived trade-off between present and future consumption.

Just because a project has an ERR that exceeds its consumption rate of interest, this does not automatically mean that the project will be accepted and implemented. It does mean that the project represents an efficient use of resources, given acceptance of the consumption rate of interest as being the relevant one. 3/ However, there is always the possibility that other uses of a limited budget can provide higher rates of return than the project being studied.

1/ Some pocket calculators now available will calculate the ERR directly.

2/ If NPW > 0, then ERR > i used;
If NPW = 0, then ERR = i used;
If NPW < 0, then ERR < i used.
Where "i" equals the discount rate used.

3/ Assuming that the other two conditions for economic efficiency are met.

Table 9.4

ECONOMIC RATE OF RETURN (ERR) - PHILIPPINE PROJECT

	YEARS															
	0	1	2	3	4	5	6	7	8	9	10	11	12	13	14	15
1. Net Benefits (costs)	(1163)	(1163)	(1163)	(1163)	(100)	(100)	(100)	5286	5887	5887	6523	6523	7147	7147	7759	5887
2. Present value of Benefits (costs) discounted at 32%	(1163)	(881)	(667)	(506)	(33)	(25)	(19)	757	639	484	406	309	255	193	159	92
3. NPW at 32% 1/	0															

1/ i.e., sum of the present values of net benefits (costs) discounted at 32% per year.

The above two measures of economic worth can be used to analyse the economics of project components and entire projects. When NPW is used, the usual approach - as discussed in Chapter 2 - is to analyse components first, making sure that all separable components ending up as part of a project "package" have NPW's at least equal to zero. Once a set of economically efficient project components has been assembled into a project, then exactly the same approach can be used in calculating the NPW or ERR for the total project. As mentioned, the least cost condition for economic efficiency does not involve calculation of a NPW or an ERR. Rather, the costs of alternatives are compared directly to find the least cost alternative. Some analysts prefer to treat the costs avoided by undertaking the project instead of the least cost known alternative as the "benefits" of the project alternative being analysed. These "benefits" are then used in calculating a NPW for the project alternative being analysed. If it is positive then this shows that it is the least cost alternative among the known set of alternatives. If the NPW is zero, then the least cost alternative to the project has costs exactly the same as the project being analysed. If the NPW is negative, then the alternative to the project being analysed has lower costs. While there is nothing conceptually wrong with this approach, it can become confusing; thus, it is recommended that costs of alternatives are compared directly. (Confusion can arise in cases where the project has to be compared with other entirely different projects which are competing for the same budget. In point of fact, the costs avoided by undertaking one alternative rather than another to achieve a given output do not necessairly represent a true measure of benefits.)

9.5.4 Relationships between NPW and ERR

NPW and the ERR represent alternative means of presenting the relationship between costs and benefits. In mathematical terms the relationship between the two is as follows:

$$\text{Net present worth} = \sum_{t=0}^{n} \left[\frac{(B_t - C_t)}{(1+i)^t} \right]$$

Economic rate or return is that discount rate d such that

$$\sum_{t=0}^{n} \left[\frac{(B_t - C_t)}{(1+d)^t} \right] = 0$$

where

B_t = benefits in each year t

C_t = costs in each year t

n = number of years to end of project

i = discount rate or consumption rate of interest (CRI)

d = the internal economic rate of return (ERR).

From these definitions, the following relationship holds: When NPW = zero, then the ERR = CRI, or the consumption rate of interest (or the discount rate used in calculating the NPW). Given the definitions and the above relationship between the two measures, what can be said about the information provided by each of them in terms of the three conditions for economic efficiency mentioned in Chapter 2 ? 1/

Neither of the two measures of project worth tell anything about the least cost (or third) condition for economic efficiency. This condition has to be studied in a separate analysis undertaken in the design and preparation stages of the project. (See Chapter 11).

Both measures do provide information related to whether PV of benefits are less than, equal to, or greater than the PV of costs for a project component and the total project. In point of fact, they both provide exactly the same answer to the question of whether or not a project or project component is economically efficient in terms of these first two conditions. If a project is accepted as being efficient in terms of one measure (i.e., NPW ≥ 0), it will also be acceptable in terms of the other measure (i.e., ERR ≥ CRI) and vice versa.

So far in the discussion, it can be seen that either of the two measures could be used equally well to determine whether a project is economically efficient (assuming no lower cost means to achieve the project objectives is known to exist). Thus, the choice of which of the two to calculate and use is unimportant in terms of this basic question, although the analyst obviously has to calculate the measure commonly used by the institution for which he is carrying out the analysis.

Each of the two measures provides additional information that the other does not provide. The NPW measure, in contrast to the ERR, provides information on the absolute value or magnitude of the present value of net benefits of a project. Yet it tells nothing about how large the cost will be to achieve the NPW. Thus, there could be a project with a NPW of $1 000 which costs $2 million or one with the same NPW that costs $5 000. Both would have the same NPW. On the other hand, the ERR is a relative measure of project worth, which gives information on the returns per unit of cost and thus provides more relevant information for comparing the benefits which can be expected from alternative uses of a limited budget. Therefore, it is more useful for ranking independent project alternatives when it is not possible for budget or other reasons to undertake all projects that meet the basic economic efficiency conditions.

1/ These are -

(1) Total present value of project benefits must be equal to or greater than total present value of project costs.

(2) Each separable project component must have PV of benefits at least equal to PV of costs.

(3) There is no lower cost means of achieving the project benefits.

This latter point brings up an important consideration. Because a project alternative is shown to be economically efficient, using either of the two measures of economic efficiency or worth, that does not mean that the project will be undertaken. Budget considerations as well as many other factors will enter the picture and the decision-making process. If a number of alternative uses (projects) exist for a limited budget, some system of ranking projects in terms of their economic worth has to be developed to provide guidance in choosing from the set of alternatives that has been shown to meet the basic efficiency conditions. The two measures of project worth do not necessarily provide the same order of ranking for projects. In point of fact, the NPW measure provides no guidance in terms of ranking projects that are not mutually exclusive.[1]

The whole problem of choice among alternative uses of a given budget is complex and goes beyond the confines of an economic analysis of a given project alternative. The choice of ranking system is often made on the basis of political objectives. Even in terms of the economic efficiency objective, one or the other of the two measures is often chosen for ideological reasons or because it is "simpler to understand".

While consideration of choices among projects for a limited budget, i.e., decision-making, is beyond the subject of EAFP the following points are emphasized:

(i) Neither of the two measures of project worth discussed provide information related to the third condition for economic efficiency; namely, that there is no known lower cost means to achieve the project benefits. This condition needs to be tested by other means - generally through a cost comparison as described in Chapter 11.

(ii) Both measures of project worth provide the same answer to the question of whether or not a given project alternative or project component is economically efficient in and of itself (assuming that the third condition is met).

(iii) Since the NPW provides an absolute measure of project worth, while the ERR only provides a relative measure of average expected returns per unit of cost (and no information on the absolute size of the net benefits), it is recommended that the analyst should calculate both the NPW and the ERR for projects. Regardless of which measure will be used by decision-makers in making project decisions, information on both absolute and relative economic efficiency is useful.

(iv) Finally, while calculation of the NPW requires that the analyst has an estimate of the appropriate discount rate (or CRI) in hand, the discount rate is not required in order to calculate the ERR. Still, in order to make use of a calculated ERR, i.e., in order to determine whether or not the project (or project component) being analysed represents an economically efficient use of resources, some estimate of the CRI is needed, since the ERR only has meaning in the context of the other possible uses for resources.

---

[1] Mutually exclusive project alternatives are those for which only one of a set of alternatives can be undertaken. For example, two projects which envisage using the same area of forest land are mutually exclusive alternatives. Only one of the two can be undertaken at a given time.

Chapter 10

TREATMENT OF UNCERTAINTY

10.1 INTRODUCTION

Mention has been made that (a) identification and valuation of costs and benefits for any project involve looking into the future, (b) estimates of future values are subject to uncertainty, and (c) the analyst needs to recognize and to treat explicitly the uncertainty surrounding his forecasts of future events and values. This chapter considers how to treat uncertainty in a project analysis. The main technique suggested is sensitivity analysis, or the testing of the sensitivity of the chosen measure(s) of project worth to alternative assumptions about values of inputs and outputs and various technical relationships, i.e., how will the value of the NPW or ERR change if the assumed value(s) of a given parameter (group of parameters) is (are) changed?

Uncertainty refers to the fact that the analyst cannot be sure today about anything that is going to happen in the future. Or, because of inadequate information, he cannot be sure about past and present events which he needs as a basis for forecasting future conditions. Using available information relating to past events he makes estimates (or guesses) of what is likely to happen - what future demand for pine sawnwood will be, what the cost of labour will be, how natural hazards will affect a plantation project, etc. However, he is never certain how close his estimates will be to what actually will happen. There is always some uncertainty involved.

The analyst may feel more confident about some estimates than others, probably because he has more experience (more accurate observation of past events and trends) on which to base the estimates. In some cases he may even have enough quantitative information on past occurrences to be able to estimate the statistical probability of occurrence of some future event. A situation where this is possible is often referred to as a situation of "risk". In contrast, when there is little or no basis for deriving quantifiable probabilities, there is a situation of "uncertainty".

While this distinction between risk and uncertainty is useful in conceptual discussions, it may merely serve to confuse the analyst dealing with a real project, since in reality he is dealing with a continuum from one extreme where probabilities of occurrence can be quantified (e.g., in cases where actuarial evidence is available) to the other extreme where no information is available on which to base probability estimates. In most cases, the forecasting problems faced in project planning fall somewhere between situations of risk and total uncertainty.

10.2 PURPOSE OF TREATING UNCERTAINTY

Using information which has been generated with time and funds available for the analysis, the analyst identifies and then values the inputs and outputs associated with the project being analysed (Chapters 4 through 8). The resulting "expected" values, i.e., those considered to be most likely to occur, are then used in the initial calculation of the chosen measure(s) of project worth (Chapter 9). To make a complete and useful economic analysis the analyst also has to provide some idea of what would happen to the chosen measures of project worth or efficiency if the actual values of various inputs and/or outputs turn out to be different from the expected values used in the analysis. If a "reasonable" change in the assumption about the expected value for a given parameter (or value of a combination of parameters) is "critical" in terms of the expected measure of project worth or efficiency, he generally will want to take some steps to reduce the uncertainty. The term "reasonable" in this context refers to an estimate of what the possible values are for a parameter around the expected value used in the basic analysis. The term "critical" generally refers to the point where the measure of project worth or efficiency moves from positive to negative (or vice versa) in terms of the relevant decision criterion. [1]

As an example, assume a 20-year plantation project where labour is a major cost component. The expected value of labour used in calculating the NPW of P1 200 for the project is P2 per day. This value (shadow price) for labour is based on a reduction of 50 percent in the current actual wage to account for high unemployment in the project area. The P2 per day figure is used for the entire project period. However, looking at developments in the project region over the past 10 years, and considering planned developments in the region, it is felt that, even if the project being analysed were not undertaken, unemployment may be reduced gradually over the project life. Thus it is reasonable to test the sensitivity of the project NPW to an assumption that labour value will gradually increase to P4 (the actual wage level) by year 10 and then continue at that level until the end of the project (10 more years). Note that no quantitative basis exists on which to estimate how the wage rate will change in the future, with or without the project. The different wage rate assumption used in the sensitivity analysis is considered "reasonable" on an intuitive basis. Most such judgements have to be made on an intuitive basis. The analyst may want to test several other wage rate assumptions in addition to the P4 per day, for example, an increase beyond P4/day for the last ten years of the project. That will depend on his judgement, the time and funds available for the analysis, and the results of the sensitivity analysis using the initial "reasonable" assumption concerning possible changes in labour value (from P2 to P4/day). If the project outcome is not sensitive to this assumption, then it will not be sensitive to a change in the expected labour value that is less extreme than P4 per day. Thus, there will be no need to test other, less extreme values. However, more extreme values can be tested.

---

[1] For example, when the NPW moves from positive to negative, using the guiding discount rate to discount costs and benefits, or when the ERR falls below the discount rate used for evaluating public projects.

10.3 GUIDELINES FOR TREATMENT OF UNCERTAINTY

The following practical systematic approach to analysing uncertainty is recommended. It involves three steps which are explained in more detail later:

(i) identify likely major sources of uncertainty for the project being analysed and for each source establish some estimate of a reasonable range of values for the parameters involved;

(ii) carry out a sensitivity analysis for the project using various combinations of different assumptions concerning the values of the parameters associated with the major sources of uncertainty. Analyse in more detail the parameters for which changes in value assumptions are critical in terms of project outcome;

(iii) determine appropriate ways of changing the design of the project or modify it to eliminate or reduce the major sources of uncertainty which are critical in terms of project outcome.

An underlying rationale for this approach is avoidance of unnecessary expenditure of funds on detailed analysis of parameters which do not appear to have much influence on the outcome of the project decision. The sensitivity analysis provides a low cost means to identify project parameters in order to design a sound, workable project, and to understand and reduce the uncertainty surrounding the project outcome.

The degree to which further information is generated on various parameters to which the project outcome is sensitive will ultimately depend on the budget available for project preparation and appraisal, the estimated impact of uncertainty on project outcome, and the particular orientation of the institution undertaking the analysis. The steps suggested provide a logical framework for the process, regardless of the funds and effort devoted to the analysis and the orientation of the analysis.

10.4 IDENTIFYING LIKELY MAJOR SOURCES OF UNCERTAINTY

From the planner's point of view, a useful distinction can be made between controllable and uncontrollable uncertainty. It may be possible to assess and to account for uncontrollable uncertainty in the appraisal of, and decision on, a project. However, nothing can be done within a given project framework to alter the underlying conditions which cause it. Controllable uncertainty, on the other hand, relates to factors which can be changed within the design of the project itself.

From a practical point of view, the analyst and decision-maker are mainly interested in how uncontrollable uncertainty could and should affect the decision whether or not to undertake a particular project and how controllable uncertainty can and should be handled in project design.

Uncertainty is associated with the availability and timing of most inputs and outputs, relationships between inputs and outputs (production functions), their prices (or values), and even the objectives of the project. However, it is obviously difficult and expensive to deal with uncertainty associated with every factor involved in a project.

Thus, a first step is to identify systematically the likely major categories of uncertainty associated with a proposed project and to make an initial assessment of their potential importance to the decision on a particular project being studied.

In forestry projects some of the main types of uncertainty which may be important relate to:

(i) Natural factors such as wind, rain, fire, insects, diseases, natural variation between species and in a given species grown in different locations. These elements of uncertainty are often particularly important for plantation projects since the period between investment and return (harvest) can be long. (In some cases these factors can be analysed in terms of probabilities.)

(ii) Technology and productivity factors related to processing different types of wood, input-output relationships in tree growing, processing yields, effects of alternative technologies (including silvicultural systems) on non-wood values derived from forests, labour productivity, transportation systems, etc.

(iii) Financial and economic factors related to values assumed for inputs and outputs, availability and cost of capital, etc.

(iv) Human factors related to labour availability and cost, the ability of man to forecast future events (wood volume availability, markets, etc.), and, most important, management capability.

The potential importance of any of these sources of uncertainty will depend on the circumstances surrounding the particular project being analysed. Theoretically, the analyst could test the sensitivity of project outcome to changes in assumptions concerning any input or output parameter or combination of such. In practice, the sensitivity analysis will be limited to a few major potential sources of uncertainty for any given project. The analyst has to use his own judgement in deciding on which parameter values he will test in the sensitivity analysis, given his time and budget constraints. If he is particularly uncertain about future labour values, for example, and labour is an important input item in the project, then he would likely carry out a sensitivity analysis for alternative assumptions concerning future labour value (see previous example). Similarly, he also should analyse the impact on project worth or the chosen measure of economic efficiency of changes in assumptions concerning output values, since generally these have the greatest impact on project outcome. There are no rules which can be made for choosing the parameters or combinations to be tested. The FAO case studies provide some examples of choices of items tested in a variety of actual forestry project situations.

In general, if an acceptable NPW and/or ERR is obtained for a project, using the initial estimates of parameter values (the "expected" values), then the analyst will be interested in testing alternative value assumptions that are less favourable in terms of project outcome, i.e., higher cost assumptions and/or lower benefit assumptions. The results provide some indication of how large unexpected cost increases or benefit value reductions would have to be to have a critical effect on the chosen measure(s) of project worth (see previous definition of "critical").

To summarize, the analyst first assesses what the main elements of uncertainty and risk are likely to be for the proposed project. This type of assessment may uncover some common problems, e.g., delay in start-up, potential factor cost increases, wood supply bottlenecks, market uncertainties, etc. Such information provides the analyst with a first approximation of the factors which should be tested in the sensitivity analysis. The analyst then looks at the relative magnitude and timing of various input and output items (which can be identified from the value flow tables for the project being analysed) and lists all those which represent a significant part of project benefits or costs. He then makes an initial estimate of a range of values which could reasonably be expected for each, relying on past experience and projected trends. At this stage, he should err on the side of making the range too broad, rather than too narrow - narrowing can occur in later stages of the analysis. He also makes some estimate of the interdependence of the values of the input and output factors, e.g., the extent to which lower or higher prices for some inputs and outputs are associated with lower or higher prices for other inputs and outputs.

In practice, he generally ends up with a limited number of major parameters which will be tested in the sensitivity analysis. As mentioned, the case studies cited in Appendix A provide some examples of practical sensitivity analyses for some actual forestry projects.

10.5 THE SENSITIVITY ANALYSIS

Using the list of parameters and estimates of the reasonable range of values for them (as developed in the previous step), the analyst then carried out the sensitivity analysis. A number of computer programmes are available for handling the calculations. However if systematically organized, it is comparatively simple to carry out the analysis using a hand calculator. There are also programmable hand calculators which can easily handle the complex calculations involved in a sensitivity analysis. If time permits, it is better to include a number of sensitivity analyses rather than a few, since sometimes it is not easy to anticipate the factors to which the project outcome is sensitive.

In addition to an analysis of alternative parameter values, the analyst may also want to test the sensitivity of results to (a) delays in implementation, and (b) changes in assumptions which reflect different objectives. This latter type of sensitivity analysis is relevant in cases where objectives include redistribution of income, environmental quality, increased employment, etc., in addition to the economic efficiency objective.

10.5.1 Using net present worth measures for sensitivity analysis

It is usually desirable to test the sensitivity of project outcome to a combination of changes in input and/or output value assumptions and different levels of changes in the values for given inputs or outputs. If this is the case, then it is usually easier to work with NPW rather than ERR. The effort and time involved generally will be less, as indicated below. However, the sensitivity analysis can be carried out using either NPW or ERR.

Table 10.1 shows the sensitivity analysis results for a fuelwood project in the Republic of Korea.[1] Using a 12 percent discount rate, the project had a NPW of 102 500 Won/ha. The table shows the sensitivity of NPW to a 20 percent change in any of the major cost and benefit elements shown in Column 1.

The entries in the body of the table are interpreted as follows (using planting cost as an example):

- if planting cost were 20 percent higher than expected, then the NPW (column 2) would be Won 8 400/ha lower, other assumptions remaining as before;

- if planting cost were 20 percent lower than expected, then NPW would be Won 8 400/ha higher.

In other words, the table can be used to estimate changes in NPW due to increases or decreases in the value of any given item.

In addition to these basic interpretations estimates of sensitivity of measures of project worth can also be derived from:

- Different magnitudes of changes for a given parameter value. For example, a 40 percent increase in planting cost would result in a W 16 800 (W 8 400 x 2) decrease in NPW. Similarly, a 30 percent increase would result in a W 12 600 (8 400 x 1.5) decrease in NPW.

- Combinations of changes in input/output values. For example, suppose all costs except harvesting cost were assumed to be 20 percent higher. The cumulative effect on NPW would be to reduce it by W 31 570/ha, or ((0.72 + 14.20 + 8.40 + 0.58 + 2.10 + 4.07 + 1.50) x 1 000). Since the "expected" value of the NPW was Won 102 500/ha, the project would still be considered economically efficient since the NPW would still be positive (Won 102 500 - Won 31 570). Any other combination of changes and magnitudes of changes could be tested in the same way. Thus it is an extremely flexible and inexpensive approach to testing the outcome of the project (NPW) to a great variety of value assumptions.

It should be noted that the table does not tell the analyst anything about the interaction between factors, i.e., which combinations and magnitudes could likely be expected. That still remains as a judgemental task of the analyst. But once he has settled on likely combinations, he can assess their impacts by using the sensitivity table. Further, the effects of changes in a parameter value are assumed to be linearly related to the measure(s) of project worth (i.e., the NPW in this case).

It is recognized that in some cases an ERR is used instead of NPW as the measure of economic efficiency. A sensitivity analysis using the ERR measure involves recalculation of the ERR for each change in assumption or combination of assumptions.

---

[1] See Case Study No. 2. FAO 1979.

Table 10.1

KOREA FUELWOOD CASE STUDY - SENSITIVITY ANALYSIS ('000 Won/ha.)

A 20 percent change in:	Causes changes as follows in the NPW [1]
	(12.00 Percent Discount Rate)
1. Seedlings	14.20
2. Planting	8.40
3. Fertilizing	2.10
4. Supervision	4.07
5. Miscellaneous - Tools	1.50
6. Harvesting	32.65
7. Fuelwood	79.58

[1] Net present worth (NPW) at 12 percent 102.55

Source: See Case Study No. 2 FAO, 1979

The sensitivity analysis using NPW as a basis can also provide some critical information concerning sensitivity of ERR to changes in input or output parameter values. This follows from the definitions discussed in Chapter 9, where it was pointed out that when NPW is zero the ERR is equal to the discount rate used in calculating the NPW. Thus, in the NPW sensitivity analysis, when costs are increased (benefits decreased) to the point where NPW is zero, then the ERR is equal to the discount rate used. This "breakeven" point is of interest to decision-makers. (See Section 10.5.2)

If the analyst wants to test the sensitivity of ERR to specified changes in parameter values (other than those which result in a NPW of zero), he will have to recalculate the ERR each time for each change in value. If a computer is available, it is a simple matter to run through a great number of different combinations in a short time. If a desk calculator is used, it is equally simple in terms of process, but more cumbersome in terms of the time and steps involved. (It should be pointed out that even at this point in time there are some relatively inexpensive calculators which can handle this type of sensitivity analysis in a short time and in a relatively simple manner.)

10.5.2 Breakeven analysis

One common type of sensitivity analysis is the breakeven (BE) analysis. Given the fact that the calculated measure(s) of project worth are primarily used as an aid in deciding whether or not a project will be economically efficient, it is natural that decision-makers are interested in how much less favourable parameter values can be before a calculated positive measure of project worth falls below the criterion (or criteria) for acceptability, i.e., how much higher can costs be and/or how much lower can benefits be before the NPW drops below zero or the ERR drops below the accepted discount rate? Similarly, for projects where use of expected values for parameters produce negative NPW's or ERR's below the guiding rate, the decision-maker will be interested in seeing how large parameter value changes (decreases in costs or increases in benefits) have to be in order to make the project acceptable in terms of the chosen economic efficiency criteria. This type of BE analysis provides useful information particularly in cases where the decision on a project will be based on a number of considerations in addition to economic efficiency.

Strictly speaking, BE analysis is usually carried out by varying the value of only one parameter, with all others taking on their expected values (i.e., holding other values constant). However, it can also be carried out for a general change in costs or benefits, e.g., by determining what percentage change in all costs is needed to reach the breakeven point (where NPW = zero, or ERR = the accepted discount rate).

The values of parameters being tested which make the NPW = 0 or the ERR = accepted discount rate are called "switching" values, i.e., the values which switch the decision on a project (based on these criteria) from a "yes" to a "no", or vice versa.

In cases where uncertainty about future values or benefits is particularly high, the analyst can use a "cost-price" approach. In this case, he calculates the price or value of the output which would make benefits equal to costs when both are discounted at the accepted discount rate. Thus, this is merely a variation on the basic BE analysis. The following example of calculation of cost price illustrates the approach.

A plantation project is being planned. The analyst is fairly certain about the costs involved - $250 for establishment in year zero and $10/ha/a starting in year 1. Technical personnel are fairly sure about their estimates of average yields and optimum rotation. The expected yield is 428 m^3 on a 15-year rotation. Present stumpage value is $\$5/m^3$, but there have been fluctuations and the expectation is that demand pressure on the limited supply will push up the stumpage price in the future. The analyst is uncertain about his estimate of a stumpage value 15 years from now. (He used an expected value of $\$7/m^3$ based on projection of past trends in real prices.) Given this uncertainty, one useful piece of information would be the stumpage value which would make NPW equal to zero at the relevant discount rate of 10 percent. The task for the analyst is to calculate this value, which is called the "cost-price".

He can approach the task dealing with future values (in year 15) or present values. Since it is easier (one less step) and makes more sense to deal with the future, he approaches it by compounding values instead of discounting them. 1/ He uses the following basic equation:

Establishment cost (C) compounded to year 15

\+ Annual costs (Ai) compounded to year 15

= Price (P) x Yield (Y)

Since he is solving for P, he arranges the equation as follows:

$$P = \frac{C(1+i)^{15} + A\frac{(1+i)^{14} - 1}{i}}{Y}$$

$$P = \frac{(1.10)^{15}(250) + \$10\frac{(1.10)^{14} - 1}{.10}}{428\ m^3}$$

$$P = \$3.1/m^3$$

(Note: The compounded annual payment factor comes from Appendix B).

What this cost-price of $\$3.1/m^3$ means is that with other values as assumed, the project could afford to return as little as $\$3.1/m^3$ and still break even at 10 percent. Since the analyst and decision-maker are quite certain that the price will be at least at the current level of \$5/m3, they accept the project as having a good chance of obtaining at least the 10 percent return required for this type of project.

If the cost-price had turned out to be around \$6/m3 (i.e., higher than the present price but lower than the analyst's estimated \$7/m3), then the analyst might want to take a closer look at the project, treating it in one of the ways suggested in the following sections.

1/ Compounding is the exact opposite of discounting.

The cost-price approach has further application in cases where a project involves non-market priced goods and services, e.g., environmental effects. It provides the decision-maker with information on what such goods or services have to be "worth" if the project is going to breakeven in terms of the relevant social rate of discount. While the decision-maker may not be able to decide on a specific value for some non-market priced output, he may be able to say to himself: "It is at least worth that much, therefore, the project is acceptable from an economic point of view." Alternatively, if the cost-price is very high, he may say: "I cannot justify the value implied by the cost-price calculation. Therefore I will not accept the project as being acceptable in economic terms and I will reject it, or attempt to redesign it to reduce costs."

10.6 DEALING WITH CRITICAL FACTORS IDENTIFIED IN THE SENSITIVITY ANALYSIS

Where a reasonable change in the assumption about the expected value for a given parameter (or values of a combination of parameters) is critical in terms of the expected outcome of a project, it is desirable to generate additional information about the parameter(s), if this is possible.[1] This may involve statistical estimation of probabilities of occurrence of different values using sampling techniques and available data, or it may merely involve developing subjective probabilities, or a number of other less formal approaches to increasing the knowledge about the likelihood of occurrence of the values that are critical to the project outcome.

Technical personnel and available literature can be consulted to obtain estimates of parameter values and ranges of such under varying conditions and more detailed effort can be spent on market surveys. Further, project planners can often find a wealth of information on species characteristics and other properties of woods available from national or international wood testing laboratories. Such information should be used to full advantage. The same general comment can be made about biological production function information, information related to insect and disease problems, etc. In most cases, data on which to base an objective probability analysis are lacking and cannot be generated in a short period of time. Yet considerable usable information is often available for use in developing subjective probabilities.

If further information on the critical parameter(s) indicates that there is a reasonable chance (1 in 20, or whatever is chosen) that the parameter(s) could indeed take on values which would influence the decision regarding a project, then the alternatives for further treatment of them in project planning fall into three categories which are not mutually exclusive:

- change project design
- build in contingencies and safeguards
- adjust the decision criteria used.

---

[1] Again, "reasonable" here refers to an initial estimate of what the possible range in values might be. "Critical" in terms of project outcome refers to the point where a factor's value reaches its "switching value", i.e., where the NPW moves from positive to negative or the ERR falls below the guiding rate of interest.

The first two of these possibilities are discussed below. The third relates to the broader issues surrounding project decision-making and is outside the scope of EAFP.

10.6.1 Changing the project design

Controllable uncertainty may be reduced by redesigning the project, e.g., changing its scale, changing factor proportions, integrating it with further processing or with raw material production, etc. Flexibility may also be built into the project by staging various project activities in a different way and with a different time schedule than initially planned, or by redesigning it to include more flexibility in terms of choice of factor inputs or outputs after implementation, etc.

Some examples will help illustrate how redesign can reduce uncertainty. In the case of scale, if an initial project design is for a scale of project that would fully meet an estimated future market demand which is somewhat uncertain, then the project possibly can be scaled down so that its capacity is near a lower estimate of market demand. This would reduce market uncertainty effects on the project. At the same time, if economies of scale are involved, it may increase costs. In this case the project planner has to weigh reduced uncertainty against higher costs. In the case of phasing of project activities, it might be possible to redesign so that the project starts with a smaller capacity sawmill or plantation and gradually builds up in phases as estimated future market conditions, factor availabilities, etc., become less uncertain. For example, investment in some of the fixed infrastructure, such as roads and buildings, could be delayed until the situation regarding future conditions became more certain. The potential impact of an uncertain market for one specific product could be reduced by expanding a forest industry project to include a more diversified output mix. For example, a sawmilling project could have a moulding production unit attached to it so there would be some flexibility in terms of shifting production from sawnwood to moulding as market conditions warranted. Diversification in plantation projects could also help to reduce uncertainty. For example, planting more than one species could help to reduce the risk of insect and/or disease problems in monoculture plantations. Species diversification could also reduce uncertainty with regard to markets, if the planted species have some overlap in characteristics and uses but also some unique characteristics which permit placing them in different markets as conditions warrant it. An example of a project which explicitly included this type of flexibility is the Korean fuelwood project; part of the area planted included "dual purpose" species which could be used for either fuel or timber, depending on how future market conditions developed for fuelwood.

A few words of caution are needed concerning redesign. In most cases, if the initial project design was based on thorough analysis of alternatives, then it was likely considered to be an optimum design in terms of the criteria for judging project worth and contribution. If redesign is undertaken, it is likely that expected costs will be increased and/or expected returns reduced over the initial optimum design. What this amounts to is a need for consideration of trade-offs between lower levels of uncertainty and lower levels of project worth (as compared with the expected return for the initial optimum design project). While the project planner can attempt to calculate and point out some of the trade-offs involved,it remains a matter of judgement as to the choice between alternatives. There are no general rules which can be made since it is difficult to quantify a decision-maker's subjective weighting of uncertainty.

Redesign is not the answer to all problems of uncertainty and should be approached cautiously. In many cases, redesign may not be desirable, and it will be necessary to resort to other methods of treating uncertainty. In cases of uncontrollable uncertainty, redesign may not be possible in the context of the project objective. In such cases other approaches can be used to take uncertainty into account.

10.6.2 Building safeguards into a project

Safeguards may be built into projects, including insurance on various elements of the project (which increases the project cost but reduces risk to the project entity); providing for physical contingencies (really a form of self-insurance); adding a premium to the discount rate used in calculating the NPW of the project, or arbitrarily lowering the output values and/or increasing the input cost estimates in calculations of the ERR or the NPW.

These approaches may not be sensitive to the uncertainties identified. For example, adding a premium to the discount rate penalizes future costs and benefits more than present or early costs and benefits, and this is not necessarily related to where the main uncertainties exist. On the other hand, an arbitrary increase in costs (e.g., contingency or insurance) and/or decrease in benefits would, for any given discount rate, suggest that uncertainty concerning future values is less important than uncertainty concerning present or early values. This may not be in keeping with the levels and timing of uncertainty identified. Despite their shortcomings the approaches suggested are used widely as a convenient way to reduce the chance of failure or a lower than expected rate of return. It essentially amounts to the same thing as saying that the acceptance criterion is made more strigent, i.e., a project has to show better than marginally acceptable performance. Adding a contingency allowance for physical uncertainty is likely to be the preferable way to treat the problem, since it does not tend to hide what is being done from the decision-maker. [1]

Projects can be designed with specific contingencies in mind. For example, in the case of an industrial plantation project planned for Tanzania, it was recognized that a principal uncertainty facing the project would be that the yet-to-be built pulp and paper mill, which would use the wood would not be built. Contingency plans for the project, in the unlikely event that the mill was not built, were (a) gradually to scale down the planting programme and stop it after five years, and (b) to grow the trees planted on a 25 year sawlog rotation instead of the shorter planned pulpwood rotation. The project analysis showed that there would be an acceptable market for the resulting volumes of sawlogs. The same type of contingency planning was included in the Korea fuelwood plantation programme, by planting a part of the area with "dual purpose" trees, i.e., ones that could be used both for fuel and timber.

Two additional points should be mentioned about uncertainty. First, uncertainty is often associated with the objectives for a project and the appropriate criteria for measuring the contribution of a project toward meeting objectives. This topic is not discussed here, mainly because it fits better in a discussion of sector planning, i.e., it is a question that transcends the subject of planning a particular project, given an

---

[1] See Gittinger 1972, pp. 100-104 for further discussion of contingency allowances.

objective. Objectives for a given project should be derived from a more general evaluation of the present condition of the sector and what goals it should be moving toward. The main problem with objectives at the project level relates to lack of definition. There is no sense in planning projects and project alternatives if objectives are not first defined explicitly. Criteria follow logically if objectives are defined. However, there are cases where criteria are poorly specified, mainly because objectives conflict or are loosely defined. The uncertainty in such cases is related to the lack of specified trade-off functions for the various conflicting objectives. Sensitivity analysis can contribute information on which decision-makers can base subjective judgements regarding trade-offs. The uncertainty involved really relates to uncertainty concerning the relative values placed on various objectives by society or decision-makers.

Second, a logical question is "How much should be spent on reducing uncertainty?" In general, the amount spent depends on the nature of the project and the available budget. In some cases, slight additional effort/expenditure can result in a marked reduction in uncertainty. In other cases, substantial expenditure will have little impact on reduction of uncertainty. Judgement based on past experience and knowledge about information availability and cost of information will provide some idea of the particular cost/benefit relationship facing the analyst. How much reduction of uncertainty is worth to the decision-maker is a judgemental question which has to be answered for each case separately.

For example, in the case of plywood production expansion project in the upper Amazon, the project analysts and sponsors decided that the substantial uncertainty surrounding estimates of total wood availability in the region was not significant to project viability. Ample volume was known to be available for the project at acceptable cost, and even the lower limit estimate indicated an available volume large enough to provide an ample margin of safety for the project. On the other hand, in the case of an integrated sawnwood and pulp and paper project currently being designed in Honduras, a large amount of money is being spent for detailed inventories so the project sponsors can be more certain that an adequate volume of wood is available at acceptable cost before they decide on the scale of the processing facilities and commit large sums to plant, equipment and infrastructure. In this case, uncertainty surrounding wood supply and cost is considered a critical factor by decision-makers.

PART II

USES OF ECONOMIC ANALYSIS IN FORESTRY PROJECT PLANNING

Chapter 11

USE OF ECONOMIC ANALYSIS IN PROJECT DESIGN

11.1 INTRODUCTION

This Chapter discusses the main uses of economic analysis in project identification and design. As pointed out earlier, many of the most critical decisions in the planning process are made at the early stages of identification and design. If economic efficiency considerations are absent, then opportunities might be lost to explore better alternatives.

Even the most simple projects can be designed in alternative ways. Thus, even for such projects, the economic analyst may have a role to play in project preparation and in the development of the most economically efficient project designs.

The process of design should encompass a number of project dimensions, such as scale, location, technology and timing for the various identifiable components of a given project. Once separable components have been identified and analysed, the analyst should look at the economic efficiency of alternative groups of components in terms of the project dimensions which will lead to the desired objective(s). The purpose is to arrive at the most economically efficient overall design, taking into account technical options, uncertainty, and interactions between separable components. Since the potential combinations of design components can be numerous, some technical judgement and experience are desirable in choosing the alternatives to be analysed.

Only one design for a given component can be chosen. For example, an analyst might be considering three alternative technologies or designs for the logging component of a particular project - one using mainly labour and cheap hand tools, one using power saws and less labour, and one using heavy machinery. Only one of them can be chosen for a given harvest and area at any one time. Similarly, only one overall project design can be chosen for a given situation.

When a situation exists where only one alternative out of several can be chosen at any given time to use a given resource or to meet a specific goal or objective it is referred to as a situation of mutually exclusive alternatives. The use and interpretation of indicators of project or component worth will vary according to whether projects or components are mutually exclusive or independent (not mutually exclusive).

Once the most efficient design for each mutually exclusive project component has been determined, the next task is to integrate them back into a total package, thus arriving at a design for the whole project. There may be several total project alternatives to compare. Therefore, once the analyst has chosen among mutually exclusive designs for a component, he still faces the choice between mutually exclusive projects. The approaches discussed in this chapter are relevant whether dealing with a component or an overall project.

When consideration has been narrowed down to one alternative project, then the final appraisal for the total project is developed. This process is discussed in Chapter 12, applying the general principles defined in Part I.

11.2 IDENTIFYING ALTERNATIVE PROJECTS AND THEIR COMPONENTS

There is no formula for determining the number and type of project alternatives which should be considered in a given situation. The process should strike a balance between covering the range of alternatives available and design costs, but no guidelines can be laid down regarding how to obtain an appropriate balance. This remains a matter of judgement based on circumstances. Technical competence and experience are important ingredients. However, some general considerations can be mentioned.

Sources for alternative designs are many. In many developing countries the range of project and project component alternatives considered are often moulded by political commitments, crises and experience. Project ideas might surface as a result of an analysis of the forest-based sector where the purpose is to develop a programme (an interrelated set of projects) to accomplish some broad objectives for the sector which fit within the general framework for national development. For example, in a large South American country an exercise was carried out in which estimated future demands for forest products were compared with the existing and expected industrial and resource capacities in order to determine the areas where important gaps existed or were more likely to materialize in the future. This exercise suggested possibilities for investment projects to attain a high level of self-sufficiency, which was one of the main policy objectives of the government. Ideas were advanced for several potential industrial expansion projects and for increasing the existing industrial plantation targets to levels consistent with the proposed industrial projects.

Most project ideas are identified in a less systematic fashion, without reference to an overall sector strategy for development. A forest service field officer may seek an opportunity to undertake a plantation project in his region ; a rural development official may see the potential for including a forestry component in a regional employment programme for a given region. An industry specialist may come up with an idea for a processing project which could make better use of a given forest resource, and so on.

While the integrated and systematic approach might seem more attractive because of its internal consistency, in practice both ways of generating ideas are desirable and complementary. It has been repeatedly observed that one of the main obstacles to forestry development in many countries is the dearth of constructive project ideas. Therefore, from a practical perspective, in many cases it is better to have good independent ideas than no project ideas at all. The way in which a project idea surfaces is not as important as the way in which the project idea is defined relative to an objective or set of objectives.

Given a project idea (or several ideas) the project analyst is concerned with two things initially. First how does the project idea relate to a definable and acceptable objective or set of objectives? Second, to what extent are there alternative ways in which that objective or set of objectives could be met? In answering the latter question, the project planner will want to take a given project concept and look at alternative ways in which the concept could be implemented. For example, the initial concept presented may be to utilize and manage a given unexploited 200 000 ha forest area. A plywood mill has been suggested initially. The project planner will want to ask: Is it possible that some other alternative - perhaps an integrated utilization complex that will use a greater number of species from the area - might be better? Perhaps the initial project idea should be expanded to include an integrated management and utilization scheme for a larger or smaller area, with conversion of the cutover forest to plantations which will, in turn, have a different use. Perhaps instead of one large plywood mill, it may be better to develop five or six small sawmills either alone or in combination with a larger central resaw unit that can provide further elaborated sawnwood products.

Some considerations can be suggested for avoiding passing over good project components and overall project ideas.

First, to the extent possible, the forestry project planner should be aware of the different developments taking place in the sector and he should understand thoroughly the policies and objectives set forth for sector development. If these are considered in a systematic fashion, the planner will have a better perspective on the relative merits of alternative approaches available for a given idea. Since one person is seldom an expert in all areas of forestry, it is generally better to develop project ideas using an interdisciplinary team which can consider project concepts and objectives from a number of different perspectives. In any case, a great deal of discussion of project ideas is desirable before focussing attention on any one idea.

Second, the project planner should follow the dictum that there are always alternative ways to design a project to meet a given objective or set of objectives. If all but one alternative are initially eliminated, then the project planner has in effect made a major project decision before he has even started the task of designing and analysing projects. That, in general, should not occur. Options should be developed so the decision maker has some choice.

One working objective in project planning is to avoid unnecessary expenditure of time and effort on analysis of project alternatives which at an early stage can be determined to be inferior (for a variety of reasons). Thus, the advantages of generating alternatives must be balanced against the costs of doing so. Only realistic alternatives should be considered.

Determination of whether or not a given alternative is "realistic" involves consideration of a number of factors other than financial and economic ones. First, and most obviously, the project alternative has to be technically sound and feasible. Second, it has to be commercially feasible. Inputs have to be available when needed, i.e., sources of supply have to be ascertained. Outputs have to be considered in terms of whether they will be purchased, in the case of market goods/services, or whether they will be required and used, if non-market goods or services. Third, the organizational and managerial feasibility of the project has to be ascertained, i.e., does it fit within the present legal and administrative framework, or can the administrative framework be changed so that it will fit? Will there be personnel available to manage the project appropriately? These three considerations are beyond the scope of EAFP. However, they need to be mentioned in the context of the present discussion, since they are central to a definition of realistic alternatives which will be subjected to economic (and financial) analysis.

In cases where multiple conflicting objectives exist - for example, where the objectives of employment of unskilled labour and economic efficiency conflict - it is particularly important that the project planner consider a range of alternative designs. In such cases a clear, quantifiable objective function cannot be defined and the decision-maker will want a number of alternatives to consider in making his subjective judgement on the trade-offs between different objectives.

The project identification stage is one of the most important stages in the project planning process. It often is not given enough systematic attention. It does not matter how good the design and appraisal of a project alternative is if it is not the "right" alternative to meet the objective. Although adjustments can be made at later stages, it is quite often true that after considerable effort has been spent appraising a given alternative there is a reluctance on the part of planners to abandon what they have been working on and to admit that the wrong alternatives were chosen. Thus, the obvious suggestion is to spend enough time and thought initially at the identification stage to ensure as far as possible that the right alternatives for a given objective and situation have been identified.

When a range of alternative project and project component designs have been defined, the economist can commence his task of analysing the economic efficiency of alternative designs for components as well as the overall economic efficiency associated with alternative projects made up of a number of separable and nonseparable components. These types of analyses are the subject of the remainder of this chapter.

The basic questions asked are: (a) What is the most economically efficient design for a given separable project component? and (b) Is it worthwhile adding a component (designed in the most efficient manner) to the total project? These are discussed in Sections 11.3 and 11.4.

11.3 USE OF ECONOMICS IN DESIGN OF SEPARABLE PROJECT COMPONENTS[1]

Assume that a limited number of separable project components have been identified initially for a total project package that will meet certain objectives. Each of the components can be designed in alternative ways by varying the technology, the scale, the location and the timing of the component. The question here is: "What guidance can the economist give in terms of identifying and determining the design which is most economically efficient?"

The question of alternative designs for a project component (or a total project) involves analysis of mutually exclusive alternatives. The basic steps involved in analyses of mutually exclusive alternatives are exactly the same as discussed in Part I. Inputs and

---

1/ If there are no separable components associated with a given project, then what follows applies directly to the project as a whole.

and outputs for the alternatives are identified and economic values are estimated for the identified inputs and outputs to arrive at estimates of costs and benefits. Then the costs and benefits associated with each alternative are compared to find the most economically attractive alternative among the ones being analyzed.

Depending on the relationship between the mutually exclusive alternative designs being considered, one of the following two types of comparisons will be relevant (see also Section 2.2.4).

(i) If the mutually exclusive alternatives produce the same benefit streams but involve different inputs and/or cost streams, then the PV of costs of the alternatives have to be compared, and that alternative that has the lowest cost chosen. This follows from the fact that if benefits are the same for all alternatives, then the alternative with the lowest cost (in PV terms) has the highest NPW. This would be the case, for example, if several alternative technologies for producing the same output are being investigated. Thus, three alternative logging technologies that could be used in harvesting the given volume of wood required for the project processing component might be identified. The lowest cost alternative would be chosen in terms of the economic efficiency objective. (Remember, though, that indirect costs that differ among alternatives also have to be considered.)

(ii) If the mutually exclusive alternatives being analyzed were to produce somewhat different benefit and cost streams, then the NPW's of the alternatives can be compared directly . This would be the case, for example, if the analyst was looking at alternative scales for a project or for a separable component, or if he was comparing alternative final processing activities for using a given volume of project wood output, or looking at the possibilities for adding on a purpose to a project (e.g., adding soil or watershed protection purpose to a primary purpose of producing fuelwood or other roundwood).

In the remainder of Section 11.3 these approaches are applied to four main design elements, namely, technology, scale, location and timing.

11.3.1 Technology alternatives

Most project components can be undertaken using different technologies (combinations of labour, land, and capital). For example, alternative sawmill designs are available that use different relative amounts of labour and capital (machinery). Plantations can be developed using different species and different establishment and maintenance practices or different combinations and intensities of inputs (e.g., fertilizer). Processing facilities can utilize different types of power. Transport of wood products can involve different combinations of equipment (and roads). Logging of a given output can be carried out with much labour and simple hand tools or with sophisticated machinery and few men.

If the output (or benefits) associated with different mutually exclusive technology options will be the same regardless of technology chosen, then a simple comparison of the PV's of costs of the alternatives can provide the appropriate information for choosing among them. That option with the lowest PV of costs is the most economically efficient to produce the given output.

From an economic efficiency point of view, the choice of technology to produce a given output will depend on the relative factor costs (e.g., the relation between the cost of labour, land, and machinery). If labour is relatively cheap, then labour intensive technologies will generally have lower average costs per unit output than capital intensive ones; if land is inexpensive (or has a low opportunity cost) then intensive management to maximize growth per acre will be less attractive than in a situation where the opportunity cost of land is high; if the cost of fertilizer is high, the less likely it is that intensive fertilization of forest plantations will provide a positive contribution to a project's NPW, other conditions being the same. The following example illustrates this point.1/

A forestry project is proposed in Tunisia. One component involves clearing land for a plantation. There are 400 ha to be cleared and a five year period for clearing is considered appropriate. Two technology options have been proposed. One involves manual clearing and the other mechanical clearing. Costs for these options are shown in columns 2, 3 and 4 of Table 11.1. Given the existing relation between costs of labour and capital, the mechanical option has a lower PV of costs if 10 percent is the discount rate chosen. However, if labour costs were lower relative to the costs of machines, say 20 percent lower, then the labour intensive alternative would have the lower PV of costs, given the same rate of discount. While the concept is intuitively evident, certain technologies which are geared to an intensive use of productive factors that are abundant in developing countries are often dismissed by project designers and preference given to "modern", usually capital-intensive, technologies developed in advanced countries with radically different resources endowments. This is a point which should be kept in mind at the earliest stages of forest-based project design in developing countries.

Another consideration of relevance in the analysis of alternative technologies is that they will, in all probability, generate different costs streams over time. If this is so, then it is likely that choice among alternative technologies will be affected by the discount rate applied to the analysis. In the example above considering the original labour cost assumption, the mechanical option has the lower PV of costs, if the rate of discount is 10 percent. If this rate were 20 percent then the labour intensive option is cheapest in PV terms. This is because the labour intensive option has a comparatively larger proportion of costs towards the end of the period, while the mechanical option involves a larger initial capital expenditure with lower operation and maintenance costs in the following years. A higher "discounting" of future costs then favours that alternative which has a greater proportion of its costs occurring in the future.

By following an iterative process, the analyst arrives at the "cross-over discount rate" of 14 percent. This is the discount rate at which a pair of alternatives with different cost-streams have equal NPW's. Its comparison with the social discount rate provides a straightforward decision rule, concerning mutually exclusive alternatives. Thus, if the relevant social discount rate is below 14 percent, then the mechanized alternative is more efficient because the PV of its costs will be lower. If the appropriate rate is above 14 percent, then the labour intensive option is the best choice from an economic point of view.2/ The results of this type of analysis can be presented in graphical form (Figure 6).

1/ Adopted from Gittinger 1972.

2/ As Gittinger points out, there may be various social reasons why the labour intensive option is preferred even if the appropriate social discount rate is below 14 percent. In this case, the choice involves objectives and criteria quite separate from those associated with the economic efficiency analysis.

Table 11.1

CHOICE BETWEEN MECHANICAL AND MANUAL LAND CLEARING ALTERNATIVES - TUNISIA

(US$)

Manual clearing

(1) Year	(2) Wages	(3) Other costs	(4) Total costs
1	44 050	3 800	47 850
2	44 050	3 800	47 850
3	44 050	3 800	47 850
4	44 050	3 800	47 850
5	44 050	3 800	47 850
Total	220 250	19 000	239 250

Mechanical clearing

Year	Equipment cost	Operation & Maintenance	Total costs
1	90 700	21 586	112 286
2	-	25 134	25 134
3	-	25 134	25 134
4	-	26 227	26 227
5	-	26 227	26 227
Total	90 700	124 308	215 008

Present value 10 percent discount rate:	manual clearing	US$ 181 399
Present value 10 percent discount rate:	mechanical clearing	175 905
Present value 20 percent discount rate:	manual clearing	143 101
Present value 20 percent discount rate:	mechanical clearing	148 759
Present value 10 percent discount rate, labour costs 20 percent lower:	manual clearing	133 586

Source: based on Gittinger 1972. p. 124.

Figure 11.1

TUNISIA: Choice between Mechanical and Manual Land Clearing Alternatives Illustrating Graphical Derivation of Crossover Discount Rate

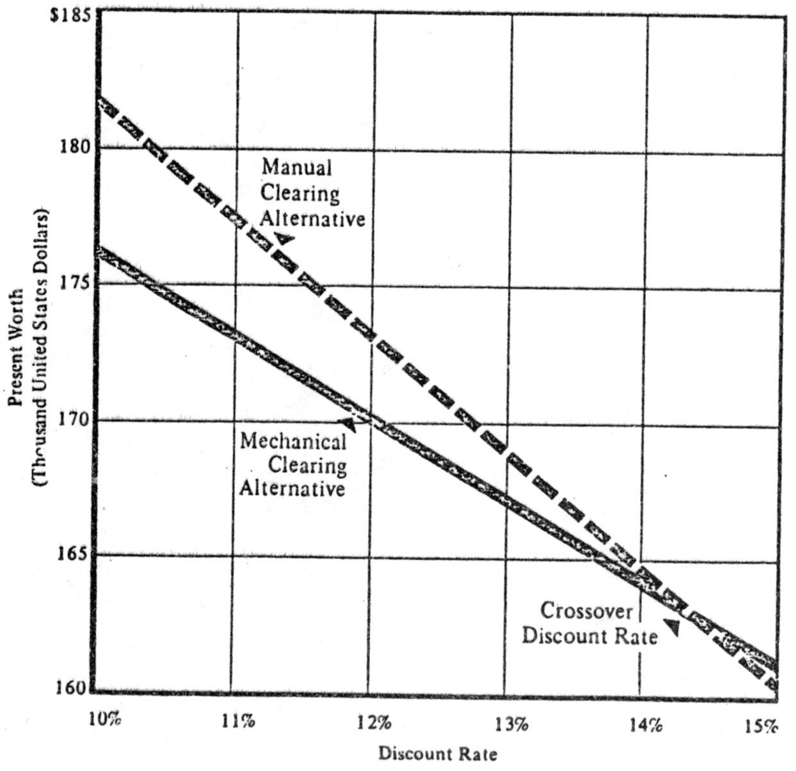

Source: Table 15 and Gittinger 1972 p. 125.

In some cases, different technologies for a component will involve different total benefits. While the primary purpose benefits (e.g., the value of wood produced) may be the same, there can be some secondary benefits that differ. For example, an alternative that involves use of locally produced simple machinery can have different secondary benefits for an economy than an alternative that involves use of imported heavy machinery to produce the same primary output.

In these cases the NPW's of the alternatives need to be compared rather than just the PV's of their costs.

Another important consideration in the analysis of technological options relates to the quality of the output and questions related to the value of the output. For example, the quality of output of a mechanized operation can be considerably more uniform than the quality of output from small labour intensive operations. Even if the output volume is the same, the value of the output (the benefits) associated with the alternatives may be different because of quality differences.

Similarly, the analyst has to be certain that he has adequately defined the project purpose and final output in making technology comparisons. For example, in forestry there is a common tendency to think of a faster growing species as being superior to a slower growing one (i.e., the "objective" is taken to be maximum volume yield per unit area per unit time). While in many cases this is the relevant criterion to use in choosing among alternative species, there are also many cases where it is not true. The analyst has to keep in mind that what matters in an economic analysis is the increase in value and that this increase might or might not be closely related to physical increases in volume per unit area per unit of time.

The technology options considered for a component or project will depend heavily on the experience and knowledge of the technical personnel involved in planning the project. Choice of the most efficient (best) alternative will also depend on the decisions regarding other project components. Thus, in some cases, the analyst will be presented with alternative packages of technologies for all project components and asked to analyse these alternatives as a whole without separating out the components for separate analysis. The same approaches as suggested above for analysing separable components can be used in this case.

11.3.2 Scale options

For many types of activities, substantial "economies of scale" in certain ranges of production can be encountered. Economies of scale refer to the variation in average per unit costs that can be achieved by varying the scale of operations. The economist may be asked to analyse the economics of alternative scales of production in cases where the project or component output level is not set by the market (e.g., in the case of some types of export oriented projects, or in the case where a project could be designed to include several small production units or one larger unit).

The most practical approach is to define several mutually exclusive alternative scales - perhaps combined with several alternative technologies - and then to analyse each separately, picking the one with the highest NPW as the optimum scale for the project. There are no new conceptual problems involved in this process. This question of scale can be treated in exactly the same manner as the question of technology. Indeed, it is often the case that the two are analysed together.

The main problem encountered in dealing with the question of scale is the general lack of adequate information on the variations in costs which will occur with different scales of output. For some types of processing activities, there are fairly good estimates available. However, such figures often reflect factor cost relationships in countries other than the project country. The analyst has to use caution in adapting such information directly to his analysis. For most forestry activities there is very little specific, empirical information available on economies of scale. Rough estimates have to be used, based on experience and judgement of technical personnel.

In cases where economies of scale exist, it is sometimes of interest to calculate the breakeven size for a given operation, or that scale of operation at which the PV of costs equals the PV of estimated benefits. At scales larger than this breakeven level NPW would be positive and at scales below it the NPW would be negative. Breakeven analysis is also used in sensitivity analyses, as explained in Section 10.5.2, where the breakeven price (or cost-price) calculation was used as an example. The approach to calculating a breakeven size for a given operation is exactly the same, except output quantity is being solved as the unknown.

11.3.3 Location alternatives 1/

In some cases, the analyst will be asked to analyse alternative locations for a given project or project component. The analytical work can be reduced by looking only at those cost and benefit elements which vary with location. For example, in the case of alternative plantation locations, the analyst generally need look only at effects of site, transport cost and land value differences. Information on these differences can be presented in simple tabular form to give the decision-maker a clear picture of the costs and benefits associated with alternative locations.

Quite often taxes will vary by location, as will subsidies given for development in relatively backward regions of a country (e.g., in the case of Brazil). These differences will not affect the economic analysis in terms of NPW, but will affect the financial analysis as discussed in Chapter 9.

1/ The term "location" as used here can refer to either the region or area in which the project will be located as well as the specific site within a region for locating production facilities.

In **analysing** mutually exclusive alternatives, the nature of the comparison must first be established **i.e., in terms of what factors are the alternatives mutually exclusive or** what is the limiting factor being considered? In some cases it will be a given input (e.g., the land area); in other cases it will be an output constraint (e.g., the size of the project market, etc.). Constraints have to be clearly specified before a meaningful economic analysis of mutually exclusive alternatives can be undertaken.

11.3.4 Timing alternatives

Some of the most common questions faced in forestry related projects concern timing of components of a project. Of particular importance is the question of rotation or felling age determination. There are also other questions which relate to timing of activities. Many of the critical timing problems encountered in project analyses relate to timing of a given component in relation to other components. These problems are discussed in Section 11.4, which deals with interactions and choice of component alternatives to include in a total project package.

11.3.4.1 Rotation determination or appropriate felling age

The basic question from an economic point of view is - given a species, site conditions, and values for costs and benefits associated with a given situation, what rotation length or growing period maximizes NPW?

The determination of the optimum economic rotation or felling age raises no new problems which are different from those involved in analyses of other aspects of mutually exclusive alternatives. NPW's are calculated for alternative felling ages and the one that results in the highest NPW is chosen as the preferred option from an economic point of view. If yield and unit value information by years is available, then it becomes a straightforward process to find the rotation with the highest NPW.

As shown in Table 11.2, the analyst estimates the total benefits (col. 3) and the total costs (col. 4) which would occur if the stand was held for each of n years. He then discounts these values and subtracts PV of costs (col. 6) including land opportunity cost from PV of benefits (col. 5) for each alternative rotation to get a NPW (col. 7) for each. That rotation with the highest NPW is then chosen. In this case, the peak of NPW is reached in years 20 and 21.

In order to arrive at the optimum rotation it is not necessary to calculate the NPW for all years. The analyst can quickly arrive at the optimum rotation length by calculating NPW's for a few years spread apart, and then concentrating on the **years where the NPW is near its maximum and then starts to drop.**

Alternatively, as shown in column 8, the analyst can estimate the marginal rate of return (MRR) (marginal ERR) on holding the stand another year, which is the same as the rate of increase in the net current benefits from one year to the next. The last year for which the MRR is above or equal to the relevant discount rate (in this case 7 percent is assumed) is the rotation length that will maximize NPW. In this **example the MRR drops to 7 percent between years 20 and 21.** Thus, again the optimum rotation length is between 20 and 21 years.

Table 11.2

ROTATION DETERMINATION (PER HA BASIS)

(1) Year	(2) Yield 1/ (m^3r)	(3) Benefits	(4) Costs 2/	(5) PV of Benefits (7%)	(6) PV of Costs 2/ (7%)	(7) NPW (7%)	(8) Rate of increase in current net benefits (7%)
0	0	0	30	0	30	(30)	
1-4	0	0	2	0	36		
5	7.7	11	2	8	45		
6	17.1	24	2	16	46		
7	27.0	38	2	24	48		
8	37.7	53	2	31	49		
9	49.2	65	2	35	50		
10	61.4	86	2	44	51	(7)	
11	74.5	104	2	49	52		
12	88.5	124	2	55	53	2	
13	103.3	145	2				
14	119.2	167	2				
15	136.2	191	2				
16	154.4	216	2	73	56	17	
17	173.7	243	2	77	56	21	12
18	193.2	270	2	80	57	23	10
19	212.6	298	2	82	57	25	10
20	231.3	324	2	84	58	26	8
21	248.9	348	2	84	58	26	7
22	265.6	372	2	84	59	25	6
23	281.3	394	2	83	59	24	5
24	296.1	414	2				
25	310.0	434	2				
26	323.1	452	2				
27	335.6	470	2				

1/ Based on Gregory, 1972, data for pine yields in Lota, Chile.

2/ Establishment costs, \$29; annual management cost starting in year 1, \$1, annual opportunity cost for land, \$1.

The MRR approach is probably the easiest to use. For each year, the total benefits are estimated as shown in column 3. To get the MRR the analyst subtracts the benefits in one year from those in the succeeding year, then subtracts any costs which would occur during that year, (i.e., col. 4) and divides the result (the net current benefit increase) by the benefit which would be obtained if the stand were cut instead of being held for another year. For example, the benefit in year 20 ($324) is subtracted from the benefit in year 21 ($348). The result equals $24. Then the $2 annual management cost is subtracted from the $24 to give $22. This is the net benefit gain which could be obtained if the stand were held during year 20 (until year 21) instead of being cut at the beginning of year 20 (or end of year 19). This net benefit is then divided by the benefit which could be obtained at the beginning of year 20, i.e., $324, and multiplied by 100 to give a MRR of 7 percent.

The logic of this approach is quite simple. As long as benefits are growing at a rate higher than the discount rate, it is worth leaving the stand since NPW is increasing. This approach applied to analyses based on market values is basically the "financial maturity" concept familiar to foresters.

It is not necessary to calculate MRR's for all years in order to arrive at the appropriate felling age. The MRR was calculated first for years 16-17 and found to be above 7 percent (the assumed discount rate). It was then calculated for year 23 and found to be below 7 percent. At years 20-21 it was found to be 7 percent.

When two or more mutually exclusive alternatives that involve different time periods are being compared (e.g., alternative rotation lengths), adjustments have to be made to take into account the different time periods involved. In the example, this adjustment was made by including the annual opportunity cost of land in its best alternative use. For example, if a 20 year rather than a 15 year rotation was chosen, 5 years of net value which could be obtained from the land ($5, or $1 per year) if it was harvested in year 15 and used for the ensuing 5 years in its best alternative use would be foregone.

A common way that foresters take unequal time periods into account in choosing between rotation ages is to calculate what is called the land or soil expectation value (SEV). The one with the highest SEV is then chosen as the optimum rotation from an economic point of view.

The SEV is essentially a financial measure, equal to the present value of an infinite number of equal periodic net returns of $R received every r years. In forestry, r is identified with the rotation age and $R with the net returns from one rotation. In the calculation of SEV, the land value is not included, and therefore the SEV essentially indicates what could be afforded for land and still breakeven (have value of benefits equal to value of costs when both are discounted back to the present using the relevant discount rate). Further details on the SEV and how it is calculated are shown in Appendix B.

If the opportunity cost of land as used is correctly estimated, then the NPW for one rotation, the MRR and SEV approaches should all give the same answer with regard to which rotation length or felling age is optimum in economic terms. As mentioned the MRR approach is probably the easiest to use. Once the optimum rotation of felling age has been determined, then the NPW and/or ERR can be calculated. 1/

11.3.4.2 Other timing considerations

In addition to rotation determination, optimum scheduling of other project activities will have to be considered. For example, what would be the impacts on NPW of scheduling investments in plant and equipment over longer periods than initially envisaged, i.e., phasing project build-up? Or, how should investments in infrastructure be phased, i.e., when should roads be built? These questions may or may not be relevant depending upon the assignment of the analyst and the project being analysed. In many cases, the technical personnel set the initial timing of various activities. If other questions of timing are relevant to the economic analysis, then the analyst would again define, with the help of the technical personnel, several alternative timings. Using the value flow tables for each alternative, he would analyse each as a separate alternative, comparing the NPW's to find that alternative with the highest NPW. If amounts and timings of outputs (benefits) remain the same for all alternatives, then the analyst can compare the PV's of costs and pick the one with the lowest value as being the relatively most economically efficient one.

11.3.5 Comments on design choices for separable components

Alternative designs of components in terms of the appropriate approaches to analysing mutually exclusive alternatives have been discussed. While the assumption of mutual exclusivity holds for any given situation involving the same resource (e.g., land area) or output constraint, it is also possible that a detailed analysis will indicate that two or more designs should be incorporated in the same project for different segments of the project. For example, two or more different logging systems can be employed in the same project if different conditions exist for parts of the total project area. Similarly, different management intensities can be used for different parts of the project area, or two or more different technologies can be used to produce the total project output, depending on specific conditions encountered in the project environment. The point still remains that for any given segment of project area, or for any given portion of the planned project output, only one or another design can be chosen. Thus, an analysis of disaggregated sub-components can be dealt with, but the relevant consideration for each still relates to which of the mutually exclusive alternatives identified has the highest NPW for each subcomponent.

Table 11.3 summarizes the appropriate approaches to analysing mutually exclusive alternatives for different design elements.

1/ The MRR only relates to the difference between holding or harvesting a given stand in year t or year t+1. A NPW or the interval rate of return (financial or economic) still has to be calculated to indicate whether in fact planting the trees is worthwhile in the first place.

Table 11.3

APPROACHES TO ANALYSING MUTUALLY EXCLUSIVE ALTERNATIVES FOR DIFFERENT DESIGN ELEMENTS

Design Elements	Alternative Designs Have the Same Timing and Value for Benefits	Alternative Designs Have Different Timings and/or Values for Benefits 1/
Technology	For each alternative calculate PV of costs that differ between alternatives. Pick the one with lowest PV of costs.	Compare NPW's of alternatives; pick one with highest NPW.
Scale	N.A. 2/ (benefits will vary with scale for any given component being analysed).	Compare NPW's of alternatives; pick the one with highest NPW. Or analyse increments in scale using MRR approach, and pick that size where MRR drops down to the discount rate.
Location	For each alternative, calculate PV of costs that differ with location. Pick the one with the lowest PV of costs.	Compare NPW's of alternative locations; pick the one with highest NPW.
Timing	N.A. (timing of outputs and thus benefits will vary).	Compare NPW's of alternatives; pick the one with highest NPW. Or use MRR approach, e.g., in the case of rotation or felling age determination. (Make sure to adjust for time differences, e.g., by adding in land opportunity costs.)

1/ If two (or more) alternatives have a number of cost and/or benefit elements in common, then the net difference in PV of costs and benefits that differ between alternatives can be compared. If the net difference between alternative i (considered the base alternative) and j (the one being compared) is positive then j is preferable. If the difference is negative, then i is still preferable. This approach holds for all design elements. It is essentially the MRR approach discussed for rotation or felling age determination.

2/ N.A. = not applicable.

11.4 INTERACTIONS BETWEEN DESIGN ELEMENTS AND SEPARABLE PROJECT COMPONENTS

The four design elements discussed - technology, scale, location and timing - are generally interrelated for a given component. The discussion in Section 11.3 treated the four separately in order to clarify some of the differences in approach which can be applied in arriving at the most economically efficient design for each element, holding other design elements constant. It is often practical to separate the elements initially when the analysis is started. However, at some point the analyst has to consider the interactions between elements i.e., allow for variation in several elements at the same time.

The variety of interactions which theoretically may be considered for different project situations is virtually limitless, while those that practically can be considered in any given project analysis are generally severely limited by budget and time constraints imposed on the project planning effort. Thus most analyses will start with a few alternative designs for a component (and a project) which include specified technology, scale, location and timing combinations. In this case, the economist develops for each of the specified alternatives a value flow table and a NPW and then compares the NPW's of the limited number of mutually exclusive alternative designs being considered.

If the situation calls for it, he may also make some more detailed investigations of alternative timings for a component, or suggest looking at alternative scales not included in the limited alternatives provided by the technical personnel. He might also look at a few technology options not initially considered. For example, if data were available on the response of several potential species to different management intensities (e.g., thinning regimes and application of fertilizers) he might analyse the benefits and costs associated with such alternatives to arrive at a more efficient technology for the particular project situation. In this case, he would be looking at the interaction between technology, timing and location elements for the plantation component, but also the implications for other project components. An example will illustrate this.

Assume an integrated wood production-processing project is being considered. The scale and output level of the processing component have been fixed by the estimated requirements for the output at the time it becomes available. Thus, given the processing technology and output level, the volume of wood input is also fixed. The wood production technology envisaged would require 300 ha of land to supply the wood input requirements of the processing component on a sustained basis, starting 6 years after the first planting. Thus, the processing component investment has been timed to come on stream in 6 years, the minimum period required to generate the wood raw material needed.

Although fertilization of the plantation is not considered in the initial design, there is information available which provides some indication of the response of the chosen species to fertilization. Thus, the economist could undertake a partial analysis of the economics of fertilization. Fertilization could have several impacts in terms of the overall project. First, less land would be required to grow a given volume of wood required. Second, this would have implications not only in terms of land costs, but also in terms of silvicultural management, logging and transportation costs.

Third, the trees might reach optimum economic rotation at an earlier age. The impacts of these differences might be felt in terms of other input requirements, location of plantations and timing of the processing investment. (Since wood would become available before 6 years, the processing component could come on stream earlier than initially planned.) Therefore, these changes could result in significant differences in the total project NPW's with and without the fertilizer application. A partial analysis that only considers effects on yield changes on an average per unit area per year basis would not capture all the implications of fertilization in terms of the overall project. There are implications for other design elements and for other components included in the project.

Another more complex situation arises when substantial economies of scale exist for a given activity (component), but the volume of output required to take full advantage of the potential economies of scale is above estimated requirements for the output of the project, at least in the initial years. In such cases, the analyst will want to look at the economics of capacity utilization. It may be that the economies of scale are so great that the larger design of the component should be undertaken, even though its capacity would not fully be utilized for several years. The physical flow and value flow tables can be prepared for several alternative combinations of investment, operating and output assumptions, and the NPW's of the alternatives can be compared to arrive at the most economically efficient one, given the constraint identified for requirements. For example, one alternative would be to build initially a pulp and paper mill with 150 000 tons per year capacity and operate it below capacity for the first five years until requirements reach 150 000 tons per year. Another alternative could be, for example, to build initially a mill that produces 110 000 tons (the assumed initial requirements) and then put another mill on the stream in 5 years that would produce an additional 60 000 tons (the assumed minimum economic capacity) and have a reduced excess capacity only from year 5 to year 7 where requirements are expected to reach 170 000 tons.

This particular problem involves considering both scale and timing elements in the same partial analysis. It also involves technology considerations in arriving at the relevant cost estimates for the two alternatives. Location considerations may also enter the picture in terms of location of the two, phased mills against the one larger mill. Thus, here is an example of a case where four design elements are closely interrelated.

Even in the case of an analysis of a single separable project component, consideration of many alternatives with regard to technology, scale, location, and timing can become a major task in terms of the computations involved. Once it is recognized that there are interactions between components, the task becomes even more complex, if many such interactions are considered. Take the example shown in Table 11.4, which only includes four components and a few limited designs for each. Since two locations are being considered for the processing component and two for the plantation component, there are 4x2x2x8 or 128 possible combinations. If the assumptions regarding fixed elements are relaxed slightly, over 500 different alternatives could be obtained.

Naturally, in most cases, the number of alternatives considered will be limited in the technical analysis stage to considerably fewer than 128. If a computer is readily available and fairly good data exist on which to base physical input-output relationships for the alternatives, then it is simple to run through a great number of alternative designs and to arrive at the optimum design in terms of economic efficiency. If computer facilities (and the expertise needed to use them) are not readily available, then the project planners will probably want to reduce the number of alternatives to a few, using their judgement and experience concerning which ones are most desirable in the particular project context.

Table 11.4

CONSIDERING COMBINATIONS OF COMPONENTS IN A TOTAL PROJECT PACKAGE

Project: integrated plantation, harvesting and processing project to produce sawnwood for local market.

Constraints: output is fixed (i.e., scale of lumber production).

Processing of sawnwood:

Alt. 1 labour intensive sawmill
Alt. 2 capital intensive sawmill

For each of these alternatives, two locations are being considered; scale is fixed by market; timing will depend directly on when first harvestable wood will become available.

Transport of wood to mill:

Alt. 1 large trucks with road improvements
Alt. 2 smaller trucks utilizing existing roads

For each alternative, location considerations are fixed depending on sawmill location and location decided on for plantation; scale is fixed by wood volume needed (actually grown) to meet mill requirements; timing is fixed within limits by when wood becomes available.

Harvesting of wood:

Alt. 1 labour intensive with cheap hand tools
Alt. 2 capital intensive technology with machinery and less labour

For each alternative, timing depends on rotation or harvest age set for plantations (from 12 years to n years); scale is fixed by volume requirements and volume actually grown; location is fixed by location of plantations.

Growing of wood:

Alt. 1 species X with fertilization
Alt. 2 species X without fertilization
Alt. 3 species Y with fertilization
Alt. 4 species Y without fertilization

For each alternative, two locations are being considered, timing of planting depends on calculation of optimum rotation (most profitable age for harvest), scale is fixed for any given alternative by volume requirements at mill and area available.

11.5 COMPARING HORIZONTALLY RELATED COMPONENTS

In some cases there are several horizontally related components which are being considered for the same project, and the components are not mutually exclusive, i.e., one or both could be included in the same project. The approach to analysing such alternatives can be illustrated with a simple example.

Assume a project which is initially designed to produce plywood and sawnwood from a given and limited raw material base. The two components are independent in terms of processing and marketing, but they both depend on the same raw material, which is considered the limiting factor in this example. The economic analyst may be asked to provide information on the relative efficiency of producing one or the other or both of the products in combination. The question really boils down to one of the optimum allocation of a scarce resource, in this case the wood. Assume the simplest case, where both products can be produced from the same raw material and the market analysis indicates that if the raw material were utilized solely for either produce the resulting output could be fully marketed without influencing prices. Given the above, the analyst would want to provide information on the net benefit if the wood were put into (a) plywood production, and (b) sawnwood production. If he finds that one gives a higher NPW than the other, then the project may be redesigned to include production of only the one with the highest NPW.

This example represents the simplest situation. If only some of the wood is suitable for plywood, while all of it can be used for sawnwood, or if the market capacity for one or the other or both of the products is limited within the range of possibilities offered by the available raw material, then the constraints change and it is possible that some combination of components will provide the maximum NPW possible.

In determining that combination which gives the maximum NPW, the analyst can be guided by net benefit estimates per unit of input or output for each product and the various constraints identified. Table 11.5 provides an example of a simple analysis of optimum product mix, given expected market prospects and other constraints. In this case, plywood gives a much higher estimated net return than sawnwood per m3 (r), i.e.,$40 vs. $20, and is thus the best product from an efficiency point of view when wood is a scarce resource and taken to be the limiting factor. However, the market potential for plywood is only 20 000 m3. Thus, the first step is to allocate all the wood needed to produce 20 000 m3 of plywood, or 40 000 m3 (r). The remainder of the wood is then allocated to sawnwood production. This amounts to 110 000 m3 which can produce 68 750 m3 of sawnwood. As can be seen, both constraints - market and wood availability - enter the analysis at different stages.

The same approach could be taken if the alternative uses of a plantation output were being investigated. For example, the output might be used for sawnwood or plywood or for pulp, paper or fibreboard. In this case, the alternatives are compared using different value assumptions for the wood output, depending on the particular use being analysed. For each alternative, the benefits would be calculated in terms of the value assumptions for each alternative product or use, and the costs would be based on the costs of producing the wood or the opportunity cost of the wood, whichever is highest. For example, assume that the cost of wood production in the project has a PV of $150/ha; the value of the wood from the project if used for sawnwood has a PV of $200/ha, and the value of the wood from the project if put into pulp production has a PV of $180/ha. If the sawnwood alternative is

Table 11.5

DETERMINING OPTIMUM PRODUCT MIX

Plywood:

Market constraint:	20 000 m^3/a
Wood required:	40 000 m^3 (r)/a (2m^3 (r)/m^3 plywood)
Average PV of net return per m3 (r):	$40

Sawnwood:

Market constraint:	100 000 m^3/a
Wood required:	160 000 m^3 (r)/a (1.6m^3 (r)/m^3 sawnwood)
Average PV of net return per m3 (r):	$20

Wood available:

Total:	150 000 m^3 (r)/a
Usable for plywood:	50 000 m^3 (r)
Usable for sawnwood:	All

Design project to produce:

20 000 m^3 plywood from 40 000 m^3 (r)

68 750 m^3 sawnwood from 110 000 m^3 (r)

NPW (maximum possible, given market constraint)

40 000 x $40	=	1 600 000
110 000 x $20	=	2 200 000
Total		$3 800 000

being analysed $200/ha will be used as the measure of benefits and, if land is the limiting factor, then $180/ha will be used as the measure of opportunity cost, since this is the value given up by using the wood for sawnwood rather than for pulpwood. If land is not the limiting factor - i.e., the pulpwood could be produced in another location for $150, then the relevant opportunity cost would be based on production costs and would be $150. This follows from the fact that, by using the wood for sawnwood, an extra cost of $150 would be incurred to get the equivalent amount of wood for pulpwood.

11.6 ADDING ON A PROJECT PURPOSE

In some cases, the analyst may want to look at the economics of adding on a secondary purpose to the initially conceived project purpose(s). For example, he may be considering a plantation project to produce fuelwood for a local community. The suggestion is made that with slight additional expenditure, the project design could be modified so it would produce significant soil protection benefits (valued in terms of crop losses avoided). How would he determine whether it is worthwhile adding on this project purpose?

The difference here from the case of a separable component being considered for inclusion is that the costs of the two purposes - fuelwood and soil protection - are for the most part shared costs. Both purposes share the major expense, namely the basic plantation establishment and maintenance costs. The approach in this case is to compare the present value of the incremental costs required to add on the purpose with the incremental benefits associated with the add-on purpose. Put another way, the NPW of the differences between the value flows with and without the soil protection component can be calculated. If it is positive, then it is worth adding on the purpose. If it is negative, then the additional benefits do not justify the additional costs. This approach essentially parallels that suggested in Table 11.3 for dealing with mutually exclusive alternatives for specific design components. The MRR approach could also be used in this case, where the MRR associated with adding on a purpose would be calculated.

The difference shown in Table 11.6 (col. 4) in the cost streams for the two alternatives (fuelwood purpose alone and fuelwood/soil protection combined) is due to the higher cost associated with shifting the plantation activity to a steeper more critical area that can produce protection benefits, modification in the planting rate per unit area, and modification in the maintenance and harvesting approaches. The differences in benefit streams are due to the inclusion of the soil protection benefits (valued in terms of crop losses avoided).

As noted in Table 11.6, the NPW of the difference between the two value flows is a positive $75 770 which means that it is worthwhile to add on the soil protection purpose, given the discount rate of 5 percent.

If the analyst is uncertain about the value of the benefits associated with an add-on purpose, then he can estimate the minimum value which such add-on benefits must have in order to justify the incremental costs associated with producing such benefits. To do this, he calculates a cost-price such as discussed in Chapter 10.

Table 11.6

DETERMINING WHETHER A PROJECT PURPOSE SHOULD BE ADDED TO THE MAIN PURPOSE(S)

		(Values in $ '000)		
Item	Fuelwood	Combined w/soil Protection	Difference between the two	PV of the difference (at 5 percent)
(1)	(2)	(3)	(4)	(5)
Costs (by years)				
0	150	180	30	30.00
1	30	50	20	19.05
2	20	30	10	9.07
3	20	25	5	4.32
4-15	15	18	3	20.5 1/
Total	-	-	-	82.94
Benefits (by years)				
2	-	10	10	9.07
3	-	15	15	12.96
4-15	90	110	20	136.68 1/
Total	-	-	-	158.71

NPW of difference (at 5 percent) equals
158 710 - 82 940 = $75 770

1/ The formula for the PV of a series of equal annual payments was used to obtain the value of the series expressed in year 4 value terms. That value was then discounted back an additional 4 years to arrive at the PV in year zero (see Appendix B for formula).

Chapter 12

USE OF ECONOMIC ANALYSIS IN THE APPRAISAL OF A FORESTRY PROJECT: A SUMMARY [1/] [2/]

12.1 INTRODUCTION

This chapter provides a review of the main elements which should be included in a financial and economic appraisal of a project at a fairly advanced stage in its preparation[3/]. It is essentially a summary of the steps discussed in Part I. As it is preferable and usual that the economic and financial appraisal results be presented in the same document, and in an integrated fashion, the discussion which follows includes both types of information.

There is no unique or "best" way to present the information needed by decision makers to evaluate the financial and economic worth of a forestry project. The amount of detail required depends on, among other things, the specific nature and size of the project, its technical complexity and scope, as well as the particular requirements or standards of the institution for which the analysis is being prepared. However, though there will thus be some variations from project to project, every appraisal report should contain at least summaries of the following basic elements:

A. Inputs into the analysis:

(i) Direct physical relationships, presented in the form of Physical Flow Tables, showing inputs and outputs and their relationships over time and by categories of inputs and activities as required by the objectives of the analysis. These relationships are obtained from the engineering and technical studies for the project and from financial analysis documents (see Chapter 4).

(ii) Indirect physical relationships. These are usually not included in financial appraisals but must be included in the economic analysis. As explained in Chapter 4, the nature and magnitude of these relationships are generally identified from sources other than the technical and engineering studies required to carry out financial appraisals,

---

1/ Because so many tables are used in this chapter and referred to in a number of places, they are all grouped together at the end of the chapter.

2/ See the case studies in FAO, 1979, and Appendix A for other examples of the suggested format.

3/ i.e., the implicit assumption is that a number of studies of alternative designs, project scopes, timing, etc., have been carried out in arriving at the project design which is being subjected to appraisal (see Chapter 11).

although these are a logical point of departure from which to investigate indirect effects. These effects can be incorporated either directly in shadow prices or, preferably in the form of an Indirect Physical Effects Table and/or Statement. (Chapter 4).

(iii) Market prices for financial inputs and outputs including their estimated changes over time net of inflation. This information is presented in a Financial Unit Value Table. (Chapters 5 and 6).

(iv) Shadow price for economic inputs and outputs (including those for indirect effects if available) and their expected changes over time. This information is presented in an Economic Unit Value Table. (Chapters 5-8).

B. Outputs of the Analysis:

(i) Financial effects of the project over time presented in Financial Total Cash Flow Tables for the total project, and for important separable project components. (See Chapter 9).

(ii) Economic effects of the project over time displayed in Economic Value Flow Tables for the total project and for major separable components. (Chapter 9).

(iii) Estimates of economic and financial measures of project worth. (Chapter 9).

(iv) Tests of the sensitivity of the measures of project worth to changes in assumptions about input/output relationships and unit values assumed in the basic analyses, i.e., explicit treatment of project uncertainty. (Chapter 10).

(v) Conclusions/recommendations (if required as part of the assignment).

This approach fits with the earlier recommendation that a financial analysis be performed prior to, or simultaneously with, an economic evaluation. The results of both should be presented together.

The following sections illustrate each of the above steps and their interrelationships, using as an example an afforestation project in a tropical country.

12.2 INPUTS INTO THE ANALYSIS

The first step in deriving and organizing information for an economic or a financial analysis involves the identification and measurement of physical input-output relationships, both direct and indirect. The second step consists of developing financial and economic values for the inputs and outputs and other values needed in the financial and economic analyses.

12.2.1 Direct physical input/output relationships

The technical design of a project involves the processing of a great deal of data on the physical dimensions of the proposed project. Input-output relationships have to be quantified and total input requirements to meet output goals need to be tabulated, generally by input types and activities and by the years in which inputs and outputs occur. In some cases inputs are further grouped by source, foreign or domestic, if foreign exchange is a relevant concern. Similarly, outputs are subdivided in terms of destination, foreign or domestic. Inputs and outputs are broken down into these categories in order to facilitate shadow pricing. Other groupings of inputs and outputs may also be used for other purposes.

Quite often, physical input and output data can be estimated on an "average" unit basis - e.g. average per ha input requirements and average per ha output for the project. This approach is typical where information is not available to break the analysis down by sub-areas/site classes and by locations within the project region. In cases where data are available on which to base a disaggregated analysis, the analyst can develop separate input and output tables for each different type of area or condition identified. (See Chapter 4).

In the case of the afforestation project used as an example in this Chapter, the analyst used Tables 12.1 and 12.2 to present the output conditions assumed in the analysis. The project is aimed at establishing 18 000 ha of plantations over a period of six years to provide wood for an industrial complex. The relevant project period, based on when yields will occur was set at 35 years (see Chapter 4).

The species proposed are *Pinus* spp. and *Eucalyptus* spp. A land area equivalent to the project's requirements and close to the main industrial market is already available as a public Forest Reserve. The project will be implemented by the Forest Service and financed by a loan obtained by the Service which carries an interest rate of 7 percent. According to the design of the project, pines will be planted at a rate of 2 000 ha per year and eucalypts at a rate of 1 000 ha per year to reach a total planted area of 12 000 ha of pines and 6 000 ha of eucalypts at the end of six years.

It is planned to grow both species over rotations designed to produce both sawlogs and small diameter roundwood to be used as poles and for pulp. The pine plantations will be thinned at ages 6, 8, 12 and 21 years with the final cut being made at 30 years. Eucalypts will be thinned at ages 3, 5 and 6 years with the final cut carried out at 8 years. Species trials and silvicultural treatment experiments carried out in similar ecological conditions in neighbouring countries provided the basic information for the choice of species and silvicultural treatment. Also, earlier commercial plantings and subsequent research in the country have allowed satisfactory and reliable practices to develop.

Detailed prescriptions have been worked out to protect plants at the nursery from fungus, insects and pests. Eucalypts will be treated with insecticide in the nursery as a protection against damage from termites after planting in the field. Land will be cleared by tractor followed by burning, ploughing and disking. Plantations will be weeded manually and mechanically during the first and second year. Controlled burning will be carried out early in the dry season to reduce fire hazard. It was considered that highly mechanized operations should be used because of the scale of operations and because the timing of operations was a critical factor. Large-scale use of labour would not be feasible from a logistical point of view. It is estimated that yields after deducting for mortality, breakage, fire and other losses will be those shown in Table 12.1. These multiplied by the appropriate area under plantation in a given year generate total output flows by type of products over time as displayed in Table 12.2.'

On the input side the analysts generated separate estimates of input streams for each main input. Thus, for example, Table 12.3 shows the pattern of estimated labour requirements over time. Similar physical flow tables were prepared for other physical inputs but they are not reproduced here for the sake of brevity.

12.2.2 Indirect physical relationships

While the project is aimed at increasing wood supplies, there are several indirect effects which will likely be generated. The following were identified by the project team:

- Reduced erosion will lower the cost of maintenance of a reservoir downstream of the pine plantations. The pine plantations will eventually cover 12 000 ha, a major part of the watershed. It is estimated that soil erosion under these plantations would be about 0.5 m3/ha/a while under present land use erosion rates are 5 times higher. The timing of these effects needs detailed examination. Soil erosion will diminish gradually when crown and litter cover increases and ends when the plantations are clear-felled. Due to lack of more precise studies of the relationships over time between forest cover and erosion rates, it was assumed that between years 5 and 15 erosion will be reduced from 2.5 to 1.5 m3/ha/a and that from year 16 to the end of the project period the full protective effect of plantations will take place reducing the erosion rate further from 1.5 to 0.5 m3/ha/a. The results are displayed in Table 12.4.

- Training and experience will be obtained in the implementation of the afforestation project and could eventually benefit other similar projects in the country. Due to the difficulty in estimating quantitatively the probable impact of this effect over time, the analyst limited his assessment to a qualitative judgement of the new skills which will become available as a result of the project. (This qualitative judgement was presented as a statement in the project appraisal document.)

- The increased economic activity in the project region will generate a stimulating effect on the depressed local economy by increasing employment and use of resources previously idle beyond increased direct use of labour and other resources in the project. Local impact studies suggest that net indirect benefits derived from increased use of resources which would remain idle without the project are roughly equivalent to 80 percent of local monetary wages in the project and therefore this coefficient was used in the calculations. 1/

1/ Due to the uncertainty surrounding this figure, appraisal results were presented both with and without including this indirect effect.

12.2.3 Unit values

Unit values used in pricing the project's output are displayed in Table 12.5. Stumpage prices existed in the market and they were estimated to correspond to economic values in the case of small roundwood. Therefore they were used in both the financial and economic analyses. However, in the case of sawlogs, due to the existence of government subsidized prices, the economic value was estimated at 1.25 times the market value. In connection with indirect effects, the value of erosion avoided was estimated on the basis of reduced maintenance costs of the reservoir reaching $1 per cubic metre. As explained in Section 12.2.2 the value of the increased local economic activity generated by the project was estimated to be equal to 80 percent of the project wages. On the input side, land was valued at zero in the financial analysis since it was already owned by the Forest Service (see Table 12.5). However, based on land demand projections for the project region its economic opportunity cost was estimated to be positive and rising over time. For purposes of economic analysis land was valued at $2 per ha per year through year 8, $3 per ha per year between years 9 and 15 and $4 per ha per year from then on. Also due to heavy unemployment in the project area, which to a certain extent is expected to persist in the future, all labour costs were valued at 60 percent of financial costs in the economic analysis. All other unit values are assumed to be the same in the financial and economic analyses.

Finally, a 7 percent discount rate was used in the financial analysis since this was the rate to be applied to the loan used to finance the project. The Central Planning Office of the country has determined that a 9 percent discount rate should be used in all project evaluations and this was, therefore, the rate used in carrying out the economic analysis.

12.3 OUTPUTS OF THE ANALYSIS

This section shows how the basic data collected and processed in the form described above was utilized to develop information for decision-making.

12.3.1 Financial and economic value flows

Information provided in Table 12.2 and Table 12.5 was combined to generate the financial benefit flow displayed in the first four rows of Table 12.7. Information on physical input requirements and input unit prices was used to generate an estimate of total costs per activity and per hectare, as displayed in Table 12.6 which in turn, multiplied by the relevant number of hectares yielded the financial estimates of total project costs over time shown in rows 5 to 8 of Table 12.7.

Because of the great uncertainty surrounding the estimation of indirect benefits, the economic analysis was divided into two parts. The first part of the analysis includes only direct effects. Second, quantifiable indirect effects were incorporated to calculate the total (although less certain)estimated economic impact of the project. The results of these estimations are displayed in the economic value flow Tables 12.8 and 12.9 respectively.

In addition, because the project has two components which are separable (pine and eucalypts), it was necessary to explore the worthiness of each of these in both financial and economic terms. (See Chapter 4 for guidelines concerning separability) . The procedure used in producing the flow tables for each component is exactly the same as that used to generate the aggregate flow table for the whole project. The results of this exercise are displayed in Tables 12.10 and 12.11 for the eucalypt component and 12.12 and 12.13 for the pine component, comprising the corresponding financial flows and economic flows net of indirect effects. In each case the relevant project component period is equal to the number of years which is necessary to materialize total physical yields. Finally, since the quantifiable indirect effects will have a differential effect on both components, the analyst also prepared Tables 12.14 and 12.15 which display the economic value flows including indirect effects for the eucalypt and pine component respectively. This completed the basic financial and value flow estimates for the project and its components.

With the basic value flow tables available, the next stage is the estimation of measures of project worth.

12.3.2 Project Worth

As mentioned in Chapter 9, several measures of project worth can be calculated. The most common measures are the financial internal rate of return (FRR) and NPW and the economic internal rate of return (ERR) and NPW.

The FRR, ERR, and the two NPW measures for the project can be derived directly from Tables 12.7 and 12.8 which present the financial cash and economic value flows for the project, excluding indirect effects. The procedure for calculating these measures of project worth and efficiency was discussed in Chapter 9. Similarly the ERR and economic NPW for the project, including indirect effects, can be estimated from Table 12.9. Tables 12.10 through 12.15 set out the calculations of the measures of worth for each project component.

It can be observed from Tables 12.7 and 12.8 that the project as a whole is financially and economically profitable as both the FRR and ERR are superior to the financial and economic discount rates respectively. Also, although the project is economically viable on the basis of its direct effects alone, the inclusion of indirect effects produces a significant positive change in the ERR as displayed in Table 12.9.

The analysis of the project's worth also shows that each of the project's components is financially and economically profitable. (See tables 12.10 through 12.13), and that their economic worth is higher when indirect effects are included (Tables 12.14 and 12.15). Finally the analysis indicates that the eucalyptus component is more profitable, both financially and economically, than the pine component and therefore suggests the idea that more land might be dedicated to eucalyptus plantations provided market constraints permit. Thus, the economic analysis of the project could lead eventually to a re-examination of the project and possibly to alternative designs.

12.3.3 Sensitivity Analysis

The nature of the sensitivity analysis and how it should be developed have been discussed in Chapter 10. There are a number of elements which might have different magnitudes than those assumed in the analyses. Thus, it is desirable to recalculate the project value flows and worth in order to take into account possible changes in parameter values. A number of parameters which are of key importance could be tested to analyse the sensitivity of the project to changes in their values. In this case the following were chosen:

(i) Market variables - Throughout the analysis of the afforestation project it was assumed that the project's output of pulpwood would be entirely used to supply the additional requirements of a planned expansion of an existing pulp and paper mill. However, the market studies indicated that the domestic market might not absorb the additional production of paper during the first years of operation of the proposed expansion of the mill and that it is unlikely that any excess could be profitably exported because of the small amounts involved.

The analyst noted that there were practically no alternative outlets for the pulpwood-sized wood from the project. This wood is not suitable for poles and posts and is likely to be too costly to compete either with indigenous woods as firewood, or with wood residues as a raw material for a planned particleboard plant. There is, therefore, some possibility that part of the project's output of pulpwood would not be needed for some years. In the sensitivity analysis carried out to explore this possibility, it was assumed, based on alternative estimates of possible demand developments, that the planned expansion of pulp and paper would be postponed from year 6 until year 14 and that, therefore, all pulpwood produced before then will not be used. As indicated in Table 12.16, under these conditions both components still remain financially and economically viable although the pine component is clearly in a critical position with FRR and ERR close to the financial and economic cut-off rates (7 and 9 percent respectively).

(ii) Yields - In the case of the eucalypt component there was a question of whether pest control treatments will be entirely effective as no previous large scale plantations of this species existed in the country. In these circumstances the analyst assumed that an additional large amount of wood in the sawlog size would be damaged by termites and that the net effect would be to reduce the usable output of sawlogs by as much as 30 percent. 1/ The recalculation of this component's financial and economic worth indicates that even in these extreme circumstances the eucalypt component would remain financially and economically viable (Table 12.16).

(iii) Location - Finally, the analyst was aware that at the time of the study the Forest Service was simultaneously considering the alternative idea of dedicating most of pine component land area to a wildlife reserve and locating the pine plantations in other available land with basically the same opportunity cost but situated 40 miles further away from the market. This would mean additional transportation costs equal to $0.074/m3 per mile and therefore a reduction of the economic unit value of the component output (stumpage value) equal to $2.96/m3. The calculations carried out under this assumption indicate, as shown in Table 12.16, that the pine component would not be financially or economically sound.

1/ Note that this particular sensitivity analysis could also provide information on other possible causes of effective yield reductions, e.g. fire damage, over-estimation of growth, etc.

12.4 CONCLUSION

Many other changes in assumptions could have been introduced in the analysis of this project. However, the analyst felt that with the studies carried out so far, certain key elements had been identified which gave a clear perspective of the economic worthiness of the project.

First, the project and its component were both financially and economically viable if the conditions orginally assumed materialized.

Second, the eucalypt component was substantially more attractive than the pine component on both financial and economic grounds.

Third, given the financial and economic strength of the eucalypt component, it is very likely that this component will remain viable even if adverse conditions materialize. The same cannot be said of the pine component, which would not be financially or economically sound, if it is located in the alternative site considered and only marginally viable if the market does not develop as rapidly as assumed in the original analysis.

Fourth, therefore, consideration needs to be given either to redesigning the project or to closer scrutiny of the variables influencing the development of the pine component.

This example illustrates both the procedures used in carrying out an economic analysis of a project and the use of the economic information derived. It also highlights the importance of economic analysis in identifying key elements in design and exploring areas of uncertainty as basic factors in the decision-making process.

Table 12.1

AFFORESTATION PROJECT

Yields of Thinnings and Final Harvest
(m3/ha)

	Year	Sawlogs	Small Poles	Pulpwood
Eucalyptus				
First thinning	3	3	14	12
Second "	5	21	19	17
Third "	6	32	7	11
Final felling	8	67	5	16
Pines				
First thinning	6	-	-	8
Second "	8	4	-	19
Third "	12	23	-	18
Fourth "	21	79	-	15
Final felling	30	297	-	39

Table 12.2

AFFORESTATION PROJECT. TOTAL OUTPUT

('000 m^3)

Product	YEARS 0-2	3	4	5	6	7	8	9	10	11	12	13	14	15	16	17	18-20	21-26	27-29	30-35
Sawlogs	0	3	3	24	56	56	131	128	128	107	121	121	46	46	46	46	-	158	-	594
Small Poles	0	14	14	33	40	40	45	31	31	12	5	5	-	-	-	-	-	-	-	-
Pulpwood	0	12	12	29	40	40	110	98	98	81	90	90	36	36	36	36	-	30	-	78
TOTAL OUTPUT	0	29	29	86	136	136	286	257	257	200	216	216	82	82	82	82	0	188	0	672

Table 12.3

AFFORESTATION PROJECT

LABOUR REQUIREMENTS
('000 man days)

Year	Land Clearing/ Road Construction	Planting Activities
-1	15.47	6.74
0	15.47	57.72
1	15.47	71.90
2	15.47	84.13
3	15.47	93.97
4	15.47	103.82
5		111.69
6		70.55
7		66.22
8		68.60
9		65.31
10		62.03
11		53.99
12		50.71
13		47.43
14		39.39
15		39.39
16		39.39
17		39.39
18-20		39.39
21-26		39.39
27-29		39.39
30		39.39
31		32.83
32		26.26
33		19.70
34		13.13
35		6.57

Table 12.4

AFFORESTATION PROJECT

REDUCED EROSION EFFECT

Years	Number of ha under plantation ('000)	Reduced erosion effect (m3/year)
0	2 000	-
1	4 000	-
2	6 000	-
3	8 000	-
4	10 000	-
5-15	12 000	12 000
16-29	12 000	24 000
30	10 000	20 000
31	8 000	16 000
32	6 000	12 000
33	4 000	8 000
34	2 000	4 000
35	-	-

Table 12.5

AFFORESTATION PROJECT

MAIN UNIT VALUES

Output	Financial Analysis	Economic Analysis
Pine Sawlogs	$15.08/m^3$	$18.85/m^3$
Pine Pulpwood	$ 5.69/m^3$	$ 5.69/m^3$
Eucalypt sawlogs	$12.11/m^3$	$15.14/m^3$
" poles	$ 9.64/m^3$	$ 9.64/m^3$
" pulpwood	$ 4.70/m^3$	$ 4.70/m^3$
Reduced erosion	-	$ 1.00/m^3$
Increased local economic activity	-	80% of project wages
Main Inputs		
Land	0	-$2/per hectare per year years 1 to 8 -$3/per hectare per year years 9 to 15 -$4/per hectare per year years 16 to 35
Labour	$0.5 man-hour	$0.3 man-hour

All other Inputs: financial unit value = economic unit value

Other Analytical Parameters		
Discount rate	7%	9%

Table 12.6

AFFORESTATION PROJECT

PLANTATION ACTIVITIES AND FINANCIAL COSTS PER HECTARE

($/ha)

YEAR

ACTIVITY	-1		0		1		2-4		5		6-7		8		9-30	
	Labour	Other	Labour	Other	Labour	Other	Labour	Other	Labour	Other	Labour	Other	Labour	Other	Labour	Other
Land Clearing	29.20	335.76														
Road Construction	12.04	97.62														
Land Preparation			6.92	95.14												
Nursery			21.80	9.34												
Planting			16.96	44.00												
Fertilizing			8.65	7.44												
Singling			23.18													
Weeding			43.77	46.70	18.82	23.72										
Pruning									19.02				9.51			
Fire Control							7.26	4.50	7.26	4.50	7.26	4.50	7.26	4.50	7.26	4.50
Road Maintenance			1.04	6.22	1.04	6.22	1.04	6.22	1.04	6.22	1.04	6.22	1.04	6.22	1.04	6.22
Station Overheads	9.34	21.80	9.34	21.80	9.34	21.80	9.34	21.80	9.34	21.80	9.34	21.80	9.34	21.80	9.34	21.80
Division Overheads	8.62	20.10	8.62	20.10	8.62	20.10	8.62	20.10	8.62	20.10	8.62	20.10	8.62	20.10	8.62	20.10

Year 0 = Year of Planting

Table 12.7

AFFORESTATION PROJECT

FINANCIAL CASH FLOW

('000 US$)

YEAR

	-1	0	1	2	3	4	5	6	7	8	9	10	11	12
BENEFITS														
Sawlogs	-	-	-	-	36.33	36.33	290.64	678.16	678.16	1610.17	1573.84	1573.84	1319.53	1625.69
Small poles	-	-	-	-	134.96	134.96	318.12	385.60	385.60	433.80	298.84	298.84	115.68	48.20
Pulpwood	-	-	-	-	56.40	56.40	136.30	279.04	279.04	570.45	514.06	514.06	434.16	496.26
TOTAL BENEFITS	-	-	-	-	227.69	227.69	745.06	1342.80	1342.80	2614.43	2386.74	2386.74	1869.37	2170.15
COSTS														
Land Clearing/ Road Construction														
Labour	61.86	61.86	61.86	61.86	61.86	61.86								
Other	650.07	650.07	650.07	650.07	650.07	650.07								
PLANTING														
Labour	26.94	230.87	287.60	336.50	375.89	415.28	446.75	282.21	264.87	274.38	261.25	248.12	215.97	202.84
Other	62.85	447.14	554.90	633.83	712.76	791.69	807.77	502.41	473.58	473.58	447.27	420.96	394.65	368.34
TOTAL COSTS	801.72	1389.94	1554.43	1682.26	1800.58	1918.90	1254.52	784.62	738.45	747.96	708.52	669.08	610.62	571.18
NET BENEFITS	(801.72)	(1389.94)	(1554.43)	(1682.26)	(1572.89)	(1691.21)	(509.46)	558.18	604.35	1866.47	1678.22	1717.66	1258.75	1598.97

	13	14	15	16	17	18–20	21–26	27–29	30	31	32	33	34	35
BENEFITS														
Sawlogs	1625.69	693.68	693.68	693.68	693.68	-	2382.64	-	8957.52	8957.52	8957.52	8957.52	8957.52	8957.52
Small poles	48.20													
Pulpwood	496.26	204.84	204.84	204.84	204.84	-	170.70	-	443.82	443.82	443.82	443.82	443.82	443.82
TOTAL BENEFITS	2170.15	898.52	898.52	898.52	898.52	-	2553.34	-	9401.34	9401.34	9401.34	9401.34	9401.34	9401.34
COSTS														
Land Clearing/ Road Construction														
Labour														
Other														
PLANTING														
Labour	189.71	157.56	157.56	157.56	157.56	157.56	157.56	157.56	157.56	131.30	105.04	78.78	52.52	26.26
Other	342.03	315.72	315.72	315.72	315.72	315.72	315.72	315.72	315.72	263.10	210.48	157.86	105.24	52.62
TOTAL COSTS	531.74	473.28	473.28	473.28	473.28	473.28	473.28	473.28	473.28	394.40	315.52	236.64	157.76	78.88
NET BENEFITS	1638.41	425.24	425.24	425.24	425.24	(473.28)	2080.06	(473.28)	8928.06	9006.94	9085.82	9164.70	9243.58	9322.46

NPW = 6158

FRR = 10.4%

Table 12.8

AFFORESTATION PROJECT

ECONOMIC VALUE FLOW

('000 US$)

	YEAR																											
	-1	0	1	2	3	4	5	6	7	8	9	10	11	12	13	14	15	16	17	18-20	21-26	27-29	30	31	32	33	34	35
BENEFITS																												
Sawlogs	-	-	-	-	45.41	45.41	363.30	847.70	847.70	2012.71	1967.30	1967.30	1649.41	2032.11	2032.11	867.10	867.10	867.10	867.10	-	2978.30	-	11196.90	11196.90	11196.90	11196.90	11196.90	11196.90
Small Poles	-	-	-	-	134.96	134.96	318.12	385.60	385.60	433.80	298.84	298.84	115.68	48.20	48.20													
Pulpwood	-	-	-	-	56.40	56.40	136.30	279.04	279.04	570.46	514.06	514.06	434.16	496.26	496.26	204.84	204.84	204.84	204.84	-	170.70	-	443.82	443.82	443.82	443.82	443.82	443.82
TOTAL BENEFITS	-	-	-	-	236.77	236.77	817.72	1512.34	1512.34	3016.97	2780.20	2780.20	2199.25	2576.57	2576.57	1071.94	1071.94	1071.94	1071.94	-	3149.00	-	11640.72	11640.72	11640.72	11640.72	11640.72	11640.72
COSTS																												
Labour	53.28	175.64	209.68	239.02	262.65	286.28	268.05	169.33	158.92	164.63	156.75	148.87	129.58	121.70	113.83	94.54	94.54	94.54	94.54	94.54	94.54	94.54	94.54	78.78	63.02	47.27	31.51	15.76
Land	6.00	12.00	18.00	24.00	30.00	36.00	36.00	36.00	36.00	36.00	51.00	48.00	45.00	42.00	39.00	36.00	36.00	48.00	48.00	48.00	48.00	48.00	48.00	40.00	32.00	24.00	16.00	8.00
Other	712.92	1097.21	1204.97	1283.90	1362.83	1441.76	807.77	502.41	473.58	473.58	447.27	420.96	394.65	368.34	342.03	315.72	315.72	315.72	315.72	315.72	315.72	315.72	315.72	263.10	210.48	157.86	105.24	52.62
TOTAL COSTS	772.20	1284.85	1432.65	1546.92	1655.48	1764.04	1111.82	707.74	668.50	674.21	655.02	617.83	569.23	532.04	494.80	446.26	446.26	458.26	458.26	458.26	458.26	458.26	458.26	381.88	305.50	229.13	152.75	76.38
NET BENEFITS	(772.20)	(1284.85)	(1432.65)	(1546.92)	(1418.71)	(1527.27)	(294.10)	804.60	843.84	2342.76	2125.18	2162.37	1630.02	2044.53	2081.71	625.68	625.68	613.68	613.68	(458.26)	2690.74	(458.26)	11182.46	11258.84	11335.22	11411.59	11487.97	11564.34

NPW - 5537 (base:year zero)

ERR - 13%

Table 12.9

AFFORESTATION PROJECT

ECONOMIC VALUE FLOW INCLUDING INDIRECT BENEFITS

('000 US$)

	YEAR																											
	-1	0	1	2	3	4	5	6	7	8	9	10	11	12	13	14	15	16	17	18–20	21–26	27–29	30	31	32	33	34	35
BENEFITS																												
Direct	-	-	-	-	236.77	236.77	817.72	1512.34	1512.34	3016.97	2780.20	2780.20	2199.25	2576.57	2576.57	1071.94	1071.94	1071.94	1071.94	-	3149.00	-	11640.72	11640.72	11640.72	11640.72	11640.72	11640.72
Indirect	71.04	234.18	279.87	318.69	350.20	381.71	369.40	237.77	223.90	231.50	221.00	210.50	184.78	174.27	163.77	138.05	138.05	150.05	150.05	150.05	150.05	150.05	146.05	121.04	96.03	71.02	44.02	21.02
TOTAL BENEFITS	71.04	234.18	279.87	318.69	586.97	618.48	1187.12	1750.11	1736.24	3248.47	3001.20	2990.70	2384.03	2750.84	2740.34	1209.99	1209.99	1221.99	1221.99	150.05	3299.05	150.05	11786.77	11761.76	11736.75	11711.74	11684.74	11661.74
TOTAL COSTS	772.20	1284.85	1432.65	1546.92	1655.48	1764.04	1111.82	707.74	668.50	674.21	655.02	617.83	569.23	532.04	494.86	446.26	446.26	458.26	458.26	458.26	458.26	458.26	458.26	381.88	305.50	229.13	152.75	76.38
NET BENEFITS	(701.16)	(1050.67)	(1152.78)	(1228.23)	(1068.51)	(1145.56)	75.30	1042.37	1067.74	2574.26	2346.18	2372.87	1814.80	2218.80	2245.48	763.73	763.73	763.73	763.76	(308.21)	2840.79	(308.21)	11328.51	11379.88	11431.25	11482.61	11531.99	11585.36

NPW = 8350

ERR = 16.2

Table 12.10

AFFORESTATION PROJECT FINANCIAL CASH FLOW

EUCALYPTUS COMPONENT

('000 US$)

	YEARS														
	-1	0	1	2	3	4	5	6	7	8	9	10	11	12	13
BENEFITS															
Sawlogs	-	-	-	-	36.33	36.33	290.64	678.16	678.16	1489.53	1453.20	1453.20	1198.89	811.37	811.37
Small Poles	-	-	-	-	134.96	134.96	318.12	385.60	385.60	433.80	298.84	298.84	115.68	48.20	48.20
Pulpwood	-	-	-	-	56.40	56.40	136.30	188.00	188.00	263.20	206.80	206.80	126.90	75.20	75.20
TOTAL BENEFITS	-	-	-	-	227.69	227.69	745.06	1251.76	1251.76	2186.53	1958.84	1958.84	1441.47	934.77	934.77
COSTS															
-Land clearing/ road construction:															
Labour	20.62	20.62	20.62	20.62	20.62	20.62									
Other	216.69	216.69	216.69	216.69	216.69	216.69									
-Planting and management:															
Labour	8.98	86.61	104.52	127.16	140.29	153.42	157.57	94.07	88.29	78.78	65.65	52.52	39.39	26.26	13.13
Other	20.95	138.14	174.06	200.37	226.68	252.99	258.35	167.47	157.86	157.86	131.55	105.24	78.93	52.62	26.31
TOTAL COSTS	267.24	461.06	515.89	564.84	604.28	643.72	415.92	261.54	246.15	236.64	197.20	157.76	118.32	78.88	39.44
NET BENEFITS	(267.24)	(461.06)	(515.89)	(564.84)	(376.59)	(416.03)	329.14	990.22	1005.61	1949.89	1761.64	1801.08	1323.15	855.89	895.33

NPW = 3562
FRR = 21.3

Table 12.11

AFFORESTATION PROJECT ECONOMIC VALUE FLOW

EUCALYPTUS COMPONENT

('000 US$)

YEARS

	-1	0	1	2	3	4	5	6	7	8	9	10	11	12	13
BENEFITS															
Sawlogs	-	-	-	-	45.41	45.41	363.30	847.70	847.70	1861.91	1816.50	1816.50	1498.61	1014.21	1014.21
Small Poles					134.96	134.96	318.12	385.60	385.60	433.80	298.84	298.84	115.68	48.20	48.20
Pulpwood					56.40	56.40	136.30	188.00	188.00	263.20	206.80	206.80	126.90	75.20	75.20
TOTAL BENEFITS	-	-	-	-	236.77	236.77	817.72	1421.30	1421.30	2558.91	2322.14	2322.14	1741.19	1137.61	1137.61
COSTS															
Labour	17.76	63.74	75.08	88.67	96.55	104.42	94.54	56.44	52.97	47.27	39.39	31.51	23.63	15.76	7.88
Land	2.00	4.00	6.00	8.00	10.00	12.00	12.00	12.00	12.00	12.00	15.00	12.00	9.00	6.00	3.00
Other	237.64	354.83	390.75	417.06	443.37	469.68	258.35	167.47	157.86	157.86	131.55	105.24	78.93	52.62	26.31
TOTAL COSTS	257.40	422.57	471.83	513.73	549.92	586.10	364.89	235.91	222.83	217.13	185.94	148.75	111.56	74.38	37.19
NET BENEFITS	(257.40)	(422.57)	(471.83)	(513.73)	(313.15)	(349.33)	452.83	1185.39	1198.47	2341.78	2136.20	2173.39	1629.63	1063.23	1100.42

NPW = 4044
ERR = 26.4

Table 12.12

AFFORESTATION PROJECT FINANCIAL CASH FLOW

PINUS COMPONENT

('000 US$)

YEARS

	-1	0	1	2	3	4	5	6	7	8	9	10	11
BENEFITS													
Sawlogs	-	-	-	-	-	-	-	-	-	120.64	120.64	120.64	120.64
Pulpwood	-	-	-	-	-	-	-	91.04	91.04	307.26	307.26	307.26	307.26
TOTAL BENEFITS	-	-	-	-	-	-	-	91.04	91.04	427.90	427.90	427.90	427.90
COSTS													
Land clearing/ Road construction													
Labour	41.24	41.24	41.24	.41.24	41.24	41.24							
Other	433.38	433.38	433.38	433.38	433.38	433.38							
Planting													
Labour	17.96	145.26	183.08	209.34	235.60	261.86	289.18	188.14	176.58	195.60	195.60	195.60	176.58
Other	41.90	309.00	380.84	433.46	486.08	538.70	549.42	334.94	315.72	315.72	315.72	315.72	315.72
TOTAL COSTS	534.48	928.88	1038.54	1117.42	1196.30	1275.18	838.60	523.08	492.30	511.32	511.32	511.32	492.30
NET BENEFITS	(534.48)	(928.88)	(1038.54)	(1117.42)	(1196.30)	(1275.18)	(838.60)	(432.04)	(401.25)	(83.42)	(83.42)	(83.42)	(64.40)

	12	13	14	15	16	17	18-20	21-26	27-29	30	31	32	33	34	35
BENEFITS															
Sawlogs	814.32	814.32	693.68	693.68	693.68	693.68	0	2382.64	0	8957.52	8957.52	8957.52	8957.52	8957.52	8957.52
Pulpwood	421.06	421.06	204.84	204.84	204.84	204.84	0	170.70	0	443.82	443.82	443.82	443.82	443.82	443.82
TOTAL BENEFITS	1235.38	1235.38	898.52	898.52	898.52	898.52	0	2553.34	0	9401.34	9401.34	9401.34	9401.34	9401.34	9401.34
COSTS															
Land clearing/ Road construction															
Labour															
Other															
Planting															
Labour	176.58	176.58	157.56	157.56	157.56	157.56	157.56	157.56	157.56	157.56	131.30	105.04	78.78	52.52	26.26
Other	315.72	315.72	315.72	315.72	315.72	315.72	315.72	315.72	315.72	315.72	263.10	210.48	157.86	105.24	52.62
TOTAL COSTS	492.30	492.30	473.28	473.28	473.28	473.28	473.28	473.28	473.28	473.28	394.40	315.52	236.64	157.76	78.88
NET BENEFITS	743.08	743.08	425.24	425.24	425.24	425.24	(473.28)	2080.06	(473.28)	8928.06	9006.94	9085.82	9164.70	9243.58	9322.46

NPW = 2595

FRR = 8.4%

Table 12.13

APFORESTATION PROJECT

ECONOMIC VALUE FLOW

PINUS COMPONENT

('000 US$)

YEARS

	-1	0	1	2	3	4	5	6	7	8	9	10	11	12
BENEFITS														
Sawlogs								-	-	150.80	150.80	150.80	150.80	1017.90
Pulpwood								91.04	91.04	307.26	307.26	307.26	307.26	421.06
TOTAL BENEFITS	-	-	-	-	-	-	-	91.04	91.04	458.06	458.06	458.06	458.06	1438.96
COSTS														
Labour	35.52	111.90	134.59	150.35	166.10	181.86	173.51	112.88	105.95	117.36	117.36	117.36	105.95	105.95
Land	4.00	8.00	12.00	16.00	20.00	24.00	24.00	24.00	24.00	24.00	36.00	36.00	36.00	36.00
Other	475.28	742.38	814.22	866.84	919.46	972.08	549.42	334.94	315.72	315.72	315.72	315.72	315.72	315.72
TOTAL COSTS	514.80	862.28	960.81	1033.19	1105.56	1177.94	746.93	471.82	445.67	457.08	469.08	469.08	457.67	457.67
NET BENEFITS	(514.80)	(862.28)	(960.81)	(1033.19)	(1105.56)	(1177.94)	(746.93)	(380.78)	(354.63)	0.98	(11.02)	(11.02)	0.39	981.29

	13	14	15	16	17	18–20	21–26	27–29	30	31	32	33	34	35
BENEFITS														
Sawlogs	1017.90	867.10	867.10	867.10	867.10	0	2978.30	0	11196.90	11196.90	11196.90	11196.90	11196.90	11196.90
Pulpwood	421.06	204.84	204.84	204.84	204.84	0	170.70	0	443.82	443.82	443.82	443.82	443.82	443.82
TOTAL BENEFITS	1438.96	1071.94	1071.94	1071.94	1071.94	0	3149.00	0	11640.72	11640.72	11640.72	11640.72	11640.72	11640.72
COSTS														
Labour	105.95	94.54	94.54	94.54	94.54	94.54	94.54	94.54	94.54	78.78	63.02	47.27	31.51	15.76
Land	36.00	36.00	36.00	48.00	48.00	48.00	48.00	48.00	48.00	40.00	32.00	24.00	16.00	8.00
Other	315.72	315.72	315.72	315.72	315.72	315.72	315.72	315.72	315.72	263.10	210.48	157.86	105.24	52.62
TOTAL COSTS	457.67	446.26	446.26	458.26	458.26	458.26	458.26	458.26	458.26	381.88	305.50	229.13	152.75	76.38
NET BENEFITS	981.29	625.68	625.68	613.68	613.68	(458.26)	2690.74	(458.26)	11182.46	11258.84	11335.22	11411.59	11487.97	11564.34

NPW = 1493
ERR = 10%

Table 12.14

AFFORESTATION PROJECT

ECONOMIC VALUE FLOW INCLUDING INDIRECT EFFECTS

EUCALYPTUS COMPONENT

('000 US$)

	YEARS														
	-1	0	1	2	3	4	5	6	7	8	9	10	11	12	13
BENEFITS															
Direct	-	-	-	-	236.77	236.77	817.72	1421.30	1421.30	2558.91	2322.14	2322.14	1741.19	1137.61	1137.61
Indirect	23.68	84.98	100.11	118.14	128.73	139.23	126.06	75.26	70.63	63.02	52.52	42.02	31.51	21.01	10.50
TOTAL BENEFITS	23.68	84.98	100.11	118.14	365.50	376.00	943.78	1496.56	1491.93	2621.93	2384.66	2364.16	1772.70	1158.62	1148.11
TOTAL COSTS	251.40	422.51	471.83	513.73	549.92	586.10	364.89	235.91	222.83	217.13	185.94	148.75	111.56	74.38	37.19
NET BENEFITS	(233.72)	(337.59)	(371.72)	(395.59)	(184.42)	(210.10)	578.89	1260.65	1269.10	2404.80	2198.72	2215.41	1661.14	1084.24	1110.92

NPW = 4811
ERR = 32

Table 12.15

AFFORESTATION PROJECT
ECONOMIC VALUE FLOW INCLUDING INDIRECT EFFECTS
PINUS COMPONENT

(1000 US$)

YEARS

	-1	0	1	2	3	4	5	6	7	8	9	10	11	12
BENEFITS														
Direct	-	-	-	-	-	-	-	91.04	91.04	458.06	458.06	458.06	458.06	1438.96
Indirect	47.36	149.20	179.46	200.46	221.47	242.48	243.34	162.51	153.26	168.48	168.48	168.48	153.26	153.26
TOTAL BENEFITS	47.36	149.20	179.46	200.46	221.47	242.48	243.34	253.55	244.30	626.54	626.54	626.54	611.32	1592.22
TOTAL COSTS	514.80	862.28	960.81	1033.19	1105.56	1177.94	746.93	471.82	445.67	457.08	469.08	469.08	457.67	457.67
NET BENEFITS	(467.44)	(713.08)	(781.35)	(832.73)	(884.09)	(935.46)	(503.59)	(218.27)	(201.37)	169.46	157.46	157.46	153.65	1134.55

	13	14	15	16	17	18-20	21-26	27-29	30	31	32	33	34	35
BENEFITS														
Direct	1438.96	1071.94	1071.94	1071.94	1071.94	-	3149.00	-	11640.72	11640.72	11640.72	11640.72	11640.72	11640.72
Indirect	153.26	138.05	138.05	150.05	150.05	150.05	150.05	150.05	146.05	121.04	96.03	71.02	44.02	21.01
TOTAL BENEFITS	1592.22	1209.99	1209.99	1221.99	1221.99	150.05	3299.05	150.05	11786.77	11761.76	11736.75	11711.74	11684.74	11661.74
TOTAL COSTS	457.67	446.26	446.26	458.26	458.26	458.26	458.26	458.26	458.26	381.88	305.50	229.13	152.75	76.38
NET BENEFITS	1134.55	763.73	763.73	763.73	763.73	(308.21)	2840.79	(308.21)	11328.51	11379.88	11431.25	11482.61	11531.99	11585.36

NPW - 3539

ERR - 11.8

Table 12.16

AFFORESTATION PROJECT. SENSITIVITY ANALYSIS

Original assumption	New assumption	Eucalyptus component FRR	Eucalyptus component ERR	Pine component FRR	Pine component ERR
Original assumptions	Same	21.3	26.4	8.4	1.0
100 percent of project's pulp-wood will be needed	No pulpwood will be needed up to year 13	18.1	23.3	7.8	9.4
Original eucalypt sawlog yields	Eucalypt sawlog output 30 percent lower	16.8	21.5	-	-
Original pine plantation site	Pine plantations located 40 miles further away from market	-	-	6.3	8.6

Appendix A

FORESTRY PROJECT CASE STUDIES PREPARED BY FAO

FAO has prepared six case studies representing a variety of actual forest-based development projects. These have been published as a supplement to EAFP. 1/ The main categories of projects and countries represented are as follows:

Industrial plantations

Cases 4, 5 and 6 (Kenya and Zambia)

Smallholder tree farming

Case 1 (Philippines)

Village woodlots

Case 2 (Republic of Korea)

Natural hardwood forest utilization

Case 3 (South America)

Integrated forestry and forest industry

Cases 3 and 6 (South America and Zambia)

Taken as a group they show most of the common problems and types of analytical needs which the project planner will encounter in appraising forest projects designed to produce wood and/or wood fibres. 2/

The case studies are based on actual project appraisals which have been modified only in terms of presenting a clearer picture of how the analyst got from the "objective" to the measures of project worth and his recommendation on the project.

The six case studies are as follows:

(a) Case Study No. 1, Philippine Smallholder Tree-Farming Project.

This case concerns a forest plantation programme which forms part of a rural development effort. Loans are made available to smallholders, with an average of some 10 ha of land, to enable them to grow a fast-growing tree (*Albizzia falcataria*), that is

---

1/ FAO, 1979.

2/ See Gregersen and Brooks paper in FAO, forthcoming, for case material illustrating the application of economic analysis to forestry projects with water and soil protection outputs.

usable as pulpwood, on part of their land. The farmers sign an agreement with a local pulp and paper company which guarantees them a price and a market, but leaves them free to try and sell their output at a higher price elsewhere. The case study considers the project mainly from the point of view of the farmer and the government. The analysis is based on a typical farm unit rather than the project as a whole.

(b) Case Study No. 2, Village Fuelwood Plantations in Korea.

This case study deals with a village fuelwood programme in the Republic of Korea which constitutes a component of a large integrated rural development project. Like the Philippines project it includes a great number of small sub-projects, but in this case in different parts of the country. The overall programme is analysed as one project. Some of the problems of averaging input/output relationships for diverse elements in a large programme are brought out. The study also emphasizes the organizational aspects of this type of project, and the problems involved in dealing with shadow pricing of inputs and outputs and future demand and markets.

(c) Case Study No. 3, South American Natural Forest Utilization Project.

This case study deals with a project for expanding an integrated forest industry development based on a natural tropical hardwood resource. The emphasis is on a practical approach to appraising such a project and the elements to consider in a financial analysis, including treatment of loans and various government fees and tax incentives.

(d) Case Study No. 4, Kenya I Sawlog and Pulpwood Plantation Project.

This case examines a six-year time slice of an on-going sawlog and pulpwood plantation programme in Kenya. At the time this project was appraised, Kenya had experience growing and processing only sawlogs. This project included some of the country's first pulpwood plantations, which were established near the site of a proposed pulpmill. The mill was designed to meet Kenya's increasing domestic demands for paper products. In addition, the project financed the continuing sawlog plantation programme, which was designed to supply domestic sawmills which produce lumber for domestic use and export. The case provides an example of analysing project components separately, and illustrates problems associated with estimating import substitution and export benefits.

(e) Case Study No. 5, Kenya II Sawlog and Pulpwood Plantation Project.

This case deals with a further six-year time slice of the on-going Kenya sawlog and pulpwood plantation programme. This project continued and expanded its predecessor, the Kenya I project. At the time of the Kenya II appraisal, the pulpmill had been completed and was just starting production. As before, the project was made up of separate pulpwood and sawlog components. A comparison of this case with the Kenya I case shows the evolution of the project planning approach over time in a particular situation. For example, in Kenya II input and output prices and quantities were revised in the light of the Kenya I experience.

(f) Case Study No. 6, Zambia Industrial Forestry Project.

Like the Kenya case studies, this one deals with a large-scale government afforestation programme, and shows how to deal with a "time slice" project. It is concerned with a situation where there is still uncertainty about both yields from the plantations and about the outlets that will be available, and shows how such uncertainty can be dealt with. It also illustrates various aspects of the relationships between forestry and forest industry activities which have to be taken into account in an integrated project.

Appendix B

COMMON DISCOUNTING AND COMPOUNDING FORMULAS

As mentioned in the text, using the value flow table as a basis for NPW and ERR calculations, the analyst avoids the need for discounting and compounding formulas other than the simple present value formula. However, there are occasions where he may find it convenient to use other formulas, all derived from the basic one, which permit him to calculate in one step the present values for equal annual or periodic series of payments or to obtain an annual equivalent for a present or future value (e.g., where he wants to calculate a rental equivalent for a purchase price).

1. Calculating the present value of a periodic series of equal payments.

Table B-1 summarizes the main formulas needed to calculate the present and future values of annual and periodic payments (costs or benefits). The PV derived by using these formulas is expressed in terms of one year (period) prior to the year (period) when the first payment occurs. Thus, the analyst has to make sure that he appropriately compounds or discounts the result if he wants PV expressed in terms of a different year (period). Application of the formulas is illustrated below.

PV of equal annual payments

Assume a situation where there is an annual maintenance fee of $12 for a plantation which starts at the beginning of year 2 (the 3rd year) of the project and continues up to and including year 15. Thus, there are (15-2) + 1, or 14 equal payments of $12. How would the PV of this series of payments be calculated, if the discount rate is 8 percent?

First, applying the appropriate formula from Table B-1 (Formula 1 for a finite number of payments) the following result is obtained:

$$\$12 \quad \frac{(1.08)^{14} -1}{.08(1.08)^{14}} = \$12\ (8.24) = \$99$$

This gives the PV in year 1 of the 14 payments starting in year 2.

Second, discounting this value ($99) back one more year ($99/(1.08))the PV in year zero is $91.60.

This formula might be useful if, for example, the analyst wanted to compare the present value of two alternative equal annual cost streams. Assume that two alternative plantation management schemes were possible, one involving four equal costs of $30/ha for years 1-4 and another involving ten equal costs of $10/ha for years 2-11. The PV in year zero for the first alternative would be (using 8 percent):

$$\$30 \quad \frac{(1.08)^4 - 1}{.08(1.08)^4} = \$99.36$$

(This is already in year zero terms since payments start in year 1.) For the second alternative, the PV in year zero would be:

$$\$10 \quad \frac{(1.08)^{10} - 1}{.08(1.08)^{10}} = \$67.1 = \text{PV in year 1}$$

$$\$67.1/(1.08) = \$62.13 = \text{PV in year zero}$$

Thus, the analyst can see that in PV terms the second alternative provides the lowest cost, assuming that the relevant discount rate is 8 percent.

Present value of a series of equal periodic payments

If payments (costs or benefits) occur every t years instead of every year for a specified period of time, then formulas 5 and 6 in Table B-1 to obtain PV's can be used. For example, suppose there is a situation where fertilizer will be applied to a stand every 5 years, starting 5 years from now and lasting during the entire rotation of 50 years except for year 50. This means that there would be 9 equal applications starting in year 5 and ending in year 45. Assume that the cost each time is estimated to be $20/ha. How would the PV of these payments be estimated? Looking at Table B-1 formula 5 would be used for a finite number of periodic payments. The PV would be calculated as follows, assuming a discount rate of 8 percent, t = 5, and N = 9:

$$\$20 \quad \frac{(1.08)^{45} - 1}{(1.08)^{45} \left[(1.08)^5 - 1\right]} = \$41.28$$

If there were also an application of fertilizer at the time of establishment, that amount would have to be added to the PV obtained above. The most common use in forestry of formulas for calculating the PV of series of equal periodic payments is in calculation of the SEV. This is explained and illustrated below.

Soil Expectation Value

The SEV gives an estimate of the present value of land if it were put into forestry and produced an infinite number of net returns of $R every r years (where r is the rotation length).

To estimate the SEV, the net benefit of forestry production at the end of the first rotation R is calculated, without taking actual land cost into account and then the NPW of a future periodic series of net benefits of $R is computed beginning with $R received at the end of the first rotation. Thus, for example, assume a situation for a plantation as follows:

Establishment cost	\$250
Rotation	11 years
Annual cost	\$10 starting one year from now
Stumpage value at rotation	\$1 000
Discount rate	8 percent

The compounded value of the establishment cost at the end of the first rotation (year 1) is:

$$\$250\ (1 + 0.08)^{11} = \$583$$

The compounded value in year ten of the ten equal annual costs (\$10 each year between years 1 and 10, both inclusive) can be calculated by using formula 2, Table B-1:

$$\$10\ \frac{(1 + 0.08)^{10} - 1}{0.08} = \$145$$

which must be compounded for one additional year:

$$\$145\ (1 + 0.08) = \$157$$

Therefore, total costs at the end of the first rotation (year 11) are \$583 + 157 = \$740 and net benefits at rotation age are \$1 000 - \$740 = \$260.

The present value of an infinite series of payments of \$260 received every 11 years, or the SEV of this forestry management alternative, can be calculated by using formula 5 in Table B-1, for an infinite number of periods:

$$\text{SEV} = 260\ \frac{1}{(1 + 0.08)^{11} - 1} = \$195$$

What does this SEV of \$195 mean? It has several meanings. Most commonly in forestry it is used to determine what amount could be paid for the land to breakeven, i.e., have PV of costs equal PV of benefits, using a discount rate i (in this case 8 percent). More generally it indicates the PV of the productive capacity of the land, given the values assumed and the assumption that the land could continue to produce timber in perpetuity at the given rate.

Table B - 1

ANNUAL AND PERIODIC PAYMENT FORMULAS

	(1)	(2)
	Payments Begin One Year (or Period) from Present	
	Finite Number of Payment	Infinite Number of Payments
1. DISCOUNTED ANNUAL PAYMENT FACTOR	$\left[\frac{(1+i)^n - 1}{i(1+i)^n}\right]$	$\left[\frac{1}{i}\right]$
2. COMPOUNDED ANNUAL PAYMENT FACTOR	$\left[\frac{(1+i)^n - 1}{i}\right]$	n.a.
3. ANNUAL CAPITAL RECOVERY FACTOR	$\left[\frac{i(1+i)^n}{(1+i)^n - 1}\right]$	n.a.
4. ANNUAL SINKING FUND FACTOR	$\left[\frac{i}{(1+i)^n - 1}\right]$	n.a.
5. DISCOUNTED PERIODIC PAYMENT FACTOR	$\left[\frac{(1+i)^{nt} - 1}{(1+i)^{nt}\left[(1+i)^t - 1\right]}\right]$	$\left[\frac{1}{(1+i)^t - 1}\right]$
6. COMPOUNDED PERIODIC PAYMENT FACTOR	$\left[\frac{(1+i)^{nt} - 1}{(1+i)^t - 1}\right]$	n.a.

i = rate of interest (discount) in decimal form

n = number of years or periods until last payment starting with 1 year from now

t = number of years between periodic payments

2. Annual Equivalency Formulas

Formulas 3 and 4 in Table B-1 are used to calculate annual equivalents of given amounts of PV of costs or benefits. The formulas are merely the inverse of formulas 1 and 2. Assume, for example, that two alternative incentive programmes for tree farmers are being compared. One alternative is to give them a lump sum today of $100. The other alternative considered is to provide them with five equal payments over 5 years, starting one year from now. For the latter incentive to be effective, the annual amount should equal the $100 of PV using their relevant discount rate. In this case it is assumed to be high - 30 percent - since they value present income considerably higher than future income. To find the annual payments necessary, formula 3 for a finite number of payments is applied. The annual amount that would have to be paid, starting one year from now, to make the farmers indifferent between $100 now and the five equal payments, would thus be:

$$\$100 \; \frac{.30(1.30)^5}{(1.30)^5 - 1} = \$41$$

In other words, given their relevant discount rates (or their trade-off rates between present and future income) they would have to be paid $41 per year for five years to make them indifferent between the two payment forms.

Appendix C

HOW TO CALCULATE THE ECONOMIC RATE OF RETURN (ERR)

Although several relatively inexpensive hand calculators contain programmes (or can be programmed) for rate of return calculations, the analyst might be faced with situations in which the computation of ERR would have to be based on more rudimentary methods. There is no formula for calculating the ERR when more than one cost and/or benefit is involved. Therefore, a trial and error technique has to be used. The approach is as follows:

(a) First calculate a NPW using a rate which is estimated to be in the neighbourhood of the expected ERR. If the NPW is negative, then the ERR must be lower than the rate of discount used. If the NPW is positive, then the ERR must be higher than the discount rate adopted.

(b) If the first NPW calculated is negative, then reduce the discount rate up to a point where the calculated NPW is positive and vice versa if the first NPW calculated is positive. The ERR must now lie between the two rates of discount used in generating the positive and negative values of NPW.

(c) Estimate the ERR by using the following formula:

$$\text{ERR} = \left(\begin{matrix}\text{low rate of}\\ \text{discount}\end{matrix}\right) + \left(\begin{matrix}\text{difference between}\\ \text{both rates of discount}\end{matrix}\right) \times \left(\frac{\text{Positive NPW}}{\begin{matrix}\text{absolute difference}\\ \text{between positive and}\\ \text{negative PNW's}\end{matrix}}\right)$$

(d) Repeat steps (a) - (c) for a more precise result, if needed.

The following example, which uses the figures of the Philippine tree-farm project, illustrates the use of this technique:

Table C-1 shows in the first row the net benefits (costs) of the Philippine tree-farm project (from Table 9.1). The second row contains the PV of each annual flow discounted at 20 percent. The NPW, using this discount rate is positive and equal to P 4638 and, therefore the ERR must be higher than 20 percent. A further discounting attempt at 30 percent generated a still positive NPW equal to P 453, as shown in row 3 of Table 35. Therefore a still higher discount rate of 35 percent was tried, which rendered a negative NPW of - P 543. The ERR must then lie between 30 and 35 percent. Using the formula from step (c) above, the ERR of this project is estimated as follows:

$$\text{ERR} = 30\% + 5\% \quad \frac{453}{996} \quad = \quad 32.27 \text{ percent}$$

This is rounded off to 32 percent.

A further interpolation using a narrower range of 31 and 33 percent would have produced NPW's equal to P 215.6 and P 198.5, respectively. Using these two new values, a second estimate of ERR would be 32.04 percent. But since the result is being rounded off to the nearest whole percentage point, this additional refinement is unnecessary.

Table C1

CALCULATING THE ERR - PHILIPPINE PROJECT

Years

		0	1	2	3	4	5	6	7	8	9	10	11	12	13	14	15
1.	Net Benefits (costs)[1]	(1163)	(1163)	(1163)	(1163)	(100)	(100)	(100)	5286	5887	5887	6523	6523	7147	7147	7759	5887
2.	Present value of net benefits (costs) discounted at 20% per year	(1163)	(969)	(808)	(673)	(48)	(40)	(34)	1475	1369	1141	1054	878	802	668	604	382
3.	Present value of net benefits (costs) discounted at 30% per year	(1163)	(895)	(688)	(529)	(35)	(27)	(21)	842	722	555	473	364	307	236	197	115
4.	Present value of net benefits (costs) discounted at 35% per year	(1163)	(861)	(638)	(472)	(30)	(22)	(16)	647	534	395	324	240	195	144	116	64

1/ From Table 9.1.

Appendix D

CALCULATION OF THE BENEFIT-TO-COST (B/C) RATIO

In addition to NPW and ERR, some institutions use B/C as a measure of project worth. It is generally expressed in terms of the ratio of present value of total benefits over present value of total costs.[1]

The calculation of a B/C ratio involves discounting the total benefit and cost rows separately. This is shown in Table D-1 for the Philippine project (see Table 9.1). A 5 percent discount rate has been employed in the calculations. The B/C ratio is equal to 5.8.

What does a B/C ratio of 5.8 indicate? It measures the amount of benefits, expressed in present value terms, that the project generates per dollar of resources used in the project, also expressed in present value terms. Put another way, at the discount rate assumed, the present value of all consumption benefits gained is 5.8 times the present value of all costs (or consumption benefits foregone) due to the project.

---

[1] Several alternative forms of the B/C ratio are in use. Here the most common one is used, which is a ratio of the present value of all benefits to the present value of all costs. Another in common use is the net B/C ratio, which is the ratio of all benefits minus operating costs to total investment cost. See paper by McGaughey in FAO, forthcoming.

Table D1

THE B/C RATIO - PHILIPPINE PROJECT. (5% DISCOUNT RATE)

	Years 0	1	2	3	4	5	6	7	8	9	10	11	12	13	14	15
1. Benefits	-	-	-	-	-	-	-	5523	6174	6174	6810	6810	7634	7434	8046	6174
2. Present value of benefits	-	-	-	-	-	-	-	3925	4179	3980	4181	3982	4140	3942	4064	2970
3. Costs	1163	1163	1163	1163	100	100	100	237	287	287	287	287	287	287	287	287
4. Present value of Costs	1163	1108	1055	1005	82	78	75	168	194	188	176	167	160	152	145	138

Note: From the above the sum of present values of benefits = 35363
sum of present values of costs = 6052

Thus, B/C ratio = $\frac{35363}{6052}$ = 5.8

Appendix E

REFERENCES

Chisholm, R.K. & Whitaker, G.R. Forecasting Methods. Homewood, Ill., Irwin.
1971

FAO. Guide for Planning Pulp and Paper Enterprises. Rome. FAO Forestry and Forest
1973 Products Studies No. 18.

FAO. An Introduction to Planning Forestry Development. Rome.
1974

FAO. Economic Analysis of Forestry Projects: Case Studies. Rome. FAO Forestry
1979 Paper No. 17, Supp. 1.

FAO. Economic Analysis of Forestry Projects: Readings. Rome. FAO Forestry Paper
forthcoming No. 17, Supp. 2.

Gittinger, J. Price. Economic Analysis of Agricultural Projects. Baltimore, Johns
1972 Hopkins University Press for the World Bank.

Gregersen, H.M. Effect of Inflation on Evaluation of Forestry Investments. Journal of
1975 Forestry, Vol. 73, No. 9, pp. 570-572.

Gregory, G.R. Forest Resource Economics. New York, Ronald Press.
1972

IDB. Social and Economic Dimensions of Project Evaluation. Washington, D.C.
1977

IUFRO. Forecasting in Forestry and Timber Economy. Helsinki, Institutum Forestale
1971 Fenniae, Folia Forestalia 101.

Little, I.M.D. & Mirrlees, J.A. Project Appraisal and Planning for Developing Countries.
1974 New York, Basic Books.

Lundgren, A.G. Tables of Compound-Discount Interest Rate Multipliers for Evaluating
1971 Forestry Investments. Washington, D.C., USDA Forest Service Research
Paper No. NC-51.

Mikesell, R.F. The Rate of Discount for Evaluating Public Projects. Washington, D.C.,
1977 American Enterprise Institute for Public Policy Research.

OECD. Manual of Industrial Project Analysis in Developing Countries, Vol. I and Annex to
1968 Vol. I. Paris, Development Centre of the OECD.

Squire, L. & van der Tak, H.G. Economic Analysis of Projects. Baltimore, Johns Hopkins
1975 University Press for the World Bank.

United Nations & FAO. Forest Products Statistics - Price Series, 1950-1976. Timber
1977 Bulletin for Europe, Vol. XXX, Supplement 6. Geneva.

UNIDO. Guidelines for Project Evaluation. New York, United Nations. Project
1972 Formulation and Evaluation Series No. 2.

UNIDO. Guide to Practical Project Appraisal: Social Benefit-Cost Analysis in Developing
1978 Countries. New York, United Nations. Project Formulation and Evaluation
Series No. 3.

UNIDO. Manual for the Preparation of Industrial Feasibility Studies, New York,
1978 United Nations.

United States Department of Agriculture, Economic Research Service. The Consumer's
1977 Surplus Issue. Agriculture Economics Research, Vol. 29, No. 4.

Ward, W.A. 'Calculating Import and Export Parity Prices', Course Note CN-3, Economic
1976 Development Institute. Washington, D.C., World Bank.

Ward, W.A. 'Economic Valuation Decision Tree', Course Note CN-61, Economic Development
1978 Institute. Washington, D.C., World Bank.

World Bank. Forestry: Sector Policy Paper. Washington, D.C.
1978

ECONOMIC ANALYSIS OF FORESTRY PROJECTS: FAO FORESTRY PAPER 17,

CASE STUDIES SUPPLEMENT 1

CASE STUDY NO. 1

PHILIPPINE SMALLHOLDER TREE-FARMING PROJECT

FOOD AND AGRICULTURE ORGANIZATION OF THE UNITED NATIONS

Rome, 1979

Preface

This is one in a series of case studies of forestry and forest industries projects that has been prepared by the Food and Agriculture Organization of the United Nations (FAO) in order to demonstrate methods of preparing and appraising projects in the forest sector. The programme of case studies has been made possible by a special allocation for this purpose from the Swedish International Development Authority (SIDA). This case study has been prepared by FAO with the assistance of the College of Forestry at the University of Minnesota.

The project reported on in this case study is a project of the Government of the Philippines and the Paper Industries Corporation of the Philippines, which has been supported by a loan from the World Bank. The case study is based largely on material contained in an appraisal of the project carried out by the World Bank (Report No. 424a-PH, 5 May 1974). Additional information was generated during a field visit to the project site. The case study describes the project as it was perceived at the time of the appraisal in 1974. The material has been released to FAO by the World Bank and the Government of the Philippines exclusively for the purposes of teaching forestry planning methods. It may not be quoted or reproduced in part or in whole without permission.

ABBREVIATIONS

DBP	=	Development Bank of the Philippines
FRR	=	Financial rate of return
P	=	Philippine Peso
PICOP	=	Paper Industries Corporation of the Philippines

CURRENCY EQUIVALENTS

P 1	=	US$0.148
US$1	=	P 6.7585

TABLE OF CONTENTS

INTRODUCTION

1. This project forms part of an on-going Agro-Forestry Development Plan initiated in the late sixties by the Paper Industries Corporation of the Philippines (PICOP) with the objectives of ensuring raw material supplies for its mill complex at Bislig Bay in Eastern Mindanao and improving the socio-economic conditions of smallholders in the area. This is a "time slice" project.

2. Under the terms of the project, loans will be granted by the Development Bank of the Philippines (DBP) to some 1 300 smallholders in the Bislig Bay area to assist them in planting tree crops and in developing crop and/or livestock production. The present case study examines only the financial and economic appraisal of the tree crop element.

3. Smallholders will plant about 10 400 ha of Albizzia falcataria over a seven-year period. Trees will be grown on an eight-year rotation. PICOP will provide the seedlings at cost and technical assistance in exchange for first rights to the mature pulpwood at the prevailing market price. Table 1 shows the total expected magnitude and timing of the project.

4. The total cost of the project is estimated at about US$1.5 million with foreign exchange costs of approximately US$150 000.

5. The case is divided into four sections: Section A presents background on the project; Section B identifies the costs and benefits involved; Section C values the financial and economic costs and benefits; Section D shows the derivation of the financial and economic rates of return and looks at the sensitivity of results to changes in assumptions.

6. This case study records an example of a successful forestry project designed to involve and benefit a segment of the rural poor. One use of the study can thus be to illustrate the institutional arrangements and technical provisions that helped make it possible[1]. Issues of substance and concept that the user may wish to consider include the size of the incentives provided to the smallholder. Points of appraisal technique that arise include the structure of the appraisal (on the basis of a typical farm unit rather than for the project as a whole), and difficulties in establishing economic values for the inputs and outputs and the use of sensitivity analysis to explore the implications of uncertainty about these values.

A. BACKGROUND

Forest Resources

7. About one-third of the 30 million hectares of land in the Philippines is presently classified as commercial public forest. These forests are being reduced at a rate of 200 000 ha a year, while reforestation averages only about 17 000 ha a year. About 40 percent of the forest destruction is due to shifting cultivators (*kaingineros*), illegal logging and accidental fires. On the Island of Mindanao, which has the largest concentration of commercial public forest, the forest resource has been decreasing by about 90 500 ha a year, nearly all of which occurs in the valuable dipterocarp forest.

8. If this trend continues and demand increases as forecast, the Philippines is likely to be a net importer of logs by the year 2000. To offset this situation, the Government has developed policies aimed at conserving the forest resource; principal among these are: (i) the phase-out of log exports, (ii) a reforestation programme, and (iii) stricter control on all forms of forest destruction.

1/ See also Case Study No. 2 - Village Fuelwood Plantations in Korea, for an example of another, but quite different, successful project for rural community development.

Table 1 - Participating Farmers and Area Planted

Year	Number of farm Sub-loans	Year 0	1	2	3	4	5	6
		(hectares planted)						
First	200	400	400	400	400			
Second	300		600	600	600	600		
Third	400			800	800	800	800	
Fourth	400				800	800	800	800
Hectares planted/yr		400	1,000	1,800	2,600	2,200	1,600	800
Cumulative ha planted		400	1,400	3,200	5,800	8,000	9,600	10,400

PICOP

9. The Paper Industries Corporation of the Philippines (PICOP) is the sole domestic producer of newsprint and the major supplier of industrial paper in the country. The Government has granted PICOP a preferred and pioneer industry status under which the company receives a number of incentives to development.

10. PICOP could advantageously meet 100 percent of its groundwood requirements from Albizzia but insufficient supplies have limited PICOP's consumption of this species to only 15 percent of total requirements. In the late 1960's the Company launched its Agro-forestry Development Plan. Under the plan a participating farmer devotes 20 percent of his land holding to food and livestock production and 80 percent to fast-growing pulpwood trees (mainly Albizzia) grown on an eight-year rotation. At the time of this analysis (1974), 868 tree farms with a total planted area of 1 816 ha have been established and 127 smallholders have received loans from DBP.

11. The present project calls for an expansion of the Plan. Tree farms would be sited within a 100 km radius of the PICOP mill at Bislig, the maximum economic distance for extending technical assistance. The total alienable and disposable land in the project area is estimated at 481 000 ha with about 160 000 ha suitable for tree-farming. Appendix 1 shows the location of the project.

12. To qualify for a loan under the project a farmer must have a signed marketing agreement with PICOP. The typical smallholder would plant 8 ha of Albizzia. The loan is for 15 years at 12 percent interest per annum with the first interest payment due in year 7 and with principal to be repaid in years 9-14.

13. Smallholders would be required to contribute a minimum of 25 percent of project costs in own labour. Smallholders would be eligible if they occupy a minimum of 5 ha, including those who have occupied alienable and disposable land for at least 10 years, but who do not have legal title. There would be no restriction on the annual rate at which the land may be developed.

14. Reliable estimates of the number of farmers who would qualify and apply for DBP loans are not available, but preliminary data from the Bureau of Lands show that over 16 000 smallholders have applied, or will apply, for land titles in the project area.

B. IDENTIFYING COSTS AND BENEFITS

Technical Considerations

15. Albizzia falcataria is indigenous to the Moluccas and was introduced to the Philippines from Java in about 1940. It is a fast-growing tree which develops a clean bole. Response to fertilizers is good and regeneration by seed and coppice is easy. Seeds, which are abundant from the third year, store well and germinate easily. The wood is soft, light in both weight and colour and is satisfactory for veneer, furniture corestock, blockboard and pulp. The most detailed yield figures available come from the Nasipit Lumber Company near Butuan City in northeastern Mindanao, which has Albizzia plantations up to 15 years old (see Appendix 2). These are the most reliable figures available and, although they are considered conservative, they provide the only reasonable basis on which to make yield predictions for this project.

16. Yield estimates for the present project are based on Nasipit site index 30, which shows yields of about 200 m^3/ha after eight years. 1/

17. Individual tree farms would be managed on a sustained yield basis using area control and an eight-year rotation. Farms would be planted initially at any rate a farmer is capable of, but the minimum rate is assumed to be 2 hectares/year. One-eighth of the total area planted would be harvested each year. During the first rotation the first cutting would be advanced to age 7, while some stands would be carried past the rotation age in order that yields can become uniform in the second and subsequent rotation. (See Table 2, Col.5)

18. For tree farms planted at rates greater than 2 hectares/year, annual harvesting of more than 1/8 the total plantation during the first rotation may be desirable. In such cases, uniform annual harvests(by area) are delayed until the second rotation and sustained yields are not reached until the third rotation.

19. A tree farm would first be clear cut of indigenous trees and all other vegetation by hand, with burning if necessary. Lining out and digging of holes are done about a week before planting of seedlings. When fertilizers are available, a few ounces would be worked into the soil two weeks and 7 months after planting. One clear weeding and two spot weedings of about one meter radius would be done during the first year. Further maintenance is not foreseen. The total number of man-days required per hectare through the first rotation is estimated at 110, broken down as follows: land preparation - 37; lining, digging and planting - 19; replanting - 8; fertilizer application - 12.5; and weeding - 34 days. Singling, starting a year after the first harvest, requires 25 man-days per ha.

20. Seedlings would be raised in PICOP's nurseries. Seeds can be obtained locally, and no shortage is likely during the project life. Although some improvement in seed genetic quality (perhaps reflected in improved form and insect resistance) is likely through provenance testing and mass selection, a tree improvement program is not considered essential for this project.

21. Spacing on tree farms would be 4x4 m, with commercial thinnings likely after four years. No estimates of thinning response are available, and thinning costs and returns were n included in this appraisal. No severe pest or disease problems have occurred in Albizzia plantations in the Philippines. However, scattered defoliation by the larvae of a certain yellow butterfly has been noted. Fire risk in the project area is minimal.

22. Farmers would receive free support services from PICOP. The Company maintains a well organized and well-staffed Agro-Forestry Department with extension offices throughout the project area. In addition to seedlings, which would be supplied at cost, PICOP would provide the following types of technical assistance to farmers; (i) locating and/or acquiring ownership of vacant alienable and disposable lands, (ii) technical advice in subdividing agro-farms for tree growing and food crops/livestock production, and (iii) technical advice on establishing and managing tree crops. PICOP also plans to continue its program of research on insect control. In addition, DBP has budgeted for a professional forester to assist in project appraisals and to coordinate with PICOP on technical matters. This forester would be responsible for safeguarding the interests of the farmers by (i) studying the development of markets competitive with PICOP and, if necessary, organizing marketing cooperatives; (ii) periodically scaling tree farm harvests to check PICOP volume estimates at mill site;(iii) appraising sites and their suitability for tree farming before loans are approved; and (iv) providing other technical advice that might be required.

1/ Site index is defined as the average total height in meters of dominant and co-dominant trees at 10 years of age. The yield table is reproduced as Appendix 2.

Summary of Inputs and Outputs

23. From the above technical description, a summary of the magnitudes and timing of inputs and outputs (costs and benefits) can be derived. Due to lack of more precise information on variability among project farms, all inputs and outputs and costs and benefits are estimated on an "average" farm basis, assuming an annual planting rate of 2 ha per year for the first four years. All estimates are based on PICOP's planting experience to date. Table 2 shows the number of ha involved each year for an "average" 10 ha farm, and Table 3 shows the magnitudes and the timing of inputs and outputs for such a 10 ha farm.

C. VALUING COSTS AND BENEFITS

Financial Costs and Benefits

24. Unit costs by input type are shown in Col. 3 of Table 4. Combining the cost figures from Table 4 with the information on timing and magnitude of inputs in Table 3, the financial cost figures for tree farming for the typical 10 ha farm are estimated in Table 5. In addition, the cost to the farmer of the DBP loan are also given in Table 5. The loan, given at 12 percent compound interest with a seven-year grace period on repayment of interest (i.e., first repayment in year 7) and nine years on repayment of principal covers 75 percent of the total cost to the farmer. The loan repayments are derived as shown in Appendix 3.

25. On the benefit side, wood produced by project farms would be easily absorbed by PICOP. Market demand for paper and paperboard is expected to grow considerably in the Philippines and in Japan, the principal export market. Total demand in the Philippines is expected to be 800 000 tons in 1985 compared with 420 000 tons in 1975. Production is expected to increase more slowly from 375 000 tons in 1975 to 420 000 tons in 1985. It is estimated that approximately 500 000 m^3 per year of Albizzia wood could be utilized to meet PICOP's present capacity. The first harvest from project tree farms would not take place until 1981 andyield will be about 284 000 m^3 by 1985.

26. PICOP actually pays P60 per cubic metre of wood and is willing to "subsidize" the hauling costs another P10. For the typical farm at a hauling distance of 75 km, total hauling costs are estimated at P27 per cubic metre. Harvesting costs are estimated at P13 per cubic metre. Therefore the resulting average stumpage value is P30/m^3 [i.e.(P60 + P10) - (P27 + P13)].

27. All financial benefits are assumed to materialize at the time of final harvest. Revenues from commercial thinnings and alterations in residual growth as a result of thinnings are excluded from the analysis. Combining outputs given in Table 3 with a unit value equal to P30 per cubic metre, financial benefits can be estimated. They are shown on line 2, Table 5.

Economic Costs and Benefits

28. The economic value of benefits was taken to be equal to the financial value of P30 per m^3, as derived above.[1] The value is based on what PICOP is actually willing to pay for wood at the time of this appraisal. There was no basis on which to estimate the potential real value increase that might take place during the project period. Thus

[1] Ideally, since the paper to be produced with project wood will substitute for imports, the economic value of the benefits should be based on the import substitution value, or the foreign exchange saved, appropriately shadow priced to account for over or under-evaluation of foreign exchange. However, data to derive such a value for the wood component are not available. Thus, benefits were estimated on the basis of PICOP's willingness to pay for the wood, which was taken as an approximation of the real value derived from import substitution. It is considered a conservative figure.

Table 2 - Physical Inputs/Outputs, 10 ha Farm

Years	Establishment Inputs 1/	Singling Input	Weeding Input	Harvest Output 2/
	Activities			
		Second & Later Roatations		
	number of ha			
0	2			
1	2			
2	2			
3	2			
4				
5				
6				
7			1	1 (7)
8		1	1	1 (8)
9		1	1	1 (8)
10		1	1	1 (9)
11		1	1	1 (9)
12		1	1	1 (10)
13		1	1	1 (10)
14		1	1	1 (11)
15 and on 3/				

1/ Planting input includes land preparation, purchase of seedlings, lining, digging, planting, replanting, fertilizing twice, and three first year weedings.

2/ Numbers in parenthesis indicate age of stand being harvested.

3/ The second and third rotations (at least) are based on coppicing.

Table 3 - Timing and Magnitudes of Physical Inputs and Outputs - for

Assumed "Average" 10 ha Farm

Item	Units/year	0	1	2	3	4	5	6	7	8	9	10	11	12	13	14	15 and on
Inputs																	
Land preparation lining, digging and planting	m.d.	74	74	74	74												
labour	m.d.	38	38	38	38												
seedlings[1]	no.	1200	1200	1200	1200												
Replanting																	
labour	m.d.	16	16	16	16												
seedlings[2]	no.	300	300	300	300												
Fertilization																	
labour	m.d.	25	25	25	25												
fertilizer	kg.	4	4	4	4												
Weeding	m.d.	68	68	68	68				34	34	34	34	34	34	34	34	34
Singling	m.d.									25	25	25	25	25	25	25	25
Output																	
Pulpwood stumpage[3]	$m^3(r)$								184.1	205.8	205.8	227.0	227.0	247.8	247.8	268.2	205.8

1/ Seedling production involves 13 m.d. of labour. Rest of cost is materials.

2/ Assumed that 20 percent would have to be replanted on the average.

3/ Based Appendix 3, using site index 30 and ages of stands harvested as shown in Table 2.

Table 4 - Unit Costs Used in
the Financial and Economic Analyses
(pesos/unit)

(1) Item	(2) Units	(3) Financial Value/Unit	(4) Economic Cost
Labour	m.d.	8	4 a/
Fertilizer	kg.	25	25 b/
Seedlings	100	8.7	5.2 c/
Output	m^3(r)	30	30 d/

a/ derived on the basis of estimated production for home consumption and alternative employment given up by a farm family in order to establish and maintain 8 ha of plantations.

b/ Assumed to be the same as the financial cost

c/ Seedling production costs are 80 percent labour and 20 percent materials. Thus, 80 percent of the P8.7 financial cost, i.e. P6.7, was reduced by half to reflect the shadow wage of P4 instead of the financial wage of P8. The shadow price for seedlings was thus calculated as follows: P8.7(.80)(P4/P8) + P8.7(.20)=P5.22

d/ The shadow price for output (stumpage) was considered to be close to the financial price, given the information available. Thus the financial price was used.

Table 5 Financial Cash Flow, 10 ha. Plantation

(value in constant pesos)

	YEARS																			
	0	1	2	3'	4	5	6	7	8	9	10	11	12	13	14	15	16	17	18	19
Cash Receipts																				
1. Bank loans	1500	1500	1500	1500	-	-	-	-	-	-	-	-	-	-	-	-	-	-	-	-
2. Sales	-	-	-	-	-	-	-	5523	6174	6174	6810	6810	7434	7434	8046	5174	6174	6174	6174	6174
Total	1500	1500	1500	1500	-	-	-	5523	6174	6174	6810	6810	7434	7434	8046	6174	6174	6174	6174	6174
Cash Disbursements																				
3. Land preparation	590	590	590	590	-	-	-	-	-	-	-	-	-	-	-	-	-	-	-	-
4. Purchase seedlings	130	130	130	130	-	-	-	-	-	-	-	-	-	-	-	-	-	-	-	-
5. Lining/Dig/Plant	300	300	300	300	-	-	-	-	-	-	-	-	-	-	-	-	-	-	-	-
6. Replanting	130	130	130	130	-	-	-	-	-	-	-	-	-	-	-	-	-	-	-	-
7. Fertilizing	300	300	300	300	-	-	-	-	-	-	-	-	-	-	-	-	-	-	-	-
8. Weeding	550	550	550	550	-	-	-	275	275	275	275	275	275	275	275	275	275	275	275	275
9. Singling	-	-	-	-	-	-	-	-	100	100	100	100	100	100	100	100	100	100	100	100
Total	2000	2000	2000	2000	-	-	-	275	375	375	375	375	375	375	375	375	375	375	375	375
Loan Payments																				
10. Accum. interest	-	-	-	-	-	-	-	3200	3050	-	-	-	-	-	-	-	-	-	-	-
11. Princ. & interest	-	-	-	-	-	-	-	-	-	1459	1459	1459	1459	1459	1459	-	-	-	-	-
Total	-	-	-	-	-	-	-	3200	3050	1459	1459	1459	1459	1459	1459	-	-	-	-	-
Cash Balance After Loan Payments	(500)	(500)	(500)	(500)	-	-	-	2048	2749	4340	4976	4976	5600	5600	6212	5799	5799	5799	5799	5799

benefits were assumed to remain constant in real terms per unit of output over the project life. Benefits from trees sold from the land clearing were excluded from project benefits, since this value could be obtained with or without the project. Since there was no basis on which to estimate potential net benefits from thinnings, they were not included in the analysis. Thus, the total project benefit figure can be considered as being conservative. Total economic benefits are shown on line 2 of Table 6, which is the basic value flow table for the "average" 10 ha farm.

29. Also, due to the lack of data, all economic costs, except labour costs, are estimated on the basis of their financial values. Labour is shadow-priced at P4/day. Derivation of real labour costs for a 10 ha farm is shown in Table 4.

30. Total economic costs for the average 10 ha farm are shown in Table 6. These include the costs of extension services provided through half-day visits per year per tree farm at P50 per visit.

D. COMPARING COSTS AND BENEFITS

Financial and Economic Rates of Return

31. On the basis of the assumptions above the financial rate of return for a 10 ha tree farm over 20 years would be about 39 percent. The economic rate of return for a period of 16 years is about 32 percent.

Sensitivity Analysis

32. Several of the original assumptions of the financial and the economic analyses were based upon uncertain data or are administratively determined. It is likely that prices and quantities of inputs and outputs will turn out to be different than originally assumed. It is also possible that some of the administrative assumptions (e.g. the fixed loan amount and the four year initial planting schedule) may be changed in the future. In order to test the sensitivity of the financial and economic rates of return to such possibilities, a sensitivity analysis was conducted. Table 7 presents the results of such analysis.

33. First, the period of time used for both financial and economic analyses was lengthened to 31 years, which corresponds to two coppice rotations in addition to the first rotation. 1/ Since the original financial and economic analyses were conducted for different periods of time (19 years and 15 years respectively), the extended period of analysis allows for more meaningful comparisons between financial and economic rates of return for a given set of assumptions. The financial rate of return is only slightly increased by the extended period of analysis (0.4 percent), but the economic rate of return is more significantly increased (by about 1%) since the period of economic analysis was substantially shorter originally (See Table 7). All further sets of assumptions used in the sensitivity analysis are also analyzed for the extended (31 year) period. 2/

34. Second, the four year initial planting period of two hectares per year was changed to an eight year sustained planting period of one hectare per year (Assumption Set 2). The two hectare per year planting rate caused initial costs to be carried over longer periods of time than necessary without providing significant benefits any earlier, which

---

1/ The 31 year period of analysis corresponds to two coppice rotations only when harvests begin in the ninth year and continue annually, assuming an eight year rotation. This is the case for all sets of assumptions used in the sensitivity analysis, except for one assuming a six year rotation. Also, for the original set of assumptions, the 31 year period corresponds to one year more than two coppices, since harvesting was advanced to the eighth year.

2/ All calculations were also carried out for 23 years (one coppice rotation), but there was little difference between 23 and 31 years. Therefore the analysis uses only the latter period.

Table 6. Economic Value Flow, 10 ha Plantation (value in constant pesos)

	Years															
	0	1	2	3	4	5	6	7	8	9	10	11	12	13	14	15
Benefits																
1. Thinning	-	-	-	-	-	-	-	-	-	-	-	-	-	-	-	-
2. Harvest	-	-	-	-	-	-	-	5523	6174	6174	6810	6810	7434	7434	8046	6174
TOTAL	-	-	-	-	-	-	-	5523	6174	6174	6810	6810	7434	7434	8046	6174
Costs																
3. Land Preparation	295	295	295	295	-	-	-	-	-	-	-	-	-	-	-	-
4. Purchase of seedlings	78	78	78	78	-	-	-	-	-	-	-	-	-	-	-	-
5. Lining/Dig/Plant	150	150	150	150	-	-	-	-	-	-	-	-	-	-	-	-
6. Replanting	65	65	65	65	-	-	-	-	-	-	-	-	-	-	-	-
7. Fertilizing	200	200	200	200	-	-	-	-	-	-	-	-	-	-	-	-
8. Weeding	275	275	275	275	-	-	-	137	137	137	137	137	137	137	137	137
9. Singling	-	-	-	-	-	-	-	-	50	50	50	50	50	50	50	50
10. Administrative	100	100	100	100	100	100	100	100	100	100	100	100	100	100	100	100
TOTAL	1163	1163	1163	1163	100	100	100	237	287	287	287	287	287	287	287	287
Net Benefits (costs)	(1163)	(1163)	(1163)	(1163)	(100)	(100)	(100)	5286	5881	5881	6522	6522	7147	7147	7559	5887

therefore depressed both the financial and economic rates of return. 1/ Adopting the sustained planting assumption, the financial rate of return increased by about 2% and the economic rate by almost 1%. Since this sustained planting assumption drops the administrative assumption of planting two hectares per year and harvesting only one, it measures the actual rate of return for each hectare assuming an eight year rotation. Therefore, this set of assumptions was used as the base for the sensitivity analysis.

35. In order to assess changes in financial profitability of an average farm induced by different alternative loan policies, the first changes in assumptions were applied to the loan the smallholders received from DBP. In Assumption Set 3 (see Table 7) the loan was assumed to be half as large (to cover 3/8 of establishments costs) as the original loan, which caused the financial rate of return to drop by 10 percent to 31.4 percent. When the loan was assumed to be eliminated entirely in Assumption Set 4, the rate of return dropped by a further 4 percent, to an FRR of 27.1 percent.

36. One factor which by definition must be variable is transportation cost. The economic hauling distance varies from zero to 100 kilometers. The financial and economic rates of return for an "average" plantation would be highest if it was located adjacent to the mill (zero hauling distance) and lowest if it was located 100 kilometers from the mill (the maximum distance considered). To test the maximum sensitivity to hauling cost it was necessary only to calculate the rates of return for the extreme distances. Translated into benefit (revenue) terms, a plantation located adjacent to the mill would have a 23.33% revenue increase in the financial analysis and a 56.67% benefit increase in the economic analysis. 2/ Thus, for a plantation located adjacent to the mill the financial rate of return would increase by 5% over the rate of a plantation located 75 kilometers from the mill, and the economic rate of return would increase by about 7%, (as shown in Assumption Set 5). For a plantation located 100 kilometers from the mill both the financial revenue and economic benefits would be decreased by 9.3%. 3/ In this case, the financial rate of return dropped by 2.3% and the economic rate by 1.5% (see Assumption Set 6).

37. The sensitivity analysis considers two other decreased benefit (revenue) assumptions, which could be caused by one or more of the factors discussed earlier. Assumption Set 7 assumes that benefits (revenues) are 25% lower, which results in a decline in the financial rate of return of 7% and in the economic rate of 4%. Assumption Set 8 assumes that benefits (revenues) are 50% lower. Under this assumption the financial rate of return drops by 18% and the economic rate by 10%.

1/ Under the original two hectare per year planting rate, harvests were begun one year earlier (age 7 or year 8), but this was merely an administrative decision which could also have been made with the sustained (8 year) planting assumption. If it was necessary for PICOP to obtain supplies of Albizzia earlier than in the ninth year, it would perhaps have been better to initially pay higher delivered prices for Albizzia, thereby encouraging earlier planting and shorter initial rotations. In that case it probably would have been necessary for smallholders to plant their entire eight hectares earlier and to vary their first rotation cutting in order to meet PICOP's immediate needs, but higher prices could then justify such planting and harvesting.

2/ Any other single factor or combination of factors (including transportation cost) which would cause benefits (revenue) to increase by these amounts would have the same effect on the rates of return. Thus, this sensitivity analysis could be used to test a number of changes.

3/ An increase in transportation cost of P 2.79/m^3 is a 9.3% increase in a base price of P 30/m^3.

38. Three cost changes were tested for their effects on the rates of return. First, in Assumption Set 9 labor cost was assumed to be 25% higher than originally, which could be due to higher wages (shadow wages) and/or lower productivity. The financial rate of return under this assumption was 10% lower, and the economic rate was 3% lower. Second, in Assumption Set 10 labor cost was assumed to be 50% higher than originally, which lowered the financial rate of return another 5% and lowered the economic rate of return an additional 2%. 1/ And third, it was assumed in Assumption Set 11 that fertilizer cost was twice as high, which decreased the financial rate by 2% and the economic rate by 1%.

39. Even though most of the assumption changes had a large effect on both rates of return, and especially the financial rate, no assumption change reduced either rate below 20%. This apparently suggests that the project is acceptable, both economically to the country and financially to the smallholder, under a wide range of alternative assumptions concerning costs and returns.

Table 7 - Sensitivity Analysis

	Original assumption	New assumption set	Internal Rate of Return (%) Financial Analysis (31 years)	Economic Analysis (31 years)
1.	Original assumption	Same	39.4	33.2
2.	2 ha per year for 4 years	Sustained Planting (SP) (1 ha/year for 8 years)	41.8	33.9
3.	P 1500 loan for each of first 4 years	Sustained Planting (SP)	31.4	-
4.	Same	SP - no loan	27.1	-
5.	Hauling distance 75 km	SP - hauling distance 0 km	47.2	41.3
6.	Same	SP - hauling distance 100 km	39.5	32.4
7.	Original revenues (benefits)	SP - Revenues (benefits)	34.5	29.5
8.	Same	SP - Revenues (benefits) 50 percent lower	23.7	23.5
9.	Original labour costs	SP - labour costs 25 percent higher	31.8	31.2
10.	Same	SP - labour costs 50 percent higher	26.4	29.1
11.	Original fertilizer cost	SP - fertilizer cost doubled	39.0	32.8

1/ Increased labor cost assumptions had a greater effect on the financial rate of return because financial labor cost was twice as high as the (50% shadow priced) economic labor cost.

APPENDIX 1

Map of Project Area

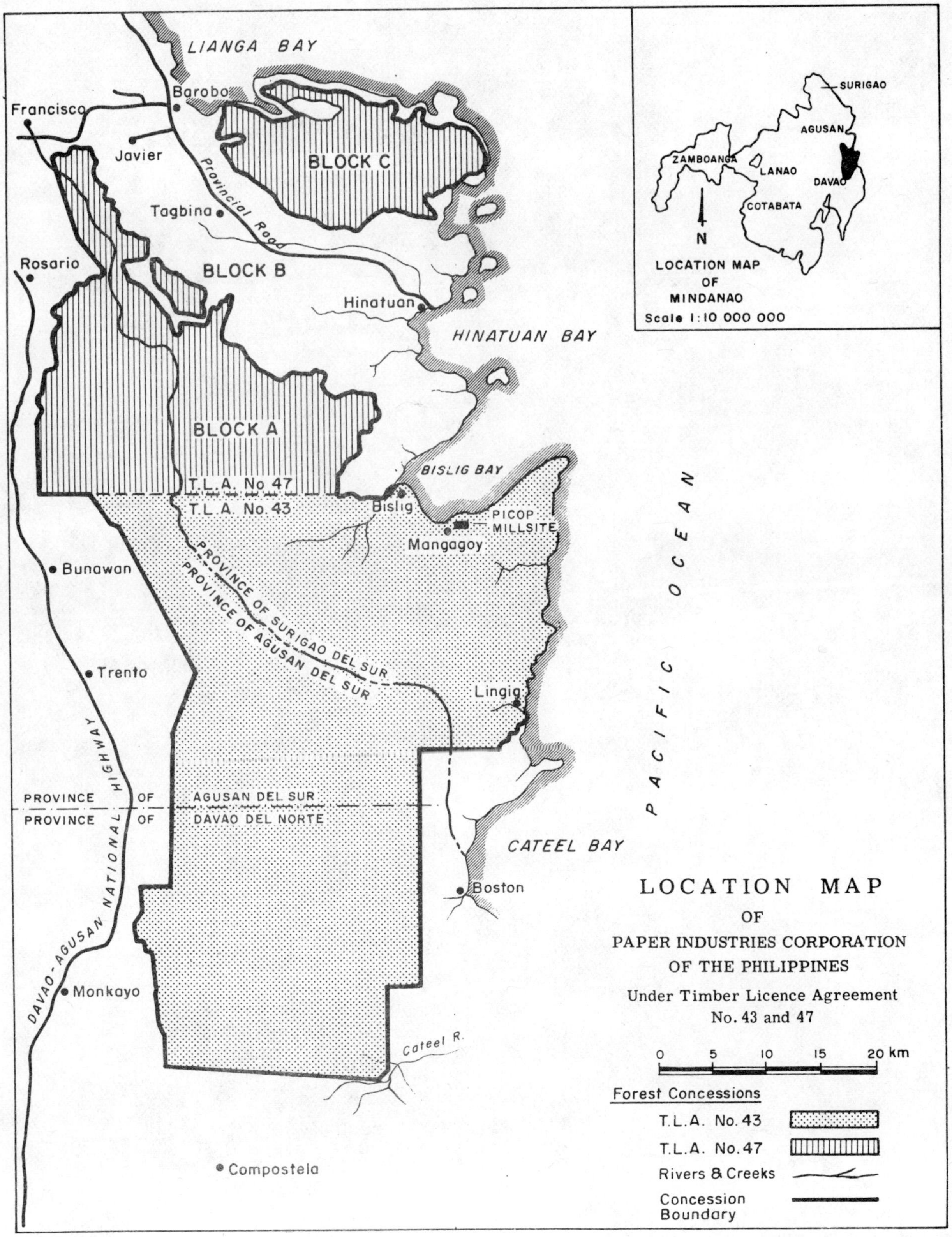

APPENDIX 2

Yields* by Site Index of Albizzia Falcataria
Nasipit Lumber Co. Plantations (m^3/ha)

Age	18	26	28	30	32	34	50
2	22.3	48.0	56.1	64.8	74.2	84.3	189.2
3	31.2	67.3	78.6	90.9	104.0	118.1	265.2
4	39.6	85.6	99.9	115.5	132.2	150.1	337.1
5	47.7	103.0	120.4	139.1	159.3	180.8	406.0
6	55.5	119.9	140.1	161.9	185.4	210.5	472.6
7	63.0	136.4	159.3	184.1	210.8	239.4	537.4
8	70.5	152.4	178.1	205.8	235.6	267.6	600.7
9	77.8	168.2	196.4	227.0	259.9	295.2	662.6
10	84.9	183.6	214.5	247.8	283.7	322.2	723.4
11	91.9	198.8	232.2	268.2	307.2	348.9	783.2
12	98.9	213.7	249.6	288.5	330.3	375.1	842.1
13	105.7	228.5	266.9	308.4	353.1	401.0	900.2
14	112.4	243.0	283.9	328.1	375.6	426.5	957.6
15	119.0	257.4	300.7	347.5	397.9	451.8	1014.0

* up to 10 cm top diameter inside bark not including branches, 2 x 3 m initial spacing, weeded first and second year, no thinning.

APPENDIX 3

Derivation of Farmer Loan Repayments

Year	Loan Amount 1/	Interest Due At End of Year 7 (Beginning of year 8)
0	P 1,500	P 1816
1	P 1,500	P 1461
2	P 1,500	P 1143
3	P 1,500	P 860
	Total	P 5280

Interest due beginning year 7	P 5,280
Amount paid in year 7	P 3,200 2/
Remaining interest due	P 2,080
Principal outstanding (4x$1 500)	P 6,000
Total loan outstanding	P 8,080
One year's interest 12%	P 970
Total interest due 3/ in year 8	P 3,050

The annual repayments of the loan for years 9 - 14 (line 11, Table 5) are derived as follows:

$$A = P\ 6{,}000 \left[\frac{.12(1.12)^6}{(1.12)^6 - 1} \right]$$

$$= P\ 6{,}000\ (0.2432) = P\ 1459$$

The factor in parentheses is called the "capital recovery multiplier", and is given as follows:

$$\left[\frac{i(1+i)^n}{(1+i)^n - 1} \right]$$

The value of this multiplier can be found in most tables of compound and discount factors.

1/ Assuming 2 ha planted per year at a total annual cost equal to P 2000 (Table 2). 75 percent of this amount, or P 1500, is covered by the Bank loan.

2/ Arbitrarily set at this level in order to keep interest payments in years 7 and 8 approximately equal.

3/ P 2080 plus P 970.

ECONOMIC ANALYSIS OF FORESTRY PROJECTS: CASE STUDIES

FAO FORESTRY PAPER 17, SUPPLEMENT 1

CASE STUDY NO. 2

VILLAGE FUELWOOD PLANTATIONS IN KOREA

FOOD AND AGRICULTURE ORGANIZATION OF THE UNITED NATIONS

Rome, 1979

Preface

This is one in a series of case studies of forestry and forest industries projects that has been prepared by the Food and Agriculture Organization of the United Nations (FAO) in order to demonstrate methods of preparing and appraising projects in the forest sector. The programme of case studies has been made possible by a special allocation for this purpose from the Swedish International Development Authority (SIDA). This case study has been prepared by FAO with the assistance of the College of Forestry at the University of Minnesota.

The project reported on in this case study is a project of the Government of Korea which has been assisted by a loan from the World Bank. The case study is based largely on material contained in an appraisal of the project carried out by the World Bank (Report No. 958a - KO). A field visit to the project provided further information. The case study describes the project as it was perceived at the time of appraisal in late 1975. This material has been released to FAO by the World Bank and the Government of Korea exclusively for the purpose of teaching forestry planning methods. It may not be quoted or reproduced in part or in whole without permission.

ABBREVIATIONS

ERR	=	Economic rate of return
VFA	=	Village Forestry Associations

CURRENCY EQUIVALENTS

US$1.00	=	Won 485
Won 1,000	=	US$2.06

TABLE OF CONTENTS

INTRODUCTION

1. The fuelwood project discussed here is part of an integrated rural development programme in Korea. The overall programme was conceived as a means of improving the living conditions and income earning potentials of some 15 000 villages within the Saemaul Undong (New Community Movement), a community development programme in which villages set their own development priorities and contribute, in kind or cash, to the selected projects.

2. In addition to the fuelwood plantations project analysed here, the integrated rural development programme includes 66 irrigation projects, ranging from 50 to 700 ha in size, 35 upland development areas, 850 km of village roads, 200 bridges, water supplies for 2 000 villages, and 2 400 rural electrification schemes. The programme also includes several supporting research components. The total cost of the fuelwood project discussed here is $14 million, of which an estimated 5 percent would be foreign exchange.

3. The project would significantly reduce local shortage of fuel by establishing about 11 000 village fuelwood blocks, covering about 127 000 ha. Sub-projects would vary from 3 to 50 ha in size and be implemented in all Provinces, but concentrated in those which are more densely populated where fuel shortages are most severe. The project would be an extension of an on-going programme to expand fuelwood plantations in Korea, which in 1975 had already successfully covered more than half a million hectares.

4. Plantations would be established by communal labour, mostly on private land with few alternative uses. Seedlings, fertilizer and technical assistance would be provided by government agencies. Species used would typically contain a mixture of fuelwood and timber-producing species with the proportion of timber species not exceeding 35 percent. The project includes support for fuelwood research (vehicles, equipment and materials). The research would address problems such as species selection, optimum spacing and fertilizer use.

5. The present case study looks in detail at the economic appraisal of the Korean fuelwood project. The case is divided into five sections. Section A provides some general background on the project and the appraisal of it. Section B gives a description of the technical forestry elements of the project and input requirements. Section C analyses project costs, while Section D discusses output and benefits. Finally, Section E provides a comparison of costs and benefits and looks at the sensitivity of the economic analysis and the economic rate of return (ERR) to changes in assumptions.

6. This case study records an example of a successful forestry project designed to involve and benefit the rural poor. One use of the study can thus be to illustrate the conditions and arrangements that helped make it possible and helped to ensure its successful implementation. 1/ Particular issues of substance and concept that the user may wish to consider include local organizational arrangements, coordination of village-level organization with the broader regional and national elements of organization needed to implement the overall project, and the concept of - and problems involved with - the dual purpose (fuelwood and timber) plantations that were made part of the project. Points of technique and organization of the analysis include the development of shadow prices, particularly for labour and fuelwood output, questions of timing and project period (final harvest output assumptions), use of "average" per ha values in the analysis of the entire 127 000 ha included in the project, and use of sensitivity analysis to determine the sensitivity of project results to changes in assumptions regarding input and output values.

A. APPRAISAL BACKGROUND

The Need for Fuelwood

7. Basic to the forestry problem in Korea is the long period of extreme winter cold, when the mean daily temperature is below freezing for almost three months. The heating system commonly in use (the "ondol" system) requires large quantities of fuel (averaging 4.2 tons per annum for a household of 6 persons) and in rural areas has led to the almost total impoverishment of those areas of forest land which are adjacent to agricultural communities. All areas of forest land have traditionally been regarded as freely available for fuelwood gathering and therefore, control of harvest has been almost impossible to

---

1/ See also Case Study No.1, "Philippine smallholder tree-farming project", for an example of another, but quite different, successful project for rural community development.

impose. The supply of fuelwood is now quite inadequate to meet requirements and thus leaves, grass, and forest litter are raked up for fuel. Rice straw, máize stalks and similar agricultural residues are also used in large quantities. The removal of forest litter has led to serious erosion and flooding problems and also to the lowering of soil fertility levels, whilst the use of agricultural residues as fuel deprive the individual farmers of a potential source of income, and the country of valuable raw materials (for composting, thatching and the manufacture of rope and bags).

8. Recognizing the seriousness of the problem, the Government introduced early in 1973 the following series of measures:

(a) The Office of Forestry was placed under the Ministry of Home Affairs, thus effectively strengthening its authority and ability to enforce forest protection. At the same time, the status of the Provincial and County Forestry Offices was raised and their staffs strengthened;

(b) A strong extension and publicity campaign was mounted, to inform the rural population of the vital necessity of restoring forest cover to the land, and to induce them to adopt the following fuel-saving methods:

(i) Conversion of cooking and heating units to coal burning, where coal supplies are adequate;

(ii) Improvement of the design of cooking stoves, for more efficient conversion of fuel;

(iii) Replacement (under the Saemaul Movement) of thatched roofs by tile, asbestos, and corrugated iron, thereby reducing the consumption of rice straw for thatching;

(iv) Elimination of the practice of cooking all feed for working oxen; and

(v) The introduction of methane gas producers (using animal manure) and kerosene heaters;

(c) Enforcement of regulations forbidding the raking of leaves, grass and litter within forest areas; and

(d) The introduction of a national reforestation scheme to create a well managed fuelwood plantation for each village, the reforestation work to be carried out by the villagers themselves.

9. Due in part to these new measures, standards of forest management in Korea improved in the last years. The annual rate of reforestation with fuelwood plantations built up to more than 30 000 ha per annum, ground vegetation began to develop in the absence of litter-raking, and materials other than thatch for roofing gained widespread use. The forestry plantations were well laid out and survival of trees is high. Silvicultural treatment and logistical facilities were also adequately planned. There do not seem to be any problems of supervision either at the technical advisory level or at the village foreman level.

Forest Land Tenure and Use

10. Of the land classified as forest, 73% is privately owned, 20% is in National Forests, while the remaining 7% is Public Forest (communally owned).

11. The National Forests are located mostly in Gengweon Province, which is the more mountainous and sparsely settled part of Korea. Therefore, the proportion of privately-owned forest land is correspondingly higher in the other, largely agricultural, provinces of the country. The private forest-land holdings are for the most part very small, as shown in Table 1.

12. Private forest-land is used mostly for the gathering of fuel and is therefore usually almost bare of trees or else carries impoverished pines less than 2 or 3 m high. Such land is frequently most valued for its use as a family burial ground. Given the high proportion of privately-owned forest, it is clear that private forest land has to be used for the establishment of village fuelwood plantations. As they are mostly small farmers, the owners are usually not able to bear the cost of planting the land themselves. Under the Forest Development Law of 1973, the Village Forestry Associations (VFA's) are empowered to take over the planting and management of private forest land under an agreement whereby the owner retains the ownership of the land and is entitled to a 10% share of the fuelwood produced from the plantations established upon it. The duration of the contract with the landowner equals the nominal economic life of the species to be planted. This arrangement has resulted in compact blocks of land being available for fuelwood plantations.

Table 1: Ownership Pattern of Private Forest Land

Holding Size (ha)	Number of Holdings	% of Holdings	Total Area (ha)	% of Area
Less than 1.0	985,000	56.0	387,000	8.4
1.0 - 10.0	699,000	39.7	2,275 000	49.2
10.0 - 100.0	74,000	4.2	1,519,000	32.9
More than 100.0	2,000	0.1	441,000	9.5
Total	1,760,000	100.0	4,622,000	100.0

Source: World Bank Report 958a - KO, Feb. 15, 1976.

B. TECHNICAL CONSIDERATIONS

Proposed Project Magnitude and Objectives

13. The establishment of village fuelwood plantations on a total of 127 000 ha in all nine provinces of Korea is proposed for the project (Table 2)

14. Arising out of a recent survey, the fuelwood needs of most villages have been assessed using a standard Fuelwood Forest Record. Based on the number of households and fireplaces in the village, total fuel demand is calculated. The forest resources of the village are assessed and, based on the availability of suitable land and of alternative fuels (such as coal and kerosene), a judgement is made by the district Office of Forestry staff on the size of fuelwood plantation needed. Priority in allocating government staff time and materials to particular villages depends largely on the competence and enthusiasm of the VFA but also on the village's own priorities, as expressed through its Saemaul Committee. Priority is also given to villages with particularly acute shortages of fuel, which is the reason for the concentration of the programme in the more densely populated provinces

Table 2: Provincial Distribution of the Fuelwood Component

Province	No. of sub-projects	Area to be Planted (a)		
		1976	1967	Total
Gyeong-gi	1,220	2,400	7,000	9,400
Gang-weon	943	2,300	7,000	9,300
Chungcheong Bug	897	3,400	5,000	8,400
Chungcheong Nam	1,674	8,100	7,900	16,000
Geonra Bug	1,077	4,400	7,000	11,400
Jeonra Nam	2,240	10,100	15,000	25,100
Gyeongsang Bug	1,796	8,600	14,000	22,600
Gyeongsang Nam	1,231	10,200	13,600	23,800
Jeju	51	500	500	1,000
Total	11,129	50,000	77,000	127,000

Source: World Bank Report 958-KO, Feb. 15, 1976

15. The survey showed that new or supplementary fuelwood blocks are required in about 11 100 villages and this is the basis for the proposed fuelwood project. Sub-projects for each village would vary from 3 ha to about 50 ha, the smaller ones being intended to supplement previously established plantations. The average sub-project size would be 11 ha. While fuelwood production would be the main purpose, about 30% of the trees planted would be intended primarily for timber production (mainly for local house building and, in some cases, for commercial uses), with fuelwood as a by-product. 1/ The main objectives of the fuelwood project are as follows:

(a) To stabilize a sustained supply system of fuelwood for the rural areas by the eventual production of 640 000 tons of additional fuelwood annually;

(b) To supply fuel for both forest owners and non-forest owners;

(c) To protect the forest through self-policing by Village Forestry Association members;

(d) To enrich the national forest resource by the planting of timber-producing species, as an admixture to the fuelwood species;

1/ The percentage for each village block and region would vary depending on needs and location.

(e) To build the productivity of the remaining forest land through the prevention of litter raking; and

(f) To boost the income of farm households through alternative uses for agricultural by-products.

16. In view of the size and success of the 1975 programme (40 000 ha), and the satisfactory establishment and protection of previous years' plantings, the targets of 50 000 ha in 1976 and 77 000 ha in 1977 seem well within practical attainment. Nursery stocks, transport, and technical supervision also appear quite adequate for the proposed project and, most important of all, it appears to have the full understanding and support of the villagers themselves.

Planting sites

17. The sites selected are on moderate to steep land (all other land being given over to agricultural crops), close to the villages. Sites may be either areas classified as "forest", which are preferred for larger plantings, or consist of small pieces of unused "non-forest" land such as roadsides, streambanks, strips adjacent to cultivated land, or risers between bench terraces. As far as possible, communal land is selected for the plantation but, as most forest land is privately owned, arrangements are made for such land to be managed by the Village Forestry Association (VFA), whereby the VFA assumes all management responsibilities (including the organization of planting) and the owner receives 10% of the yield of the plantation. The inclusion of timber-producing species (which have a cash value) eases any problems in obtaining the consent of the landowner and, in nearly all cases, there is no need to resort to the powers of eminent domain provided by the Forest Development Law.

Site preparation

18. In most cases, the areas selected for planting carry only a light growth of shrubs and grass. Clearing is therefore easy and can be carried out with simple hand tools. There is burning before planting and the planting pits, 30 cm in diameter, are made at the time of planting.

Seed supplies and nursery stocks

19. All seeds are produced in Korea. Seed collection is organized by the VFA's. Transplants are produced from a variety of well-organized nurseries, including: (a) state nurseries, (b) VFA nurseries at province, county and village levels, and (c) private commercial nurseries. The largest production is from the VFA nurseries. Stock from these nurseries is sold at a small profit to cover costs and to give a return to the VFA members.

Transport

20. The country has an adequate road network and therefore primary transport of seedlings by truck is no problem, with secondary distribution being carried out by bullock cart and by human portage. There are no serious problems of access to the plantation areas. Project supervision, by the Office of Forestry and the VFA unions, is presently hampered by a lack of suitable vehicles. Under the project, 140 pick-up trucks, one for each county, would be procured.

Planting

21. Planting takes place in March and April. To obtain maximum early yields, planting density is high, with 4 000 plants/ha proposed for "forest" areas, and 10 000 plants/ha for "non-forest" areas, where only low-growing shrubby species are employed, in order to avoid shading of adjacent crops. While research results show that these spacings are optimal from a yield point of view, they may not give sufficient weight to cost factors and to soil and site conditions. There may be additional benefits from less dense spacing

and therefore various spacings would be field tested under the project. Planting is in straight lines up and down the hillsides, to facilitate weeding and later extraction of the produce. Open-rooted planting is used and NPK fertilizer (at a rate which varies from zero to 80 g per plant, with an average of 6 g) is applied at the time of planting.

Species selection

22. The selection of a particular species (or species mixture) for an individual site is governed by its location in the country, soil fertility, and other site characteristics. The main fuelwood species used are Black Locust (Robinia pseudoacacia), Alder (Alnus firma and A. hirsuta) and the leguminous shrubs Bush Clover (Lespedeza spp.) and Indigo Bush (Amorpha sp.). The main timber species are Poplar (Populus alba x glandulosa) and Larch (Larix leptoledis) while Pine (Pinus rigida and P. thunbergii) may be considered dual-purpose. An interesting development is the use of Poplar in alternating lines with Bush Clover on relatively dry mountain sites. This is the so-called "Sunchon" method of fuelwood production. The Lespedeza gives fuelwood in the first year, plus a cash crop from seed production, whilst the Poplar gives fuelwood from the pruning of side branches, and can produce pulpwood at nine years old, or saw timber on longer rotations. Wherever site conditions permit, plantations of mixed timber-producing and fuelwood-producing species are created. For example, Poplar (for pulpwood production) is planted with Lespedeza or Robinia. Pinus rigida (for fuel and timber production) may be planted with Black Locust on low-quality sites, where the nitrogen-fixing capacity of the Black Locust is valuable for site improvement, whilst the Pine is tolerant of the low site quality.

23. While the mixing of fuel and timber-producing species may not be ideal from an economic viewpoint, there are good sociological reasons for its use. The main timber species are well understood by the villagers and produce significant amounts of fuel from pruning and thinning. High-yielding fuel-producing species are more novel and may have problems, such as the thorniness of the Black Locust. Under the project, about 30% of the project area would be planted to dual-purpose species (such as pines), 30% would be devoted to Bush Clover and Indigo Bush in "non-forest" areas, while the remaining areas would grow Black Locust and Alders (Table 3).

Maintenance

24. The plantations are weeded for two or three years after planting, until the trees are free from grass competition. Insect pests pose some problems but control methods are well understood. If necessary, fire breaks are constructed around the plantation. Losses from fire have been low in recent years.

Inputs and outputs

25. The physical inputs required for a typical or "average" ha of fuelwood plantations are shown in Table 4, together with the estimated output (discussed in par. 38). A 20-year rotation is assumed for the fuelwood component.

Organization

26. A Village Forestry Association has been formed within every village in the country. The head of each VFA is elected by his fellow villagers for a five-year term. He is directly responsible for activities of the VFA, such as the production of transplants from the village nursery, oak mushroom cultivation, and so on. He obtains technical guidance in his work from the foresters of the VFA Union at county level, and the district officers of the Office of Forestry (see Figure 1).

27. The Office of Forestry and the VFA field organizations are fully staffed at all levels, with the Office of Forestry having approximately 40 technically-trained foresters in each provincial office and 25 in each county office. These numbers are impressive and represent a very adequate forestry organization for carrying out the project.

Table 3 - Fuelwood Component: Species Distribution

Species	% of Area
"Forest" Areas - fuelwood species	
Pobinia pseudoacacia	22
Alnus Firmus) Alnus hirsuta)	18
- dual-purpose species	
Pinus rigida	29
Pinus thunbergii	2
"Non-Forest" areas - fuelwood species	
Lespedeza sp) Amorpha sp)	29
	100

Note: The above distribution is based on preliminary surveys only and has been used for cost estimating purposes only. A more detailed breakdown will be available after final designs are completed and may include additional minor species.

Table 4 - Physical Inputs/Outputs by Years per Ha. (Average) 1/

	Units	YEARS 0	1	2	3	4	5	6	7-19	20
Fuelwood Output	Tons	-	-	-	0.5	1.0	2.0	4.0	5.0	125.0
Inputs										
Labor (local)	Man	48.7	-	-	4.1	8.3	16.7	33.3	41.7	2083.3
Site Prep	Days	3.1	-	-	-	-	-	-	-	-
Planting	"	36.5	-	-	-	-	-	-	-	-
Fertil. Appl.	"	9.1	-	-	-	-	-	-	-	-
Weeding 4/	"	..	..	..	-	-	-	-	-	-
Harvesting	"	-	-	-	4.1	8.3	16.7	33.3	41.7	2083.3
Supervision 2/	"	0.1	0.1	0.1	0.1	0.1	0.1	0.1	0.1	0.1
Misc. (tools,etc.) 3/		..	-	-	-	-	-	-	-	-
Plants	No.	5944	-	-	-	-	-	-	-	-
Fertilizer	Kg.	42	-	-	-	-	-	-	-	-

1/ These are weighted averages assumed for 85,836 ha of "forest" land and 41,164 ha of "non-forest" land (Total: 127,000 ha)

2/ Includes only office of Forestry input

3/ Physical breakdown not available

4/ No figures give, but weeding cost is assumed to be offset by fuelwood value of grass collected

Figure 1 - Korea Rural Infrastructure Project
Organization Chart 1/ Fuelwood

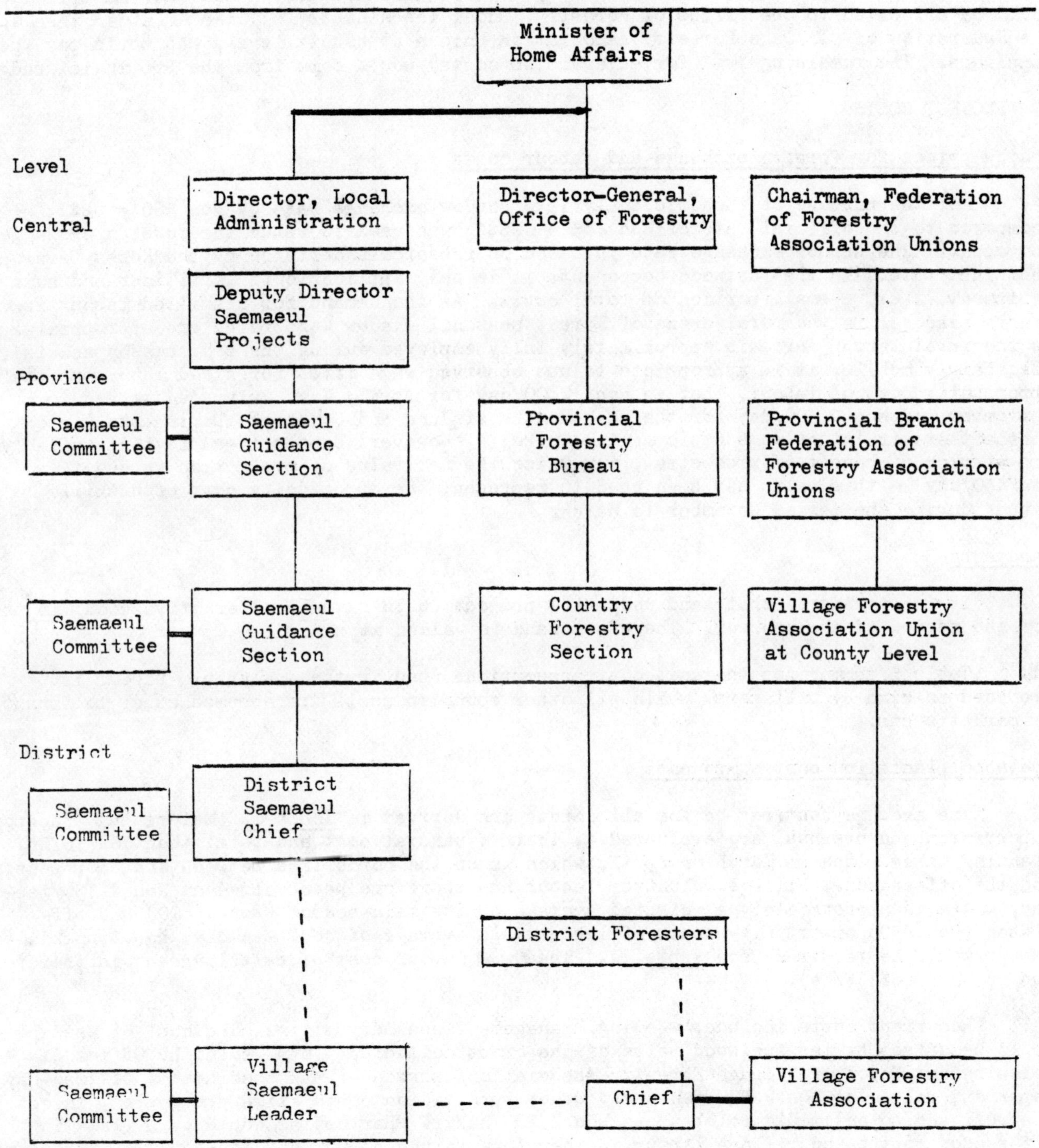

1/ Principal project units only shown.

28. For the fuelwood component, the Government would fully subsidize materials (mostly seedlings and fertilizer) and supervision. Labour to establish and maintain the plantations would be provided by the villagers. About 90% of required project funds (for materials) would be allocated to the Office of Forestry, under the Ministry of Home Affairs, then to the Federation of Village Forestry Association Unions at county level, who would pay the suppliers. The remaining 10% (for supervision costs) would come from the Provincial budget.

C. PROJECT COSTS

Shadow prices for foreign exchange and labour costs

29. For the purpose of economic analysis a shadow exchange rate of Won 560 = US$1 (compared to the official rate of Won 485 = US$1) was used to value the foreign exchange component. The shadow exchange rate is based on research undertaken by the Korea Development Institute. In the fuelwood component, it is only applicable to fertilizer and some equipment, i.e., a small portion of total costs. As the demand for unskilled labour is highly seasonal in the rural areas of Korea, seasonal shadow wage rates are appropriate. As the rural labour force is approximately fully employed during the crop season and is also fairly mobile, it is appropriate to use observed wage rates for hired labour as the opportunity cost of labour, that is Won 1,800/day for April, May, July, August and September and Won 1,900/day for the peak months of June and October. During the winter months there is little wage employment. There is, however, some self-employment, notably the weaving of bags from rice straw, for which the net value of production is about Won 450/day - this value has been used to represent the opportunity cost of unskilled labour during the period November to March.

Land cost

30. It is considered that land under the project would have no alternative economic use for the period of the project. Therefore land is valued at zero.

31. Table 5 summarizes the unit cost assumptions used in the analysis. Labour is provided in kind by villagers, while all other economic costs are assumed equal to financial or market costs.

Fuelwood plantation cost components

32. The average costs per ha for all inputs are derived in Table 6. Materials, transport, supervision and overhead are evaluated at their financial cost and total Won 76 376/ha. Planting takes place in March or April, which is on the borderline between the crop season and the off-season. Village volunteer labour has therefore been valued at Won 1 150/man-day, which is approximately a weighted average of the main-season (Won 1 800) and off-season (Won 450) opportunity costs of labour. An average of 48.7 man-days/ha of unskilled labour would be required (see Table 4). The total labour cost of establishment is therefore Won 56 000 (US$115/ha).

33. Recurrent costs include: weeding, management and harvesting. The cost of weeding would be offset by the fuelwood value of the grass collected. Management by Office of Forestry and Union of Village Forestry Associations personnel has been set at 0.1 man-day/ha/yr and valued at Won 2,500/man-day. Other supervision costs (fire prevention, insect control, and so on) would total about Won 2 200/ha/yr. Harvesting would take place in the off-season months and village labour is therefore valued at Won 450/man-day. It is assumed that one man can harvest and transport (by A-frame) 120 kg/day. Thus the harvesting cost would be Won 3 750 (US$7.75)/ton. Final harvest costs (year 20) are equal to Won 937 500 per ha as derived in Table 6, Footnote 2. Costs per ha by years are summarized in Table 7.

Table 5 - Unit Costs

	Units 4/	Cost W/Unit
Labor		
Harvesting 1/	m.d	450
Other 2/	m.d	1,150
Supervision		
OOF Pers.	m.d	2,500
Other	ha/yr	2,200
Transport 3/	ha	2,000
Fertilizer	kg	77.2
Seedlings	plant	11.6

1/ Harvesting takes place during the off months, therefore the minimum shadow wage was used

2/ Establishment starts at the end of the peak season for agricultural activity. Therefore a weighted average of off season and high season labour opportunity costs was used.

3/ Transport of materials (seedlings, fertilizer).

4/ m.d. = man days

Table 6 - Derivation of costs per hectare

	Units	"Forest"	"Non-forest"	Avg. 1/
SEEDLINGS				
Plants	(No./ha)	4,000	10,000	-
cost/plant	(W)	11.6	11.6	
cost/ha	(W/ha)	46,400	116,000	69,000
transport to site	(W/ha)	1,300	3,200	2,000
TOT. (Del)	(W/ha)	47,700	119,200	71,000
FERTILIZER				
Application	(g/plant)	6	6	6
Cost	(W/g)	0.077	0.077	0.077
Cost/Plant	(W/plant)			
plants/ha	(Plants/ha)	4,400	10,000	-
cost/ha	(W/ha)	1,853	4,600	2,743
transport to site	(W/ha)	100	200	133
TOT. (Del)	(W/ha)	1,953	4,800	2,876
ESTABLISHMENT (LABOR)				
Planting				
man-days	(M.D./ha)	24.6	61.5	36.5
value	(W/ha)	28,300	70,700	42,000
Fertilizing				
man-days	(m.d./ha)	6.2	15.4	9.1
value	(W/ha)	7,100	17,700	10,500
Site Prep.				
man-days		4.6		3.1
value		5,300	-	3,600
Weeding		(assumed to be offset by grass fuel collected)		
SUPERVISION				
Office of Forestry (0.1 m.d./ha/yr x W2,500/m.d.				W 250
Other (fire prevent, insect/disease control, etc.)				W 2,200
Total				W 2,450
HARVESTING				
120 kg./m.d or 8.3 m.d/ton and W 450/m.d. (Idle season labor cost) Cost/ton = W450 x 8.3 m.d. = W3750				
Cost/ha (by years)				
Years	Yield (ton/ha)			
0-2	0			0
3	0.5			1,875
4	1.0			3,750
5	2.0			7,500
6	4.0			15,000
7-19	5			18,700
20	125			937,500 2/
MISCELLANEOUS (tools, etc.)				7,500

1/ Weighted avg. 85,836 ha of "Forest" and 41,164 ha of "non-forest"

2/ 60 Kg/man/day can be harvested. Thus the cost is

$$\frac{125{,}000 \text{ kg/ha}}{60 \text{ kg/man day}} \times \text{W450/man day} = \text{W937,500/ha}$$

Table 7 - Economic Analysis of Average Fuelwood Plantation (one ha basis) 1/

W 1000/ha (constant prices)

	0	1	2	3	4	5	6	7	8	9	10	11	12	13	14	15	16	17	18	19	20
BENEFITS																					
Fuelwood	-	-	-	6.5	13.0	26.0	52.0	65.0	65.0	65.0	65.0	65.0	65.0	65.0	65.0	65.0	65.0	65.0	65.0	65.0	1,625.0
COSTS																					
Establishment:																					
Site Prep.(labor)	3.6																				
Seedlings	71.0																				
Planting (labor)	42.0																				
Fertilizer (del/site)	2.9																				
Fertilizing (labor)	10.5																				
Weeding (labor) 2/	-	-	-																		
Supervision	2.4	2.4	2.4	2.4	2.4	2.4	2.4	2.4	2.4	2.4	2.4	2.4	2.4	2.4	2.4	2.4	2.4	2.4	2.4	2.4	2.4
Misc. (tools, etc.)	7.5																				
Harvesting				1.9	3.7	7.5	15.0	18.7	18.7	18.7	18.7	18.7	18.7	18.7	18.7	18.7	18.7	18.7	18.7	18.7	937.5
Total Cost	139.9	2.4	2.4	4.3	6.1	9.9	17.4	21.1	21.1	21.1	21.1	21.1	21.1	21.1	21.1	21.1	21.1	21.1	21.1	21.1	939.9
NET BENEFIT (cost)	(139.9)	(2.4)	(2.4)	2.2	6.9	16.1	34.6	43.9	43.9	43.9	43.9	43.9	43.9	43.9	43.9	43.9	43.9	43.9	43.9	43.9	685.1

1/ Original World Bank Appraisal version corrected for consistency

2/ Weeding cost assumed to be offset by grass fuels obtained during operation.

D. PROJECT BENEFITS AND OUTPUT

34. The principal benefits of the fuelwood component would be:

(a) an increase in farmers' incomes, as a reduction in the use of agricultural residues for heating purposes would allow them to be used for composting and other productive purposes;

(b) conservation of Korea's scarce mineral energy resources (principally coal);

(c) creation of permanent timber-producing forests, as a secondary purpose of the fuelwood component;

(d) a reduction in soil erosion and flood damages; and

(e) institution building at the village level, leading to increased self-reliance.

35. At attempt is made to quantify benefits (a) and (b) by shadow pricing fuelwood as shown below. Benefits (c), (d), and (e) will be discussed only in qualitative terms. As discussed earlier, many of the fuelwood plantations would have an admixture of timber species. As the returns from timber production are known to be at least as attractive as those from fuelwood, only the latter type of plantation has been evaluated. This should provide a minimum estimate of the rate of return for the project.

Valuation of fuelwood

36. A precise estimation of the benefits of the fuelwood component is difficult, since at present fuelwood is not traded and therefore there are no data on its market value. Consequently, it is necessary to value the fuelwood in terms of the value of the fuels that it would replace, namely, agricultural residues and coal.

37. Field observations suggest that 80 percent of the additional fuelwood production would replace agricultural residues and 20 percent would replace coal. The calorific value of agricultural residues is about 90 percent that of fuelwood, while that of Korean coal (which is very low grade) is about equal to fuelwood. Thus 1 kg of fuelwood would replace 0.9 kg of agricultural residues plus 0.2 kg of coal. The price of both fuels, taken to represent their opportunity cost, is about Won 12/kg and thus the opportunity cost of fuelwood is about Won 13 000 (US$27)/ton. 1/ There is no foreign exchange effect since both substitute materials are produced and sold locally.

Project output

38. Assuming plantation establishment in Year 0, annual harvest would reach 5.0 ton/ha/y in year seven for a typical fuelwood stand. Yields would build up as follows:

Year	Yield (ton/ha)	% of full development
0-2	0	0
3	0.5	10
4	1.0	20
5	2.0	40
6	4.0	80
7-19	5.0	100

1/ i.e., 1.1 kg of other fuels is worth 1.1 x Won 12/kg or Won 13.20, which is rounded off to Won 13/kg or Won 13 000/ton.

39. It is assumed that the average plantation would be clear-cut in Year 20, yielding 125 tons/ha of fuelwood. Average yield over the rotation would be 9.4 tons/ha/yr. During Years 0 to 2, there would be a certain amount of grass and weeds produced but it is assumed that the economic value of this would be merely equal to the opportunity cost of labour used in collecting it.

Direct project benefits

40. Using the above schedule of outputs and the opportunity cost of fuelwood (estimated above as Won 13 000 per ton) we arrive at the benefit estimates shown in Table 7.

Indirect benefits

41. Besides the direct benefits of the fuelwood component, which are evaluated above, there are several indirect and intangible benefits, for which few data are available. Concentration of fuelwood production close to the village would ease the pressure on the natural "forest" around the village as a source of fuelwood and would enhance its potential for commercial timber production. Leaving forest litter on such areas would retard erosion and help to reduce flood peaks.

42. A major intangible benefit of the fuelwood component would be its contribution to the strengthening of village institutions. That is one reason why fuelwood projects are often among the first to be undertaken by "basic" villages under the Saemaul Movement. Fuelwood plantations, while technologically simple, give a very tangible demonstration to the villagers of the benefits of cooperative action.

E. COMPARING COSTS AND BENEFITS

43. The economic cost and benefit streams are shown in Table 7. The economic rate of return (ERR) is estimated to be 17 percent, which indicates that the investment is attractive.

Sensitivity analysis

44. To test the sensitivity of the ERR to different input and output assumptions, certain changes were tested. These include changes in values for costs such as fertilizer and seedling costs and benefits such as yields and price of fuelwood. In addition, alternative assumptions were tested regarding labour costs. The results are shown in Table 8.

45. As indicated in that table, the ERR is sensitive to an annual change in labour cost and to an annual change in fuelwood value, if introduced separately. In combination, a 4 percent annual increse in labour cost and a 2 percent increase in fuelwood value cancel each other out, giving an ERR of 17 percent, the same as the original ERR. This is because of the different timing of the impacts of these elements.

45. The ERR is also sensitive to other labour cost changes. The original analysis assumed that only half as much fuelwood could be cut per man day in the final harvest (year 21) as in the earlier thinnings. 1/ If it is assumed that the same amount could be cut in the final harvest, the ERR increases by 1.2 percent. The original analysis assumed that the shadow wage rate for off-season labour was Won 450 per man-day, as compared to Won 1800-1900 per man-day during the peak season. If this off-season shadow

---

1/ 60 kg/day vs. 120 kg/day

Table 8 - Sensitivity Analysis 1/

Original Assumption	New assumption 2/	New ER (percent)
1. Constant real labour cost for project period.	1. 4 percent annual increase in rural labour cost.	negative
2. Constant shadow price for fuelwood for project period.	2. 2 percent annual increase in real value of fuelwood	21
3. Same as 1 and 2 above.	3. Combination of 1 and 2 above.	17 (no change)
4. Seedling cost of Won 71 000 per ha.	4. Seedling cost of Won 35 500 per ha.	20
5. Fertilizer cost of Won 2 900/ha.	5. Fertilizer cost of Won 6 000/ha.	17 (no change)
6. Final harvest of 125 tons/ha in year 20.	6. Final harvest of 29 tons/ha each in years 20-24.	16.7
7. Final harvest of 125 tons/ha in year 20.	7. No final harvest. Year 20 "thinning" of 5 tons/ha.	14.4
8. Final harvest of 125 tons/ha in year 20.	8. No final harvest. Years 20-25 annual "thinnings" of 5 tons/ha.	15.2
9. Final harvest of 125 tons/ha in year 20.	9. No final harvest. Years 20-30 annual "thinnings" of 5 tons/ha.	15.5
10. Annual thinnings of 5 tons/ha in years 16-19. Final harvest of 125 tons/ha in year 20.	10. No thinnings or harvest beyond year 15.	12.6
11. Annual thinnings of 5 tons/ha in years 13-19. Final harvest of 125 tons/ha in year 24.	11. No thinnings or harvest beyond year 12.	10.0
12. Yields as shown in Table 6.	12. Yields 80% of those shown in Table 6.	14.7
13. Yields as shown in Table 6.	13. Yields 60% of those shown in Table 6.	11.9
14. Final harvest: 60 kg/md.	14. Final harvest: 120 kg/md.	18.2
15. Off-season labour shadow wage = 450W/md.	15. Off-season shadow wage = W600/md.	14.3
16. Off-season labour shadow wage = W450/md.	16. Off-season shadow wage = W675/md.	12.7
17. Fuelwood value of W 13 000/ton.	17. Fuelwood price of W 10 400/ton (80%).	12.9
18. Fuelwood value of W 13 000/ton.	18. Fuelwood price of W 7 800/ton (60%).	5.6

1/ Original ERR was 17 percent.

2/ All other original assumptions remaining the same.

wage is raised to Won 600, the ERR falls nearly 3 percent, and if it is raised to Won 675, it falls by over 4 percent.

46. Reducing seedling cost by 50 percent increases the ERR by 3 percent, while a doubling of fertilizer cost has no effect on the ERR because it is relatively insignificant to begin with.

47. Changes in the yield and timing of output also affected the ERR When yields for each year were assumed to be 80% of the original amounts, the ERR dropped by more than 2%, and when yields were assumed to be only 60%, the ERR decreased by more than 5%. If the final harvest was spread equally over five years, due to demand deficiency or other reasons, the ERR would remain about the same. 1/ To test further for demand deficiency, it was assumed that no more than 5 tons per hectare per year could be used in the community, and that demand would drop to zero after a specified year. For a demand of 5 tons per hectare per year through year 12, the ERR dropped to 10%. For demand cut-offs after years 15, 20, 25, and 30, the ERR were 12.6%, 14.4%, 15.2%, respectively. 2/

48. Finally, if fuelwood values fell to 80% of the originally assumed value, the ERR dropped to 12.9%, and if values decreased to 60%, the ERR fell to 5.6%.

1/ This assumes that instead of harvesting the total volume of 125 tons/ha in year 20, 20 tons/ha/year are harvested from year 20 to year 24, which is a total of 145 tons over 5 years (includes 20 tons of growth over the four years from year 20-24).

2/ For demand assumptions through years 25 and 30, this assumes that the fuelwood plantations established in year 0 continue to yield 5 tons/ha/year (at the same harvest and supervision costs) through that final year of demand.

ECONOMIC ANALYSIS OF FORESTRY PROJECTS: CASE STUDIES

FAO FORESTRY PAPER 17, SUPPLEMENT 1

CASE STUDY NO. 3

SOUTH AMERICAN NATURAL FOREST UTILIZATION PROJECT

FOOD AND AGRICULTURE ORGANIZATION OF THE UNITED NATIONS

Rome, 1979

Preface

This is one in a series of case studies of forestry and forest industries projects that has been prepared by the Food and Agriculture Organization of the United Nations (FAO) in order to demonstrate actual methods used in preparing and appraising projects in the forest sector. The programme of case studies has been made possible by a special allocation for this purpose from the Swedish International Development Authority (SIDA). This case study has been prepared by FAO with the assistance of the College of Forestry at the University of Minnesota.

The present case study is based on a real project in a South American country. Certain key data have been changed to avoid disclosure of proprietary information. However, none of the realism of the original project has been lost. The case study describes the project as it was perceived at the time the original project appraisal was carried out, in 1973.

ABBREVIATIONS

ERR	=	Economic rate of return
FOB/f.o.b.	=	Free-on-board
FRR	=	Financial rate of return
HP	=	Horse power
IPT	=	Industrial production tax
NPW	=	Net present worth
TT	=	Trade tax

TABLE OF CONTENTS

INTRODUCTION

1. This is a case study of a project to expand wood supply and processing capabilities for an existing wood processing company. It is an integrated project involving harvest and utilization of natural tropical hardwoods. The proposed project would add plywood, lumber, moulding and sliced veneer capacity and expand significantly the company's ability to supply its raw material needs. The project involves increased commercialization of native species. The proposed project involves an investment of some US$13 million.

2. The Company was founded about 25 years ago as a sawmilling and log export firm. A plywood mill was added about 15 years ago and began operations with a single peeler, which has been added to at intervals so that the mill now has a capacity of about 20 000 m^3 plywood per annum. There is also a small sawmill with a capacity of about 2 800 m^3 per annum.

3. Although the plywood mill started operations almost 15 years ago, it is only in recent years, since the Company received positive incentives to encourage exports, that the mill began to operate under profitable conditions. Log exports have contributed significantly to financing the development of the current mill operation.

4. During the last few years a firm market has been established in Europe for plywood products, based mainly upon one species. From this base other species have been successfully introduced on the market. The present policy is to progressively introduce more native wood species onto the international market. This will broaden the base of raw material supply, which will improve the yield of utilizable timber per hectare.

5. This policy promises to be rewarding, particularly in the longer run, since the region's forests remain one of the few large untouched resources of tropical hardwoods left in the world. In light of growing world demand and limited expansion in other tropical zones, due to physical, economic, and in some cases political constraints, progressively more favourable market conditions for the region's forest products should develop.

6. The present mill is located on a major river, with many tributaries that supply logs to the mill. The mill site is on a small bay out of the main current - which provides a fairly still lagoon for log storage and handling. Undeveloped land on either side of the present mill site could be acquired if future expansion is warranted.

7. While the standing volume of timber in the region is large, much of it is at present inaccessible by water, which is almost the only transport means available. Further, much of the accessible timber has no known commercial market. Thus, while the region certainly could support additional processing capacity, the lack of adequate information on delivered wood cost, availability of presently marketable woods, shortage of skilled labour, infrastructure, etc., has resulted in a slow growth of the wood processing sector.

8. Since the project region is still largely undeveloped, the economic, financial and social conditions which are normally needed for a successful processing operation are not generally available. Therefore, this project raises interesting questions connected with the uncertainty surrounding the proposed operation and its treatment in an economic appraisal. Other issues in this case study include the analytical treatment of separable components, and the differences involved in financial vis-à-vis economic analyses.

9. Section A provides a description of the project, including technical considerations, availability of required inputs, market for outputs and unit prices. Section B provides estimates of project costs and benefits. Costs and benefits are compared in Section C. Finally, Section D explores the sensitivity of the project to changes in key assumptions.

A. THE PROJECT

Outputs and Inputs

10. The project includes a new plywood mill which will be added to the existing one. A new sawmill and moulding plant will also be built and the existing sawmill will be closed down. In addition, high capacity slicing equipment will replace the existing veneer slicer. A logging system new to the region will be introduced.

11. The proposed output mix of the project is structured so as to make use of a variety of species available in the region. Output volumes and types were set at levels that would not overburden the Company's marketing capacity.

Technical Considerations - Processing Components

12. The planned production mix of the industrial complex at full project development (fifth year) is summarized in Table 1. Table 2 shows build-up of production for the main products from year 0 to 4, when full production is reached for all project components.

13. The production of the existing plywood mill will be raised from about 13 000 m^3 per annum to 20 000 m^3 per annum with some minor alterations and additions and improvement in operating skills. (See Appendix 1.)

14. A new plywood mill (plymill) is proposed as the core of the industrial complex. A 3-line mill appears the most appropriate choice to take advantage of the variability of the species that occur in the region, since it greatly increases operational flexibility with no proportional increase in cost over a two-line mill. This is also important for plant maintenance and in case of breakdowns, since the operation is relatively isolated from centres of engineering and spare part supplies. (See Appendix 2 for a description of the plywood mill.)

15. With regard to the proposed sawmill, recommendations about main items of equipment and levels of mechanization have been based on the levels of skilled labour and maintenance and spare parts availabile in the area of the project. A "medium level technology" sawmill is proposed. (Details on plant, equipment and production estimates are shown in Appendix 3.)

16. A pre-drier will be installed to bring all timber for kiln seasoning down to a uniform moisture content of 30 percent. The timber will then be passed to two seasoning kilns for drying to moisture equilibrium (12-14 percent).

17. A moulding plant consists of three 6-cutter planers and moulders and one 4-cutter, together with a 3-saw trimmer. Appropriate bundling equipment is provided for, together with a packaging and covered storage area for the finished products. About 11 000 m^3 of mouldings will be processed per annum. These will be composed of a variety of products in varying proportions according to special orders. Basically two types of mouldings will be produced, i.e. wall panelling and semi-decorative mixed mouldings. (See Appendix 3).

18. The veneer slicer will consist of high capacity equipment, 4 m^3 per hour. Production will be about 500 m^3 per year. Decorative veneer has been included in the production programme in order to optimize local utilization of high grade logs suitable for veneer slicing which in the past have been exported in log form. It is estimated that such logs will not exceed one percent of total supplies.

19. Also a steam plant will be added for the new plywood mill and sawmill. It is estimated that the combined energy requirement of both mills can be adequately satisfied with the available wood waste.

Table 1 - Planned Project Output[1/]

		(m³ per year)
Old plywood mill	plywood	20 000
New plywood mill	plywood	45 000
New sawmill/moulding plant	green lumber	30 000
	dried lumber	13 000
	mouldings	10 800
New veneer slicing plant	sliced veneer	500
	Total	119 300

Table 2 - Phasing of Output

($1\ 000\ m^3$)

	Year 0	1	2	3	4-14
Plywood Old Mill (additional)[2/]	4.6	7.0	7.0	7.0	7.0
Plywood New Mill	-	4.4	26.0	45.0	45.0
Green Lumber	-	2.3	15.5	28.5	30.0
Dried Lumber	-	1.0	6.7	12.3	13.0
Mouldings	-	.8	5.9	10.3	10.8
Sliced Veneer	-	-	0.2	0.4	0.5
Total	4.6	15.5	61.3	103.5	106.3

1/ See Appendices 1-3 for additional details.

2/ The old mill will eventually be producing 20 000 m^3, i.e. 7 000 m^3 in addition to 13 000 m^3 being produced now.

Markets for Outputs and Price Assumptions

20. World demand for semi-decorative plywood is growing rapidly. The Company presently has difficulties satisfying foreign orders, and prices on the international market have increased substantially over the last few years. Also, several new species have been successfully tested in export markets. In local markets, prices have been lower on average than for exports but demand is also increasing steadily. It is expected that about 10 percent of the Company's production will be lower grade plywood for local markets.

21. For the purpose of economic evaluation, f.o.b. prices of $\$170/m^3$ and $\$150/m^3$ are assumed for exported and locally marketed plywood respectively.

22. The market for mouldings will be mainly the USA. Market prices vary considerably according to the type of mouldings and wood species used. A conservative average price of US$125 f.o.b. per cubic metre is used for economic calculations. This price reflects the situation on the low price US market for semi-decorative mouldings.

23. In view of the growing sawnwood deficit in Europe, there is no problem expected in placing the small planned volume of 13 000 m^3 kiln dried lumber in this market. In addition, the static but very substantial US market could be approached. The development of sawnwood exports implies the necessity to adopt suitable timber grading and seasoning methods. Given the experience of other companies in the region, no problems are expected in marketing sawnwood output.

24. The principal market for the production of green lumber will be an adjoining wood scarce region. Despite a low per capita consumption of wood products in this region, the absolute size of the market is substantial with an estimated consumption of about 400 000 to 500 000 m^3 per annum. Planned output of 30 000 m^3 should be readily absorbed at full project development in year 4, taking into account the output of competing sawmills. F.o.b. prices assumed in the economic analysis correspond to present levels, i.e., US$$80/m^3$ for exported and $\$50/m^3$ for locally marketed lumber.

25. With regard to decorative veneer no problems are foreseen in finding markets for the modest output envisaged. An f.o.b. price of $\$300/m^3$ has been assumed as a conservative estimate for sliced veneer.

Raw Material Requirements and Supply

26. Raw material requirements will rise from an estimated 80 000 m^3 per year at present to 325 000 m^3 per year at project maturity: plywood production will require about 162 500 m^3 per year (including some 40 000 m^3 per year for the existing mill operation); sawnwood and mouldings will require 120 000 m^3; veneer will require 2 500 m^3 and log exports will absorb an additional 40 000 m^3 per year. (See Table 3 for phasing of additional requirements.) To secure this volume on a regular year-round basis, a departure from exclusive reliance on the traditional logging system will be necessary. At present logs are obtained entirely from a supply system based on the independent operations of many individual suppliers. Local workers fell the trees by axe on a narrow strip of flood plain land along the river shores upstream from the mill. As water levels rise, the logs are floated out onto the river. The logs are then bought and collected by local operators and are in turn purchased by company agents at recognized collection points. Logs are assembled into larger rafts and floated downstream to the mill, using Company tugboats for guiding the rafts.

27. Some major drawbacks of the traditional wood supply system are the following:

(1) Seasonal fluctuation of supplies as a function of seasonal flooding and the resulting necessity to operate substantial log storage or suffer irregular capacity utilization;

Table 3 - Phasing of Additional Raw Material Requirements
(m^3 (r))

	Year				
	0	1	2	3	4-14
Old Plymill (additional)[1]	4 000	10 000	10 000	10 000	10 000
New Plymill	-	11 000	65 000	112 500	112 500
Sawmill/Moulding	-	9 000	62 000	114 000	120 000
Veneer Slicing	-	-	1 000	2 000	2 500
Total	4 000	30 000	138 000	238 500	245 000

[1] Wood requirements for the increased output from the old plymill will be covered by greater efficiency in use of the wood already coming into it plus the additional amounts shown here (which are new wood requirements).

(ii) limitation of usable species due to restricted accessible logging areas on low lands along river banks and necessity of floating the logs. This leads to little concentrated exploitation over large catchment areas, difficult logistics and high losses in transit.

28. To secure year-round log supplies, which cannot be provided by the traditional system, it is essential to introduce some form of mechanical logging which does not depend on seasonal water level variations.

29. To determine the logging methods to be adopted and to estimate in detail the investment and operating cost necessary to establish a complete system to guarantee wood supplies, as well as to provide inventories of selected areas to be logged mechanically, a forest inventory/logging survey will be carried out. The system envisioned would include a gradual and selective introduction of mechancial equipment in flood plain logging coupled with optimization of the traditional system, supplemented by dryland logging in particularly suitable areas.

30. Estimates made by the Company indicate that, by perfecting the present labour intensive log supply system, it should be possible to approximately double log supplies to about 155 000 m^3 per annum for the next 10-15 years. This would correspond to roughly one half of the project's total wood requirements at full project development. The means by which this increase is expected are better organization of assembling, improved log barging and use of new collection equipment (tugboats, radio communication). The possibility of providing about 75 000 m^3 per annum of additional logs through the traditional system and the balance (170 000 m^3 per annum) by mechanized logging is taken as the most likely assumption. The mechanized system assumed here will require investments spread over 3 years reaching production capacity of 170 000 m^3 in the fifth year. Other logging system assumptions are tested in the sensitivity analysis.

Labour and Technical Staff

31. With regard to labour, there is an adequate supply to meet project requirements, but it is generally lacking the needed skills. Absenteeism is about 10 percent per day and turnover about 10 percent annually. The lack of skills points to the need to establish in-service training and to improve the general working conditions in the plant so as to attract and retain labour. Although substantial industrial development is taking place in the region, the inflow of people from outlying areas and other parts of the country should be sufficient to ensure a readily available supply of direct production labour.

32. Since the Company's plant management in the past has shown certain deficiencies, particular emphasis will be laid on broadening the management structure and improving its quality. The number of staff in leading and supervisory functions will be increased about threefold and a number of new departments created. Initial inquiries indicate that additional technical, supervisory and executive staff can be attracted from forest industries in other parts of the country.

B. PROJECT COSTS AND BENEFITS

Costs

33. Needed investment to achieve the project's envisaged processing capacity is shown in Table 4, and phasing of the investment is shown in Table 5. Appendices 1-3 show detailed breakdown of investment costs for the processing component of the project.

34. Estimated operating costs for the processing stage at full development are shown in Table 6 and their phasing up to full development (the fifth year) is shown in Table 7. Appendix 4 shows further details for derivation of operating costs.

Table 4 - Summary of Investment Costs for Processing Stage

(US $1,000)

	Equipment	Buildings	Other Civil Works	Engineering Design[1/]	Freight Insurance[2/]	Installation[3/]	Contingencies[4/]	Total
Existing Plymill	167	-	-	5	25	33	17	247
New Plymill	3,497	1,124	37	140	524	699	466	6,487
Sawmill and Moulding Plant	928	201	28	35	138	185	115	1,630
Veneer Slicing Plant	275	-	-	8	41	55	28	407
Steam Plant	640	60	-	21	96	128	70	1,015
Wood Assembly Equipment[5/]	105	-	-	-	15	-	11	131
Transport Equipment, Plane, Cars	100	-	-	-	15	-	10	125
Sub-Total	5,712	1,385	65	209	854	1,100	717	10,042
Total[6/]	...........7,157.........							

1/ 3% of equipment, buildings and other civil works costs

2/ 15% of equipment cost

3/ 20% of equipment cost including electrical installations

4/ 10% of equipment, buildings and other civil works costs

5/ tugboats, radio communication

6/ details of the totals are shown in Appendices 1-3

Table 5 - Phasing of Investment - Processing Stage

(US $ 1,000)

	Year									
	0	1	2	3/4	5	6-9	10	11	12/13	14
Existing Plymill	247	-	-	-	-	-	-	-	-	-
New Plymill	-	6,487	-	-	-	-	-	-	-	(330)
Sawmill and Moulding	-	1,630	-	-	-	-	-	-	-	(50)
Veneer Slicing Plant	-	-	407	-	-	-	-	-	-	-
Steam Plant	-	1,015	-	-	-	-	-	-	-	(20)
Log Assembly Equipment	131	-	-	-	-	-	131	-	-	-
Transport Equipment	100	25	-	-	80	-	100	25	-	-
Total	478	9,157	407	-	80	-	231	25	-	(400)[1/]

[1/] Approximately 25% of civil works and structures used as salvage value and entered as a benefit/revenue in cash flow tables. See Table 3.

Table 6 - A Summary of Annual Operating Cost at full development processing stage[1/]

	Plymill[2/]	Sawmill/Moulding Slicing Plant	Services[3/]	Total
	(US$ '000)			
Direct production labour	678	333	195	1,206
Supervisoty & Administrative labour	144	160	588	893
Auxiliary materials	1,417	200	83	1,700
Repair and maintenance	144	48	31	222
Electricity	485	193	Negligible	678
Sales cost 5%[4/]	439	202	-	641
Reaforestation cost[4/]	31	31	-	61
Total cost[5/] (excluding logs)	3,338	1,166	897	5,400

[1/] Details provided in Appendix 4

[2/] Including additional costs of existing plymill

[3/] Including additional cost of log assembly under traditional supply system

[4/] Attributed according to sales and intake volumes

[5/] Totals may not add due to rounding

Table 7 - Phasing of Operating Cost During Buildup

(1000 U$)

	Year				
	0	1	2	3	4 2/
Additional Log intake (000 m^3)	4	30	138	239	245
Additional Sales ($ 000)	777	2,221	7,685	12,605	12,828
Direct Production Labor					
- Plymill	-	166.6	500.0	678.3	678.3
- Sawmill	-	83.3	166.6	250	332.8
- Services	-	195.2	195.2	195.2	195.2
Supervisory and Administrative Labor	422.3	666.6	892.5	892.5	892.5
Auxiliary Materials					
- Plymill	125	309.6	896.5	1,416.2	1,416.2
- Sawmill	-	15.3	103.3	190	200
- Services	-	83.3	83.3	83.3	83.3
Repair and Maintenance					
- Existing Plymill	6.2	6.2	6.2	6.2	6.2
- New Plymill	-	137.6	137.6	137.6	137.6
- Sawmill Moulding	-	36.7	36.7	36.7	36.7
- Veneer Slicing	-	-	10.2	10.2	10.2
- Steam Plant	-	24.3	24.3	24.3	24.3
- Log Assembly equipment	3.3	3.3	3.3	3.3	3.3
- Transport Equipment	2.7	3.2	3.2	3.2	3.2
Reforestation	1	5	34.5	59.6	61.3
Electricity					
- Sawmill	-	48	96.2	144.3	192.5
- Plymill	-	121.1	363.6	485	485
Sales Costs	38.8	111	384.3	630.9	646.3
Total cost (excluding logs)	599.3	2,016.8	3,937.8	5,247.0	5,400.6
Service of existing enterprise 1/	(87.5)	(87.5)	(87.5)	(87.5)	(87.5)
Supplementary Cost (rounded)	512	1,929	3,850	5,159	5,313

1/ It represents sunk costs in old mill that can be used in new mill without reducing output of old mill.

2/ Operating costs in years 5-14 are the same as in year 4.

Table 8 - Phasing of Investment and Operating Costs for Logging System

	Year 0	1	2	3	4	5	6	7	8	9	10	11	12	13	14
Physical Flows ($1000\ m^3$)															
Traditional system wood supply ($1000m^3$) 1/	4	30	75	75	75	75	75	75	75	75	75	75	75	75	75
Mechanized logging wood supply	-	-	63	164	170	170	170	170	170	170	170	170	170	170	170
Total	4	30	138	239	245	245	245	245	245	245	245	245	245	245	245
Costs ($US 1000)															
Traditional system 3/	60	450	1125	1125	1125	1125	1125	1125	1125	1125	1125	1125	1125	1125	1125
Mechanized system investment	650 2/	650 2/	1000	-	-	250	250	1000	-	-	250	250	1000	-	-
Mechanized system operation 4/			803	2091	2167	2167	2167	2167	2167	2167	2167	2167	2167	2167	2167
Total	710	1100	2928	3216	3292	3542	3542	4292	3292	3292	3542	3542	4292	3292	3292

1/ Assumed upper limit of $75,000m^3$ additional wood per year.

2/ Pilot operation with no significant production. $400,000 of cost is non-recurring development costs; $250,000 are for equipment with five year life

3/ $\$15/m^3(r)$ delivered

4/ $\$12.75/m^3(r)$ of operating costs in addition to investment costs.

35. With regard to the cost of raw material supplies it is unlikely that, given increased future demand in the region, the cost of logs at factory gate, supplied through the traditional system, will remain at the present level. A long-term cost of logs supplied by the traditional system of US$15/$m^3$ at factory site is the estimate used here. This is almost twice the present cost.

36. The mechanized system of logging proposed in the project and its exact cost will depend on the result of the logging/inventory survey. For purposes of this analysis it has been assumed that the system will require an estimated initial investment of $2.3 million. Of the initial investment, $800 000 represents non-recurring development costs. The remaining $1.5 million has a five-year life. Table 8 summarizes log cost assumptions using a combination of traditional and mechanized logging.

Benefits

37. Combining data about project output over time shown in Table 2 and unit price assumptions as specified earlier, the total value of output can be derived. Estimates of benefits from year 0 up through full development are shown in Table 9.

C. COMPARING COSTS AND BENEFITS

Economic Analysis

38. For the purposes of this evaluation, market values for all inputs and outputs were taken to reflect economic costs and benefits. Thus, no adjustments were made to market prices.

39. It was considered that wood essentially had an opportunity cost of zero on the stump. Even log exporting is not expected to be affected by the project. There are no apparent structural (long term) unemployment problems in the mill area for the types of labour used in the mill, so the wages and salaries paid were taken to represent economic cost as well. In the case of the labour used in logging and initial transport to collection points, the situation is more difficult to analyse. In the absence of further information on this element (which is not large in any case relative to other input costs), it was assumed that the return to this labour also adequately reflected economic costs. With regard to foreign exchange and capital, shadow prices for both should be established on a macro level. Since none were provided to the analysts by the Government, the market exchange rate and the nominal interest rate were used as measures of economic value. The flow of estimated economic costs and benefits over the life of the project (15 years) is shown in Table 10.[1/]

40. Using these value estimates, the economic rate of return (ERR) of the project is 27 percent. Its net present worth (NPW), using a 12 percent discount rate, is US$9 824 000.

41. It is of interest to the Government to know the impact of the project on the balance of payments (i.e., the net foreign exchange effect). This effect is summarized in Tables 11 and 12.

42. Annual net foreign exchange earnings at full project development, taking into account the foreign exchange component of logging costs, would be about $6 million. As indicated in Table 12, the effect will vary from year to year, with a net outflow of some US$ 7.6 million during the first two years.

1/ The costs and benefits for the log export component were not calculated since log exports were not part of the project, i.e. they did not require project investment, and would have taken place with or without the project.

Table 9 - Phasing of Revenues

($US 1,000)

Product	Price (FOB) US $/m^3	Year				
		0	1	2	3	4 and on
Plywood - Export (90%)	170	704	1,744	5,049	7,956	7,956
- Local (10%)	150	69	171	485	780	780
Green Lumber	50	-	115	775	1,425	1,500
Dried Lumber	80	-	80	536	984	1,040
Mouldings	125	-	100	737	1,288	1,350
Sliced Veneer	300	-	-	60	120	150
Total		773	2,210	7,642	12,553	12,776

Table 10 Economic Analysis of Integrated Operation

(US $1000)

		Year														
		0	1	2	3	4	5	6	7	8	9	10	11	12	13	14
	Costs															
1	Processing - investment 1/	478	9157	407	-	-	80	-	-	-	-	231	25	-	-	-
2	Processing - operations 2/	512	1929	3850	5159	5313	5313	5313	5313	5313	5313	5313	5313	5313	5313	5313
3	Logging - investment 3/	783	650	1000	-	-	250	250	1000	-	-	250	250	1000	-	-
4	Logging - operations 4/	60	450	1928	3216	3292	3292	3292	3292	3292	3292	3292	3292	3292	3292	3292
5	Total	1833	12186	7185	8375	8605	8935	8855	9605	8605	8605	9086	8880	9605	8605	8605
	Benefits															
6	Plywood	773	1915	5534	8736	8736	8736	8736	8736	8736	8736	8736	8736	8736	8736	8736
7	Sawnwood/moulding	-	295	2048	3697	3890	3890	3890	3890	3890	3890	3890	3890	3890	3890	3890
8	Veneer	-	-	60	120	150	150	150	150	150	150	150	150	150	150	150
9	Salvage value 5/															400
10	Total	773	2210	7642	12553	12776	12776	12776	12776	12776	12776	12776	12776	12776	12776	13176
11	Net Benefits (Costs)	(1060)	(9976)	457	4178	4171	3841	3921	3171	4171	4171	3690	3896	3171	4171	4571

1/ Details in Table 5 and Appendices 1-3. Includes logging survey cost of $133,000 in year 1.

2/ Details in Table 7 and Appendix 4.

3/ Details in Table 8.

4/ Details in Table 8. Includes cost of logs from traditional supplies.

5/ Assumed at 25 percent of civil works and structure values

Table 11 - Balance of Payment Effects at Full Development

(US $ 1000)

Estimate of foreign exchange components

	Total	Foreign Exchange
Investment[1/]		
existing plymill	247	247
new plymill	6,487	5,176
sawmill and moulding plant	1,630	1,366
veneer slicing plant	407	407
steam plant	1,015	947
transport equipment	256	256
	10,037	8,399
Operating cost at full development[2/]		
direct production labor	1,206	-
supervisory and administrative labor	893	-
auxiliary materials	1,700	1,400
repair and maintenance	222	100
electricity	678	500
sales cost	641	300
reforestation cost	61	-
	5,413	2,300
Logging investment[3/]	2,300	2,070
Logging operating cost at full development[4/]	3,292	1,975
Sales returns[4/]		
plywood	8,736	7,956
lumber	2,540	1,040
mouldings	1,350	1,350
sliced veneer	150	150
	12,776	10,496

1/ See Table 4
2/ See Table 6
3/ See Table 8
4/ See Table 9

Table 12 - Phasing of Balance of Payments Effect

(US $ 1000)

	Years														
	0	1	2	3	4	5	6	7	8	9	10	11	12	13	14
Foreign Exchange Outflow 1/															
Plant - investment	478	7506	407	-	-	80	-	-	-	-	198	-	-	-	-
Plant - operating cost	198	798	1600	2196	2300	2300	2300	2300	2300	2300	2300	2300	2300	2300	2300
Logging - costs	594	715	1903	2090	2139	2302	2302	2789	2139	2139	2302	2302	2789	2139	2139
Sub-total	1270	9019	3910	4286	4439	4682	4602	5089	4439	4439	4800	4602	5089	4439	4439
Foreign Exchange inflow															
Sales	704	1924	6382	10348	10496	10496	10496	10496	10496	10496	10496	10496	10496	10496	10496
New foreign exchange inflow (outflow)	(566)	(7095)	2472	6062	6057	5814	5894	5407	6057	6057	5696	5894	5407	6057	6057

1/ Estimates based on detailed analysis of materials, equipment and service requirements

Financial Analysis

43. The financial analysis of this project was carried out from the point of view of the Company. Basic project costs and benefits are taken to be the same as in the economic analysis above. However several adjustments are needed as follows:

- The proposed loan and terms of the loan must be considered and entered in the cash flow.

- Taxes and fiscal incentives must be considered and included.

44. These two adjustments are summarized in Table 15. Derivations of the two adjustments are presented below.

45. (i) The proposed loan and its terms

The proposed loan would be for US$11.5 million at 10 percent interest, with a grace period for repayment of principal of 3 years after the year of the first disbursement and a repayment period of 5 years. The loan and its repayment is shown in Table 15 (lines 18-20 and 6-8). Derivation of interest and principal payments is shown in Appendix 5.

46. (ii) Taxes and fiscal incentives

Due to tax legislation, the impact of corporate income tax is slight. However, the impacts of fiscal incentives for export are significant. The nature of these impacts is indicated below and their magnitudes are shown in Table 15.

47. Corporate income tax - The present operation is exempt for nine more years. If a new enterprise for the extended operations were to be implemented before one year, a 10-year tax exemption period would apply from the date of establishment. Thereafter, the tax would be 33 percent of net revenue according to present legislation.

48. Industrial production tax (IPT) - This is a tax which must be paid on every industrial product processed by a firm. In the case of export sales a cash bonus of the amount of the industrial production tax is credited to the industry. Since the Company will export more than it will sell in the local market, it will receive a net credit of about US$1.2 million per annum when the project is at full development. (See Table 13). The tax rates are the following:

Sawn timber	3	percent	of	value
Mouldings	10	"	"	"
Plywood	14.5	"	"	"
Veneer	14.5	"	"	"

49. Trade tax (TT) - The trade tax rates are the same as for industrial production tax. In the case of export sales, a bonus of the same amount is credited to the industry. This credit cannot be taken in the form of cash but serves to off-set trade taxes due on local sales. Since the Company exports more than it sells in the local market, it will not have to pay out any trade tax.

50. Rural fund - Two percent of the value of logs at the point of transfer from the producer to the project (i.e. delivered cost excluding shipping cost) is due to be paid into a rural fund for development. An average price of $\$10/m^3$ at the point of takeover is assumed for the purposes of calculation. (See Table 14).

51. Financial Rate of Return (FRR) - The information needed to calculate the FRR is summarized in Table 15. Applying standard discounting procedures to the bottom row of this table (net return (expenditure) after taxes) a FRR of 187 percent is obtained. It should be remembered that this is the FRR from the Company's point of view. It is extremely high, mainly because of the very favourable loan conditions.

Table 13 - Expenditures and Receipts Due to Tax Legislation During Buildup Period

(US $ 1000)

	Year				
	0	1	2	3	4 and on
Local Sales 1/					
Plywood	69	171	485	780	780
Sawnwood	-	115	775	1,425	1,500
Export Sales 1/					
Plywood	704	1,744	5,049	7,956	7,956
Sawnwood	-	80	536	984	1,040
Mouldings	-	100	737	1;288	1,350
Veneer	-	-	60	120	150
Industrial Production and Trade Taxes due 2/					
Plywood (local sales)	10.1	24.8	71.8	113.1	113.1
Sawnwood (local sales)	-	3.4	23.2	42.7	45.0
Total Due	10.1	28.2	95.0	115.8	158.1
Industrial Production and Trade Tax Credits 2/					
Plywood export	102.1	252.9	732.1	1,153.6	1,153.6
Sawnwood export	-	2.4	16.1	29.5	31.2
Mouldings export	-	10.0	73.7	128.8	135.0
Veneer export	-	-	8.7	17.4	21.7
Total received	102.1	265.3	830.6	1,329.3	1,341.5
Net amount received 3/	92.0	237.1	735.6	1,173.5	1,183.4

1/ From Table 9.

2/ Tax rates in percent of value:

Sawnwood	3.0
Plywood and veneer	14.5
Mouldings	10.0

paid on local sales/credited to company on export sales.

3/ Total received minus total due.

Table 14 - "Rural Fund" Payment Due on Logs

	Years				
	0	1	2	3	4 and on
Total logs purchased ($1000m^3$)	4	30	138	238.5	245.0
Value of logs at point of takeover 1/	40	300	1380	2385	2450
Rural fund due 2/	0.8	6.0	27.6	47.7	49.0

1/ Assumed value is $10/m^3 at point of takeover

2/ 2% of value of logs at point of takeover

Table 15 - Financial Analysis of Integrated Operation

(US $ 1000)

		Years															
		0	1	2	3	4	5	6	7	8	9	10	11	12	13	14	Sources/Comment
1	Expenditures																
2	Processing - investment	478	9157	407	-	-	80	-	-	-	-	231	25	-	-	-	Table 10
3	Processing - operations	512	1929	3850	5159	5313	5313	5313	5313	5313	5313	5313	5313	5313	5313	5313	" "
4	Wood supply - investment	783	650	1000	-	-	250	250	1000	-	-	250	250	1000	-	-	" "
5	Wood supply - operations	60	450	1928	3216	3292	3292	3292	3292	3292	3292	3292	3292	3292	3292	3292	" "
6	Loan																
7	interest due - grace period	-	70	1010	1150	-	-	-	-	-	-	-	-	-	-	-	Appendix 5
8	principal and interest repay	-	-	-	-	3034	3034	3034	3034	3034	-	-	-	-	-	-	" "
9	Rural fund payment	1	6	28	48	49	49	49	49	49	49	49	49	49	49	49	Table 14
10																	
11	Total	1834	12262	8223	9573	11688	12018	11938	12688	11688	8654	9135	8929	9654	8654	8654	
12																	
13	Receipts																
14	Sales																
15	Domestic	69	286	1260	2205	2280	2280	2280	2280	2280	2280	2280	2280	2280	2280	2280	Tables 9, 12
16	Export	704	1924	6382	10348	10496	10496	10496	10496	10496	10496	10496	10496	10496	10496	10496	" " "
17	Salvage value	-	-	-	-	-	-	-	-	-	-	-	-	-	-	400	Text
18	Loan																
19	Plant	450	9150	400	-	-	-	-	-	-	-	-	-	-	-	-	Appendix 5
20	Logging	250	250	1000	-	-	-	-	-	-	-	-	-	-	-	-	" "
21	Net IPT/TT benefits[1/]	92	237	736	1173	1183	1183	1183	1183	1183	1183	1183	1183	1183	1183	1183	Table 13
22	Total	1565	11847	9778	13726	13959	13959	13959	13959	13959	13959	13959	13959	13959	13959	13959	
23																	
24	Net Receipts (expenditures)																
25	Before taxes	(269)	(415)	1555	4153	2271	1941	2021	1271	2271	5305	4824	5030	4305	5305	5705	-
26																	
27	Corporate taxes	-	-	-	-	-	-	-	-	-	-	1592	1660	1421	1751	1883	Text
28	Net Receits (expenditures) after																
29	taxes	(269)	(415)	1555	4153	2271	1941	2021	1271	2271	5305	3232	3370	2884	3554	3822	-

1/ Benefits not exempt from corporate taxes.

D. SENSITIVITY ANALYSIS

52. The sensitivity of the measures of project economic worth to changes in assumptions concerning values of a number of key parameters is shown in Table 16.

53. It is evident that only one alternative (No. 4) in Table 16 brings the ERR down below 12 percent. This is the case when it is assumed that all costs are 20 percent higher than the original assumption. If all costs were 15 percent higher (No. 5 in Table 16), the project would just break even at 12 percent.

54. Since prices used to value outputs in the analysis were conservative estimates, the sensitivity of the ERR and NPW were also tested for output price increases. In Case No. 2, operating costs were increased by 4 percent and revenues by 2 percent on a compound basis. As shown, this results in a slight increase in the ERR and NPW over results achieved in the original analysis. Similarly, in Case No. 6 all costs and revenues with the exception of salvage value, were increased by 10 percent over the values of the original assumptions. The result is a slight increase in both ERR and NPW. More optimistic changes (increases) in revenues would raise the ERR above 28 percent and the NPW at 12 percent above US$11 million.

55. There is considerable uncertainty surrounding the costs for wood that will result with the new mechanized logging system. Thus the sensitivity of the project's NPW and ERR were analysed for alternative, higher operating costs for the mechanized system. The results are as follows:

Delivered wood cost (excl. investment) (US$/$m^3$(r))	ERR (percent)	NPW (US$ 1 000)
12.75 (orig. assumption)	27	9 824
25.50 (100% increase)	9	(1 463)
19.13 (50% increase)	19	4 155

56. As indicated, in the unlikely event that operating costs for the mechanized system were twice the expected cost used in the analysis, the ERR would drop to 9 percent and the NPW would become negative at 12 percent. With operating costs one and a half times the original estimate, the ERR would only drop to 19 percent and the NPW would remain positive.

57. In sum, using 12 percent as the relevant discount rate or cutoff rate, the project remains viable under a variety of reasonable potential cost increase assumptions.

58. In connection with the financial sensitivity of the project, the financial cash flow is the same as the economic cash flow, with the exception of the loan, fiscal incentives and tax components. The difference between the ERR and FRR is due entirely to these components. If the loan were halved, the FRR would be 51 percent. If the loan were removed, the FRR would drop to 35 percent. Thus, even if the loan were not secured, the project would appear to be viable for a private enterprise with an alternative rate of return below 35 percent.

Table 16. Sensitivity of Measures of Project Economic Worth to Changes in Assumptions[1/]

No.	New Assumption	New ERR (percent)	New NPW using 12 percent discount rate ($US 1000)
1	operating costs for processing phase increase by 4 percent per year on a compound basis	14	674
2	operating costs and sale revenues increase by 4 and 2 percent per year respectively on a compound basis	28	10,760
3	all costs 10 percent higher than in Table 10	ERR >12%	3,622
4	all costs 20 percent higher than in Table 10	ERR <12%	-2,580
5	all costs 15 percent higher than in Table 10	ERR ≈12%	521
6	all costs and revenues 10 percent higher than in Table 10	ERR >27%	10,797

[1/] Original ERR = 27 percent and NPW = $US 9,824,000. The original value flow data are shown in Table

APPENDIX 1

Present Plywood Plant - Recommended New Equipment

Item		Cost US$
1 Waste hog		21 000
1 Green veneer pneumatic clipper		19 000
1 Torwegge type veneer composer complete with Diehl type jointer and glue applicator		30 000
6 Sets Carboloy sizing saws - 3 Rip 3 Cutoff		3 000
2 8 000 lb loft trucks		25 000
1 Dust extractor system - Cyclone		15 000
1 Plywood pre-press		15 000
1 Dump truck for hog fuel		8 000
2 Lay up hoists for spreaders		6 000
	Total	142 000

Other Installations

Item		Cost US$
Lathe waste conveyor to hog		2 000
Infeed and outfeed to hog		1 000
Hog conveyor to fuel storage bin or boilers		1 500
Direct hook up to clippers		7 500
Rebuild transfer infeed to clippers		2 500
Rebuild and extend green chain offbearing tables		2 500
Build hog fuel storage bin		2 000
Relocation of sizing saws		3 000
Plywood pre-press transfer rolls		3 000
	Total	25 000
	Grand Total	167 000

APPENDIX 2

New Plymill and Veneer Slicing Operations

Plymill - Plant Type, Capacity and Investment Cost

1. The proposed plywood plant would have a productive capacity of 45 000 m^3 of finished plywood per annum on the basis of two ten hour shifts, operating 288 days per annum.

2. The plant is designed to manufacture phenolic and urea resin bonded plywood from four principal species.

3. Increasingly, other species can be utilized with improved log supply.

4. Recovery is based on 40% of cubic input to finished plywood. This is low by international standards but has been taken as a conservative estimate in the light of technical skills at present available.

5. The plywood would have a main building area of 11 - 12 000 m^2. The buildings would be of steel frame and other materials not subject to damage by insect or climatic conditions in the area, and have reinforced concrete floors. Foundations for the major equipment would be on piling.

6. Peeler log diameters would range from 45 cm to 110 cm.

7. Veneer will be peeled on high production lathes fully equipped with dual spindles, block charger and back pressure roll attachment. Veneer from the lathe will go direct to the veneer clippers where usable widths and grades would be sized for further processing.

8. The green veneer would be dried in steam heated veneer dryers; moisture meters would be installed at the outfeed end of the dryers to ensure maximum production and proper moisture content for quality of bond in later process.

9. The veneer preparation department would consist of dry veneer recovery clippers, manually operated, veneer jointers and veneer taping machines to process the random width veneers to full sheet sizes for the gluing operations.

10. Two glue spreaders for each line would be provided for the lay up of veneer prior to pre-pressing and hot pressing in 5 ft by 8 ft 20-opening hot presses, equipped with press loader and unloader. Glue mixing facilities and storage for adhesives and components would be adjacent to the main building at ground level, with the mixed glue being pumped to the spreaders. The finishing department would consist mainly of a panel sizing unit, putty patching facilities, wide belt speed sanders, crating, steel strapping, storage and shipping accommodation. Refuse conveyors would be arranged for the disposal of waste materials and a dust collection system would be installed. Peeler cores would not be fired in the boilers as presently but sawn to produce packaging crates or low grade lumber for local sales. Storage areas for green veneer, dry veneer and finished plywood would be provided. The movement of materials within the plant would be by lift trucks and dollies.

11. Total installed power would be about 4 000 HP.

12. The loading or shipping area would be in line with the storage area for finished products.

13. At this stage the equipment is listed only in such detail that the major equipment items, capital cost and manufacturing cost estimates can be outlined.

14. Detailed equipment comparisons and design development work will be required to make final decisions on such items as plant layout and equipment selection and have been accounted for in the investment cost estimate.

15. Investment costs of new plymill are shown in Appendix 2, Table 1. Table 2 shows equipment needs and capital cost estimates for the veneer slicing operation.

APPENDIX 2

Table 1 - Investment Cost - New Plymill

		US$
A.	**Equipment**	
	Debarker	90 000
	Blocksaw	25 000
	Front end loader	135 000
	Bark hog	17 000
	Lathe chargers	195 000
	8' veneer lathes	435 000
	Green veneer clippers	54 000
	Green veneer core saws	30 000
	Veneer waste hog	15 000
	Veneer driers	525 000
	Veneer moisture meters	33 000
	Veneer jointers	90 000
	Veneer tapers	36 000
	Torwegge veneer composer	45 000
	Glue spreaders	60 000
	Lay up scissor hoists	18 000
	Plywood pre-presses	45 000
	20-opening hot presses	375 000
	Hot press loader and unloaders	330 000
	Glue mixers, tanks, scales etc.	30 000
	Sizing and edging saws	105 000
	Sub-total	2 688 000
	2 Drum (1 top and 1 bottom) wide belt speed sanders	150 000
	Cyclone dust collection system	35 000
	Air compressors	45 000
	Veneer carts	9 000
	Waste conveyors, waste hog machines	200 000
	Fork lifts	100 000
	Strapping tools	1 000
	Maintenance equipment	25 000
	Sharpening and grinding equipment	20 000
	Spare parts	75 000
	Total	3 347 000
B.	Buildings including floor, lighting, sprinklers, 11 240 m^2 at 100 $\$/m^2$	1 124 000
C.	Logyard 5 000 m^2	17 000
	Hoists, log-lifting equipment	150 000
	Log sorting	20 000
	Total investment (excluding freight, installation, etc.)	4 658 000

APPENDIX 2

Table 2 - New Veneer Slicing Operation

Major Equipment and Capital Cost Estimate

Item		Cost Installed $ US
1	Overhead hoist and monorail	3 000
	Motors, starters, wiring	4 000
1	8 ft vertical type of veneer slicer complete with carriage and controls	85 000
1	Single line endless bed veneer drier	100 000
1	Veneer guillotine jointer	30 000
2	Diehl veneer tapers complete with motors and controls	25 000
1	8 000 lb lift truck	10 000
25	Veneer dollies for transport of veneer within the plant	3 000
2	Sunken steam or hot water vats	10 000
1	20 litre glue mixer and other miscellaneous supplies	5 000
	Total Slicer Operation - Equipment	275 000

APPENDIX 3

Sawmill and Moulding Plant

A. INTRODUCTION

1. The general concept in the design of the sawmill is to provide equipment and a process flow which is to be operated and maintained according to the required high production capacity and to the given special conditions of the raw material and the technological level of the available manpower.

2. The raw material consists of logs of different tropical species (up to 30 and more) with a density range from 400 kg/m^3 to 800 kg/m^3. The diameters vary from 40 cm to about 1.5 m with an average of about 60 cm.

3. The presently available labour force and supervisory staff shows a medium level technological expertise but it seems to be able to advance with the future requirements of the production.

4. Accordingly a band sawmill has been selected having an intermediate level of mechanization but which can be more highly mechanized at a larger stage.

B. SAWMILL AND MOULDING PLANT: OUTLINE OF PRODUCTION PROCESS

5. (i) Basic technical data:

Input	120 000 m^3 logs/yr	
Operating period	2 x 10 hr. shifts/day	
	288 days/yr	
Average log size	- diameter	: 0.43 m
	- length	: 4.00 m
	- volume	: 0.57 m^3
Recovery	sawn lumber	: 55%
	all products	: 45%
Output products	30 000 m^3 green lumber	
	13 000 m^3 dried lumber	
	8 640 m^3 net, moulding for panelling walls and ceilings	
	2 160 m^3 net other mouldings	
Total products for sale	53 800 m^3 = 45% net recovery	
Waste	62 800 m^3 (600 kg/m^3 density absolutely dry)	
	32 680 tons absolutely dry	
	All products	: 53 800 m^3
	Waste	: 62 800 m^3
	Shrinkage lost	: 3 400 m^3
	Total	120 000 m^3

6. (ii) Primary log breakdown:

Two log bandsaws with carriages of heavy duty, mechanised with hydraulic clips, log turner ("nigger"), log loader ("flipper") and reception arms.

The log-bandsaws and carriages would remove the outside slabs and cut the centre of the logs into cants and side pieces.

In case of a breakdown at the secondary equipment (re-saws) the log-bandsaws would do more cutting; they would also produce lumber for edging.

7. (iii) Secondary breakdown equipment:

Following the log bandsaws and carriages, the secondary equipment would consist of 3 band re-saws for re-sawing the cant to raw boards for mouldings and 2 twin-edgers for edging the side pieces and the lumber cut for sale.

An undertable crosscut saw should trim the side pieces if it is necessary.

8. (iv) Cutting procedures:

The 3 re-saws should be interconnected by transport facilities in a way that the cants, not having been cut completely in the first passing, can return to the second or to the first re-saw (merry-go-round). The transport of cants, slabs and boards has to be mechanized with longitudinal and transversal conveyors.

9. (v) Impregnation, trimming and grading of green lumber for sale:

At the end of the cross transfer green chain, the green lumber for sale should be classified, impregnated, trimmed and marked.

10. (vi) Drying:

In the pre-drier the lumber should be dried from green to a uniform moisture content of about 30%. In this manner the kiln-drying equipment could be used more effectively and overall costs would be reduced as compared with a system of exclusive kiln drying.

Two kiln-driers should establish the final moisture content of the dried product at about 12%.

At the lumber yard packages sorted by species should be collected to complete loadings for the pre-drier. All transport should be made with fork-lift trucks.

11. (vii) Moulding plant:

The dried lumber should be moulded, trimmed, bundled and packed. The dried lumber for sale should be trimmed and classified in the same plant.

Major Equipment Items

12. **Sawmill**

		HP	Investment Cost
(1) Cutting equipment			
(a)	2 log bandsaws 20 m/min. each 65 ft/min. each	400	
(b)	3 re-saws 22 m/min. each 72 ft/min. each	300	
(c)	2 twin-edgers 34 m/min. each 110 ft/min. each	100	
(d)	1 undertable cross-cut saw	5	
(e)	1 three-saw trimmer	10	
(2) Conveyor equipment			
(a)	1 log-cutter	15	
(b)	1 log-haul	10	
(c)	3 log-kickers	6	
(d)	2 live log-decks	10	
(e)	153 m. roller conveyor)	100	
(f)	24 m. re-saw feeding c.)		
(g)	54 m. green chain transfer	40	US$ 480 000
(3) Other sawmill equipment			
(a)	saw-dust conveyor)		
(b)	waster transfer)		
(c)	waste hog)	175	
(d)	compressor, etc.)		
(e)	saw sharpening equipment)		
			US$ 115 000
(4) Internal transport			
2	front-fork lift trucks 3 tons capacity each		US$ 25 000
			US$ 620 000

		HP	Investment Cost
13.	**Drying equipment**		
	(1) 1 preseasoner $9 \times 2 \times 40\ m^3 = 375\ m^3$ Board cap.	50	US$ 60 000
	(2) 2 drying kilns $7 \times 4.5 \times 20\ m^3$ c.o. = $300\ m^3$ each	150	US$ 200 000
14.	**Moulding equipment**		
	(1) 3 six cutter planer and moulder 15 m/min. each 50 ft/min. each	135	US$ 35 000
	(2) 1 four cutter planer and moulder 11 m/min. 35 ft/min.	16	US$ 7 000
	(3) 1 three-saw trimmer	15	US$ 5 000
	(4) bundling equipment		US$ 1 000
	Total investment for equipment and machinery	1 552 HP	US$ 928 000

Electricity consumption : 1 552 HP = 1 130 kw = 1 614 KVA x 20 hr x 288 days x 0.8 (utilization factor) = 7 436 800 Kwh/year

APPENDIX 4

Estimated Operating Costs - Processing Stage

Estimate of cost of auxiliary materials at full project development, including cost of glues in plywood production

		US$ p.a.
(A) Plymill		
(a) Glues		987 786.60
(b) Other auxiliary material (US$8.30/$m^3$ produced)		433 333.30
	Total	1 421 119.90
	Rounded	1 420 000
(B) Sawmill/Moulding/Slicing Plant		
(a) Knives, chemicals, etc.		
US$ 2 m^3 sawn wood		132 000
US$ 1 m^3 mouldings		10 800
(b) Packaging material		
3% of f.o.b. value of mouldings and sliced veneer		45 000
	Total	187 800
	Rounded	190 000
(C) Services		
- Mainly hooks, cables US$ 0.17/m^3		40 833.30
- Fuel for boats (additional)		22 000.00
- Other		16 666.60
	Total	79 499.90
	Rounded	80 000

Estimate of Costs of Glues in Plywood Production

Ingredient	Quantity Consumed per m^3 of Plywood (kg)	Price (US$)	Total Cost (per m^3)
A. Phenolic Glues			
Phenol Resin	37.0	0.4	14.8
Hardener HP4	15.0	0.215	3.225
Albapin	1.5	1.06	1.59
Flour	1.5	0.206	0.309
		Total	19.924
		Rounded	20

Ingredient	Quantity Consumed per m3 of Plywood (kg)	Price (US$)	Total Cost (per m^3)
B. Urea Formol Glues			
Urea Resin	27.0	0.235	6.345
Hardener FH-40D	1.0	0.315	0.315
Flour	16.0	0.206	3.296
Emulsion TF	3.0	2.10	6.3
		Total	16.256

Total additional cost of glues at full development:

40 000 m^3 phenolic	x	20	=	800 000
12 000 m^3 urea formol	x	16.25	=	195 000
				US$995 000

Cost of the Reafforestation Programme

Year	1	2	3	4	5
Logs utilized ('000 m^3)	4	20	138	239	245
Annual cost of reafforestation (US$ 1 000)	1	5	34.5	60	61.3

Estimate of Electricity Costs

	Sawmill/ Moulding	Plymill	Total
Installed power (hp)	1 600	4 000	5 600
Installed power (kw)	1 176	2 940	4 116
Installed power (kva)	1 680	4 200	5 880
Operating Hours p.a.	* * * * * * * * * *	5 760	* * * * * * * * * *
Utilization factor of installed electrical equipment	-	0.8	-
KWH consumed p.a.	7.7 million	19.4 million	27.1 million
Cost per KWH	-	$0.25	-
Electricity cost p.a.	US$ 192 500	US$ 485 000	US$ 677 500

Estimate of Management and Labour Cost

		Per Month (US$)	Per Annum (US$)
1	Managing Director	1 666.60	20 000
4	Directors	133.30	64 000
14	Managers	833.30	140 000
1	Chief Accountant	833.30	10 000
1	Company Doctor	833.30	10 000
2	Foresters	833.30	20 000
	Total		264 000
	Social Charges 60%		158 400
			422 400
24	Superintendents, supervisors	300	
2	Engineers	300	
1	Saw Doctor	300	97 166.60
1	Staff Nurse	166.60	
12	Buyers	166.60	
5	Lagoon Foremen	166.60	
19	Plymill Foremen	166.60	
6	Sawmilling/Moulding Foremen	166.60	
1	Slicer Plant Foreman	166.60	
8	Maintenance Foremen	166.60	
30	Secretarial Staff, Clerks, Drivers, etc.	166.60	169 000.00
	Total		261 166.60
	Social Charges 80%		208 933.30
			470 099.90
	Workers		
182	Sawmilling/Moulding	66.60	
26	Veneer Slicing	66.60	
20	Lagoon	66.60	
424	Plymill	66.60	
50	Boat Crews	66.60	
20	Power/Boiler	66.60	
12	Maintenance Sawmill	66.60	
20	Maintenance Plymill	66.60	
	Total		603 200
	Social Charges 100%		603 200
			1 206 400
	Total Labour Cost	US$	2 098 899
	Expatriate Staff during Running-in Period (1 year)		
1	Technical Director Plymill	5 000	60 000
1	Technical Director Sawmill	5 000	60 000
1	Production Manager Plymill	4 000	48 000
			168 000

Estimate of Repair and Maintenance Cost

(Supplementary)

	Civil Works and Buildings[1/] (US$ '000)	Equipment[2/] (US$ '000)	Total (US$ '000)
Existing Plymill (new investment)	-	6.2	6.2
New Plymill	13.8	123.9	137.7
Sawmill and Moulding Plant	2.6	34.2	36.8
Veneer Slicing Plant	-	10.2	10.2
Steam Plant	0.7	23.7	24.4
Transport Equipment	-	3.1	3.1
Wood Assembly Equipment	-	3.3	3.3
Total	17.1	204.6	221.7

1/ 1% of investment cost p.a.

2/ 2.5% of investment cost p.a.

APPENDIX 5

Loan: Derivation of Interest and Principal Repayment

Conditions: 3-year grace period on principal repayment and 10% interest with full repayment by year 9.

Year	New Loan Given (beginning year)	End of Year Total Loan Outstanding	Beginning year payments Interest	Principal	Total
		(US$ 1 000)			
1	700	700	-	-	-
2	9 400	10 100	70	-	70
3	1 400	11 500	1 010	-	1 010
4	-	11 500	1 150	-	1 150
5	-	9 616	1 150	1 884	3 034
6	-	7 544	962	2 072	3 034
7	-	5 264	754	2 280	3 034
8	-	2 757	526	2 507	3 034
9	-	-	276	2 757	3 034

Note: Repayment of principal (and interest) for the last five years could also bave been obtained by using the capital recovery multiplier formula,

$$\left[\frac{i(1+i)^n}{(1+i)^n - 1}\right]$$

, where i = 10% and n = 5 years. Looking this factor up in standard tables and multiplying it (0.2638) times the principal outstanding ($11 500 000) we arrive at the annual payment of $3 034. Note also that interest is not paid until the year after the loan is received (see years 1-3).

ECONOMIC ANALYSIS OF FORESTRY PROJECTS: CASE STUDIES

FAO FORESTRY PAPER 17, SUPPLEMENT 1

CASE STUDY NO. 4

KENYA I SAWLOG AND PULPWOOD PLANTATION PROJECT

FOOD AND AGRICULTURE ORGANIZATION OF THE UNITED NATIONS

Rome, 1979

Preface

This is one of a series of case studies of forestry and forest industries projects that has been prepared by the Food and Agriculture Organization of the United Nations (FAO) in order to demonstrate methods of identifying, preparing and appraising projects in the forest sector. The programme of case studies has been made possible by a special allocation to FAO for this purpose from the Swedish International Development Authority (SIDA). This case study has been compiled by FAO with the assistance of the College of Forestry at the University of Minnesota.

The project reported on in this case study is a project of the Government of Kenya which has been supported by a loan from the World Bank. The case study is based largely on material contained in an appraisal of the project carried out by the World Bank (Report No. 706-KE). Additional information was gathered through a visit to the project. The case study describes the project as it was perceived at the time of appraisal in 1974. This material has been released to FAO by the World Bank and the Government of Kenya exclusively for the purpose of teaching forestry planning methods. It may not be quoted or reproduced in part or in whole without permission.

ABBREVIATIONS

ERR	=	Economic rate of return
F.O.B.	=	Free on board
FRR	=	Financial rate of return

CURRENCY EQUIVALENTS

Currency Unit	=	Kenya Shilling (KSh)
US$ 1	=	KSh 7.14
KSh 1	=	US$ 0.14
K£ 1	=	US$ 2.80
	=	KSh 20

TABLE OF CONTENTS

INTRODUCTION

1. This case study is based upon a project which is part of Kenya's continuing industrial sawlog and pulpwood production programme[1]. It is a "time slice" project covering plantation activities during the years 1970-1975. A continuation project, started in 1975, is covered in a separate case study[2].

2. The project involves the establishment and maintenance of about 19 400 ha of sawlog plantations and 9 350 ha of pulpwood plantations located in the Kenyan highlands. The pulpwood plantations will be established near the proposed site of Kenya's first pulpmill. The mill is expected to be built during the first years of the plantation project and will have an annual output of 50 000 metric tons, almost all of which will be consumed domestically. Local sawmills are expected to process all of the sawlog output. Half of the sawnwood produced is expected to be consumed domestically and half exported, mostly to the Middle East and Europe.

3. In addition to plantations, the project includes roads, buildings, vehicles and equipment for Kenya's continuing plantation programme. The project expenditure over the six-year period of plantation establishment totals approximately US$ 4.0 million.

4. The case study is divided into five sections. Section A provides background to the project. Section B gives the specific technical details of the project, including estimates of input requirements and output levels. Section C analyses project costs, while Section D discusses project benefits. Section E provides a comparison of costs and benefits and looks at the sensitivity of the economic analysis and the economic rate of return (ERR) to changes in assumptions concerning input and output values.

5. The present case study raises many of the issues commonly associated with appraising industrial afforestation projects. These issues include a number of shadow pricing points, relating to the value of labour, forest land and the project outputs. It also provides background for discussion of a number of forest management issues (including questions of thinning regimes and separate or joint production of sawlogs and pulpwood)[3]. Finally, in combination with the case study of the followup project, it provides useful insights into how information availability and thinking about a major plantation programme evolved over time.

A. BACKGROUND

6. Kenya's population was estimated to be 9.9 million in 1967. The northern and north-eastern parts of the country, comprising three-fifths of the whole, are arid. Northwest from a low-lying tropical coastal plain, the country rises to the Kenya Highlands - a block of alternating plateaux and mountain ranges at elevations of up to 4 000 metres. The principal forest areas are located in these areas of high elevation and favourable rainfall.

7. Forestry and forest industries account currently for less than 2 percent of GDP. Forestry employs between 8 000 and 9 000 permanent workers and forest industries a further 1 500. Kenya's indigenous forest resources cover some 17 275 km^2 over approximately 3 percent of the country's total land area. Over the last 50 years, these forests have been systematically exploited, mainly for the production of sawn timber. The economically accessible reserves of the main commercially valuable species (*Podocarpus gracilior*, *Podocarpus milanjianus*, *Juniperus procera*, *Ocotea* spp., *Olea* spp.) have been largely cut out.

1/ Policies and practices described in this case study are not necessarily those actually in force in Kenya, either at present or during the period covered by the project.

2/ See Case Study No. 5 "Kenya II Sawlog and Pulpwood Plantation Project".

3/ A feature of this particular case is that the outlets for its products were well defined, as were nearly all questions about species choice and treatment. For an industrial afforestation project where these aspects presented a much greater degree of uncertainty, see Case Study No. 6 "Zambia Industrial Forestry Project".

8. Regeneration with indigenous species is difficult and growth rates for these species are low. Conversely, regeneration with exotic, fast-growing pine, cypress and eucalypt species has proved very successful. Favourable climatic conditions in the Highlands - a rainfall satisfactory in both amount and distribution, and the lack of any extremes of temperature to cause an annual period of dormancy or slow growth - coupled with fertile deep volcanic soils, make the area comparable with the world's best softwood growing areas. Consequently, since 1945, the Forest Department has been engaged in a systematic reafforestation programme on the cut-over indigenous forest areas using these exotic species. By 1968 nearly 80 000 ha of pine and cypress and 8 000 ha of eucalypt had been planted. The plantations have recently begun to yield substantial volumes of timber (164 000 m^3 of roundwood in 1967 compared to 42 000 m^3 in 1958).

9. Headed by a Chief Conservator of Forests, the Forest Department consists of 35 senior staff (Graduate Forest Officers), 104 intermediate staff (Senior Foresters and Foresters), 142 Rangers and 897 Forest Guards. The total permanent labour force is about 8 200, including miscellaneous staff such as drivers and clerical staff.

10. The Government's 1965-70 Development Plan provided for a continuation of the present planting programme at a yearly rate of about 6 000 ha of sawlog and pulpwood plantations. This would entail Government investment until about 1980; thereafter, the softwood programme should be self-supporting financially. The planting programme is spread over more than 40 Forest Districts in 11 geographic Forest Divisions.

11. On the industrial utilization side the majority of sawmills require modernization, although some improvement for selected mills and associated logging operations has been made in recent years. A factory to produce low-cost prefabricated wooden houses has begun production at Nakuru. An expansion of this activity would foster sawnwood sales. Until 1968, Kenya imported all her plywood and veneer requirements from Uganda, Tanzania and overseas. (See Net Trade Balance in Forest Products, Appendix 1). A new plywood mill at Elburgon in Kenya, using coniferous plantation wood, commenced production in 1968, and another mill at Elgeyo will start operation in early 1969.

12. All Kenya's pulp and paper consumption is presently based on imports. Plans are now under consideration for a pulp and paper mill to be situated on the Nzoia River at Webuye (Broderick Falls) in the Western Province, 300 km west of Nairobi (see map, Appendix 2). The mill would have two paper machines with total rated capacity of 45 000 metric tons of paper per annum. A first expansion to 75 000 metric tons is anticipated by about 1977, but will be dependent upon development of the market. The mill would produce both bleached and unbleached paper products including kraft paper, paperboard, carton board, and uncoated printing and writing paper, but not newsprint.

13. Under current plans, the mill would be commissioned in early 1972 and would be operating at full capacity by 1974/75. The expansion to the 75 000 ton capacity in 1977 would require temporary exports of limited quantities, but by 1982 domestic demand would again catch up with the mill's output. The pulpwood requirements would be of the order of 400 000 m^3 annually by 1982/83, and by that time the first deliveries from the proposed pulpwood plantings could be expected. Until pulpwood plantings are ready for cutting, the mill would obtain its pulpwood requirements from plantations originally established for sawnwood and situated within distances of 100-150 km from the mill.

14. An existing import duty of 15 percent on industrial paper is waived for most imports at the present time. However, to encourage the new mill, the Government of Kenya has agreed to impose a 40 percent ad valorem tariff on all paper products and pulp imports. A financial return of about 19 percent is estimated for the proposed mill, based on estimated prices taking the tariff into account.

B. DESCRIPTION OF THE PLANTATION PROJECT

15. The project consists of the establishment and maintenance of 19 425 ha of sawlog plantations and 9 350 ha of pulpwood plantations. The sawnwood plantations would be located near the road and rail axis of Nairobi-Eldoret-Uganda, (see map, Appendix 2) and would constitute about two-thirds of Kenya's total sawlog plantings during the period 1970-75. The pulpwood plantations would be located close to the proposed Webuye pulpmill. Project plantations would be established in the six years 1970-1975 and output would be used to meet domestic demand for sawn timber and plywood and as raw material for the proposed Webuye pulp and paper mill.

Sawnwood Plantations

16. Cypress (Cupressus lusitanica) and pine (Pinus patula) would be the two principal sawlog species planted. Both would be grown on a 30-year rotation. Cypress has given uniformly successful results in the project areas and its timber is of excellent quality both as sawnwood and as plywood peeler logs. The spacing used would correspond to a stand of 1 680 trees per ha. The average yield of commercial timber for this species in the project area is estimated to be 560 cubic metres per ha over a 30 year rotation. This is made up of approximately 70 cubic metres per ha for each of two thinnings at 15 and 20 years, and about 420 cubic metres per ha when the trees are clear-felled at 30 years (see Table 1 for total sawlog yields). While pine yields are anticipated as being higher, sound and conservative management dictates that within the six years of the planting programme a balance should be maintained between plantings of the well-tried cypress and the more newly-introduced pine. In addition, the plantings of two species will reduce the risk of extensive damage caused to the plantations by possible spreading disease. It is estimated that about 80 percent of sawnwood plantings would be of cypress. Sufficient seeds are available from local seed orchards, and seedlings would be raised at District nurseries. Planting would be made after clearing of indigenous forests, at the beginning of the long rains (March to June). Plantation maintenance would include weeding, pruning and thinning.

17. In all sawlog plantation operations, the "taungya" or "shamba" system would be used. Under this system, after the removal of commercial species of timber, a plot of indigenous forest is temporarily handed over to a resident worker. The resident worker receives a monthly wage, now 50 shs, which is paid only during nine months of each year during which time he works for the Department to do plantation work including nursery work, weeding, pruning, housing and road construction. During the three months in which he is not paid, he is required to clear his plot on which he is then free to grow agricultural crops during that and the following two years. The crops he produces are his recompense for clearing the area. In the third year, the agricultural crops are interplanted with trees.

Pulpwood Plantations

18. Some 3 000 ha of the pulpwood circle have been planted in the last three years at Turbo, near the Webuye pulpmill site. During the project planting period, 1970-1975, the planting rate would reach about 1 600 ha annually, and a total of 9 350 ha would be planted, giving about 12 300 ha of planted pulpwood by 1975 (see Table 5). A total of about 18 500 ha of new plantations would be required to supply the proposed mill on a sustained production basis.

19. In addition to the 3 000 ha already planted about 7 000 plantable hectares of Government owned land are available for use by the Forest Department at Turbo. Some additional 2 350 ha are thus required by the Forest Department for pulpwood plantings during the first five years of the project. The Government foresees no problems in obtaining this area of land by public acquisition and intends to have acquired this land by 31 December 1971.

Table 1. Planting - Harvesting Schedule for Sawlogs

Year Planted	Hectares Planted	Year of Thinnings: 1984/85	1985/86	1986/87	1987/88	1988/89	1989/90	1990/91	1991/92	1992/93	1993/94	1994/95	...	Year Harvested: 1999/00	2000/01	2001/02	2002/03	2003/04	2004/05
1969/70	3,238	3,238	-	-	-	-	3,238	-	-	-	-	-	...	3,238	-	-	-	-	-
1970/71	3,238	-	3,238	-	-	-	-	3,238	-	-	-	-	...	-	3,238	-	-	-	-
1971/72	3,238	-	-	3,238	-	-	-	-	3,238	-	-	-	...	-	-	3,238	-	-	-
1972/73	3,238	-	-	-	3,238	-	-	-	-	3,238	-	-	...	-	-	-	3,238	-	-
1973/74	3,238	-	-	-	-	3,238	-	-	-	-	3,238	-	...	-	-	-	-	3,238	-
1974/75	3,238	-	-	-	-	-	3,238	-	-	-	-	3,238	...	-	-	-	-	-	3,238
Total Hectares	19,425.0	3,238	3,238	3,238	3,238	3,238	6,475.0	3,238	3,238	3,238	3,238	3,238		3,238	3,238	3,238	3,238	3,238	3,238
Total Volume (thousand m^3)		226.5	226.5	226.5	226.5	226.5	453.0	226.5	226.5	226.5	226.5	226.5		1,359	1,359	1,359	1,359	1,359	1,359

Table 2. Planting - Harvesting Schedule for Pulpwood

Year Planted	Year Harvested: 1983/84	1984/85	1985/86	1986/87	1987/88	1988/89	1989/90	Total
				hectares				
1969/70	1,210	40	-	-	-	-	-	1,250
1970/71	-	1,340	280	-	-	-	-	1,620
1971/72	-	-	1,220	400	-	-	-	1,620
1972/73	-	-	-	1,090	530	-	-	1,620
1973/74	-	-	-	-	950	670	-	1,620
1974/75	-	-	-	-	-	790	830	1,620
Total Hectares	1,210	1,380	1,500	1,490	1,480	1,460	830	9,350
(Age 14)	1,210	1,340	1,220	1,090	950	790	-	6,600
(Age 15)	-	40	280	400	530	670	830	2,750
Total Volume (thousand m^3)	313.1	359.3	409.9	410.2	410.8	410.5	250.2	2,564.0

20. The areas for planting with pulpwood would be flat and free of forest. The main species planted would be Pinus patula, but during 1970/71-1974/75, about 120 ha annually would be planted with Pinus radiata. P. radiata is higher yielding than P. patula, but is susceptible to Dothistroma Blight (a fungus disease of the needless). More experimentation is needed on the control of this disease before the Forest Department can contemplate large P. radiata plantings. Plantings would be contiguous to facilitate aerial spraying. Pulpwood plantings would be at a density of 1 330 trees per ha and would be grown on a 15-year rotation. Yields at 15 years would be about 290 and 400 cubic metres per ha respectively for P. patula and P. radiata (see Table 2 for expected total pulpwood yields). No thinnings are contemplated.

Future Supply-Demand of Wood Raw Material

21. Projections of requirements for sawnwood from project plantations are developed in Appendix 3. The forecast of sawnwood and plywood requirements equivalent to 2.4 million m3 of roundwood in the year 2000 has been based on an estimate that domestic demand would reach about 1.2 million m3 and on an assumption that a further 1.2 million m3 could be exported mainly to other East African countries and the Near East region. Total requirements would increase yearly five times from the 1967 level of about 0.5 million m3 of production. The expected production from project and earlier plantings, supplemented by thinnings from later plantings and by exploitation of indigenous forests, would meet the estimated sawnwood requirements.

22. The project output of about 400 000 m3 annually of pulpwood would meet the anticipated requirements of the Webuye pulpmill. Since no alternative source of raw material exists for the pulpmill within an economic distance for their transport to Webuye, apart from limited supplies of sawnwood residues and smallwood, it may be confidently assumed that project pulpwood would be used by the proposed mill.

23. In the unlikely event that the pulpmill proposal should fail to materialize, it would be possible to maintain the trees under a longer rotation for use as sawlogs. If this should occur, however, a moderate surplus of sawnwood would probably be created, resulting in some sales being delayed. As a consequence, the return on investment in any pulpwood plantings used in this way would be lower than on regular sawnwood plantings. (See sensitivity analysis, Section E).

C. PROJECT COSTS

24. Project economic costs are assumed to equal the financial costs. Individual cost items are explained in Table 3. The cost estimates are based on Forest Department experience with appropriate modifications in light of experience of plantation developments in other countries.

25. Table 4 describes labour requirement per hectare needed to carry out silvicultural activities. These requirements multiplied by the number of hectares associated with each silvicultural activity (Table 5 and Table 7 for sawlog and pulpwood plantations respectively)

equal total labour requirements. The latter are found in Table 6 (sawlogs) and Table 8 (pulpwood). In order to arrive at total labour costs, labour requirements are then multiplied by the daily labour cost (Table 9) 1/. The schedule of all costs is found in the relevant financial cash flow tables for the sawlog and pulpwood components (Table 10 or Table 12). The cost of the land was assumed to be zero.

26. As it is difficult to establish costs under the present accounting systems, a physical contingency allowance of 15% of total costs has been included in the financial (economic) analysis.

D. PROJECT BENEFITS

Financial Revenue

27. Financial revenue equals the price the Government of Kenya expects to receive for a unit of stumpage multiplied by the volume thinned/harvested. For sawlogs the Ministry of Natural Resources plans to raise stumpage prices by 1975 to KSh 17.66, KSh21.19, and KSh 26.49 per cubic metre measured under bark for the first and second commercial thinnings and final harvest, respectively. These rates are used in the project revenue calculations. Sawlog thinnings occur during years 17 to 27, and final harvests occur during years 32 to 37 (see Table 1). For pulpwood there would be no thinnings, and harvests would occur during years 15 to 21 (see Table 2). Project revenue calculations assume that the pulpwood stumpage price is KSh 21.19 per cubic meter measured under bark. This price has been calculated back from the price that the pulpmill could afford to pay (KSh 106) and at the same time provide a financial return of 19 percent, which is considered sufficient to attract investors.

Economic Benefits

28. For pulpwood the stumpage price of KSh 21.19 per m3 is assumed to be the economic price as well as the financial price, and therefore pulpwood economic benefits are the same as the financial benefits.

29. For sawlogs the output is shadow priced at a higher value. It is assumed that half of the sawlog output will be consumed domestically and substitute for imports, and half will be exported as sawnwood. Both halves of the output are shadow priced based upon the savings/earnings generated, as explained below.

30. The sawnwood value of the 50% of the Kenyan sawlog output which is assumed to substitute for imports is valued at the approximate world F.O.B. sawnwood price of KSh 243.71/m3 (s) in the United Kingdom plus KSh 95.36 for transportation from the United Kingdom to Nairobi, minus KSh 158.94 for the cost of converting standing Kenyan timber to sawnwood delivered to Nairobi, which results in a stumpage price (sawnwood measure) of KSh 180.13/m3. The 50% of the sawlog output which is exported is valued at the approximate world sawnwood price of KSh 247.24/m3 sawn F.O.B. at point of export from Kenya which after netting out KSh 194.26 for logging, sawmilling, and freight to the export point leaves a residual stumpage price (sawnwood measure) of KSh 52.98/m3. An average stumpage price (sawnwood measure) is arrived at as the mean of the import substitution and export values (KSh 110.56). Stumpage prices in roundwood measure are then computed by applying conversion factors (to convert sawnwood measure to roundwood measure) of 3.0, 2.5, and 2.0 for the first thinning, the second thinning, and the final harvest respectively. The average unit economic benefit values calculated in this way are KSh 38.85 and KSh 46.62

1/ In the case of nursery cost only, there is an additional non-labour cost per hectare which must be multiplied by the associated hectares and the result added to the labour cost to get the total nursery cost.

Table 3. Explanation of Costs for Kenya I Analysis

Cost Item	Sawlogs	Pulpwood
Nursery	Labor, plus 10Ksh per ha for fertilizer tools, equipment, and stores.1/	Labor, plus 7.9 KSh per ha for fertilizer, tools, equipment and stores, plus 9.4 KSh per ha for polyethylene sleeves, plus 16.7 KSh for filling sleeves by piecework. (Total cost per ha in addition to labor equals 34.0 KSh.)
Stumping	n.a.	Cost equals KL 5.00, and is applied to 364.2 ha in year 0, 404.7 ha in year 1, and 364.2 ha in yr. 2.
Plowing & Harrowing	n.a.	Cost equals KL 6.2 per ha planted in the year of planting.
Demarcation	All labor	n.a.
Clearing	All labor	n.a.
Pitting	All labor	n.a.
Staking Out	All labor	All labor
Planting	All labor	All labor
Plantation Surveys	All labor	All labor
Beating Up	All labor	All labor
Weeding(s)	All labor	All labor
Pruning(s)	All labor	All labor
Fire Protection	All labor	All labor
Other Protection	All labor	All labor
Dothistroma Control	n.a.	Applies to 121.4 ha of P. radiata planted annually in years 2 - 6.2/ Nursery treatment just prior to planting (age 0) costs KSh 24.7/ha. Field treatment just after planting (also age 0) costs Ksh 37.1/ha. Aerial treatments twice a year at Ksh 49.4/ha each begins at age 1 and are continued annually through age 6 (6 years of 2 treatments each year).
Vehicle (& Tractor) Purchase	Costs given in cash flow table.	Costs given in cash flow table.
Vehicle (& Tractor) Running Cost	Costs given in cash flow table.	Costs given in cash flow table.
Agricultural Equipment	Cost equals Ksh 12.4 per ha planted during the planting period, and 1.03 KSh annually per ha established after the planting period.3/	n.a.
Tools, Stores, (& Other Equipment)	Cost equals 37.1 Ksh per ha planted during the planting period, 3.1 KSh annually per acre established after the planting period up through project year 27, and 1.03 KSh annually per ha established for the remainder of the project.	Cost equals 30.9 Ksh per ha planted during the planting period, and 2.1 KSh annually per ha established after the planting period.
Building Construction	Spread equally over the 6 planting years: 10 forest ranger houses at KL 400 each, 15 forest guard houses at KL 100 each, 425 village houses at KL 75 each, 4 water supply installations at KL 500 each, and six offices and stores at KL 500 each, making a total of KL 42,375 (7.1 Thousand KL per year).	Spread equally over the 6 planting years: 2 forester houses at KL 1000, eight forest ranger houses at KL 400, 10 forest guard houses at KL 100, 200 village houses at KL 100, and 2 water supplies at KL 500, making a total of KL 27,200 (4.5 thousand KL per year.)
Building Maintenance	An annual cost (beginning the year following construction) which equals 10% of the cumulative construction cost during the planting period, and which equals 10% of total construction cost times a factor of (number of cumulative established ha divided by 19,425) after the planting period.	An annual cost (beginning the year following construction) which equals 10% of the cumulative construction cost during the planting period, and which equals 10% of total construction cost times a factor of (number of cumulative established ha divided by 9,350) after the planting period.
Road Construction	One km of access road per 100 ha planted built during the year of planting at KL per km (KL 1.55 per ha planted). Roads are rebuilt at the same cost after 10 years.	One km of access road per 125 ha planted is built during the year of planting at KL 155 per km (KL 1.24 per ha planted).
Road Maintenance	An annual cost (beginning the year following construction) which equals 10% of the cumulative construction cost during the planting period, & which equals 10% of the total construction cost times a factor of (number of cumulative established ha divided by 19,425) after the planting period.	An annual cost (beginning the year following construction which equals 10% of the cumulative construction cost during the planting period, and which equals 10% of the total construction cost times a factor of (number of cumulative established acres divided by 9,350 after the planting period.
Staff Requirements	Costs given in cash flow table.	Costs given in cash flow table.
Office Running Costs	Cost equals 35.21 KSh/ha planted during planting period & 2.93 annually/ ha established after the planting period.	Cost equals Ksh 37.1/ha planted during the planting period and 3.21 KSh annually per ha established after the planting period.
Thinning, Marking Out	All labor	n.a.

1/ Labor requirements outlined in Tables 5 and 6, and compiled in Tables 8 and 10.

2/ See Table 9.

3/ Established ha equal the total number of ha planted minus the total number of ha harvested.

KENYA I PLANTATION PROJECT

Table 4 - Silvicultural Labour Input

Requires
Labour (md/ha)

	Sawlogs	Pulpwood
Nursery - Planting	18.48	9.34
Nursery - Beating Up	2.77	1.40
Demarcation	3.21	n.a.
Clearing	19.77	n.a.
Pitting	1.23	n.a.
Staking Out	4.94	2.47
Planting	7.41	4.94
Plantation Surveys	2.47	2.47
Beating Up	1.11	0.74
Weeding (s)	4.94	4.94
Pruning - First	9.88	9.88
Pruning - Second	12.36	n.a.
Pruning - Third	14.83	n.a.
Pruning - Fourth	9.88	n.a.
Fire Protection - Initial	4.45	2.47
Fire Protection - Final	2.47	1.24
Other Protection - Initial	1.24	1.24
Other Protection - Final	0.74	0.74
Pre-commercial Thinning	14.83	n.a.
Marking Out (Commercial Thinning)	2.47	n.a.

Table 5. Hectares Associated with Plantation Activities - Sawlogs

YEARS

Item/Activity	0	1	2	3	4	5	6	7	8	9	10	11	12	13	14	15	16	17	18
Hectares Planted	-	-	3,238	3,238	3,238	3,238	3,238	3,238	-	-	-	-	-	-	-	-	-	-	-
Hectares Harvested	-	-	-	-	-	-	-	-	-	-	-	-	-	-	-	-	-	-	-
Cumulative Established ha.	-	-	3,238	6,475	9,713	12,950	16,188	19,425	19,425	19,425	19,425	19,425	19,425	19,425	19,425	19,425	19,425	19,425	19,425
. . . .																			
Nursery-Planting	-	3,238	3,238	3,238	3,238	3,238	3,238	-	-	-	-	-	-	-	-	-	-	-	-
Nursery-Beating Up	-	-	3,238	3,238	3,238	3,238	3,238	3,238	-	-	-	-	-	-	-	-	-	-	-
Demarcation	3,238	3,238	3,238	3,238	3,238	3,238			-	-	-	-	-	-	-	-	-	-	-
Clearing	3,238	3,238	3,238	3,238	3,238	3,238	-	-	-	-	-	-	-	-	-	-	-	-	-
Pitting	-	-	3,238	3,238	3,238	3,238	3,238	3,238	-	-	-	-	-	-	-	-	-	-	-
Staking Out	-	-	3,238	3,238	3,238	3,238	3,238	3,238	-	-	-	-	-	-	-	-	-	-	-
Planting	-	-	3,238	3,238	3,238	3,238	3,238	3,238	-	-	-	-	-	-	-	-	-	-	-
Plantation Survey	-	-	3,238	3,238	3,238	3,238	3,238	3,238	-	-	-	-	-	-	-	-	-	-	-
Beating Up	-	-	-	3,238	3,238	3,238	3,238	3,238	3,238	-	-	-	-	-	-	-	-	-	-
Weeding	-	-	-	3,238	3,238	3,238	3,238	3,238	3,238	-	-	-	-	-	-	-	-	-	-
Pruning-First	-	-	-	-	-	-	3,238	3,238	3,238	3,238	3,238	3,238	-	-	-	-	-	-	-
Pruning-Second	-	-	-	-	-	-	-	-	-	3,238	3,238	3,238	3,238	3,238	3,238	-	-	-	-
Pruning-Third	-	-	-	-	-	-	-	-	-	-	-	3,238	3,238	3,238	3,238	3,238	3,238	-	-
Pruning-Fourth	-	-	-	-	-	-	-	-	-	-	-	-	-	-	3,238	3,238	3,238	3,238	3,238
Fire Protec.-Initial	-	-	3,238	6,475	9,713	12,950	16,188	-	-	-	-	-	-	-	-	-	-	-	-
Fire Protec.-Final	-	-	-	-	-	-	-	19,425	19,425	19,425	19,425	19,425	19,425	19,425	19,425	19,425	19,425	19,425	19,425
Other Protec.-Initial	-	-	3,238	6,475	9,713	12,950	16,188	-	-	-	-	-	-	-	-	-	-	-	-
Other Protec.-Final	-	-	-	-	-	-	-	19,425	19,425	19,425	19,425	19,425	19,425	19,425	19,425	19,425	19,425	19,425	19,425
Pre-Commer'l Thinning	-	-	-	-	-	-	-	-	-	-	-	-	3,238	3,238	3,238	3,238	3,238	3,238	-
Marking Out(Commer'l Thinning)	-	-	-	-	-	-	-	-	-	-	-	-	-	-	-	-	-	3,238	3,238
Road Construction	-	-	3,238	3,238	3,238	3,238	3,238	3,238	-	-	-	-	3,238	3,238	3,238	3,238	3,238	3,238	-

Item/Activity	19	20	21	22	23	24	25	26	27	28	29	30	31	32	33	34	35	36	37
Hectares Planted	-	-	-	-	-	-	-	-	-	-	-	-	-	-	-	-	-	-	
Hectares Harvested	-	-	-	-	-	-	-	-	-	-	-	-	-	3,238	3,238	3,238	3,238	3,238	3,238
Cumulative Established ha.	19,425	19,425	19,425	19,425	19,425	19,425	19,425	19,425	19,425	19,425	19,425	19,425	19,425	16,188	12,950	9,713	6,475	3,238	-
. . . .																			
Nursery-Planting	-	-	-	-	-	-	-	-	-	-	-	-	-	-	-	-	-	-	-
Nursery-Beating Up	-	-	-	-	-	-	-	-	-	-	-	-	-	-	-	-	-	-	-
Demarcation	-	-	-	-	-	-	-	-	-	-	-	-	-	-	-	-	-	-	-
Clearing	-	-	-	-	-	-	-	-	-	-	-	-	-	-	-	-	-	-	-
Pitting	-	-	-	-	-	-	-	-	-	-	-	-	-	-	-	-	-	-	-
Staking Out	-	-	-	-	-	-	-	-	-	-	-	-	-	-	-	-	-	-	-
Planting	-	-	-	-	-	-	-	-	-	-	-	-	-	-	-	-	-	-	-
Plantation Survey	-	-	-	-	-	-	-	-	-	-	-	-	-	-	-	-	-	-	-
Beating Up	-	-	-	-	-	-	-	-	-	-	-	-	-	-	-	-	-	-	-
Weeding	-	-	-	-	-	-	-	-	-	-	-	-	-	-	-	-	-	-	-
Pruning-First	-	-	-	-	-	-	-	-	-	-	-	-	-	-	-	-	-	-	-
Pruning-Second	-	-	-	-	-	-	-	-	-	-	-	-	-	-	-	-	-	-	-
Pruning-Third	-	-	-	-	-	-	-	-	-	-	-	-	-	-	-	-	-	-	-
Pruning-Fourth	3,238	-	-	-	-	-	-	-	-	-	-	-	-	-	-	-	-	-	-
Fire Protec.-Initial	-	-	-	-	-	-	-	-	-	-	-	-	-	-	-	-	-	-	-
Fire Protec.-Final	19,425	19,425	19,425	19,425	19,425	19,425	19,425	19,425	19,425	19,425	19,425	19,425	19,425	16,188	12,950	9,713	6,475	3,238	-
Other Protec.-Initial	-	-	-	-	-	-	-	-	-	-	-	-	-	-	-	-	-	-	-
Other Protec.-Final	19,425	19,425	19,425	19,425	19,425	19,425	19,425	19,425	19,425	19,425	19,425	19,425	19,425	16,188	12,950	9,713	6,475	3,238	-
Pre-Commer'l Thinning	-	-	-	-	-	-	-	-	-	-	-	-	-	-	-	-	-	-	-
Marking Out(Commer'l Thinning)	3,238	3,238	3,238	3,238	3,238	3,238	3,238	3,238	3,238	-	-	-	-	-	-	-	-	-	-
Road Construction	-	-	-	-	-	-	-	-	-	-	-	-	-	-	-	-	-	-	-

Table 6. Unskilled Labour Requirements - Sawlogs (thousand man days)

YEARS

Silvicultural Activity	0	1	2	3	4	5	6	7	8	9	10	11	12	13	14	15	16	17	18
Nursery	-	59.8	68.8	68.8	68.8	68.8	68.8	9.0	-	-	-	-	-	-	-	-	-	-	-
Demarcation	10.4	10.4	10.4	10.4	10.4	10.4	-	-	-	-	-	-	-	-	-	-	-	-	-
Clearing	64.0	64.0	64.0	64.0	64.0	64.0	-	-	-	-	-	-	-	-	-	-	-	-	-
Pitting	-	-	4.0	4.0	4.0	4.0	4.0	4.0	-	-	-	-	-	-	-	-	-	-	-
Staking Out	-	-	16.0	16.0	16.0	16.0	16.0	16.0	-	-	-	-	-	-	-	-	-	-	-
Planting	-	-	24.0	24.0	24.0	24.0	24.0	24.0	-	-	-	-	-	-	-	-	-	-	-
Plantation Survey	-	-	8.0	8.0	8,0	8.0	8.0	8.0	-	-	-	-	-	-	-	-	-	-	-
Beating Up	-	-	-	3.6	3.6	3.6	3.6	3.6	3.6	-	-	-	-	-	-	-	-	-	-
Weeding	-	-	-	16.0	16.0	16.0	16.0	16.0	16.0	-	-	-	-	-	-	-	-	-	-
Pruning	-	-	-	-	-	-	32.0	32.0	32.0	72.0	72.0	120.0	88.0	88.0	120.0	80.0	80.0	32.0	32.0
Fire Protection	-	-	14.4	28.8	43.2	57.6	72.0	48.0	48.0	48.0	48.0	48.0	48.0	48.0	48.0	48.0	48.0	48.0	48.0
Other Protection	-	-	4.0	8.0	12.0	16.0	20.0	14.4	14.4	14.4	14.4	14.4	14.4	14.4	14.4	14.4	14.4	14.4	14.4
Thinning,Marking Out	-	-	-	-	-	-	-	-	-	-	-	-	48.0	48.0	48.0	48.0	48.0	56.0	8.0
Total Labor Require.	74.4	134.2	213.6	251.6	270.0	288.4	264.4	175.0	114.0	134.4	134.4	182.4	198.4	198.4	230.4	190.4	190.4	150.4	102.4
(Total Labor Cost,KSh/md)	2.354	2.354	2.354	2.9425	3.531	4.1195	4.708	4.708	4.708	4.708	4.708	4.708	4.708	4.708	4.708	4.708	4.708	4.708	4.708

Silvicultural Activity	19	20	21	22	23	24	25	26	27	28	29	30	31	32	33	34	35	36	37
Nursery	-	-	-	-	-	-	-	-	-	-	-	-	-	-	-	-	-	-	-
Demarcation	-	-	-	-	-	-	-	-	-	-	-	-	-	-	-	-	-	-	-
Clearing	-	-	-	-	-	-	-	-	-	-	-	-	-	-	-	-	-	-	-
Pitting	-	-	-	-	-	-	-	-	-	-	-	-	-	-	-	-	-	-	-
Staking Out	-	-	-	-	-	-	-	-	-	-	-	-	-	-	-	-	-	-	-
Planting	-	-	-	-	-	-	-	-	-	-	-	-	-	-	-	-	-	-	-
Plantation Survey	-	-	-	-	-	-	-	-	-	-	-	-	-	-	-	-	-	-	-
Beating Up	-	-	-	-	-	-	-	-	-	-	-	-	-	-	-	-	-	-	-
Weeding	-	-	-	-	-	-	-	-	-	-	-	-	-	-	-	-	-	-	-
Pruning	32.0	-	-	-	-	-	-	-	-	-	-	-	-	-	-	-	-	-	-
Fire Protection	48.0	48.0	48.0	48.0	48.0	48.0	48.0	48.0	48.0	48.0	48.0	48.0	48.0	40.0	32.0	24.0	16.0	8.0	-
Other Protection	14.4	14.4	14.4	14.4	14.4	14.4	14.4	14.4	14.4	14.4	14.4	14.4	14.4	12.0	9.6	7.2	4.8	2.4	-
Thinning,Marking Out	8.0	8.0	8.0	16.0	8.0	8.0	8.0	8.0	8.0	-	-	-	-	-	-	-	-	-	-
Total Labor Require.	102.4	70.4	70.4	78.4	70.4	70.4	70.4	70.4	70.4	62.4	62.4	62.4	62.4	52.0	41.6	31.2	20.8	10.4	
(Total Labor Cost,KSh/md)	4.708	4.708	4.708	4.708	4.708	4.708	4.708	4.708	4.708	4.708	4.708	4.708	4.708	4.708	4.708	4.708	4.708	4.708	4.708

Table 7. Hectares Associated with Plantation Activities - Pulpwood

Item/Activity	0	1	2	3	4	5	6	7	8	9	10	11	12	13	14	15	16	17	18	19	20	21
											YEAR											
Hectares Planted	-	1,250	1,620	1,620	1,620	1,520	1,620	-	-	-	-	-	-	-	-	-	-	-	-	-	-	-
Hectares Harvested	-	-	-	-	-	-	-	-	-	-	-	-	-	-	-	1,210	1,380	1,500	1,490	1,480	1,460	830
Cumulative Established	-	1,250	2,870	4,490	6,110	7,730	9,350	9,350	9,350	9,350	9,350	9,350	9,350	9,350	9,350	8,140	6,760	5,260	3,770	2,290	830	-
. . . .																						
Nursery-Planting	1,250	1,620	1,620	1,620	1,620	1,620	-	-	-	-	-	-	-	-	-	-	-	-	-	-	-	-
Nursery-Beating Up	-	1,250	1,620	1,620	1,620	1,620	1,620	-	-	-	-	-	-	-	-	-	-	-	-	-	-	-
Staking Out	-	1,250	1,620	1,620	1,620	1,620	1,620	-	-	-	-	-	-	-	-	-	-	-	-	-	-	-
Planting	-	1,250	1,620	1,620	1,620	1,620	1,620	-	-	-	-	-	-	-	-	-	-	-	-	-	-	-
Plantation Survey	-	1,250	1,620	1,620	1,620	1,620	1,620	-	-	-	-	-	-	-	-	-	-	-	-	-	-	-
Beating Up	-	-	1,250	1,620	1,620	1,620	1,620	1,620	-	-	-	-	-	-	-	-	-	-	-	-	-	-
Weedings	-	1,250	2,870	3,240	3,240	3,240	3,240	1,620	-	-	-	-	-	-	-	-	-	-	-	-	-	-
Pruning	-	-	-	-	-	1,250	1,620	1,620	1,620	1,620	1,620	-	-	-	-	-	-	-	-	-	-	-
Fire Protection-Initia	-	1,250	2,870	4,490	6,110	7,730	-	-	-	-	-	-	-	-	-	-	-	-	-	-	-	-
Fire Protection-Final	-	-	-	-	-	-	9,350	9,350	9,350	9,350	9,350	9,350	9,350	9,350	9,350	8,140	6,760	5,260	3,770	2,290	830	-
Other Protection-Initi	-	1,250	2,870	4,490	6,110	7,730	-	-	-	-	-	-	-	-	-	-	-	-	-	-	-	-
Other Protection-Final	-	-	-	-	-	-	9,350	9,350	9,350	9,350	9,350	9,350	9,350	9,350	9,350	8,140	6,760	5,260	3,770	2,290	830	-
Stumping	364	405	364	-	-	-	-	-	-	-	-	-	-	-	-	-	-	-	-	-	-	-
Plowing and Harrowing	-	1,250	1,620	1,620	1,620	1,620	1,620	-	-	-	-	-	-	-	-	-	-	-	-	-	-	-
Dothistroma Control-Nursery	-	-	121	121	121	121	121	-	-	-	-	-	-	-	-	-	-	-	-	-	-	-
Dothistroma Control-Field	-	-	121	121	121	121	121	-	-	-	-	-	-	-	-	-	-	-	-	-	-	-
Dothistroma Control-Aerial	-	-	-	121	243	364	486	607	607	486	364	243	121	-	-	-	-	-	-	-	-	-
Road Construction	-	1,250	1,620	1,620	1,620	1,620	1,620	-	-	-	-	-	-	-	-	-	-	-	-	-	-	-

Table 8. Unskilled Labour Requirements - Pulpwood (thousand man days)

Silvicultural Activit	0	1	2	3	4	5	6	7	8	9	10	11	12	13	14	15	16	17	18	19	20	21
											YEAR											
Nursery	11.7	16.9	17.4	17.4	17.4	17.4	2.3	-	-	-	-	-	-	-	-	-	-	-	-	-	-	-
Staking Out	-	3.1	4.0	4.0	4.0	4.0	4.0	-	-	-	-	-	-	-	-	-	-	-	-	-	-	-
Planting	-	6.2	8.0	8.0	8.0	8.0	8.0	-	-	-	-	-	-	-	-	-	-	-	-	-	-	-
Plantation Survey	-	3.1	4.0	4.0	4.0	4.0	4.0	-	-	-	-	-	-	-	-	-	-	-	-	-	-	-
Beating Up	-	-	0.9	1.2	1.2	1.2	1.2	1.2	-	-	-	-	-	-	-	-	-	-	-	-	-	-
Weedings	-	6.2	14.2	16.0	16.0	16.0	16.0	8.0	-	-	-	-	-	-	-	-	-	-	-	-	-	-
Pruning	-	-	-	-	-	12.4	16.0	16.0	16.0	16.0	16.0	-	-	-	-	-	-	-	-	-	-	-
Fire Protection	-	3.1	7.1	11.1	15.1	19.1	11.6	11.6	11.6	11.6	11.6	11.6	11.6	11.6	11.6	10.1	8.3	6.5	4.6	2.8	1.0	-
Other Protection	-	1.6	3.6	5.6	7.6	9.6	6.9	6.9	6.9	6.9	6.9	6.9	6.9	6.9	6.9	6.0	5.0	3.9	2.8	1.7	0.6	-
Total Labor Requirement	11.7	40.2	59.2	67.3	73.3	91.7	70.0	43.7	34.5	34.5	34.5	18.5	18.5	18.5	18.5	16.1	13.3	10.4	7.4	4.5	1.6	-
(Total Labor Cost,KS nd)	2.354	2.354	2.9425	3.531	4.1195	4.708	4.708	4.708	4.708	4.708	4.708	4.708	4.708	4.708	4.708	4.708	4.708	4.708	4.708	4.708	4.708	4.708

KENYA I PLANTATION PROJECT

Table 9. Daily Unskilled Labour Cost Schedule

Year	Wage Rate (KSh)	Seven Percent Supervision Cost (KSh)	Daily Labour Cost (KSh)
1967/68-69/70	2.20	0.1540	2.3540
1970/71	2.75	0.1925	2.9425
1971/72	3.30	0.2310	3.5310
1972/73	3.85	0.2695	4.1195
1973/74 and on	4.40	0.3080	4.7080

per m3 of sawlog stumpage for the first and second commercial thinnings, and KSh 58.28 for the final harvest.

Indirect Benefits

31. Indirect benefits were not quantified in either the economic or the financial analyses. Indirect benefits would include the promotion of secondary industries such as logging and sawmills, plywood factories and furniture factories. These would generate additional employment and income. The project itself would employ some 1 500 resident workers. Other indirect benefits would be the maintenance of permanent protective vegetative cover on watershed areas where most of the permanent rivers in Kenya originate.

E. COMPARING COSTS AND BENEFITS

32. The total cost and total benefit/revenue streams are shown in the accompanying cash flow tables. Table 10 is the sawlog financial cash flow, Table 11 is the sawlog economic cash flow, and Table 12 is the pulpwood financial/economic cash flow. 1/ The financial and economic rates of return (FRR and ERR) are as follows:

sawlog financial analysis	7.8%	(FRR)
sawlog economic analysis	12.3%	(ERR)
pulpwood financial/economic analysis	11.0%	(FRR and ERR)

Sensitivity Analysis

33. The sensitivity of the economic rate of return to different input and output assumptions is presented in Table 13.

34. (a) Sawlog sub-project - First, ERR estimates were made separately for each half of the sub-project, i.e. the import substitution and the export components. For the import substitution component the rate of return was 14.8% and for the export component it was 7.8%.

35. Given the expected employment situation in the country, it was estimated that the economic wage for unskilled labour should be somewhat lower than the financial wage rate. Also, it was considered likely that in view of the projected deficits in importing countries, prices of exported Kenyan sawnwood would increase in the future. In order to explore the impact of both changes, results were tested for an increase in the price of the export component equal to 1 percent per year, combined with a shadow wage assumed equal to 75 percent of the financial wage. The economic rate of return for the export component rose from 7.8 percent to 13.5 percent.

36. (b) Pulpwood sub-project. In this case, there was no clear indication of the direction in which prices might move in the future, as pulpwood prices would be heavily influenced by government policies regarding the level of protection granted to the pulp and paper industry. For this reason, results were tested against both increases in pulpwood stumpage prices equal to 1% per year and a total decrease in revenues of 25%. The rate of return respectively increased from 11% to 12.5% and decreased to 8.5%. The results were also tested to determine the effect of shadowpricing unskilled and silvicultural labour at 75% of the original value. This raised the rate of return from 11% to 11.5%.

1/ Pitting cost was assumed not to occur in the economic analysis since it was insignificant and only applied to 25% of the sawlog area. The rate of return is unchanged by its removal.

KENYA I PLANTATION PROJECT

Table 10. Analysis of Financial Return for Sawlog Component

THOUSAND KL/TOT PROJ

COSTS (YEAR)	0	1	2	3	4	5	6	7	8	9	10	11	12	13	14	15	16	17	18
1 NURSERY	0	8.60	9.90	12.00	14.00	16.00	18.00	2.40	0	0	0	0	0	0	0	0	0	0	0
2 DEMARCATION	1.20	1.20	1.20	1.50	1.80	2.10	0	0	0	0	0	0	0	0	0	0	0	0	0
3 CLEARING	7.50	7.50	7.50	9.40	11.30	13.20	0	0	0	0	0	0	0	0	0	0	0	0	0
4 PITTING	0	0	.50	.60	.70	.80	.90	.90	0	0	0	0	0	0	0	0	0	0	0
5 STAKING OUT	0	0	1.90	2.40	2.80	3.30	3.80	3.80	0	0	0	0	0	0	0	0	0	0	0
6 PLANTING	0	0	2.80	3.50	4.20	4.90	5.60	5.60	0	0	0	0	0	0	0	0	0	0	0
7 PLANT SURVEY	0	0	.90	1.20	1.40	1.60	1.90	1.90	0	0	0	0	0	0	0	0	0	0	0
8 BEATING UP	0	0	0	.50	.60	.70	.80	.80	.80	0	0	0	0	0	0	0	0	0	0
9 WEEDING	0	0	0	2.40	2.80	3.30	3.80	3.80	3.80	0	0	0	0	0	0	0	0	0	0
10 PRUNING	0	0	0	0	0	0	7.50	7.50	7.50	16.90	16.90	28.20	20.70	20.70	28.20	18.80	18.80	7.50	7.50
11 FIRE PROTECT	0	0	1.70	4.20	7.60	11.90	16.90	11.30	11.30	11.30	11.30	11.30	11.30	11.30	11.30	11.30	11.30	11.30	11.30
12 OTH. PROTECT	0	0	.50	1.20	2.10	3.30	4.70	3.40	3.40	3.40	3.40	3.40	3.40	3.40	3.40	3.40	3.40	3.40	3.40
13 VEH/TRAC PUR	0	0	13.20	10.90	5.50	6.20	6.70	8.20	4.70	2.70	2.30	3.50	3.50	3.50	3.50	2.30	3.50	3.50	8.50
14 VEH/TRAC RUN	0	0	12.90	13.80	14.60	15.20	17.20	18.10	8.50	7.30	7.30	8.10	8.10	8.10	9.00	8.10	8.10	8.10	7.70
15 AGRIC EQUIP	0	0	2.00	2.00	2.00	2.00	2.00	2.00	1.00	1.00	1.00	1.00	1.00	1.00	1.00	1.00	1.00	1.00	1.00
16 TOOLS/STORES	0	0	6.00	6.00	6.00	6.00	6.00	6.00	3.00	3.00	3.00	3.00	3.00	3.00	3.00	3.00	3.00	3.00	3.00
17 BUILD CONSTR	0	0	7.10	7.10	7.10	7.10	7.10	7.10	0	0	0	0	0	0	0	0	0	0	0
18 BUILD MAINT	0	0	0	.70	1.40	2.10	2.80	3.50	4.20	4.20	4.20	4.20	4.20	4.20	4.20	4.20	4.20	4.20	4.20
19 ROAD CONSTRU	0	0	5.00	5.00	5.00	5.00	5.00	5.00	0	0	0	0	5.00	5.00	5.00	5.00	5.00	5.00	0
20 ROAD MAINT	0	0	0	.50	1.00	1.50	2.00	2.50	3.00	3.00	3.00	3.00	3.00	3.00	3.00	3.00	3.00	3.00	3.00
21 STAFF REQUIR	0	0	26.10	30.70	35.30	40.00	44.00	51.30	25.70	25.70	25.70	25.70	25.70	25.70	25.70	25.70	25.70	25.70	25.70
22 OFFICE RUN	0	0	5.70	5.70	5.70	5.70	5.70	5.70	2.80	2.80	2.80	2.80	2.80	2.80	2.80	2.80	2.80	2.80	2.80
23 THIN/MARKOUT	0	0	0	0	0	0	0	0	0	0	0	0	11.30	11.30	11.30	11.30	11.30	13.20	1.90
24 SUBTOTALS	0	0	0	0	0	0	0	0	0	0	0	0	0	0	0	0	0	0	0
25 CONTINGENCY	0	0	15.70	18.20	19.90	23.10	24.40	22.60	12.00	12.20	12.10	14.10	15.40	15.40	16.70	15.00	15.70	13.80	12.00
TOTAL COST	8.70	17.30	120.60	139.50	152.80	171.00	186.80	173.40	91.70	93.50	93.00	108.30	118.40	118.40	128.10	114.90	119.80	105.50	92.00
BENEFITS																			
26 THINNING	0	0	0	0	0	0	0	0	0	0	0	0	0	0	0	0	0	200.00	200.00
27 HARVEST	0	0	0	0	0	0	0	0	0	0	0	0	0	0	0	0	0	0	0
TOTAL BENEFT	0	0	0	0	0	0	0	0	0	0	0	0	0	0	0	0	0	200.00	200.00
NET BENEFITS	-8.70	-17.30	-120.60	-139.50	-152.80	-171.00	-186.80	-173.40	-91.70	-93.50	-93.00	-108.30	-118.40	-118.40	-128.10	-114.90	-119.80	94.50	106.00

COSTS	19	20	21	22	23	24	25	26	27	28	29	30	31	32	33	34	35	36	37
1 NURSERY	0	0	0	0	0	0	0	0	0	0	0	0	0	0	0	0	0	0	0
2 DEMARCATION	0	0	0	0	0	0	0	0	0	0	0	0	0	0	0	0	0	0	0
3 CLEARING	0	0	0	0	0	0	0	0	0	0	0	0	0	0	0	0	0	0	0
4 PITTING	0	0	0	0	0	0	0	0	0	0	0	0	0	0	0	0	0	0	0
5 STAKING OUT	0	0	0	0	0	0	0	0	0	0	0	0	0	0	0	0	0	0	0
6 PLANTING	0	0	0	0	0	0	0	0	0	0	0	0	0	0	0	0	0	0	0
7 PLANT SURVEY	0	0	0	0	0	0	0	0	0	0	0	0	0	0	0	0	0	0	0
8 BEATING UP	0	0	0	0	0	0	0	0	0	0	0	0	0	0	0	0	0	0	0
9 WEEDING	0	0	0	0	0	0	0	0	0	0	0	0	0	0	0	0	0	0	0
10 PRUNING	7.50	0	0	0	0	0	0	0	0	0	0	0	0	0	0	0	0	0	0
11 FIRE PROTECT	11.30	11.30	11.30	11.30	11.30	11.30	11.30	11.30	11.30	11.30	11.30	11.30	11.30	9.40	7.50	5.60	3.80	1.90	0
12 OTH. PROTECT	3.40	3.40	3.40	3.40	3.40	3.40	3.40	3.40	3.40	3.40	3.40	3.40	3.40	2.80	2.30	1.70	1.10	.60	0
13 VEH/TRAC PUR	7.30	2.70	2.70	2.70	2.70	2.70	2.70	2.70	2.70	2.40	2.40	2.40	2.40	2.30	.80	1.20	.80	0	0
14 VEH/TRAC RUN	7.30	6.50	6.50	6.50	6.50	6.50	6.50	6.50	5.70	5.70	5.70	5.70	6.10	5.40	4.20	3.30	2.30	1.40	0
15 AGRIC EQUIP	1.00	1.00	1.00	1.00	1.00	1.00	1.00	1.00	1.00	1.00	1.00	1.00	1.00	.80	.60	.40	.20	.10	0
16 TOOLS/STORES	3.00	3.00	3.00	3.00	3.00	3.00	3.00	3.00	3.00	1.00	1.00	1.00	1.00	.80	.60	.40	.20	.10	0
17 BUILD CONSTR	0	0	0	0	0	0	0	0	0	0	0	0	0	0	0	0	0	0	0
18 BUILD MAINT	4.20	4.20	4.20	4.20	4.20	4.20	4.20	4.20	4.20	4.20	4.20	4.20	4.20	3.50	2.80	2.10	1.40	.70	0
19 ROAD CONSTRU	0	0	0	0	0	0	0	0	0	0	0	0	0	0	0	0	0	0	0
20 ROAD MAINT	3.00	3.00	3.00	3.00	3.00	3.00	3.00	3.00	3.00	3.00	3.00	3.00	3.00	2.50	2.00	1.50	1.00	.50	0
21 STAFF REQUIR	25.70	25.70	25.70	25.70	25.70	25.70	25.70	25.70	25.70	25.70	25.70	25.70	25.70	21.40	17.10	12.80	18.60	4.30	0
22 OFFICE RUN	2.80	2.80	2.80	2.80	2.80	2.80	2.80	2.80	2.80	2.80	2.80	2.80	2.80	2.30	1.90	1.40	.90	.50	0
23 THIN/MARKOUT	1.90	1.90	1.90	3.80	1.90	1.90	1.90	1.90	1.90	0	0	0	0	0	0	0	0	0	0
24 SUBTOTALS	0	0	0	0	0	0	0	0	0	0	0	0	0	0	0	0	0	0	0
25 CONTINGENCY	11.80	9.80	9.80	10.10	9.80	9.80	9.80	9.80	9.70	9.10	9.10	9.10	9.10	7.70	6.00	4.60	4.50	1.50	0
TOTAL COST	90.20	75.30	75.30	77.50	75.30	75.30	75.30	75.30	74.40	69.60	69.60	69.60	70.00	58.90	45.80	35.00	34.80	11.60	0
BENEFITS																			
26 THINNING	200.00	200.00	200.00	440.00	240.00	240.00	240.00	240.00	240.00	0	0	0	0	0	0	0	0	0	0
27 HARVEST	0	0	0	0	0	0	0	0	0	0	0	0	0	1800.00	1800.00	1800.00	1800.00	1800.00	1800.00
TOTAL BENEFT	200.00	200.00	200.00	440.00	240.00	240.00	240.00	240.00	240.00	0	0	0	0	1800.00	1800.00	1800.00	1800.00	1800.00	1800.00
NET BENEFITS	109.80	124.70	124.70	362.50	164.70	164.70	164.70	194.70	165.60	-69.60	-69.60	-69.60	-70.00	1741.10	1754.20	1765.00	1765.20	1788.40	1800.00

KENYA I PLANTATION PROJECT

Table 11 . Analysis of Economic Return for Sawlog Component,

THOUSAND KL/TOT PROJ

COSTS (YEAR)	0	1	2	3	4	5	6	7	8	9	10	11	12	13	14	15	16	17	18
1 NURSERY	0	8.60	9.90	12.00	14.00	19.00	18.00	2.40	0	0	0	0	0	0	0	0	0	0	0
2 DEMARCATION	1.20	1.20	1.20	1.50	1.80	2.10	0	0	0	0	0	0	0	0	0	0	0	0	0
3 CLEARING	7.50	7.50	7.50	9.40	11.30	13.20	0	0	0	0	0	0	0	0	0	0	0	0	0
4 PITTING	0	0	.50	.60	.70	.80	.90	.90	0	0	0	0	0	0	0	0	0	0	0
5 STAKING OUT	0	0	1.90	2.40	2.80	3.30	3.80	3.80	0	0	0	0	0	0	0	0	0	0	0
6 PLANTING	0	0	2.80	3.50	4.20	4.90	5.60	5.60	0	0	0	0	0	0	0	0	0	0	0
7 PLANT SURVEY	0	0	.90	1.20	1.40	1.60	1.90	1.90	0	0	0	0	0	0	0	0	0	0	0
8 BEATING UP	0	0	0	.50	.60	.70	.80	.80	.80	0	0	0	0	0	0	0	0	0	0
9 WEEDING	0	0	0	2.40	2.80	3.30	3.80	3.80	3.80	0	0	0	0	0	0	0	0	0	0
10 PRUNING	0	0	0	0	0	0	7.50	7.50	7.50	16.90	16.90	28.20	20.70	20.70	28.20	18.80	18.80	7.50	7.50
11 FIRE PROTECT	0	0	1.70	4.20	7.60	11.90	16.90	11.30	11.30	11.30	11.30	11.30	11.30	11.30	11.30	11.30	11.30	11.30	11.30
12 OTH. PROTECT	0	0	.50	1.20	2.10	3.30	4.70	3.40	3.40	3.40	3.40	3.40	3.40	3.40	3.40	3.40	3.40	3.40	3.40
13 VEH/TRAC PUR	0	0	13.20	10.90	5.50	8.20	6.70	8.20	4.70	2.70	2.30	3.50	3.50	3.50	3.50	2.30	3.50	3.50	8.50
14 VEH/TRAC RUN	0	0	12.90	13.80	14.60	15.20	17.20	18.10	8.50	7.30	7.30	8.10	8.10	8.10	9.00	8.10	8.10	8.10	7.70
15 AGRIC EQUIP	0	0	2.00	2.00	2.00	2.30	2.00	2.00	1.00	1.00	1.00	1.00	1.00	1.00	1.00	1.00	1.00	1.00	1.00
16 TOOLS/STORES	0	0	6.00	6.00	6.00	6.30	6.00	6.00	3.00	3.00	3.00	3.00	3.00	3.00	3.00	3.00	3.00	3.00	3.00
17 BUILD CONSTR	0	0	7.10	7.10	7.10	7.10	7.10	7.10	0	0	0	0	0	0	0	0	0	0	0
18 BUILD MAINT	0	0	0	.70	1.40	2.10	2.80	3.50	4.20	4.20	4.20	4.20	4.20	4.20	4.20	4.20	4.20	4.20	4.20
19 ROAD CONSTRU	0	0	5.00	5.00	5.00	5.30	5.00	5.00	0	0	0	0	5.00	5.00	5.00	5.00	5.00	5.00	0
20 ROAD MAINT	0	0	0	.50	1.00	1.50	2.00	2.50	3.00	3.00	3.00	3.00	3.00	3.00	3.00	3.00	3.00	3.00	3.00
21 STAFF REQUIR	0	0	26.10	30.70	35.30	40.00	44.00	51.30	25.70	25.70	25.70	25.70	25.70	25.70	25.70	25.70	25.70	25.70	25.70
22 OFFICE RUN	0	0	5.70	5.70	5.70	5.70	5.70	5.70	2.80	2.80	2.80	2.80	2.80	2.80	2.80	2.80	2.80	2.80	2.80
23 THIN/MARKOUT	0	0	0	0	0	0	0	0	0	0	0	0	11.30	11.30	11.30	11.30	11.30	13.20	1.90
24 SUBTOTALS	0	0	0	0	0	0	0	0	0	0	0	0	0	0	0	0	0	0	0
25 CONTINGENCY	0	0	15.70	18.20	19.90	23.10	24.40	22.60	12.00	12.20	12.10	14.10	15.40	15.40	16.70	15.00	15.70	13.80	12.00
TOTAL COST	8.70	17.30	120.60	139.50	152.80	171.00	186.80	173.40	91.70	93.50	93.00	108.30	118.40	118.40	128.10	114.90	119.80	105.50	92.00
BENEFITS																			
26 THINNING	0	0	0	0	0	0	0	0	0	0	0	0	0	0	0	0	0	440.00	440.00
27 HARVEST	0	0	0	0	0	0	0	0	0	0	0	0	0	0	0	0	0	0	0
TOTAL BENEFT	0	0	0	0	0	0	0	0	0	0	0	0	0	0	0	0	0	440.00	440.00
NET BENEFITS	-8.70	-17.30	-120.60	-139.50	-152.80	-171.00	-186.80	-173.40	-91.70	-93.50	-93.00	-108.30	-118.40	-118.40	-128.10	-114.90	-119.80	334.50	348.00

COSTS	19	20	21	22	23	24	25	26	27	28	29	30	31	32	33	34	35	36	37
1 NURSERY	0	0	0	0	0	0	0	0	0	0	0	0	0	0	0	0	0	0	0
2 DEMARCATION	0	0	0	0	0	0	0	0	0	0	0	0	0	0	0	0	0	0	0
3 CLEARING	0	0	0	0	0	0	0	0	0	0	0	0	0	0	0	0	0	0	0
4 PITTING	0	0	0	0	0	0	0	0	0	0	0	0	0	0	0	0	0	0	0
5 STAKING OUT	0	0	0	0	0	0	0	0	0	0	0	0	0	0	0	0	0	0	0
6 PLANTING	0	0	0	0	0	0	0	0	0	0	0	0	0	0	0	0	0	0	0
7 PLANT SURVEY	0	0	0	0	0	0	0	0	0	0	0	0	0	0	0	0	0	0	0
8 BEATING UP	0	0	0	0	0	0	0	0	0	0	0	0	0	0	0	0	0	0	0
9 WEEDING	0	0	0	0	0	0	0	0	0	0	0	0	0	0	0	0	0	0	0
10 PRUNING	7.50	0	0	0	0	0	0	0	0	0	0	0	0	0	0	0	0	0	0
11 FIRE PROTECT	11.30	11.30	11.30	11.30	11.30	11.30	11.30	11.30	11.30	11.30	11.30	11.30	11.30	9.40	7.50	5.60	3.80	1.90	0
12 OTH. PROTECT	3.40	3.40	3.40	3.40	3.40	3.40	3.40	3.40	3.40	3.40	3.40	3.40	3.40	2.80	2.30	1.70	1.10	.60	0
13 VEH/TRAC PUR	7.30	2.70	2.70	2.70	2.70	2.70	2.70	2.70	2.70	2.40	2.40	2.40	2.40	2.30	.80	1.20	.80	0	0
14 VEH/TRAC RUN	7.30	6.50	6.50	6.50	6.50	6.50	6.50	6.50	5.70	5.70	5.70	5.70	6.10	5.40	4.20	3.30	2.30	1.40	0
15 AGRIC EQUIP	1.00	1.00	1.00	1.00	1.00	1.00	1.00	1.00	1.00	1.00	1.00	1.00	1.00	.80	.60	.40	.20	.10	0
16 TOOLS/STORES	3.00	3.00	3.00	3.00	3.00	3.00	3.00	3.00	3.00	1.00	1.00	1.00	1.00	.80	.60	.40	.20	.10	0
17 BUILD CONSTR	0	0	0	0	0	0	0	0	0	0	0	0	0	0	0	0	0	0	0
18 BUILD MAINT	4.20	4.20	4.20	4.20	4.20	4.20	4.20	4.20	4.20	4.20	4.20	4.20	4.20	3.50	2.80	2.10	1.40	.70	0
19 ROAD CONSTRU	0	0	0	0	0	0	0	0	0	0	0	0	0	0	0	0	0	0	0
20 ROAD MAINT	3.00	3.00	3.00	3.00	3.00	3.00	3.00	3.00	3.00	3.00	3.00	3.00	3.00	2.50	2.00	1.50	1.00	.50	0
21 STAFF REQUIR	25.70	25.70	25.70	25.70	25.70	25.70	25.70	25.70	25.70	25.70	25.70	25.70	25.70	21.40	17.10	12.80	18.60	4.30	0
22 OFFICE RUN	2.80	2.80	2.80	2.80	2.80	2.80	2.80	2.80	2.80	2.80	2.80	2.80	2.80	2.30	1.90	1.40	.90	.50	0
23 THIN/MARKOUT	1.90	1.90	1.90	3.80	1.90	1.90	1.90	1.90	1.90	0	0	0	0	0	0	0	0	0	0
24 SUBTOTALS	0	0	0	0	0	0	0	0	0	0	0	0	0	0	0	0	0	0	0
25 CONTINGENCY	11.80	9.80	9.80	10.10	9.80	9.80	9.80	9.80	9.70	9.10	9.10	9.10	9.10	7.70	6.00	4.60	4.50	1.50	0
TOTAL COST	90.20	75.30	75.30	77.50	75.30	75.30	75.30	75.30	74.40	69.60	69.60	69.60	70.00	58.90	45.80	35.00	34.80	11.60	0
BENEFITS																			
26 THINNING	440.00	440.00	440.00	968.00	528.00	528.00	528.00	528.00	528.00	0	0	0	0	0	0	0	0	0	0
27 HARVEST	0	0	0	0	0	0	0	0	0	0	0	0	0	3960.00	3960.00	3960.00	3960.00	3960.00	3960.00
TOTAL BENEFT	440.00	440.00	440.00	968.00	528.00	528.00	528.00	528.00	528.00	0	0	0	0	3960.00	3960.00	3960.00	3960.00	3960.00	3960.00
NET BENEFITS	349.80	364.70	364.70	890.50	452.70	452.70	452.70	452.70	453.60	-69.60	-69.60	-69.60	-70.00	3901.10	3914.20	3925.00	3925.20	3948.40	3960.00

KENYA I PLANTATION PROJECT

Table 12. Analysis of Economic and Financial Return for Pulpwood Component

THOUSAND KL/TOT PROJ

(YEAR)

COSTS	0	1	2	3	4	5	6	7	8	9	10
1 NURSERY	3.50	5.10	5.70	6.20	6.70	7.30	.90	0	0	0	0
2 STUMPING	4.50	5.00	4.50	0	0	0	0	0	0	0	0
3 PLOW/HARROW	0	7.80	10.00	10.00	10.00	10.00	10.00	0	0	0	0
4 STAKING OUT	0	.40	.60	.70	.80	.90	.90	0	0	0	0
5 PLANTING	0	.70	1.20	1.40	1.60	1.90	1.90	0	0	0	0
6 PLANT SURVEY	0	.40	.60	.70	.80	.90	.90	0	0	0	0
7 BEATING UP	0	0	.10	.20	.20	.30	.30	.30	0	0	0
8 WEEDING	0	.70	2.10	2.80	3.30	3.80	3.80	1.90	0	0	0
9 PRUNING	0	0	0	0	0	2.90	3.80	3.80	3.80	3.80	3.80
10 FIRE PROTECT	0	.40	1.00	2.00	3.10	4.50	2.70	2.70	2.70	2.70	2.70
11 OTH. PROTECT	0	.20	.50	1.00	1.60	2.20	1.60	1.60	1.60	1.60	1.60
12 MOTH CONTROL	0	0	.40	1.00	1.00	2.20	2.80	3.00	3.00	2.40	1.80
13 VEHIC. PURCH	0	9.30	0	1.20	3.50	2.70	1.80	2.70	1.20	0	2.30
14 VEH. RUNNING	0	3.80	5.20	5.60	6.10	6.80	7.40	4.10	3.10	3.10	2.70
15 TOOLS/STORES	0	1.90	2.50	2.50	2.50	2.50	2.50	1.00	1.00	1.00	1.00
16 BUILD CONSTR	0	4.50	4.50	4.50	4.50	4.50	4.50	0	0	0	0
17 BUILD MAINT.	0	0	.50	.90	1.40	1.80	2.30	2.70	2.70	2.70	2.70
18 ROAD CONSTR	0	1.60	2.00	2.00	2.00	2.00	2.00	0	0	0	0
19 ROAD MAINT.	0	0	.20	.40	.60	.80	1.00	1.20	1.20	1.20	1.20
20 STAFF REQUIR	0	11.20	12.70	14.40	16.00	16.80	17.40	8.70	8.70	8.70	8.70
21 OFFICE RUN	0	2.50	3.00	3.00	3.00	3.00	3.00	1.50	1.50	1.50	1.50
22 SUBTOTALS	0	0	0	0	0	0	0	0	0	0	0
23 CONTINGENCY	0	8.30	8.60	9.10	10.40	11.70	10.70	5.30	4.60	4.30	4.50
TOTAL COST	8.00	63.80	65.90	69.60	79.70	89.50	82.20	40.50	35.10	33.00	34.50
BENEFITS											
24 HARVEST	0	0	0	0	0	0	0	0	0	0	0
TOTAL BENEFIT	0	0	0	0	0	0	0	0	0	0	0
NET BENEFITS	-8.00	-63.80	-65.90	-69.60	-79.70	-89.50	-82.20	-40.50	-35.10	-33.00	-34.50

(YEAR)

COSTS	11	12	13	14	15	16	17	18	19	20	21
1 NURSERY	0	0	0	0	0	0	0	0	0	0	0
2 STUMPING	0	0	0	0	0	0	0	0	0	0	0
3 PLOW/HARROW	0	0	0	0	0	0	0	0	0	0	0
4 STAKING OUT	0	0	0	0	0	0	0	0	0	0	0
5 PLANTING	0	0	0	0	0	0	0	0	0	0	0
6 PLANT SURVEY	0	0	0	0	0	0	0	0	0	0	0
7 BEATING UP	0	0	0	0	0	0	0	0	0	0	0
8 WEEDING	0	0	0	0	0	0	0	0	0	0	0
9 PRUNING	0	0	0	0	0	0	0	0	0	0	0
10 FIRE PROTECT	2.70	2.70	2.70	2.70	2.40	2.00	1.50	1.10	.70	.20	0
11 OTH. PROTECT	1.60	1.60	1.60	1.60	1.40	1.20	.90	.70	.40	.10	0
12 MOTH CONTROL	1.20	.60	0	0	0	0	0	0	0	0	0
13 VEHIC. PURCH	1.20	0	2.10	1.20	0	2.70	.80	0	1.50	0	0
14 VEH. RUNNING	2.60	2.60	2.60	2.80	2.50	2.20	1.70	1.20	.80	.40	0
15 TOOLS/STORES	1.00	1.00	1.00	1.00	.90	.70	.60	.40	.20	.10	0
16 BUILD CONSTR	0	0	0	0	0	0	0	0	0	0	0
17 BUILD MAINT.	2.70	2.70	2.70	2.70	2.40	2.00	1.50	1.10	.70	.20	0
18 ROAD CONSTR	0	0	0	0	0	0	0	0	0	0	0
19 ROAD MAINT.	1.20	1.20	1.20	1.20	1.00	.80	.70	.50	.30	.10	0
20 STAFF REQUIR	8.70	8.70	8.70	8.70	7.10	5.90	4.60	3.30	2.00	.70	0
21 OFFICE RUN	1.50	1.50	1.50	1.50	1.30	1.10	.80	.60	.40	.10	0
22 SUBTOTALS	0	0	0	0	0	0	0	0	0	0	0
23 CONTINGENCY	3.70	3.40	3.60	3.50	2.80	2.80	1.90	1.30	1.00	.30	0
TOTAL COST	28.10	26.00	27.70	26.90	21.80	21.40	14.60	10.20	8.00	2.20	0
BENEFITS											
24 HARVEST	0	0	0	0	331.80	381.00	434.40	434.70	435.30	435.00	245.20
TOTAL BENEFIT	0	0	0	0	331.80	381.00	434.40	434.70	435.30	435.00	245.20
NET BENEFITS	-28.10	-26.00	-27.70	-26.90	310.00	359.60	419.60	424.50	427.30	432.80	245.20

Table 13

Economic Rates of Return for Alternative Assumptions
Used in the Sensitivity Analysis

(ERR in percent)

	Sawlog	Pulpwood
1. Half Sawlog Output Substitutes for Imports	14.8	n.a.
2. Half Sawlog Output is Exported	7.8	n.a.
3. Half Sawlog Output is Exported with 1% Annual World Sawnwood Price Increase and 75% Shadow Wage Rate for Unskilled Labour	13.5	n.a.
4. 1% Annual Pulpwood Stumpage Price Increase	n.a.	12.5
5. 2% Annual Pulpwood Stumpage Price Increase	n.a.	14.0
6. 75% Shadow Wage Rate for Unskilled Labour	n.a.	11.5
7. 25% Benefit Reduction	n.a.	8.5

APPENDIX 1

KENYA I

FOREST PLANTATIONS PROJECT

NET TRADE BALANCE IN FOREST PRODUCTS 1963-1967

	1963	1964	1965	1966	1967
	K£ '000				
Fuelwood and Charcoal	+ 77.9	+ 91.8	+ 153.5	+ 134.7	+ 147.6
Sawlogs	- 0.7	- 3.5	+ 2.6	+ 20.7	+ 7.0
Poles and Posts	+ 29.8	+ 23.8	+ 57.9	+ 81.9	+ 78.5
Sawn Timber	+125.9	+218.0	+ 326.8	+ 197.7	+ 173.3
of which: Softwood	(+153.8)	(+204.8)	(+ 273.4)	(+ 193.7)	(+ 246.8)
Hardwood	(- 27.9)	(+ 13.2)	(+ 53.4)	(+ 4.0)	(- 73.5)
Boxes, Crates, etc.	-110.4	-158.9	- 143.2	- 152.5	- 98.5
Builders' Woodwork	+ 85.8	+ 69.7	+ 107.0	+ 59.1	+ 105.8
Wood Carvings	-	+168.9	+ 280.3	+ 341.7	+ 294.5
Plywood	-115.7	-158.9	- 147.7	- 205.2	- 180.6
Matches	+ 3.4	- 18.1	+ 60.8	+ 57.2	+ 34.8
Other	- 2.5	+ 2.3	+ 29.9	- 23.0	- 78.7
Sub-total Wood:	+ 93.4	+235.1	+ 727.9	+ 512.3	+ 483.7
Paper, Paperboard and Manufactures thereof	-2 292.6	-2 163.3	-2 604.0	-2 843.6	-3 046.5
Fibre Board	- 41.7	- 87.9	- 118.2	- 113.6	- 104.1
Sub-total Pulp:	-2 334.3	-2 251.2	-2 722.2	-2 957.2	-3 150.6
Wattle Bark	+ 157.2	+ 197.6	+ 118.3	+ 142.8	+ 66.9
Other	+ 50.0	+ 22.1	+ 34.5	+ 26.7	+ 38.5
Sub-total:	+ 207.2	+ 219.7	+ 152.8	+ 169.5	+ 105.4
NET TRADE BALANCE:	-2 033.7	-1 796.4	-1 841.5	-2 275.4	-2 561.5

APPENDIX 2

KENYA — SAWLOG AND PULPWOOD PLANTATION PROJECT

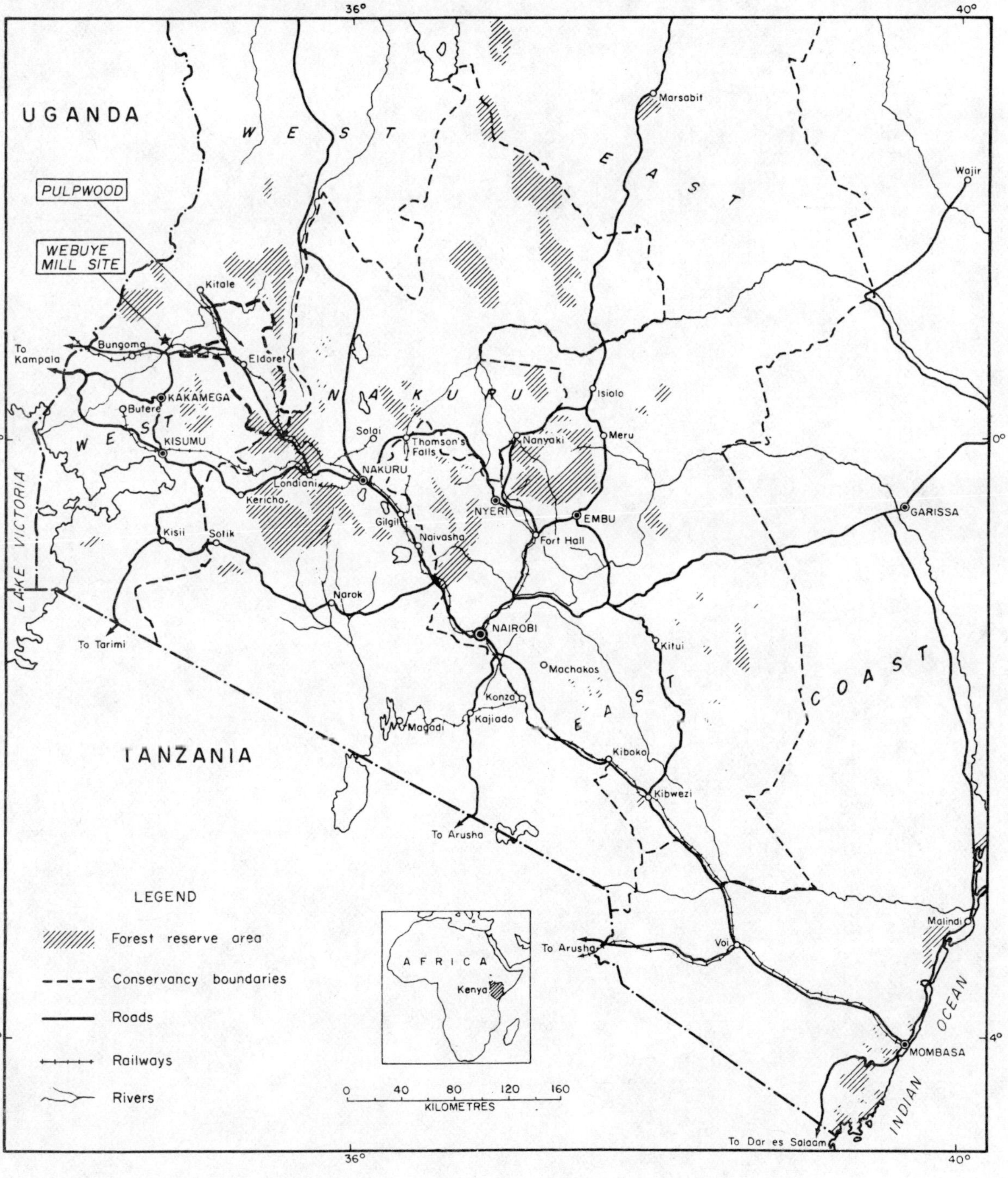

APPENDIX 3

KENYA I

FOREST PLANTATIONS PROJECT

MARKETING

A. Saw-wood

1. Output from the Saw-wood Circle would be largely of sawlogs, with smaller quantities of peeler logs being supplied to the recently established plywood industry.

2. Domestic demand projections are based on a 1962 FAO study. The FAO data have been amended by the mission in the light of up-to-date information on population growth and income per capita. Thus, in arriving at the medium forecast of demand used in the report it has been assumed that throughout the period 1960-2000, Kenya's population would increase by 3% per annum and per capita income by 3% per annum. The table below shows the projections of demand on these bases, and following this details of the estimates are discussed.

Projected Domestic Saw-wood Requirements

	1960	1967	1980	2000
Sawn Wood (1 000 m^3)	95	131	246	575
Public Sector	22	27	42	80
Commerce and Industry	21	29	57	147
Rural Households	32	47	99	249
Urban Households	20	48	48	99
Plywood (1 000 m^2)	675	835	1 860	5 575
Total Saw-wood Requirements ('000 m^3 roundwood equivalent)	243	337	566	1 217

Public Sector

3. This category includes all government, institutional, and religious buildings as well as private schools and hospitals. Capital expenditure on new buildings will be related to growth in population rather than to incomes. A rate of 3.25% has been taken as the rate of construction necessary to service a population growing at 3%.

Commerce and Industry

4. This sector will develop with the gradual emergence of cash earners in the economy. It has been assumed, however, that since efficiency in the use of wood will increase, the annual growth in commercial and industrial building will be one percentage point less than the annual growth of national income. This will give a rate of increase of about 5% per annum, and an annual consumption of sawn wood of 147 thousand m3 by the year 2000.

Rural Households

5. The use of sawn wood in private residential buildings in rural areas currently represents about a third of total estimated sawn wood consumption. Because of the rapid increase in the number of rural cash earners, it is expected that the growth of this sector will be the fastest. The use of sawn wood in rural households and associated buildings consequently is expected to reach 249 thousand m3 by year 2000, or about 43% of total domestic consumption.

Urban Households

6. This category comprises private residential buildings in all towns, settlements, and trading centres. It is assumed that the future growth rate of urban population will be 5% until 1980, a slower rate than in the past (6%) and thereafter 4% until 2000 as development of commerce and industry becomes spread more widely through the country. With the rates adopted, the use of sawn wood for urban households will reach about 99 thousand m3 in year 2000.

Plywood and Wood-based Board Products

7. Import statistics show an apparent consumption of plywood of 835 thousand m^2 in 1967, or approximately 0.1 m2 per capita per annum. Part of this is re-exported in the form of tea chests. The rate of increase in domestic use is difficult to assess since the uses of plywood overlap with those of other non-wood products and the extent to which substitution would occur is unknown. It is tentatively assumed that annual per capita apparent consumption would rise to 0.25 m2 by the year 2000, resulting in domestic sales of the order of 5.6 million m2 annually.

8. Two privately-owned plywood factories (one at Elgeyo, the other at Sokoro) started operations in 1968/69 with a total annual output capacity of 1.1 million m2. It is assumed that low production costs will permit these factories to compete with imports.

Total Domestic Roundwood Requirements

9. Assuming 2.0, instead of the present 2.5, as the conversion factor of sawn wood to roundwood in the year 2000 (thus anticipating better quality and uniformity of logs and some modernization of sawmill machinery) and a factor of 11.88 m3 of roundwood for each thousand m2 of plywood, total domestic requirements would amount to about 1.2 million m3 of roundwood in the year 2000.

Roundwood Requirements for Exports

10. A comparison of future domestic production (including the production from current planting programmes) with prospective requirements indicates a maximum deficit of about 850 thousand m3 roundwood equivalent in Tanzania and Uganda in 2000. These countries' plans to achieve self-sufficiency have so far been hampered by lack of funds. The scope of Kenyan exports to these countries would depend on the degree and the rapidity of realization of their plans. For a medium forecast, it is assumed that the deficit would be 425 thousand m3, of which 75% would be imported from Kenya. Import requirements in the Middle East are expected to reach about 3.3 million m3 roundwood equivalent of coniferous softwood by the year 2000. It is assumed that Kenya could supply 20% of these requirements. Finally, there is a market in Europe for small quantities of top grade softwood from Kenya, but it is assumed that only about 10% of Kenya's production would meet quality standards for such exports, as at present. In total, this medium forecast assumes total exports equivalent to 1.2 million m3 of roundwood annually in the early 2000's.

Projected Supply/Demand Situation

11. Adopting the preceding forecasts of domestic and export sales, the aggregate demand for roundwood would be about 2.4 million m3 in the year 2000. To assess the range of variation of total requirements, the following additional evaluations were made. In forecasting domestic demand, the income per capita growth rate was assumed to be comprised between 2% and 4.5% per annum. In forecasting exports, it was assumed, first, that Tanzania and Uganda would increase their own plantings, thus reducing their import requirements to between 25% and 75% of the level expected without an expansion of their programmes; and second, that Kenya would succeed in supplying between 15% and 25% of the coniferous wood import requirements in the Middle East. The lower and upper extremes of the range calculated with these assumptions are 1.8 million m3 and 3.0 million m3 of roundwood, respectively.

12. On the basis of conservative yield estimates, about 1.4 million m3 of roundwood annually would be available from the year 2000 from clear felling of project plantings, about 340 thousand m3 from thinning of later plantings, and at least 700 thousand m3 from clear felling of plantings made prior to the project. These sources would provide a total supply sufficient to meet the medium forecast requirement of 2.4 million m3 annually.

13. The supply and demand situation should be revised at regular intervals and planting programmes subsequent to the project adjusted accordingly. In addition, potential markets, both domestic and export, should be investigated and actual market promotion undertaken.

B. Pulpwood

14. It is assumed that, by the time project pulpwood plantings reach maturity, that is after 14-15 years, the **Webuye pulp and paper mill would be manufacturing** 75 000 m tons of paper, of which 37 500 m tons would be bleached and 37 500 m tons unbleached, requiring about 70 000 m tons of pulp. Assuming recovering rates of 42% and 47% respectively for bleached and unbleached pulp, the annual wood requirements of the pulpmill are estimated at 410 thousand m3, including an emergency reserve of about 5%. Project pulpwood plantings and earlier plantings would supply this demand as shown in the following exploitation schedule.

Projected Annual Exploitation Schedule for Pulpwood

	Plantings Made Prior to the Project		Project Plantings		
	Area (ha)	Volume ($000 \ m^3$)	Area (ha)	Volume ($000 \ m^3$)	Total Volume ($000 \ m^3$)
1981/82	630	165.9	—	—	165.9
1982/83	1 700	443.1	—	—	443.1
1983/84	400	132.2	1 210	313.1	445.3
1984/85	220	82.1	1 380	359.3	441.4
1985/86	—	—	1 500	409.9	409.9
1986/87	—	—	1 490	410.2	410.2
1987/88	—	—	1 480	410.8	410.8
1988/89	—	—	1 460	410.5	410.5
1989/90	—	—	830	250.2	250.2
Total Area Felled:	2 950		9 350		

ECONOMIC ANALYSIS OF FORESTRY PROJECTS: CASE STUDIES

FAO FORESTRY PAPER 17, SUPPLEMENT 1

CASE STUDY NO. 5

KENYA II SAWLOG AND PULPWOOD PLANTATION PROJECT

FOOD AND AGRICULTURE ORGANIZATION OF THE UNITED NATIONS

Rome, 1979

Preface

This is one of a series of case studies of forestry and forest industries projects that has been prepared by the Food and Agriculture Organization of the United Nations (FAO) in order to demonstrate methods of identifying, preparing and appraising projects in the forest sector. The programme of case studies has been made possible by a special allocation to FAO for this purpose from the Swedish International Development Authority (SIDA). This case study has been compiled by FAO with the assistance of the College of Forestry at the University of Minnesota.

The project reported on in this case study is a project of the Government of Kenya which has been supported by a loan from the World Bank. The case study is based largely on material contained in an appraisal of the project carried out by the World Bank (Report No. PA-14a). Additional information was gathered through a visit to the project. The case study describes the project as it was perceived at the time of appraisal in late 1968. This material has been released to FAO by the World Bank and the Government of Kenya exclusively for the purpose of teaching forestry planning methods. It may not be quoted or reproduced in part or in whole without permission.

ABBREVIATIONS

C.I.F./c.i.f.	=	Cost, insurance and freight
ERR	=	Economic rate of return
F.D.	=	Forest Department
F.O.B./f.o.b.	=	Free on board
FRR	=	Financial rate of return
NFW	=	Net Future Worth
NPW	=	Net Present Worth
SEV	=	Soil Expectation Value
SR	=	Soil Rent

CURRENCY EQUIVALENTS

Currency Unit	=	Kenya Shillings (K Sh)
KL	=	K Sh 20
K Sh 1.00	=	US$ 0.14
K Sh Million	=	US$ 140,056
£1.00 Sterling	=	K Sh 17.25

TABLE OF CONTENTS

INTRODUCTION

1. Kenya's forests are of fundamental importance to the country's future development. They provide a valuable economic base for industrial development and also play a crucial role in the preservation of water catchment areas and in the prevention of soil erosion. Forests help protect the country's plant and animal life and offer an important source of tourist revenue. Over the past decades, forests have continually been exploited for sawn timber, charcoal and other forest products. Also, parts of the natural forests have been given over to controlled or spontaneous settlement as a result of intense population pressure in many agricultural areas. As a result the main economically accessible reserves of commercially valuable species have been considerably diminished.

2. The Kenyan Government has over the past thirty years been replacing some of the country's forests with fast-growing exotic softwood plantations. An extensive forest development programme has been designed to meet an increasing domestic demand for forest products and to take advantage of expanding export opportunities, particularly to the Middle East and Europe.

3. Kenya's present long-term industrial forestry programme aims at the establishment of about 136 000 ha of sawlog plantations and 24 000 ha of pulpwood plantations by 1980. To date approximately 104 000 ha of sawlog plantations and 16 500 ha of pulpwood plantations have been established under the programme. In 1970 a project aimed at planting about 19 000 ha of sawlog plantations and 9 000 ha of pulpwood plantations over six years (1970-1975) was launched with the support of the World Bank (the first phase project; see Case Study No. 4). This project was completed one year ahead of schedule in 1974.

4. The project on which this case study is based (the second phase project) will, over six years (1975-1980) continue and expand the first phase project to include the total industrial plantation programme of the Kenyan Forest Department to provide raw material for sawmills and a pulp and paper plant. 1/ The project will complete the Government's target plantation programme to 1980. It will also support further development of plantations already established.

5. When fully established, the plantations would be the primary source of raw material for pulp and paper manufacturing and for the country's sawmills. Of about 46 000 ha planted during the period of implementation of the project, 38 000 ha would be used for sawlogs and the rest for pulpwood.

6. The estimated cost of the project including contingencies for the six-year plantation period is K Sh 396.5 million (US$ 55.5 million with a foreign exchange component of about US$ 12.9 million).

7. This Case Study is divided into five sections. Section A provides background for the project. Section B gives technical details of the project including estimates of input requirements and output levels. Section C analyses project costs and Section D discusses project benefits. Section E provides a comparison of costs and benefits and analyses the sensitivity of expected project performance to changes in input and output assumptions.

8. The Case Study is a self-contained example of a "time slice" project appraisal, including separate analyses of independent project components. In combination with Case Study No. 4, it provides useful insights into the evolution of a major plantation programme, and of the planning of the latter.

1/ Policies and practices described in this case study are not necessarily those actually employed in Kenya, either at present or during the period covered by the case study.

A. BACKGROUND [1/]

9. Forestry contributed about 2 percent of GNP and forest products represented less than 5 percent of total exports in 1973. Kenya remains a net importer of forest products, mainly paper and paperboard products.

10. There are about 180 licensed sawmills in Kenya, and many more unlicensed ones. Most are small operations (only five have an annual capacity of over 3 000 m^3(s)). They are designed primarily to provide sawn timber for the growing rural market, are handicapped by inefficient operations, and produce a low quality product. Total output has averaged about 100 000 m^3(s) annually over the last three years, about 10 percent of which was exported. To meet projected demand, by 1980 sawnwood production capacity would have to be increased to 210 000 m^3(s). This could be realized without major increase in new capacity if significant improvements in the efficiency of existing mills were achieved.

11. Over the past few years the Kenyan government has made increased efforts to assist mills in identifying bottlenecks and improving their efficiency, but only a limited impact has been made thus far. The government in some cases also has designed its industrial plantations in blocks of at least 4 000 ha to encourage the construction of larger, more efficient mills turning out a higher quality product, but this also has failed to produce additional capacity. The project will thus provide further technical assistance to increase the efficiency of sawmilling.

12. There are a number of mills producing plywood. Two mills, one at Elburgon and one at Elgeyo, together produced about 1 million m^2 of plywood in 1972. A third mill with a capacity of 1.5 million m^2 has been constructed at Eldoret, and mills are planned in the Mount Kenya area and the Masai Mau forest.

13. In addition there is the newly constructed pulp and paper mill at Webuye which will eventually be one of the country's largest industrial enterprises. The mill has a capacity of 54 000 air dried metric tons per annum, and its expansion to 75 000 tons by the mid-1980s is being considered. The mill will require about 300 000 m^3 of pulpwood a year at full operation at present capacity, equivalent to the output of about 24 000 ha of plantations; this would need to be increased to some 35 000 ha of plantations to meet expected increases in the mill's capacity.

14. The first phase industrial plantations project was located within four main administrative divisions west of the Rift Valley which account for about 75 percent of the Government's long-term programme. In general, the project has been carried out satisfactorily. In the early years difficulties were experienced in land acquisition, necessitating the relocation of some plantation development and in the establishment of pulpwood plantations in the Turbo area due mainly to soil deficiencies. The planting failures at Turbo had been largely overcome by 1973-1974.

15. The Kenya Government's programme for forestry development has been designed with due regard to environmental considerations. The current system of plantation development and the possible impact of fast-growing plantations on the hydrological balance in forest areas has been studied over a period of fifteen years, and there is sufficient evidence to confirm that there is no adverse impact on the water regime by changing from one form of forest crop (indigenous forest) to another (fast-growing plantations). Also, when the Kenya Government's long-term plantation programme is fully implemented, it will result in the replanting of only about 20 percent of the area of indigenous forests designated as forest reserves, leaving some 1.5 million ha untouched by the programme. Approximately 40 000 ha of this area has already been set aside by the government as Nature Reserves or National Parks.

1/ For additional information on the forest sector of Kenya, see <u>Case Study No.4: Kenya I Sawlog and Pulpwood Plantation Project</u>.

B. DESCRIPTION OF THE SECOND FORESTRY PLANTATION PROJECT

16. The proposed project (analysed here) would be implemented during the six-year period 1975-1980. The project includes:

(a) the afforestation of 32 000 ha for sawlogs and 8 000 ha for pulpwood, and the reforestation of 6 000 ha of clear-felled plantations, mainly for sawlog products (Table 1);

(b) further development of 120 000 ha of existing sawlog and pulpwood plantations;

(c) construction, up-grading and maintenance of forest roads;

(d) construction of staff housing and administration buildings;

(e) studies, technical assistance, and training.

17. The plantation activities include establishment of nurseries, land clearing and preparation, planting and weeding. About 16 million seedlings will be required per year. These will be raised in some 90 existing nurseries in addition to 16 new nurseries to be established by the project.

18. All of the sawlog planting and about 90 percent of the pulpwood planting will take place in already-exploited indigenous forest areas to be cleared using the traditional "shamba" system. 1/ The planting spacing will be 2.5 m x 2.5 m for sawlog crops and 2.7 x 2.0 m for pulpwood crops. The main pulpwood species to be used are pines (Pinus patula and Pinus elliotti) and eucalypts (Eucalyptus saligna) and the primary sawlog species is cypress (Cupressus lusitanica). Planting will be done manually. Past experience indicates that a failure rate of about 10 percent can be expected and replanting will be carried out either in the same or the next planting year.

19. Sawlogs will be grown on a 25-year rotation, with commercial thinnings at age 10 (35 m^3/ha) and age 15 (30 m^3/ha). The final harvest yield is expected to be 360 m^3/ha. Pulpwood will be grown on a 15-year rotation with a final harvest yield of 225 m^3/ha.

20. Project activities include silvicultural and forest management elements such as pruning, thinning, fire control and other services to be provided not only to the project's plantation but also to about 120 000 ha of plantations already established.

21. In addition the project provides for the construction of about 160 km of all-weather forest roads and plantation access tracks per year. Also, 20 km of service roads will be upgraded each year.

22. To meet the requirements of administration and housing, the project will include all direct and indirect administrative overhead expenses in support of the industrial plantation programme and related activities such as forestry protection. The project's staffing will comprise 36 forestry officers, 106 technical level foresters, 165 forest rangers, 164 clerks, 184 drivers and 660 forest guards.

23. To carry out the project, it is necessary to construct one new division headquarters, 4 new stations and 12 new sub-stations. The housing of some 1 300 forest labourers will be improved and some 2 000 new houses constructed at the new stations and sub-stations, making a total housing programme of some 3 300 units.

1/ See Case Study No. 4, page 3, for a description of the shamba system.

24. The project requires investment in equipment for several different project activities including afforestation, road building, administration, buildings and housing. Equipment includes bulldozers, loaders, graders, rollers, lorries, tractors, saloon cars, water pumps, and various other equipment.

25. In connection with studies and technical assistance, the project will provide for a total of 5 man-years of staff time. Staff to be recruited will include a land use planning economist, a forestry specialist, an agronomist, a hydrologist and an ecologist. Funds will also be provided for training skills to interpret data from the Earth Resources Technology Satellite (ERTS) for classifying land use categories and recording changes in land use patterns. Also, in order to improve efficiency and production of the country's sawmills, the project will include the provision of technical assistance comprising three sawmill advisers for three years and the establishment and operation of a small-scale sawmill training centre.

26. Each year 40 ha of trial plantations will be established under the project, and they will provide a basis for an evaluation of forest potential in the Arabuko-Jukoke and the Bone-Lungi semi-arid areas of the Coast Province which appear to have potential for considerable forestry plantation development.

C. PROJECT COSTS

Financial Costs

27. The total project financial cost for the six year plantation period is estimated to be Ksh 396.5 million (US$ 55.5 million). The foreign exchange component is estimated to be about US$ 12.9 million. Appendix 1 presents detailed information about the assumptions relating to physical inputs and the input unit prices which were used in the analysis. Based on these results, Table 2 displays total project financial costs, and a more detailed yearly schedule is presented in Table 3. Three plantation components - sawlog circle, pulpwood circle Eldoret Division and pulpwood circle Turbo Division - are identified and analysed separately because of geographical and/or output differences. The financial costs of the three project components are shown, on a per hectare basis, in Tables 9, 10 and 11. 1/ The cost streams in all cases include all costs of the project components which will be incurred beyond the six-year planting period since they must be incurred in order to receive the benefits. However, the financial costs of studies, technical assistance, training, etc., have been excluded from the cost flows since benefits of these activities are mostly unquantifiable and therefore the analysts preferred to treat these activities and their effects in a purely qualitative manner.

Economic Costs

28. Economic costs are also derived on a per hectare basis for the three plantation components. Individual silvicultural cost items are listed in Table 8 in Appendix 1. Silvicultural labour has been shadow priced at 75 percent of the financial wage of Ksh 12.68 per man day, i.e. at Ksh 9.51 per man day. The silvicultural costs are entered directly into the economic cash flow tables 12, 13 and 14 1/ for the three project components.

29. Non-silvicultural financial costs also have been estimated in Appendix 1 and are listed in text Table 4. 2/ Non-silvicultural economic costs are derived by shadow pricing

---

1/ The financial cash and economic value flow tables (tables 9-14) appear on pages 15-20.

2/ Non-silvicultural costs are defined to be those costs which are not directly apportioned on a per hectare basis, such as buildings, vehicles, etc. Non-silvicultural financial costs are determined for the entire plantation programme of 160 000 ha.

the foreign exchange components of these costs at 130 percent of the official foreign exchange rate.

30. These non-silvicultural costs need to be apportioned to each of the three project components on a per hectare basis for inclusion in the per hectare value flow tables. The approach used is as follows:

(a) In Table 4 each non-silvicultural cost is expressed as an average (unweighted) amount, with the foreign exchange component shadow priced as indicated above.

(b) In Table 6 each unweighted cost from Table 4 is adjusted for each project component to reflect the proportion of the total annual unskilled silvicultural labour used in that project component. Unskilled silvicultural labour was used as a basis for weighting non-silvicultural costs since it is likely to be a good indicator of the intensity of project component activity at any given time. The top half of Table 5 shows the yearly requirements of unskilled silvicultural labour for each project component, as well as the weights based on those requirements, which are used to derive the values in Table 6.

(c) Finally, each weighted total from Table 6 is split up into yearly amounts by multiplying it by the percentage of unskilled silvicultural labour used per year in that project component. These yearly percentages are found in the bottom half of Table 5. The yearly weighted non-silvicultural costs are then entered into the economic cash flow table (Tables 12, 13 and 14).

D. PROJECT BENEFITS

Financial Revenue

31. The financial revenue is calculated on the basis of the unit prices the Government of Kenya expects to obtain for stumpage multiplied by the volumes thinned and harvested. Physical yields per hectare are summarized in Table 7.

32. Financial stumpage prices expected for sawlogs are derived from expected sawnwood prices assuming certain conversion factors between sawlogs and sawnwood (which are explained in Table 8). The financial stumpage price for pulpwood is expected to be equal to Ksh 30/m^3. The results of the financial revenue calculations are displayod in Tables 9, 10 and 11.

Economic Direct Benefits

33. In calculating the economic unit value of the pulpwood output, it is assumed that if the project fails to materialize, pulpwood chips would have to be imported to run the Webuye Pulpmill. Therefore, the project's output of pulpwood will save an amount of foreign exchange which is then shadow priced at 130 percent of the financial value of that foreign exchange. The procedure followed is explained in more detail in Appendix 3 and the result is displayed in Table 8.

34. In estimating sawlog values, it is assumed, based on a demand study carried out previously, that 75 percent of the sawlog output will be consumed domestically, substituting for volumes which otherwise would have been imported in the absence of the project. The remaining 25 percent will be exported.

35. The economic benefit streams calculated using the unit values above are displayed in Tables 12, 13 and 14.

36. Inherited values and salvage values were included in the "costs" listed in Tables 4 and 6. Inherited values (costs) are the values at the start of the project of vehicles, equipment, tools, stores, buildings, and road equipment which were purchased or built during the plantation programme preceding the present project. The appropriate inherited value for these items is the value they have, if any, in their next best (highest valued) use other than the present project. Likewise, salvage values (benefits) are the values of the same types of items at the termination of the rotation, valued in terms of their best alternative use or in terms of what they could be sold for. The procedures used to estimate inherited and salvage values are explained in detail in Appendix 2.

37. Both physical and price contingencies were included in the total project costs. However, only the physical contingencies were included in project financial and economic cash flow analyses (see Appendix 1). Price contingencies are important for budgetary purposes, but by definition do not affect real (economic) rates of return.

Indirect Benefits

38. Under the project, benefits would be widely distributed among those to be employed directly in the plantation programme (who would fall within the lower rural income brackets), and in related processing industries, as well as among both rural and urban users of the final product. The project would provide continued employment of about 4 750 persons over six years, although this would fall slightly to about 4 500 persons when the main planting programme is completed. However this decrease is expected to be offset by secondary employment opportunities arising from expanding processing industries.

39. As has been mentioned before, there is also a whole array of indirect effects generated by the project which are not quantifiable. One case in point relates to the substantial efforts carried out under the project to train local technicians, and the experience which will be acquired by local project staff during the implementation of the project which could then be favourably extended to other future afforestation programmes. The project will probably provide additional incentives for rural people and will help to distribute the population of Kenya more evenly.

E. COMPARING COSTS AND BENEFITS

40. The financial and economic rates of return for each of these project components were estimated directly from the relevant financial cash and economic value flow tables. The financial and economic value streams differ only in what relates to:

- wages, which in the economic analysis are assumed to be equal to 75 percent of financial values;
- foreign exchange rate, which in the economic analysis is equal to 130 percent of the financial value;
- stumpage values, which are higher in the economic analysis.

41. The estimated financial and economic rates of return for the project components are as follows:

Component	FRR	ERR
Sawlog	6.4 %	15.5 %
Pulpwood Eldoret	2.4 %	21.7 %
Pulpwood Turbo	1.7 %	20.9 %

42. The relative importance of the shadow-priced variables in explaining the discrepancy between financial and economic rates of return was tested (Table 15). The removal of the shadow foreign exchange rate lowers each of the component's original ERR by about 1.0 to 1.4 percent. The removal of the shadow wage rate lowers each original ERR by about one-half of 1 percent. The combined effect of removing simultaneously both the foreign exchange and the wage shadow rates produces a drop of only about 2 percent in the original ERR's.

43. The sensitivity analysis shows that the major reason for the discrepancy between the FRR and ERR is the value of the final output assumed. Thus, if the pulpwood unit economic value is reduced from the original economic value of 475.8 Ksh/m^3 to the very much lower value of 70 Ksh/m^3 - which is still more than double the price assumed in the financial analysis - the ERR falls by about 12 percent.

44. For the sensitivity analysis, the sawlog component was split into its two main elements: import substitution and export. Each was analysed separately as in reality each could be implemented independently. The ERR for the output for import substitution was 16.8 percent and for the export output was 8.2 percent. Further tests of the financial and economic sensitivity of each are shown and discussed in Appendix 4.

KENYA II PLANTATION PROJECT

Table 1, Existing and Projected Planting Areas

(ha)

	Plantations Established as of 31.12.74	Planting Programme						
		Year 0	Year 1	Year 2	Year 3	Year 4	Year 5	Total
Pulpwood	16 500	2 000	2 000	1 000	1 000	1 000	1 000	8 000
Sawlogs	104 080	4 570	5 540	5 430	5 430	5 430	5 430	31 830
Total	120 580	6 570	7 540	6 430	6 430	6 430	6 430	39 830
Replanting		700	700	1 000	1 200	1 300	1 300	6 200
Total Planting plus Replanting		7 270	8 240	7 430	7 630	7 730	7 730	46 030

KENYA II PLANTATION PROJECT

Table 2, Summary of Total Project Financial Costs
(Ksh million)

	Local	Foreign	Total
Afforestation	120.1	15.3	117.4
Project Management			
Administration	81.6	8.5	90.1
Building Housing	32.9	3.9	36.8
Equipment	5.9	19.7	25.6
Studies and Technical Assistance	6.1	11.1	17.2
Sub-total	239.4	66.5	305.9
Contingencies			
Physical	2.3	2.9	5.2
Price	62.5	22.9	85.4
Sub-total	64.8	25.8	90.6
Total Project Cost	304.2	92.3	396.5

KENYA II PLANTATION PROJECT

Table 3, Project Financial Costs (KSh '000)

	Year 0			Year 1			Year 2			Year 3			Year 4			Year 5		
	Local	Foreign	Total	Local	Foreign	Total	Local	Foreign	Total	Local	Foreign	Total	Local	Foreign	Total	Local	Foreign	Total
AFFORESTATION																		
Labour	16 058	-	16 058	16 160	-	16 160	15 448	-	15 448	15 604	-	15 604	15 610	-	15 610	15 645	-	15 645
Machine Operating Cost	48	113	161	47	109	156	34	80	114	16	38	54	7	18	25	8	18	26
Vehicle Operating Cost	992	2 314	3 306	992	2 315	3 307	992	2 314	3 306	992	2 315	3 307	992	2 314	3 306	992	2 315	3 307
Fertilizer & Tools	248	187	435	248	187	435	224	169	393	230	174	404	233	176	409	233	176	409
Sub-total	17 346	2 614	19 960	14 447	2 611	20 058	16 098	2 563	19 261	16 842	2 527	19 369	14 872	2 508	19 350	16 878	2 509	19 387
FOREST ROADS																		
Road Construction:																		
Staff & Labour	1 063	-	1 063	1 064	-	1 064	1 063	-	1 063	1 064	-	1 064	1 063	-	1 063	1 064	-	1 064
Machine Operating Cost	335	783	1 118	335	783	1 118	335	783	1 118	335	783	1 118	335	783	1 118	335	783	1 118
Workshop Running Cost	33	32	65	32	33	65	33	32	65	32	33	65	33	32	65	32	33	65
Tools & Miscellaneous	11	11	22	11	11	22	11	11	22	11	11	22	11	11	22	11	11	22
Office Running Cost	35	8	43	35	8	43	35	8	43	35	8	43	35	8	43	35	8	43
Bridge & Culvert Material	122	30	152	121	31	152	122	30	152	121	31	152	122	30	152	121	31	152
Road Maintenance	185	432	617	188	439	627	193	450	643	196	458	654	201	469	670	206	480	686
Sub-total	1 784	1 296	3 080	1 786	1 305	3 091	1 792	1 314	3 106	1 794	1 324	3 118	1 800	1 333	3 133	1 804	1 396	3 150
PROJECT MANAGEMENT																		
Administration:																		
Staff & Labour	12 835	-	12 835	12 835	-	12 835	12 835	-	12 835	12 835	-	12 835	12 835	-	12 835	12 835	-	12 835
Office Running Cost	167	39	206	167	39	206	167	39	206	167	39	206	167	39	206	167	39	206
Vehicle Operating Cost	588	1 372	1 960	588	1 372	1 960	599	1 397	1 996	600	1 397	1 997	599	1 397	1 996	600	1 397	1 997
Building & Housing																		
Construction Material	4 124	815	4 939	2 927	578	3 505	2 927	578	3 505	1 109	219	1 328	1 108	219	1 327	1 109	219	1 328
Construction Labour	3 292	-	3 292	2 337	-	2 337	2 337	-	2 337	885	-	885	885	-	885	885	-	885
Maintenance Material	253	206	459	252	207	459	253	206	459	252	207	459	253	206	459	252	207	459
Maintenance Labour	1 248	-	1 248	1 248	-	1 248	1 248	-	1 248	1 248	-	1 248	1 248	-	1 248	1 248	-	1 248
Sub-total	22 507	2 432	24 939	20 354	2 196	22 550	20 366	2 220	22 586	17 096	1 862	18 958	17 095	1 861	18 956	17 096	1 862	18 958
EQUIPMENT	1 541	6 136	7 677	920	3 234	4 154	925	3 358	4 283	950	3 189	4 139	781	1 893	2 674	782	1 893	2 675
STUDIES, TECHNICAL ASSISTANCE & TRIALS																		
Land Use & Catchment Study	353	1 628	1 981	-	786	786	-	786	786	-	786	786	-	786	786	-	785	785
Socio-Economic Study	281	543	824	123	372	495	-	-	-	-	-	-	-	-	-	-	-	-
Technical Assistance:																		
Assistance to Sawmilling Industry	643	1 159	1 802	310	1 007	1 317	308	1 008	1 316	173	41	214	173	41	214	173	41	214
Training Sawmill	660	807	1 467	220	48	268	220	48	268	220	48	268	220	48	268	220	48	268
Coastal Trials:																		
Arabuko Sokoke	32	10	42	191	21	212	192	21	213	191	21	212	192	21	213	191	21	212
Boni & Lungi	256	78	334	83	19	102	120	24	144	121	28	149	121	28	149	121	28	149
Sub-total	2 225	4 225	6 450	927	2 253	3 180	840	1 887	2 727	705	924	1 629	706	924	1 630	705	924	1 629
TOTAL	45 403	16 703	62 106	41 434	11 599	53 033	40 621	11 342	51 963	37 387	9 826	47 213	37 224	8 519	45 743	37 265	8 534	45 799
CONTINGENCIES																		
Physical Contingencies	701	899	1 600	438	467	906	439	480	919	259	428	687	242	298	540	242	298	540
Price Escalation	1 198	1 640	2 838	5 298	2 525	7 823	9 103	3 701	12 804	11 715	4 338	16 043	15 685	4 837	20 522	19 525	5 866	25 391
Sub-total	1 899	2 539	4 438	5 736	2 992	8 728	9 542	4 181	13 723	11 974	4 756	16 730	15 927	5 135	21 062	19 767	6 164	25 931
TOTAL PROJECT COST	47 302	19 242	66 544	47 170	14 591	61 761	50 163	15 523	65 686	49 361	14 582	63 943	53 151	13 654	66 805	57 032	14 698	71 730

KENYA II PLANTATION PROJECT

Table 4, Summary of Unweighted Non-Silvicultural Costs and Salvage Benefits for Both Sawlog and Pulpwood Plantations

Item	Single Cost: Amount 3/ (KL) (1)	Year (2)	Area ('000 ha) (3)	Amount per ha (KSh/ha) (4)	Foreign Exchange Component (%) (5)	Amount per ha 130% Shadow F.E. Rate (KSh/ha) (6)	Annual Cost: Amount 3/ (KL) (7)	Year (8)	Area ('000 ha) (9)	Amount per ha (KSh/ha) (10)	Foreign Exchange Component (%) (11)	Amount per ha 130% Shadow F.E. Rate (12)	Total Amount per ha 130% Shadow F.E. Rate: Sawlog (KSh/ha) (13)	Pulpwood (14)
Vehicle-Equipment Purchase	-	-	-	-	-	-	135 465	-2/24 1/	160	16.933	100	22.013	594.351	374.221
Vehicle-Equipment Running	-	-	-	-	-	-	301 702	-2/24 1/	160	37.713	70	45.633	1232.091	775.761
Vehicle-Equipment Inherited	406 395	-2	160	50.799	100	66.039	-	-	-	-	-	-	-	-
Tools-Stores	-	-	-	-	-	-	22 000	-2/24 1/	160	2.750	75	3.369	90.963	57.273
Tools-Stores Inherited	33 000	-2	160	4.125	75	5.053	-	-	-	-	-	-	-	-
Building Construction	1 809 406	-2	160	226.176	10	232.961	-	-	-	-	-	-	-	-
Building Maintenance	-	-	-	-	-	-	89 100	-2/24 1/	160	11.138	10	11.472	309.744	195.024
Building Replacement	-	-	-	-	-	-	57 691	-2/24 1/	160	7.211	10	7.428	200.556	126.276
Building Inherited	721 136	-2	160	90.142	10	92.846	-	-	-	-	-	-	-	-
Road Construction-capital	202 620	-1	39.830	101.742	100	132.265	-	-	-	-	-	-	-	-
Road Construction-operating	128 920	-1	6.638	388.411	30	423.368	-	-	-	-	-	-	-	-
Road Maintenance	-	-	-	-	-	-	-	-2/24 1/	-	6.417	60	7.572	189.300	113.580
Road Equipment Inherited	141 625	-1	39.830	71.115	100	92.449	-	-	-	-	-	-	-	-
Staff Requirements	-	-	-	-	-	-	594 140	-2/24 1/	160	74.268	10	76.496	2065.392	1300.432
Office Running	-	-	-	-	-	-	44 825	-2/24 1/	160	5.603	10	5.771	155.817	98.107
Vehicle-Equipment Salvage	406 395	25 2/	160	50.799	100	66.039								
Tools-Stores Salvage	33 000	25 2/	160	4.125	75	5.053								
Buildings salvage (sawlogs only)	1 173 488	25 2/	160	146.686	10	151.087								
Buildings salvage (pulpwood only)	1 354 428	25	160	169.304	10	174.383								

1/ Year 24 is for sawlogs only. For pulpwood the ending year is year 16.

2/ Year 25 is for sawlogs only. For pulpwood the year is year 15.

3/ Columns 1 and 7 are yearly amounts for the total completed plantation area (160 000 ha) or for the total six-year project area (39 830 ha). Columns 3 and 9 give the relevant areas. Per hectare average costs (columns 4 and 10) are the result of dividing column 1 by column 3, and column 7 by column 9, respectively. In the economic analyses, foreign exchange is shadow priced at 130% of the official exchange rate. Columns 5 and 11 indicate the foreign exchange components of each non-silvicultural cost, and columns 6 and 12 give single and annual amounts, respectively, when foreign exchange has been weighted at 130%.

KENYA II PLANTATION PROJECT

Table 5, Yearly Unskilled Silvicultural Labour Requirements

	YEAR -2	-1	0	1	2	3	4	5	6	7	8	9	10	11	12	13	14	15	16	17	18	19	20	21	22	23	24	25	Total	average annual man days per ha	proportion of total plantation area (%)	weighted total cost factor
Total On-site Labour Required (man days/ha)																																
Sawlog	37.7	-	16.62	0.22	27.22	0.22	12.22	7.22	12.22	0.22	12.22	0.22	12.72	0.22	0.22	0.22	0.22	0.72	0.22	0.22	0.22	0.22	0.22	0.22	0.22	0.22	0.22	-	142.6	5.281	84.727	1.0203
Pulpwood - Eldoret	37.7	-	16.62	0.22	15.22	0.22	0.22	0.22	0.22	0.22	0.22	0.22	0.22	0.22	0.22	0.22	0.22	-											72.4	4.259	7.637	0.8750
Pulpwood - Turbo	37.7	-	25.02	7.22	7.22	0.22	0.22	0.22	0.22	0.22	0.22	0.22	0.22	0.22	0.22	0.22	0.22	-											79.8	4.929	7.637	0.9523
Yearly Proportion of Total On-site Labour (%)																																
Sawlog	26.44	-	11.65	0.15	19.09	0.15	8.57	5.06	8.57	0.15	8.57	0.15	8.92	0.15	0.15	0.15	0.15	0.15	0.15	0.15	0.15	0.15	0.15	0.15	0.15	0.15	0.15	-	99.92			
Pulpwood - Eldoret	52.07	-	22.96	0.30	21.02	0.30	0.30	0.30	0.30	0.30	0.30	0.30	0.30	0.30	0.30	0.30	0.30	-											99.95			
Pulpwood - Turbo	41.24	-	31.35	9.05	9.05	0.28	0.28	0.28	0.28	0.28	0.28	0.28	0.28	0.28	0.28	0.28	0.28	-											100.05			

KENYA II PLANTATION PROJECT

Table 6, Summary of Weighted Non-Silvicultural Costs and Salvage Benefits for Both Sawlog and Pulpwood Plantations

Cost or Benefit	Type of Cost/Benefit[1]	Year(s)	Sawlogs: Unweighted Total Amount - 130% Shadow F.E. Rate (KSh/ha)	Sawlogs: Weight	Sawlogs: Weighted Total Amount - 130% Shadow F.E. Rate (KSh/ha)	Pulpwood - Eldoret: Unweighted Total Amount - 130% Shadow F.E. Rate (KSh/ha)	Pulpwood - Eldoret: Weight	Pulpwood - Eldoret: Weighted Total Amount - 130% Shadow F.E. Rate (KSh/ha)	Pulpwood - Turbo: Unweighted Total Amount - 130 % Shadow F.E. Rate (KSh/ha)	Pulpwood - Turbo: Weight	Pulpwood - Turbo: Weighted Total Amount - 130% Shadow F.E. Rate (KSh/ha)
Vehicle-Equipment Purchase	AC	d[2]	594.351	1.0203	606.416	374.221	0.8750	327.443	374.221	0.9523	356.371
Vehicle-Equipment Running	AC	d	1232.091	1.0203	1257.102	775.761	0.8750	678.791	775.761	0.9523	738.757
Vehicle-Equipment Inherited	SC	-2	66.039	1.0203	67.380	66.039	0.8750	57.784	66.039	0.9523	62.889
Tools-Stores	AC	d	90.963	1.0203	92.810	57.273	0.8750	50.114	57.273	0.9523	54.541
Tools-Stores Inherited	SC	-2	5.053	1.0203	5.156	5.053	0.8750	4.421	5.053	0.9523	4.812
Building Construction	SC	-2	232.961	1.0203	237.690	232.961	0.8750	203.841	232.961	0.9523	221.849
Building Maintenance	AC	d	309.744	1.0203	316.032	195.024	0.8750	170.646	195.024	0.9523	185.721
Building Replacement	AC	d	200.556	1.0203	204.627	126.276	0.8750	110.492	126.276	0.9523	120.253
Building-Inherited	SC	-2	92.846	1.0203	94.731	92.846	0.8750	81.240	92.846	0.9523	88.417
Road Construction	SC	-1	555.633	1.0000[3]	555.633	555.633	1.0000[3]	555.633	555.633	1.0000[3]	555.633
Road Maintenance	AC	0/24[4]	189.300	1.0000[3]	189.300	113.580	1.0000[3]	113.580	113.580	1.0000[3]	113.580
Road Equipment Inherited	SC	-1	92.449	1.0000[3]	92.449	92.449	1.0000[3]	92.449	92.449	1.0000[3]	92.449
Staff Requirements	AC	d	2065.392	1.0203	2107.319	1300.432	0.8750	1137.878	1300.432	0.9523	1238.401
Office Running	AC	d	155.817	1.0203	158.980	98.107	0.8750	85.844	98.107	0.9523	93.427
Vehicle-Equipment Salvage	SB	25[5]	66.039	1.0203	67.380	66.039	0.8750	57.784	66.039	0.9523	62.889
Tools-Stores Salvage	SB	25[5]	5.053	1.0203	5.156	5.053	0.8750	4.421	5.053	0.9523	4.812
Buildings Salvage	SB	25[5]	151.087	1.0203	154.154	174.383	0.8750	152.585	174.383	0.9523	166.065

1/ AC = annual cost; SC = single cost; SB = single benefit

2/ "d" means that an annual cost is distributed from years minus 2 through 24 for sawlogs (through 14 for pulpwood) according to the proportion of total man days of unskilled silvicultural labour used.

3/ Road costs are unweighted since they are assumed to be equal for all hectares.

4/ Year 24 is the ending year for sawlogs only. For pulpwood the ending year is 14. This annual cost is not distributed according to silvicultural labour distribution. It is equally distributed over the rotation, beginning in year 0, which amounts to KSh 7.6 per hectare per year.

5/ Year 25 for sawlogs only. Year 15 for pulpwood.

KENYA II PLANTATION PROJECT

Table 7, Physical output per hectare (m^3)

Year	Sawnwood	Pulpwood
10	35	
15	30	225
25	360	

KENYA II PLANTATION PROJECT

Table 8, Output Unit Values

Sawlogs	Sawnwood Ksh/m^3	Conversion Factor	Stumpage Financial Price Ksh/m^3	Economic Unit Value Ksh/m^3
Sawlogs				
1st Thinning	166.67	0.30	50.00	255
2nd Thinning	166.67	0.35	58.33	297
Harvest	166.67	0.42	70.00	357
Pulpwood			30.00	475.8

See Appendix 1 for further details

KENYA II PLANTATION PROJECT

Table 9, Financial Cash Flow - Sawlog Component

KEN SHILLING/HECTARE

EXPENDIT (YEAR)	-2	-1	0	1	2	3	4	5
1 NURSERY	0	601.90	0	0	0	0	0	0
2 LAND PREPARE	478.00	0	0	0	0	0	0	0
3 PLANTING	0	0	177.50	0	0	0	0	0
4 PLANT SURVEY	0	0	12.70	0	0	0	0	0
5 BEATING UP	0	0	17.80	0	0	0	0	0
6 WEEDING	0	0	0	0	190.20	0	0	0
7 PRUNING	0	0	0	0	152.20	0	152.20	0
8 PRE-COM THIN	0	0	0	0	0	0	0	88.80
9 MARKING OUT	0	0	0	0	0	0	0	0
10 FIRE PROTECT	0	0	1.50	1.50	1.50	1.50	1.50	1.50
11 OTH. PROTECT	0	0	1.30	1.30	1.30	1.30	1.30	1.30
12 VE/EQUIP PUR	123.30	0	54.30	.70	89.00	.70	40.00	23.60
13 VE/EQUIP RUN	274.70	0	121.00	1.60	198.30	1.60	89.00	52.00
14 VE/EQU INHER	51.80	0	0	0	0	0	0	0
15 TOOLS/STORES	20.00	0	8.80	.10	14.50	.10	6.50	3.80
16 TOOLS INHER	4.20	0	0	0	0	0	0	0
17 BUILD CONSTR	230.80	0	0	0	0	0	0	0
18 BUILD MAINT	88.10	0	35.70	.50	58.60	.50	26.30	15.50
19 BUILD REPLAC	52.50	0	23.10	.30	37.90	.30	17.00	10.10
20 BUILD INHER	92.00	0	0	0	0	0	0	0
21 ROAD CONSTRU	0	490.20	0	0	0	0	0	0
22 ROAD MAINT	0	0	6.40	6.40	6.40	6.40	6.40	6.40
23 ROAD INHER	0	71.10	0	0	0	0	0	0
24 STAFF REQUIR	540.90	0	238.40	3.10	390.60	3.10	175.30	103.50
25 OFFICE RUN	40.80	0	18.00	.20	29.50	.20	13.20	7.80
TOTAL EXPEND	1937.10	1163.20	716.50	15.70	1170.00	15.70	528.70	314.90
RECEIPTS								
26 THINNINGS	0	0	0	0	0	0	0	0
27 HARVEST	0	0	0	0	0	0	0	0
28 VE/EQU SALVG	0	0	0	0	0	0	0	0
29 TOOLS SALVG	0	0	0	0	0	0	0	0
30 BUILD SALVG	0	0	0	0	0	0	0	0
TOTAL RECTS	0	0	0	0	0	0	0	0
NET RECEIPTS	-1937.10	-1163.20	-716.50	-15.70	-1170.00	-15.70	-528.70	-314.90

EXPENDIT (YEAR)	6	7	8	9	10	11	12
1 NURSERY	0	0	0	0	0	0	0
2 LAND PREPARE	0	0	0	0	0	0	0
3 PLANTING	0	0	0	0	0	0	0
4 PLANT SURVEY	0	0	0	0	0	0	0
5 BEATING UP	0	0	0	0	0	0	0
6 WEEDING	0	0	0	0	0	0	0
7 PRUNING	152.20	0	152.20	0	152.20	0	0
8 PRE-COM THIN	0	0	0	0	0	0	0
9 MARKING OUT	0	0	0	0	6.30	0	0
10 FIRE PROTECT	1.50	1.50	1.50	1.50	1.50	1.50	1.50
11 OTH. PROTECT	1.30	1.30	1.30	1.30	1.30	1.30	1.30
12 VE/EQUIP PUR	40.00	.70	40.00	.70	41.60	.70	.70
13 VE/EQUIP RUN	89.00	1.60	89.00	1.60	92.70	1.60	1.60
14 VE/EQU INHER	0	0	0	0	0	0	0
15 TOOLS/STORES	6.50	.10	6.50	.10	6.80	.10	.10
16 TOOLS INHER	0	0	0	0	0	0	0
17 BUILD CONSTR	0	0	0	0	0	0	0
18 BUILD MAINT	26.30	.50	26.30	.50	27.40	.50	.50
19 BUILD REPLAC	17.00	.30	17.00	.30	17.70	.30	.30
20 BUILD INHER	0	0	0	0	0	0	0
21 ROAD CONSTRU	0	0	0	0	0	0	0
22 ROAD MAINT	6.40	6.40	6.40	6.40	6.40	6.40	6.40
23 ROAD INHER	0	0	0	0	0	0	0
24 STAFF REQUIR	175.30	3.10	175.30	3.10	182.50	3.10	3.10
25 OFFICE RUN	13.20	.20	13.20	.20	13.80	.20	.20
TOTAL EXPEND	528.70	15.70	528.70	15.70	950.20	15.70	15.70
RECEIPTS							
26 THINNINGS	0	0	0	0	1750.00	0	0
27 HARVEST	0	0	0	0	0	0	0
28 VE/EQU SALVG	0	0	0	0	0	0	0
29 TOOLS SALVG	0	0	0	0	0	0	0
30 BUILD SALVG	0	0	0	0	0	0	0
TOTAL RECTS	0	0	0	0	1750.00	0	0
NET RECEIPTS	-528.70	-15.70	-528.70	-15.70	1199.80	-15.70	-15.70

EXPENDIT	13	14	15	16	17	18
1 NURSERY	0	0	0	0	0	0
2 LAND PREPARE	0	0	0	0	0	0
3 PLANTING	0	0	0	0	0	0
4 PLANT SURVEY	0	0	0	0	0	0
5 BEATING UP	0	0	0	0	0	0
6 WEEDING	0	0	0	0	0	0
7 PRUNING	0	0	0	0	0	0
8 PRE-COM THIN	0	0	0	0	0	0
9 MARKING OUT	0	0	6.30	0	0	0
10 FIRE PROTECT	1.50	1.50	1.50	1.50	1.50	1.50
11 OTH. PROTECT	1.30	1.30	1.30	1.30	1.30	1.30
12 VE/EQUIP PUR	.70	.70	2.30	.70	.70	.70
13 VE/EQUIP RUN	1.60	1.60	5.20	1.60	1.60	1.60
14 VE/EQU INHER	0	0	0	0	0	0
15 TOOLS/STORES	.10	.10	.40	.10	.10	.10
16 TOOLS INHER	0	0	0	0	0	0
17 BUILD CONSTR	0	0	0	0	0	0
18 BUILD MAINT	.50	.50	1.50	.50	.50	.50
19 BUILD REPLAC	.30	.30	1.00	.30	.30	.30
20 BUILD INHER	0	0	0	0	0	0
21 ROAD CONSTRU	0	0	0	0	0	0
22 ROAD MAINT	6.40	6.40	6.40	6.40	6.40	6.40
23 ROAD INHER	0	0	0	0	0	0
24 STAFF REQUIR	3.10	3.10	10.20	3.10	3.10	3.10
25 OFFICE RUN	.20	.20	.80	.20	.20	.20
TOTAL EXPEND	15.70	15.70	36.90	15.70	15.70	15.70
RECEIPTS						
26 THINNINGS	0	0	1750.00	0	0	0
27 HARVEST	0	0	0	0	0	0
28 VE/EQU SALVG	0	0	0	0	0	0
29 TOOLS SALVG	0	0	0	0	0	0
30 BUILD SALVG	0	0	0	0	0	0
TOTAL RECTS	0	0	1750.00	0	0	0
NET RECEIPTS	-15.70	-15.70	1713.10	-15.70	-15.70	-15.70

EXPENDIT	19	20	21	22	23	24	25	
1 NURSERY	0	0	0	0	0	0	0	601.90
2 LAND PREPARE	0	0	0	0	0	0	0	478.00
3 PLANTING	0	0	0	0	0	0	0	177.50
4 PLANT SURVEY	0	0	0	0	0	0	0	12.70
5 BEATING UP	0	0	0	0	0	0	0	17.80
6 WEEDING	0	0	0	0	0	0	0	190.20
7 PRUNING	0	0	0	0	0	0	0	761.00
8 PRE-COM THIN	0	0	0	0	0	0	0	88.80
9 MARKING OUT	0	0	0	0	0	0	0	12.60
10 FIRE PROTECT	1.50	1.50	1.50	1.50	1.50	1.50	0	37.50
11 OTH. PROTECT	1.30	1.30	1.30	1.30	1.30	1.30	0	32.50
12 VE/EQUIP PUR	.70	.70	.70	.70	.70	.70	0	486.00
13 VE/EQUIP RUN	1.60	1.60	1.60	1.60	1.60	1.60	0	1038.70
14 VE/EQU INHER	0	0	0	0	0	0	0	51.80
15 TOOLS/STORES	.10	.10	.10	.10	.10	.10	0	75.50
16 TOOLS INHER	0	0	0	0	0	0	0	4.20
17 BUILD CONSTR	0	0	0	0	0	0	0	230.80
18 BUILD MAINT	.50	.50	.50	.50	.50	.50	0	314.20
19 BUILD REPLAC	.30	.30	.30	.30	.30	.30	0	144.40
20 BUILD INHER	0	0	0	0	0	0	0	92.00
21 ROAD CONSTRU	0	0	0	0	0	0	0	490.20
22 ROAD MAINT	6.40	6.40	6.40	6.40	6.40	6.40	0	160.00
23 ROAD INHER	0	0	0	0	0	0	0	71.10
24 STAFF REQUIR	3.10	3.10	3.10	3.10	3.10	3.10	0	2044.70
25 OFFICE RUN	.20	.20	.20	.20	.20	.20	0	153.70
TOTAL EXPEND	15.70	15.70	15.70	15.70	15.70	15.70	0	7601.80
RECEIPTS								
26 THINNINGS	0	0	0	0	0	0	0	3500.00
27 HARVEST	0	0	0	0	0	0	25200.00	25200.00
28 VE/EQU SALVG	0	0	0	0	0	0	51.80	51.80
29 TOOLS SALVG	0	0	0	0	0	0	4.20	4.20
30 BUILD SALVG	0	0	0	0	0	0	149.70	149.70
TOTAL RECTS	0	0	0	0	0	0	25405.70	28905.70
NET RECEIPTS	-15.70	-15.70	-15.70	-15.70	-15.70	-15.70	25405.70	21103.90

KENYA II PLANTATION PROJECT

Table 10, Financial Cash Flow - Pulpwood Component - Eldoret Division

KEN SHILLING/HECTARE

EXPENDIT (YEAR)	-2	-1	0	1	2	3	4	5	6	7	8	9	10	11	12	13	14	15	
1 NURSERY	0	601.90	0	0	0	0	0	0	0	0	0	0	0	0	0	0	0	0	601.90
2 LAND PREPARE	478.00	0	0	0	0	0	0	0	0	0	0	0	0	0	0	0	0	0	478.00
3 PLANTING	0	0	177.50	0	0	0	0	0	0	0	0	0	0	0	0	0	0	0	177.50
4 PLANT SURVEY	0	0	12.70	0	0	0	0	0	0	0	0	0	0	0	0	0	0	0	12.70
5 BEATING UP	0	0	17.60	0	0	0	0	0	0	0	0	0	0	0	0	0	0	0	17.60
6 WEEDING	0	0	0	0	190.20	0	0	0	0	0	0	0	0	0	0	0	0	0	190.20
7 FIRE PROTECT	0	0	1.50	1.50	1.50	1.50	1.50	1.50	1.50	1.50	1.50	1.50	1.50	1.50	1.50	1.50	1.50	0	22.50
8 OTH. PROTECT	0	0	1.30	1.30	1.30	1.30	1.30	1.30	1.30	1.30	1.30	1.30	1.30	1.30	1.30	1.30	1.30	0	19.50
9 VE/EQUIP PUR	131.20	0	57.80	.80	52.90	.80	.80	.80	.80	.80	.80	.80	.80	.80	.80	.80	.80	0	252.30
10 VE/EQUIP RUN	292.10	0	128.80	1.70	117.90	1.70	1.70	1.70	1.70	1.70	1.70	1.70	1.70	1.70	1.70	1.70	1.70	0	560.90
11 VE/EQU INHER	44.40	0	0	0	0	0	0	0	0	0	0	0	0	0	0	0	0	0	44.40
12 TOOLS/STORES	21.30	0	9.40	.10	8.60	.10	.10	.10	.10	.10	.10	.10	.10	.10	.10	.10	.10	0	40.60
13 TOOLS INHER	3.60	0	0	0	0	0	0	0	0	0	0	0	0	0	0	0	0	0	3.60
14 BUILD CONSTR	157.90	0	0	0	0	0	0	0	0	0	0	0	0	0	0	0	0	0	197.90
15 BUILD MAINT	86.30	0	38.00	.50	34.80	.50	.50	.50	.50	.50	.50	.50	.50	.50	.50	.50	.50	0	165.60
16 BUILD REPLAC	55.90	0	24.60	.30	22.50	.30	.30	.30	.30	.30	.30	.30	.30	.30	.30	.30	.30	0	106.90
17 BUILD INHER	78.90	0	0	0	0	0	0	0	0	0	0	0	0	0	0	0	0	0	78.90
18 ROAD CONSTRU	0	490.20	0	0	0	0	0	0	0	0	0	0	0	0	0	0	0	0	490.20
19 ROAD MAINT	0	0	6.40	6.40	6.40	6.40	6.40	6.40	6.40	6.40	6.40	6.40	6.40	6.40	6.40	6.40	6.40	0	96.00
20 ROAD INHER	0	71.10	0	0	0	0	0	0	0	0	0	0	0	0	0	0	0	0	71.10
21 STAFF REQUIR	575.20	0	253.60	3.30	232.20	3.30	3.30	3.30	3.30	3.30	3.30	3.30	3.30	3.30	3.30	3.30	3.30	0	1103.90
22 OFFICE RUN.	43.40	0	19.10	.30	17.50	.30	.30	.30	.30	.30	.30	.30	.30	.30	.30	.30	.30	0	83.90
TOTAL EXPEND	2008.20	1163.20	748.50	16.20	685.80	16.20	16.20	16.20	16.20	16.20	16.20	16.20	16.20	16.20	16.20	16.20	16.20	0	4816.30
RECEIPTS																			
23 HARVEST	0	0	0	0	0	0	0	0	0	0	0	0	0	0	0	0	0	6750.00	6750.00
24 VE/EQU SALVG	0	0	0	0	0	0	0	0	0	0	0	0	0	0	0	0	0	44.40	44.40
25 TOOLS SALVG	0	0	0	0	0	0	0	0	0	0	0	0	0	0	0	0	0	3.60	3.60
26 BUILD SALVG	0	0	0	0	0	0	0	0	0	0	0	0	0	0	0	0	0	148.10	148.10
TOTAL RECTS	0	0	0	0	0	0	0	0	0	0	0	0	0	0	0	0	0	6946.10	6946.10
NET RECEIPTS	-2008.20	-1163.20	-748.50	-16.20	-685.80	-16.20	-16.20	-16.20	-16.20	-16.20	-16.20	-16.20	-16.20	-16.20	-16.20	-16.20	-16.20	6946.10	2129.00

KENYA II PLANTATION PROJECT

Table 11, Financial Cash Flow - Pulpwood Component - Turbo Division

KEN SHILLING/HECTARE

EXPENDIT	12	-1	0 (YEAR)	1	2	3	4	5	6	7	8	9	10	11	12 (YEAR)	13	14	15	
1 NURSERY	0	656.60	0	0	0	0	0	0	0	0	0	0	0	0	0	0	0	0	656.60
2 LAND PREPARE	478.00	0	0	0	0	0	0	0	0	0	0	0	0	0	0	0	0	0	478.00
3 PLANTING	0	0	177.50	0	0	0	0	0	0	0	0	0	0	0	0	0	0	0	177.50
4 PLANT SURVEY	0	0	12.70	0	0	0	0	0	0	0	0	0	0	0	0	0	0	0	12.70
5 BEATING UP	0	0	35.50	0	0	0	0	0	0	0	0	0	0	0	0	0	0	0	35.50
6 WEEDING	0	0	128.80	168.80	128.80	0	0	0	0	0	0	0	0	0	0	0	0	0	426.40
7 FIRE PROTECT	0	0	1.50	1.50	1.50	1.50	1.50	1.50	1.50	1.50	1.50	1.50	1.50	1.50	1.50	1.50	1.50	0	22.50
8 OTH. PROTECT	0	0	1.30	1.30	1.30	1.30	1.30	1.30	1.30	1.30	1.30	1.30	1.30	1.30	1.30	1.30	1.30	0	19.50
9 VE/EQUIP PUR	129.50	0	86.00	24.80	24.80	.80	.80	.80	.80	.80	.80	.80	.80	.80	.80	.80	.80	0	274.70
10 VE/EQUIP RUN	264.40	0	191.50	55.30	55.30	1.70	1.70	1.70	1.70	1.70	1.70	1.70	1.70	1.70	1.70	1.70	1.70	0	610.90
11 VE/EQU INNER	48.40	0	0	0	0	0	0	0	0	0	0	0	0	0	0	0	0	0	48.40
12 TOOLS/STORES	21.00	0	14.00	4.00	4.00	.10	.10	.10	.10	.10	.10	.10	.10	.10	.10	.10	.10	0	44.20
13 TOOLS INNER	3.90	0	0	0	0	0	0	0	0	0	0	0	0	0	0	0	0	0	3.90
14 BUILD CONSTR	215.40	0	0	0	0	0	0	0	0	0	0	0	0	0	0	0	0	0	215.40
15 BUILD MAINT	85.20	0	56.50	16.30	16.30	.50	.50	.50	.50	.50	.50	.50	.50	.50	.50	.50	.50	0	180.30
16 BUILD REPLAC	55.20	0	36.60	10.60	10.60	.30	.30	.30	.30	.30	.30	.30	.30	.30	.30	.30	.30	0	118.60
17 BUILD INNER	85.80	0	0	0	0	0	0	0	0	0	0	0	0	0	0	0	0	0	85.80
18 ROAD CONSTRU	0	490.20	0	0	0	0	0	0	0	0	0	0	0	0	0	0	0	0	490.20
19 ROAD MAINT	0	0	6.40	6.40	6.40	6.40	6.40	6.40	6.40	6.40	6.40	6.40	6.40	6.40	6.40	6.40	6.40	0	96.00
20 ROAD INNER	0	71.10	0	0	0	0	0	0	0	0	0	0	0	0	0	0	0	0	71.10
21 STAFF REQUIR	568.00	0	377.10	108.80	108.80	3.40	3.40	3.40	3.40	3.40	3.40	3.40	3.40	3.40	3.40	3.40	3.40	0	1203.50
22 OFFICE RUN	42.80	0	28.40	8.20	8.20	.30	.30	.30	.30	.30	.30	.30	.30	.30	.30	.30	.30	0	91.20
TOTAL EXPEND	2021.60	1217.90	1153.80	406.00	366.00	16.30	16.30	16.30	16.30	16.30	16.30	16.30	16.30	16.30	16.30	16.30	16.30	0	5360.40
RECEIPTS																			
23 HARVEST	0	0	0	0	0	0	0	0	0	0	0	0	0	0	0	0	0	6750.00	6750.00
24 VE/EQU SALVG	0	0	0	0	0	0	0	0	0	0	0	0	0	0	0	0	0	48.40	48.40
25 TOOLS SALVG	0	0	0	0	0	0	0	0	0	0	0	0	0	0	0	0	0	3.90	3.90
26 BUILD SALVG	0	0	0	0	0	0	0	0	0	0	0	0	0	0	0	0	0	161.20	161.20
TOTAL RECTS	0	0	0	0	0	0	0	0	0	0	0	0	0	0	0	0	0	6963.50	6963.50
NET RECEIPTS	-2021.60	-1217.90	-1153.80	-406.00	-366.00	-16.30	-16.30	-16.30	-16.30	-16.30	-16.30	-16.30	-16.30	-16.30	-16.30	-16.30	-16.30	6963.50	1602.00

KENYA II PLANTATION PROJECT

Table 12, Economic Value Flow - Sawlog Component

KEN SHILLING/HECTARE

COSTS (YEAR)	-2	-1	0	1	2	3	4	5	6	7	8	9	10	11	12
1 NURSERY	0	462.40	0	0	0	0	0	0	0	0	0	0	0	0	0
2 LAND PREPARE	388.50	0	0	0	0	0	0	0	0	0	0	0	0	0	0
3 PLANTING	0	0	133.10	0	0	0	0	0	0	0	0	0	0	0	0
4 PLANT SURVEY	0	0	9.50	0	0	0	0	0	0	0	0	0	0	0	0
5 BEATING UP	0	0	13.30	0	0	0	0	0	0	0	0	0	0	0	0
6 WEEDING	0	0	0	0	142.60	0	0	0	0	0	0	0	0	0	0
7 PRUNING	0	0	0	0	114.10	0	114.10	0	114.10	0	114.10	0	114.10	0	0
8 PRE-COM THIN	0	0	0	0	0	0	0	66.60	0	0	0	0	0	0	0
9 MARKING OUT	0	0	0	0	0	0	0	0	0	0	0	0	4.80	0	0
10 FIRE PROTECT	0	0	1.10	1.10	1.10	1.10	1.10	1.10	1.10	1.10	1.10	1.10	1.10	1.10	1.10
11 OTH. PROTECT	0	0	1.00	1.00	1.00	1.00	1.00	1.00	1.00	1.00	1.00	1.00	1.00	1.00	1.00
12 VE/EQUIP PUR	160.30	0	70.60	.90	115.80	.90	52.00	30.70	52.00	.90	52.00	.90	54.10	.90	.90
13 VE/EQUIP RUN	332.40	0	146.50	1.90	240.00	1.90	107.70	63.60	107.70	1.90	107.70	1.90	112.10	1.90	1.90
14 VE/EQU INHER	67.40	0	0	0	0	0	0	0	0	0	0	0	0	0	0
15 TOOLS/STORES	24.50	0	10.80	.10	17.70	.10	8.00	4.70	8.00	.10	8.00	.10	8.30	.10	.10
16 TOOLS INHER	5.20	0	0	0	0	0	0	0	0	0	0	0	0	0	0
17 BUILD CONSTR	237.70	0	0	0	0	0	0	0	0	0	0	0	0	0	0
18 BUILD MAINT	83.60	0	36.80	.50	60.30	.50	27.10	16.00	27.10	.50	27.10	.50	28.20	.50	.50
19 BUILD REPLAC	54.10	0	23.80	.30	39.10	.30	17.50	10.40	17.50	.30	17.50	.30	18.30	.30	.30
20 BUILD INHER	94.70	0	0	0	0	0	0	0	0	0	0	0	0	0	0
21 ROAD CONSTRU	0	555.60	0	0	0	0	0	0	0	0	0	0	0	0	0
22 ROAD MAINT	0	0	7.60	7.60	7.60	7.60	7.60	7.60	7.60	7.60	7.60	7.60	7.60	7.60	7.60
23 ROAD INHER	0	92.40	0	0	0	0	0	0	0	0	0	0	0	0	0
24 STAFF REQUIR	557.20	0	245.50	3.20	402.30	3.20	180.60	106.60	180.60	3.20	180.60	3.20	188.00	3.20	3.20
25 OFFICE RUN	42.00	0	18.50	.20	30.30	.20	13.60	8.00	13.00	.20	13.60	.20	14.20	.20	.20
TOTAL COST	2017.60	1110.40	718.10	16.80	1171.90	16.80	530.30	316.30	530.30	16.80	530.30	16.80	551.80	16.80	16.80
BENEFITS															
26 THINNINGS	0	0	0	0	0	0	0	0	0	0	0	0	8923.70	0	0
27 HARVEST	0	0	0	0	0	0	0	0	0	0	0	0	0	0	0
28 VE/EQU SALVG	0	0	0	0	0	0	0	0	0	0	0	0	0	0	0
29 TOOLS SALVG	0	0	0	0	0	0	0	0	0	0	0	0	0	0	0
30 BUILD SALVG	0	0	0	0	0	0	0	0	0	0	0	0	0	0	0
TOTAL BENEFIT	0	0	0	0	0	0	0	0	0	0	0	0	8923.70	0	0
NET BENEFITS	-2017.60	-1110.40	-718.10	-16.80	-1171.90	-16.80	-530.30	-316.30	-530.30	-16.80	-530.30	-16.80	8371.90	-16.80	-16.80

COSTS	13	14	15	16	17	18	19	20	21	22	23	24	25	
1 NURSERY	0	0	0	0	0	0	0	0	0	0	0	0	0	462.40
2 LAND PREPARE	0	0	0	0	0	0	0	0	0	0	0	0	0	388.50
3 PLANTING	0	0	0	0	0	0	0	0	0	0	0	0	0	133.10
4 PLANT SURVEY	0	0	0	0	0	0	0	0	0	0	0	0	0	9.50
5 BEATING UP	0	0	0	0	0	0	0	0	0	0	0	0	0	13.30
6 WEEDING	0	0	0	0	0	0	0	0	0	0	0	0	0	142.60
7 PRUNING	0	0	0	0	0	0	0	0	0	0	0	0	0	570.50
8 PRE-COM THIN	0	0	0	0	0	0	0	0	0	0	0	0	0	66.60
9 MARKING OUT	0	0	4.80	0	0	0	0	0	0	0	0	0	0	9.60
10 FIRE PROTECT	1.10	1.10	1.10	1.10	1.10	1.10	1.10	1.10	1.10	1.10	1.10	1.10	0	27.50
11 OTH. PROTECT	1.00	1.00	1.00	1.00	1.00	1.00	1.00	1.00	1.00	1.00	1.00	1.00	0	25.00
12 VE/EQUIP PUR	.90	.90	3.00	.90	.90	.90	.90	.90	.90	.90	.90	.90	0	605.60
13 VE/EQUIP RUN	1.90	1.90	6.30	1.90	1.90	1.90	1.90	1.90	1.90	1.90	1.90	1.90	0	1256.30
14 VE/EQU INHER	0	0	0	0	0	0	0	0	0	0	0	0	0	67.40
15 TOOLS/STORES	.10	.10	.50	.10	.10	.10	.10	.10	.10	.10	.10	.10	0	92.20
16 TOOLS INHER	0	0	0	0	0	0	0	0	0	0	0	0	0	5.20
17 BUILD CONSTR	0	0	0	0	0	0	0	0	0	0	0	0	0	237.70
18 BUILD MAINT	.50	.50	1.60	.50	.50	.50	.50	.50	.50	.50	.50	.50	0	316.30
19 BUILD REPLAC	.30	.30	1.00	.30	.30	.30	.30	.30	.30	.30	.30	.30	0	204.30
20 BUILD INHER	0	0	0	0	0	0	0	0	0	0	0	0	0	94.70
21 ROAD CONSTRU	0	0	0	0	0	0	0	0	0	0	0	0	0	555.60
22 ROAD MAINT	7.60	7.60	7.60	7.60	7.60	7.60	7.60	7.60	7.60	7.60	7.60	7.60	0	190.00
23 ROAD INHER	0	0	0	0	0	0	0	0	0	0	0	0	0	92.40
24 STAFF REQUIR	3.20	3.20	10.50	3.20	3.20	3.20	3.20	3.20	3.20	3.20	3.20	3.20	0	2100.30
25 OFFICE RUN	.20	.20	.80	.20	.20	.20	.20	.20	.20	.20	.20	.20	0	158.00
TOTAL COST	16.80	16.80	38.20	16.80	16.80	16.80	16.80	16.80	16.80	16.80	16.80	16.80	0	7800.80
BENEFITS														
26 THINNINGS	0	0	8923.70	0	0	0	0	0	0	0	0	0	0	17847.40
27 HARVEST	0	0	0	0	0	0	0	0	0	0	0	0	128501.10	128501.10
28 VE/EQU SALVG	0	0	0	0	0	0	0	0	0	0	0	0	67.40	67.40
29 TOOLS SALVG	0	0	0	0	0	0	0	0	0	0	0	0	5.20	5.20
30 BUILD SALVG	0	0	0	0	0	0	0	0	0	0	0	0	154.20	154.20
TOTAL BENEFIT	0	0	8923.70	0	0	0	0	0	0	0	0	0	128727.90	146575.30
NET BENEFITS	-16.80	-16.80	8885.50	-16.80	-16.80	-16.80	-16.80	-16.80	-16.80	-16.80	-16.80	-16.80	128727.90	138774.50

KENYA II PLANTATION PROJECT

Table 13, Economic Value Flow - Pulpwood Component - Eldoret Division

KEN SHILLING/HECTARE

(YEAR)

COSTS	-2	-1	0	1	2	3	4	5	6	7	8	9	10	11	12	13	14	15	
1 NURSERY	0	466.40	0	0	0	0	0	0	0	0	0	0	0	0	0	0	0	0	462.40
2 LAND PREPARE	358.50	0	0	0	0	0	0	0	0	0	0	0	0	0	0	0	0	0	356.50
3 PLANTING	0	0	133.10	0	0	0	0	0	0	0	0	0	0	0	0	0	0	0	133.10
4 PLANT SURVEY	0	0	9.50	0	0	0	0	0	0	0	0	0	0	0	0	0	0	0	9.50
5 BEATING UP	0	0	13.30	0	0	0	0	0	0	0	0	0	0	0	0	0	0	0	13.30
6 WEEDING	0	0	0	0	142.60	0	0	0	0	0	0	0	0	0	0	0	0	0	142.60
7 FIRE PROTECT	0	0	1.10	1.10	1.10	1.10	1.10	1.10	1.10	1.10	1.10	1.10	1.10	1.10	1.10	1.10	1.10	0	16.50
8 OTH. PROTECT	0	0	1.00	1.00	1.00	1.00	1.00	1.00	1.00	1.00	1.00	1.00	1.00	1.00	1.00	1.00	1.00	0	15.00
9 VE/EQUIP PUR	170.50	0	75.20	1.00	68.80	1.00	1.00	1.00	1.00	1.00	1.00	1.00	1.00	1.00	1.00	1.00	1.00	0	327.50
10 VE/EQUIP RUN	353.40	0	155.90	2.00	142.70	2.00	2.00	2.00	2.00	2.00	2.00	2.00	2.00	2.00	2.00	2.00	2.00	0	678.00
11 VE/EQU INNER	57.80	0	0	0	0	0	0	0	0	0	0	0	0	0	0	0	0	0	57.80
12 TOOLS/STORES	26.10	0	11.50	.20	10.50	.20	.20	.20	.20	.20	.20	.20	.20	.20	.20	.20	.20	0	50.70
13 TOOLS INNER	4.40	0	0	0	0	0	0	0	0	0	0	0	0	0	0	0	0	0	4.40
14 BUILD CONSTR	203.80	0	0	0	0	0	0	0	0	0	0	0	0	0	0	0	0	0	203.80
15 BUILD MAINT	88.90	0	39.20	.50	35.90	.50	.50	.50	.50	.50	.50	.50	.50	.50	.50	.50	.50	0	170.50
16 BUILD REPLAC	57.50	0	25.40	.30	23.20	.30	.30	.30	.30	.30	.30	.30	.30	.30	.30	.30	.30	0	110.00
17 BUILD INNER	81.20	0	0	0	0	0	0	0	0	0	0	0	0	0	0	0	0	0	81.20
18 ROAD CONSTRU	0	555.60	0	0	0	0	0	0	0	0	0	0	0	0	0	0	0	0	555.60
19 ROAD MAINT	0	0	7.60	7.60	7.60	7.60	7.60	7.60	7.60	7.60	7.60	7.60	7.60	7.60	7.60	7.60	7.60	0	114.00
20 ROAD INNER	0	92.40	0	0	0	0	0	0	0	0	0	0	0	0	0	0	0	0	92.40
21 STAFF REQUIR	592.50	0	261.30	3.40	239.20	3.40	3.40	3.40	3.40	3.40	3.40	3.40	3.40	3.40	3.40	3.40	3.40	0	1137.20
22 OFFICE RUN	44.70	0	19.70	.30	18.00	.30	.30	.30	.30	.30	.30	.30	.30	.30	.30	.30	.30	0	86.30
TOTAL COST	2039.30	1110.40	753.80	17.40	690.60	17.40	17.40	17.40	17.40	17.40	17.40	17.40	17.40	17.40	17.40	17.20	17.40	0	4820.30
BENEFITS																			
23 HARVEST	0	0	0	0	0	0	0	0	0	0	0	0	0	0	0	0	0	107055.00	107096.00
24 VE/EQU SALVG	0	0	0	0	0	0	0	0	0	0	0	0	0	0	0	0	0	57.80	57.80
25 TOOLS SALVG	0	0	0	0	0	0	0	0	0	0	0	0	0	0	0	0	0	4.40	4.40
26 BUILD SALVG	0	0	0	0	0	0	0	0	0	0	0	0	0	0	0	0	0	152.60	152.60
TOTAL BENEFT	0	0	0	0	0	0	0	0	0	0	0	0	0	0	0	0	0	107269.60	107269.60
NET BENEFITS	-2039.30	-1110.40	-753.80	-17.40	-690.60	-17.40	-17.40	-17.40	-17.40	-17.40	-17.40	-17.40	-17.40	-17.40	-17.40	-17.20	-17.40	107269.60	102449.50

KENYA II PLANTATION PROJECT

Table 14, Economic Value Flow - Pulpwood Component - Turbo Division

KEN SHILLING/HECTARE

COSTS (YEAR)	-2	-1	0	1	2	3	4	5	6	7	8	9	10	11	12	13	14	15	
1 NURSERY	0	504.50	0	0	0	0	0	0	0	0	0	0	0	0	0	0	0	0	504.50
2 LAND PREPARE	358.50	0	0	0	0	0	0	0	0	0	0	0	0	0	0	0	0	0	358.50
3 PLANTING	0	0	133.10	0	0	0	0	0	0	0	0	0	0	0	0	0	0	0	133.10
4 PLANT SURVEY	0	0	9.50	0	0	0	0	0	0	0	0	0	0	0	0	0	0	0	9.50
5 BEATING UP	0	0	26.60	0	0	0	0	0	0	0	0	0	0	0	0	0	0	0	26.60
6 WEEDING	0	0	106.60	146.60	136.60	0	0	0	0	0	0	0	0	0	0	0	0	0	389.80
7 FIRE PROTECT	0	0	1.10	1.10	1.10	1.10	1.10	1.10	1.10	1.10	1.10	1.10	1.10	1.10	1.10	1.10	1.10	0	16.50
8 OTH. PROTECT	0	0	1.00	1.00	1.00	1.00	1.00	1.00	1.00	1.00	1.00	1.00	1.00	1.00	1.00	1.00	1.00	0	15.00
9 VE/EQUIP PUR	168.30	0	111.70	32.30	32.30	1.00	1.00	1.00	1.00	1.00	1.00	1.00	1.00	1.00	1.00	1.00	1.00	0	356.60
10 VE/EQUIP RUN	349.00	0	231.60	66.90	66.90	2.10	2.10	2.10	2.10	2.10	2.10	2.10	2.10	2.10	2.10	2.10	2.10	0	739.60
11 VE/EQU INHER	62.90	0	0	0	0	0	0	0	0	0	0	0	0	0	0	0	0	0	62.90
12 TOOLS/STORES	25.80	0	17.10	4.90	4.90	.20	.20	.20	.20	.20	.20	.20	.20	.20	.20	.20	.20	0	55.10
13 TOOLS INHER	4.80	0	0	0	0	0	0	0	0	0	0	0	0	0	0	0	0	0	4.80
14 BUILD CONSTR	221.80	0	0	0	0	0	0	0	0	0	0	0	0	0	0	0	0	0	221.80
15 BUILD MAINT	87.70	0	58.20	16.80	16.80	.50	.50	.50	.50	.50	.50	.50	.50	.50	.50	.50	.50	0	185.50
16 BUILD REPLAC	56.80	0	37.70	10.90	10.90	.30	.30	.30	.30	.30	.30	.30	.30	.30	.30	.30	.30	0	119.90
17 BUILD INHER	88.40	0	0	0	0	0	0	0	0	0	0	0	0	0	0	0	0	0	88.40
18 ROAD CONSTRU	0	555.60	0	0	0	0	0	0	0	0	0	0	0	0	0	0	0	0	555.60
19 ROAD MAINT	0	0	7.60	7.60	7.60	7.60	7.60	7.60	7.60	7.60	7.60	7.60	7.60	7.60	7.60	7.60	7.60	0	114.00
20 ROAD INHER	0	92.40	0	0	0	0	0	0	0	0	0	0	0	0	0	0	0	0	92.40
21 STAFF REQUIR	585.00	0	388.20	112.10	112.10	3.50	3.50	3.50	3.50	3.50	3.50	3.50	3.50	3.50	3.50	3.50	3.50	0	1239.40
22 OFFICE RUN	44.10	0	29.30	8.50	8.50	.30	.30	.30	.30	.30	.30	.30	.30	.30	.30	.30	.30	0	94.00
TOTAL COST	2053.10	1152.50	1159.30	408.70	398.70	17.60	17.60	17.60	17.60	17.60	17.60	17.60	17.60	17.60	17.60	17.60	17.60	0	5353.50
BENEFITS																			
23 HARVEST	0	0	0	0	0	0	0	0	0	0	0	0	0	0	0	0	0	107055.00	107055.00
24 VE/EQU SALVG	0	0	0	0	0	0	0	0	0	0	0	0	0	0	0	0	0	62.90	62.90
25 TOOLS SALVG	0	0	0	0	0	0	0	0	0	0	0	0	0	0	0	0	0	4.80	4.80
26 BUILD SALVG	0	0	0	0	0	0	0	0	0	0	0	0	0	0	0	0	0	166.10	166.10
TOTAL BENEFIT	0	0	0	0	0	0	0	0	0	0	0	0	0	0	0	0	0	107288.60	107288.60
NET BENEFITS	-2053.10	-1152.50	-1159.30	-408.70	-398.70	-17.60	-17.60	-17.60	-17.60	-17.60	-17.60	-17.60	-17.60	-17.60	-17.60	-17.60	-17.60	107288.60	101935.30

KENYA II PLANTATION PROJECT

Table 15, Sensitivity Analysis

Assumption Set	Sawlog ERR (%)	Pulpwood Eldoret ERR(%)	Pulpwood Turbo ERR(%)
1. Original assumptions	15.5	21.7	20.9
2. No shadow foreign exchange rate	14.4	20.3	19.5
3. No shadow wage rate	15.1	21.1	20.4
4. Combination of 2. and 3.	13.9	19.7	18.9
5. Pulpwood price of Ksh70/m^3	-	9.8	9.0
6. Sawnwood import substitution element	16.8	-	-
7. Sawnwood export element	8.2	-	-
8. No shadow prices	6.4	2.4	1.7

See Appendix 2 for further details

APPENDIX 1

PROJECT COSTS

A. Non-silvicultural physical inputs and costs

Vehicle and Equipment Purchase Costs

1. Vehicles and equipment are assumed to have a six year useful life. A list of the total number of required vehicles and equipment for the entire plantation programme is presented below. It is assumed that equal expenditures for replacements are made each year beginning in the year of planting. This input is composed entirely of foreign exchange.

Item	Number Required	Unit Cost (KL)	Total Cost (KL)
Saloon car	1	2 650	2 650
Station wagon	1	2 450	2 450
Land Rovers	100	2 650	265 000
5-ton lorries	65	3 500	227 500
3-ton lorries	40	2 900	116 000
60-65 hp tractors	10	2 650	26 500
85 hp tractors	8	4 150	33 200
Stationary engines	60 1/	650	39 000
Fire trucks	4	5 300	21 200
Portable water pumps	12	450	5 400
Sub-totals			738 900
10 percent physical contingency			73 890
Total cost (6-year period)			812 790
Total annual cost			135 465
Annual cost per hectare (assuming 160 000 hectares)			0.8467 (KSh 16.9)

Vehicles and Equipment Running Costs

2. The total annual vehicle and equipment running cost for the entire plantation programme is KL 301 702 (see Table 1 in this appendix), which is KSh 37.7 per hectare. The foreign exchange component for running cost is 70 percent.

1/ Stationary engines have a useful life of 12 years. Therefore, only half of the required number were entered in this table.

Tools and Stores Costs

3. Tools and stores costs consist of maintenance and replacement of walkie talkie and VHF radios, uniforms for staff, hand tools and miscellaneous items. The annual cost of the above for the entire plantation programme is KL 22 000 (including a 10 percent physical contingency), which is KSh 2.8 per hectare. The foreign exchange component is 75 percent.

Building Construction Cost

4. It is assumed that one project headquarters office, one new divisional office, four new full accounting stations and 12 sub-stations will be constructed during the project period (1975-1980). Itemized costs for each are presented in Table 2 of this appendix.

5. In addition, some existing staff housing, buildings and water supplies will require replacing. The Department has commenced a programme for the replacement of forest workmen's houses, presently made from mud and wattle, with improved houses. So far about 20 percent of the housing has been replaced; the remainder are being replaced over a ten year period and the project would provide for the replacement of some 3 100 forest workmen's houses phased over the six year period. This would be in addition to some 2 000 new workmen's houses to be constructed in the new stations and sub-stations. The house building programme will be undertaken by the Forest Department using direct labour; costs are based on present day experience.

6. The proposal is to provide during the project disbursement period (1975-1980), an improved house for all permanent forestry workers (working on existing stations and sub-stations) required beyond 1980, i.e., 5 500 forestry families.

7. It is further assumed that building costs associated with each hectare are incurred in the year preceding the planting of that hectare. The foreign exchange component is 10 percent. Total building costs are obtained from the following:

New Stations	Total (KL)
Project Headquarters, Office Block Nairobi	13 650
New Division, Nabkoi	14 795
New Forest Stations (4 at KL 23 150)	92 600
New Forest Sub-Stations (12 at KL 20 250)	243 000
New Forest Workers' Houses for New Stations (2 031 at KL 210)	426 510
Sub-total (New Stations)	790 555
Existing Stations	
Replacement of Existing Forest Workers' Houses 3 116 at KL 210	654 360
Replacement of and Additions to Existing Staff Housing and Buildings	200 000
Sub-total (Existing Stations)	854 360
Sub-total Building Cost	1 644 915
10% Physical Contingency	164 491
Total Building Cost	1 809 406
	(KSh 226.2/hectare)

Building Maintenance Cost

8. Maintenance of existing buildings (see Table 3 in this appendix) and new (project) buildings will be carried out by the Forestry Department with maintenance teams located in the main stations. The foreign exchange component is 10 percent. Maintenance requirements consist of 60 teams of the following:

		KL/year
1 Mason		250
1 Carpenter		250
3 Labourers		540
Materials		310
	Sub-total team cost	1 350
	10% Physical Contingency	135
	Total team cost	1 485
Total Annual Cost of 60 Teams		89 100
		(KSh 11.1/hectare)

Road Construction Cost

9. At the beginning of the project (1975) there was one Forest Department road unit. Under this project two road units will be formed, one West of the Rift and the other East of the Rift. New road construction for both units is summarized below.

	km/year
Primary and secondary roads for new plantations to be established during project period	60
Plantation tracks	100
Service road construction to new stations and upgrading some existing roads	20
Total	180

The total amount of equipment required for both units is shown in Table 4 in this appendix. In addition, KL 3 000 per unit is required for site offices/workshops and housing. Road construction capital costs (equipment plus office/workshop and housing construction) associated with each hectare are assumed to occur in the year preceding the planting of that hectare. Road construction capital costs are composed entirely of foreign exchange.

10. Road construction operating costs consist of (1) machinery operating costs (summarized in Table 5 of this appendix), (2) bridge and culvert construction costs of KL 3 500 per unit, (3) staff and labour (summarized in Table 6 of this appendix), and (4) miscellaneous items. The foreign exchange component equals 30 percent for road construction operating costs.

11. Costs are summarized below.

Capital Costs

Equipment	KL	178 200
Buildings		6 000
Sub-total		184 200
10% Physical Contingency		18 420
Total		202 620
Total Per Hectare	KSh	101.742
		(at 39 830 "new hectares)

Operating Costs	KL/year
Machine operating costs	54 690
Bridge and culvert materials	7 000
Staff and labour	49 510
Workshop running	3 000
Tools and miscellaneous stores	1 000
Office expenses, supervision, vehicle running costs, administration, etc.	2 000
Sub-total	117 200
10% Physical Contingency	11 720
Total	128 920
(per hectare, at 6638.33 ha/year)	(KSh 388.41)
Total Road Construction Cost	KSh 490.2/ha

Road Maintenance Cost

12. Annual road maintenance operating cost per hectare is calculated as follows:

(Average Road Density) x (Machine Hours Required) x (Cost per Machine Hour)

= (.045 km/ha x (1.6 hr/km) x (KSh 50/hr) + 10%

= KSh 3.6/ha + 10% physical contingency = KSh 3.96

13. Capital road maintenance cost is based upon the assumptions that 8 graders are needed and each grader has a useful life of 7 years, which means that an average of 8/7 graders must be purchased (replaced) each year. Each grader costs KL 17 200 (including 10 percent physical contingency). Therefore, the annual capital cost per hectare is:

8/7 (KL 17 200) / (160 000)

= KL 0.12 = KSh 2.46

Total Road Maintenance Cost Per Hectare Per Year = KSh (3.96 + 2.46) = KSh 6.4. The foreign exchange component is 60 percent.

Staff Requirement Costs

14. Annual staff salaries for the entire plantation programme are itemized in Table 7 of this appendix. Ten percent is composed of foreign exchange. A summary by management unit is presented below.

	No.	Unit Cost Per Year	Total KL
Project Headquarters	1	16 000	16 000
Conservancies	3	5 150	15 450
Divisions 1/	14	4 600	64 400
Stations 2/	59	5 530	326 270
Sub-stations 3/	47	3 660	172 020
			594 140

The total annual cost of KL 594 140 is KSh 74.3 per hectare.

Office Running Costs

15. Office running costs are composed of general office expenses of KL 8 825 annually, summarized below, and annual station upkeep of KL 360 for each (100) stations. Ten percent is composed of foreign exchange.

General office expenses	KL/year	KL/year
Project Headquarters	1 000	
Conservancies (KL 300 x 3)	900	
Divisions (KL 200 x 4)	2 800	
Forest Stations (KL 50 x 59)	2 950	
Sub-stations (KL 25 x 47)	1 175	8 825
Station upkeep		36 000
Total Annual Office Running Cost		44 825 (KSh 5.6/ha)

B. Silvicultural physical inputs and costs

16. The afforestation programme will include nurseries, land clearing and preparation, planting and replanting, surveying, weeding, pruning, thinning, protection and road construction.

Nurseries

17. Pulpwood. The Turbo nursery which is capable of holding over three million plants will raise most of the plants for that division, supplemented by the Saboti Nursery.

18. Sawlog. The average annual planting programme in the sawlog area is 5 900 hectares. Twelve million plants would be needed to be met from the existing nurseries. Species to be raised in these nurseries are *Cupressus lusitanica* (80 percent of the total planting

---

1/ Includes 1 new Division to be built by project.

2/ Includes 4 new stations to be built by project.

3/ Includes 12 new sub-stations to be built by project.

programme), and Pinus patula and other species, i.e. Eucalyptus and a few hardwoods which will account for the remaining 20 percent. All seeds would be collected from the existing seed orchards. Seedlings are normally raised in the nursery for about six months before planting.

Land Clearing and Preparation

19. Pulpwood. Land is ploughed and harrowed after which maize or wheat is planted for two years before planting of pine trees is commenced.

20. Sawlog Areas. In the sawlog areas and in the pulpwood areas in Eldoret Division the land to be planted is first demarcated into 0.5 hectare plots and allocated to labourers. They clear and burn these plots, plant agricultural crops, i.e. maize, beans and potatoes, for two years, and plant trees in the third year which are intercropped for an additional three years. Felling of small trees in these plots is done by labour, while large trees are felled by the Forestry Department using power saws. Most of these areas are very fertile and therefore no fertilization is required. In the grasslands subsoiling is carried out before pitting out is done.

Planting

21. Pulpwood. Planting in the pulpwood plantations would be done by hand at a space of 2.7 m x 2.0 m (approximately 1 800 plants per hectare). The interval is designed to facilitate easy movement of harvesters while removing the maize interplanted with the trees in the first year. It is also designed to allow free movement of harrows during mechanical weeding. Planting commences immediately after sufficient moisture build-up has been formed during the rainy season (April-May); sometimes when rains are late planting may continue to mid-August. Replacement of failed seedlings (beating up) is done during the same season or the following year depending on the length of the rainy season. Normally an allowance of 10 percent of the seedlings stock is reserved for this purpose. Complete replanting during the following year is done where over 75 percent of the establishment fails. The average planting requirement is 14 man-days per hectare. This includes handling and movement of seedlings from roadside to planting site. Replanting of clearfelled areas takes place one year after felling.

22. Sawlog Areas. In the sawlog areas, planting is carried out by workmen in the allocated plots. Normal spacing is 2.5 m by 2.5 m. An average of 14 man-days is required. Replanting of the clearfelled areas takes place two years after felling. A 10 percent seedling reserve is made to facilitate beating up of failed seedlings after planting. Complete replacement is unusual in the sawlog areas.

Surveying

23. Survey of the planted areas follows immediately after planting has been completed. Survey data from the field is transferred into 1:10 000 survey sheets plotted from 1:25 000 scale aerial photographs. The survey section of the Department keeps a record of all the surveyed areas.

Weeding

24. Pulpwood Area. In the Turbo area, both manual and mechanical weeding are carried out as follows: manual weeding in the first, second and third years at seven man-days per hectare and mechanical weeding in the first, second and third year at two hours per hectare. After the rains in the Turbo area, weed growth is profuse making such heavy weeding necessary.

25. Sawlog Area and Pulpwood Areas Other Than Turbo. Since trees interplanted with agricultural crops are grown in workers' plots for the first three years, weeding is carried out by labour while tending their crops. No mechanical weeding or herbicides application is used. Continual weeding on plantations grown on grasslands is necessary for the first three years, particularly where cypress is established. However, *P. patula* can survive in grasslands so long as the grass is not taller than the trees. In these areas, weeding is particularly necessary to reduce fire danger and also to minimize rat damage to young seedlings.

Pruning

26. Pulpwood Area. No pruning and thinning is required in the pulpwood area. Where fire danger is considered high in young plantations light pruning may be carried out; otherwise, this is not necessary.

27. Sawlog Area. Pruning of trees intended to provide sawn wood is necessary so as to provide a knot-hole tree timber for better quality. To obtain this high quality timber, it is necessary to carry out pruning schedules as follows:

Cypress - five prunings over ten years

- Age 2 - all stems
- Age 4 - all stems
- Age 6 - 533 stems/hectare to 9.25 metres
- Age 8 - 533 stems/hectare to 11.25 metres
- Age 10 - 533 stems/hectare to 13.75 metres
- No other pruning is necessary after age ten.

Pines - four prunings over ten years at

- Age 4 - all the stems
- Age 6 - all the stems
- Age 8 - 711 stems/hectare to 9 metres
- Age 10 - 711 stems/hectare to 12 metres
- No further pruning necessary after age ten.

Labour requirement for pruning is based on the country's present average of 12 man-days per hectare. The average annual pruning programme will be 25 000 hectares.

28. Thinning. Thinning is not necessary in pulpwood plantations. It is, however, necessary in sawlog plantations. The purpose of thinning plantations is to increase the value of the final crop. Three thinnings would be carried out on plantations. The first thinning is hygienic and is intended to remove all dying, dead and diseased trees from the plantations. The second and third thinnings are commercial and provide a financial return to the Government while still improving the quality of the final crop. The schedule below shows the remaining crop after thinning in both pine and cypress plantations in the project.

Cypress

Age in Years	Plants Established Per Ha	Crop Remaining After Thinning	Plant Height in Metres
0	1 600	No thinning	-
5	-	533	12
10	-	355	Top height not considered
15	-	250	Top height not considered

Pines

Age in Years	Plants Established Per Ha	Crop Remaining After Thinning	Plant Height in Metres
0	1 600	No thinning	-
5	-	711	12
8	-	533	Top height not considered
13	-	356	Top height not considered

29. The average annual thinning during the project period would be 4 100 ha. Average annual man-day requirement for non-commercial thinning and marking of commercial thinning is eight/ha.

Protection

30. Protection Against Fire. Fire damage in Kenya can be very high, particularly during years of severe drought. In 1971 the country lost 19 000 hectares of forest which included over 2 000 hectares of plantations. It is therefore necessary that adequate fire protection measures are enforced during the project period. In order to ensure strong fire control, the Department established a fire control section to deal with all fire matters.

31. Firebreaks. Firebreaks would be established at a density of 40 m per ha and with a width of 20 m and maintained by early burning and all year round clearing. Firebreak maintenance in both pulpwood and sawlog plantations would be carried out both by labour and tractor with a rotary harrow. An average of 0.12 man-days per hectare would be required.

32. Protection Against Fungal Diseases. Dothistroma pinii has been the worst fungal disease. As a result, the Department has halted any further planting of P. radiata which is **particularly** susceptible to it while studies continue to find a cheaper method of control. Nursery diseases caused by Rhizoctonia solani Kuhn and pythium spp. still cause problems. Cypress canker caused by Monochaetia unicornis (Cooke and Ellis), Armillaria mellea of both pines and cypress, Diplodia pinea (Desm) have all been present in the Kenya softwood plantations. The Pathology section of the Department carries out a continuous field check on these diseases and controls them where possible. This would continue during the project period.

33. Protection Against Insect Pests. Insect pests of economic importance in Kenya are the woolly aphids of the genera Adelges and Pineus (Homoptera: Adelgidae). P. patula, P. radiata and other pines have been attacked over a wide area. The aphids suck the pine juice and retard growth, and a severe attack can cause death of the trees, although this has not yet been noticed in Kenya.

34. A summary of silvicultural inputs for each of the three project components, together with estimates of total costs per hectare, are displayed in Table 8 of this appendix.

APPENDIX 1

KENYA II PLANTATION PROJECT

Table 1 Vehicle and Equipment Running Cost

Location	Type and Number			km/vehicle/ year	KSh/ km		Annual Cost (KL)
Supervision Vehicles							
Project Headquarters	Saloon Car		1	20,000	1.1		1,100
"	Station Wagon		1	20,000	1.1		1,100
"	Land Rover	1					
Conservancies	" "	3					
Divisions	" "	14					
Main Forest Stations	" "	59					
Sub Forest Stations	" "	23 1/	100	17,000	1.1		93,500
						sub-total	95,700
General Administration Vehicles and Equipment							
Main Forest Stations and Selected Divisions	5/7 ton lorries	65		15,000	1.7		82,875
Sub Forest Stations	3-ton lorries	40 2/		15,000	1.5		45,000
Stations under construction	60-65 hp tractors	10 3/4/		1,000	19.0		9,500
Divisions, Main Station and Sub Stations	Stationary engine (water supplies)	120		1,300 hr/yr	4.0		31,200
(Location not given)	Fire trucks	4		--	--		2,000
" " "	85 hp tractors	8		1,000	20.0		8,000
						sub-total	178,575
				Sub-total all vehicles and equipment			274,275
				10% Physical Contingency			27,427
				Total Annual Cost			301,702
				Total Annual Cost Per Hectare			1.886 (KSh 37.7)

1/ Balance of 24 sub-stations would share transport with main stations and have use of 3-ton trucks.

2/ Allows for replacement of tractor units by trucks in stations where ploughing completed.

3/ Required for general carting over rough terrain during station construction and limited cultivation operations.

4/ Tractor requirements for cultivation operation in Turbo are included in afforestation estimates.

APPENDIX 1

KENYA II PLANTATION PROJECT

Table 2 Schedule of Unit Costs for New Division, Stations and Sub-Stations

	Unit Cost KL	Division		Station		Sub-Station	
		No.	Cost (KL)	No.	Cost (KL)	No.	Cost (KL)
Office	-	1	1,300	1	600	1	600
Store, general	330	1	330	1	330	1	330
Store, fuel	110	-	-	1	110	1	110
Garage	-	1	115	1	460	1	460
School	950	-	-	1	950	1	950
Dispensary	330	-	-	1	330	1	330
Shops	290	-	-	2	580	2	580
DFO/Forester House	7,250	1	7,250	1	7,250	1	7,250
Ranger/Clerk/Driver House	1,450	4	5,800	5	7,250	3	4,350
Forest Guard House	550	-	-	3	1,650	3	1,650
Water Supply	1,650	-	-	1	1,650	1	1,650
Nursery	1,450	-	-	1	1,450	1	1,450
Fire Towers	270	-	-	2	540	2	540
Labour houses 1/	210	-	-	-	1/	-	1/
Totals			14,795		23,150		20,250

1/ Calculated on basis of number of employees at cost of KL200/house/employee

APPENDIX 1

KENYA II PLANTATION PROJECT

Table 3 Schedule of Existing Forestry Department Buildings in the Project Area

(Excluding forest workmen's houses)

	HOUSING					OTHER BUILDINGS								
Division	Senior Staff	Junior Staff	Teachers	Forest Guards	Rest House	Fire Tower	Office	Store	Garage	Dispen.	School	Comm. Hall	Shop	Water Supply
Elburgon	13	69	40	129	4	11	11	12	6	11	10	5	40	12
Londiani	15	35	12	75	-	3	8	7	3	5	5	-	4	7
Nyahururu	11	66	30	103	-	6	10	13	7	7	7	1	23	11
Baringo	7	35	15	70	-	4	7	7	2	5	5	-	10	6
Eldoret/Nabkoi	11	60	33	78	2	4	11	26	3	11	15	9	39	20
Kitale	6	30	9	63	-	2	6	6	3	5	3	-	4	5
Kisumu	2	9	-	18	-	-	3	4	2	-	-	-	-	1
Turbo	11	21	-	20	-	3	5	6	3	2	-	-	-	7
Nyeri	12	65	23	104	-	5	12	12	8	10	7	3	10	11
Embu	6	28	6	50	-	2	8	6	2	2	2	-	4	6
Nairobi	10	53	21	110	-	4	9	9	4	7	7	-	14	8
Coast	5	25	-	50	-	-	5	5	3	-	-	-	-	4
Southern	3	18	-	40	2	-	3	3	2	-	-	-	-	3
TOTAL	112	514	189	920	8	44	98	116	48	65	60	18	148	101

APPENDIX 1

KENYA II PLANTATION PROJECT

Table 4 Road Construction Equipment Requirement

Item	Estimated Remaining Operational Life Per Unit	Full Operational Life	Replacement Cost or Cost (KL)	Inherited Value (KL)
Existing Equipment				
1 low loader, 20 ton	4 - 5 years	6 years	16,400	12,300
1 Massey Ferguson bulldozer (D6 equiv.)	5,000 hours	10,000 hours	28,400	14,200
1 D.6 Caterpillar	6,000 hours	10,000 hours	22,900	13,740
1 Tevex 32:40 bulldozer (D7.5 equiv.)	6,000 hours	10,000 hours	33,200	19,920
1 Wakefield motor grader (120 hp ")	8,000 hours	10,000 hours	16,000	12,800
1 Champion motor grader (Cat 150 hp ")	8,000 hours	10,000 hours	17,200	13,760
1 Cat 14 motor grader (150 hp)	6,000 hours	10,000 hours	17,200	10,320
1 Front end loader	3,000 hours	10,000 hours	19,600	5,880
1 Bomeg self-propelled road roller 7-ton	10,000 hours	10,000 hours	8,000	8,000
1 Bomeg self-propelled road roller 10-ton	6,000 hours	10,000 hours	10,000	6,000
4 Tipper lorries 5/7 ton	160,000 km	300,000 km	19,200	10,240
2 Land rovers	60,000 km	200,000 km	5,300	1,590
			Total	128,750
New Equipment (per unit)				
1 D.6 (D.7) Bulldozer with ripper & blade				28,000 1/
1 Cat 955 front end loader 115 hp				24,200
1 Motor grader, 150 hp				17,200
3 7-ton tipper lorries				14,400
2 Land Rovers LWB				5,300
			Total, 1 Unit	89,100
			Total, 2 Units	178,200

KENYA II PLANTATION PROJECT

Table 5 Machinery Operating Costs

Item	Km/hrs/year	Cost/hour or km (KSh)	Annual Cost (KL)
1 bulldozer (D6)	1,500 hours	50	3,750
1 bulldozer (D7)	1,500 hours	60	4,500
1 front end loader	1,400 hours	43	3,010
2 Motor graders	3,000 hours	49	7,350
1 road roller	1,000 hours	16	800
4 tipper lorries	70,000 kms	1.5	5,250
2 land rovers	35,000 kms	1.1	1,925
1 low loader	4,000 kms	3.8	760
		Total, 1 Unit	27,345
		Total, 2 Units	54,690

1/ 1 D.7 bulldozer will be required for one unit and one D.6 for the other unit. Cost represents average of D.6 and D.7.

APPENDIX 1

KENYA II PLANTATION PROJECT

Table 6 Staff and Labour Requirements and Costs

Staff/Labour Grade	No.	Annual Cost (KL)	Total Cost (KL)
Headquarters			
Superintending roads engineer	1	1/	-
Cheif Clerk	1	400	400
Clerk/Typists	2	350	700
Driver	1	350	350
Low Loader driver	1	400	400
			1,850
One Unit Requirements			
Road superintendent	1	1,400	1,400
Road foreman	2	850	1,700
Plant operators	7 2/	450	3,150
Plant operators helpers (dozers)	3	180	540
Plant greasers	4	180	720
Drivers	5	350	1,750
Mechanical foreman	1	1,200	1,200
Welder	1	450	450
Mechanics	4	450	1,800
Mechanics' assistants	4	200	800
Bridge building foreman	1	850	850
Bridge labour	10	180	1,800
Masons	2	350	700
Carpenters	2	350	700
Culvert labour	20	180	3,600
Chief clerk	1	400	400
Clerks	1	350	350
Storeman	1	300	300
Watchman/messenger	4	180	720
Road levellers/tracers 3/	2	450	900
		Total 1 unit	23,830
		Total 2 units	47,660
			49,510

1/ Existing post financed by bilateral agreement funds

2/ Spare driver interchangeable with lorry drivers

3/ Major surveys would be undertaken by surveyor seconded from survey branch of F.D.

APPENDIX 1

KENYA II PLANTATION PROJECT

Table 7 Staff Manning Plans and Costs of Management Units

Project Headquarters

	Grade	Annual Cost (KL)
Project Manager	M	2,600
Asst. Manager	L	2,050
Chief Accountant	L	2,050
Acct. Grade I	J	1,400
Acct. Grade III	G	850
Executive Officer	G	850
3 Secratraies	F	1,800
2 Typists	D	700
7 Clerks	D	2,450
2 Drivers	D	700
1 Storeman	D	350
1 Officer Messenger	B	200
	Total	16,000

Conservancies (3)

	Rate	Annual Cost (KL)
1 C of F	2,100	2,100
1 Executive Officer		850
1 Secratary		600
3 Clerk/Typists	350	1,050
1 Driver		350
1 Postman/Messenger		200
	Total	5,150

Divisions (14)

	Rate	Annual Cost (KL)
1 D.F.O.		1,600
1 Asst. D.F.O.		1,400
3 Clerk/Typists	350	1,050
1 Driver		350
1 Postman/Messenger		200
	Total	4,600

Main Forest Stations (59)

	Rate	Annual Cost (KL)
1 Forester		1,140
2 Rangers	500	1,000
1 Clerk/Typist	350	350
2 Drivers	350	700
1 Storeman		300
8 Forest Guards	230	1,840
1 Postman/Messenger		200
	Total	5,530

		Annual Cost (KL)
1 Forester		1,040
1 Ranger		500
1 Clerk		350
1 Driver		350
1 Storeman		300
4 FGs		920
1 Postman/Messenger		200
	Total	3,660

APPENDIX 1

KENYA II PLANTATION PROJECT

Table 8, Summary of Silvicultural Costs

Item	Sawlogs					Pulpwood - Eldoret					Pulpwood - Turbo				
	Year(s)	Labour Required (Man days)	Labour Cost 1/ (KSh/ha)	Other Cost (KSh/ha)	Total Cost (KSh/ha)	Year(s)	Labour Required (Man days)	Labour Cost 1/ (KSh/ha)	Other Cost (KSh/ha)	Total Cost (KSh/ha)	Year(s)	Labour Required (Man days)	Labour Cost 1/ (KSh/ha)	Other Cost (KSh/ha)	Total Cost (KSh/ha)
Nursery 2/	-1	44.0	418.440	44	462.440	-1	44.0	418.440	44	462.440	-1	48.0	456.480	48	504.480
Land Preparation	-2	37.7	358.527	-	358.527	-2	37.7	358.527	-	358.527	-2	37.7	358.527	-	358.527
Planting	0	14.0	133.140	-	133.140	0	14.0	133.140	-	133.140	0	14.0	133.140	-	133.140
Plantation Survey	0	1.0	9.510	-	9.510	0	1.0	9.510	-	9.510	0	1.0	9.510	-	9.510
Beating Up	0	1.4	13.314	-	13.314	0	1.4	13.314	-	13.314	0	2.8	26.628	-	26.628
Weeding: First	2	15.0	142.650	-	142.650	2	15.0	142.650	-	142.650	0	7.0	66.570	40 3/	106.570
Second	-	-	-	-	-	-	-	-	-	-	1	7.0	66.570	80 3/	146.570
Third	-	-	-	-	-	-	-	-	-	-	2	7.0	66.570	40 3/	106.570
Pruning	2,4,6,8,10	12.0	114.120	-	114.120	-	-	-	-	-	-	-	-	-	-
Pre Commercial Thinning	5	7.0	66.570	-	66.570	-	-	-	-	-	-	-	-	-	-
Marking Out	10,15	0.5	4.755	-	4.755	-	-	-	-	-	-	-	-	-	-
Fire Protection	0-24	0.12	1.141	-	1.141	0-14	0.12	1.141	-	1.141	0-14	0.12	1.141	-	1.141
Other Protection	0-24	0.1	0.951	-	0.951	0-14	0.1	0.951	-	0.951	0-14	0.1	0.951	-	0.951

1/ Labour Shadow wage rate is KSh 9.51/man day.

2/ Nursery labour required is 40 man days per hectare planted. Ten percent is added for sawlogs (and pulpwood-Eldoret) beating up, and 20 percent is added for pulpwood-Turbo beating up. Two thousand trees need to be grown in the nursery for each hectare planted in the field. (There is a 20% culling loss of seedlings for sawlog and pulpwood-Eldoret nurseries and a 10% culling loss for pulpwood-Turbo nurseries.) It costs KSh 20 per thousand seedlings grown for fertilizer, tools, boxes, etc., which amounts to KSh 44 per hectare for sawlogs (and pulpwood-Eldoret) and KSh 48 per hectare for pulpwood-Turbo.

3/ Cost represents mechanical weeding, which costs KSh 20 per hour.

APPENDIX 2

INHERITED AND SALVAGE VALUES

Both inherited and salvage values derived below are unweighted and have no shadow foreign exchange rate applied to them. Appropriate weighting and shadow pricing has been done in Tables 4 and 6 of the text.

Vehicles and Equipment

The project is assumed to inherit a stock of vehicles and equipment equal to six years of project purchases. The useful lives of vehicles and equipment are six years, so they should be, on the average, half worn out. Therefore, an average of three years of useful life is assumed to remain in the vehicles and equipment inherited. Similarly, at the end of the project the salvage value of vehicles and equipment passed on to the continuing plantation programme is assumed to be six years worth of purchases, again on the average half 'worn out'. The inherited/salvage value is, therefore, equal to the cost of three project years' purchases of vehicles and equipment, which is KSh 50.8 per hectare.

Tools and Stores

The inherited and salvage values for tools and stores are determined in the same manner as the values for vehicles and equipment, above, except that a three year useful life is assumed rather than a six year life. Assuming that the stock of tools and stores is on the average half worn out, 1½ years of useful life are inherited and salvaged, at a value of KSh 4.1 per hectare.

Building Construction

Existing buildings and their replacement costs are given in Table 1 of this appendix. Replacement cost per hectare for all existing buildings is KSh 360.6 per hectare. If on the average the buildings are half worn out, their value (for forestry purposes) is KSh 180.3 per hectare. However, it is doubtful that their value for alternative uses is as high as it is for forestry due to location and design. Assuming that their alternative value is half that of their forestry value, the proper inherited value and salvage value for existing buildings is KSh 90.14 per hectare. In addition, a salvage value of new buildings (capital cost - KSh 226.2) needs to be added. Assuming a 50 year useful life and a non-forestry value equal to half that of forestry, the salvage value of new buildings for the sawlog rotation is 25 percent of the building capital cost (KSh 56.54/ha) and for the pulpwood rotation is 35 percent of the building capital cost (KSh 79.16/ha). The salvage value for existing buildings must be added to the salvage values for new buildings to get the total building salvage values, which are KSh 146.7/ha for sawlogs and KSh 169.3/ha for pulpwood. In addition, the existing buildings, which cost KSh 360.6/ha to rebuild, will have to be replaced. Assuming a 50 year useful life for existing buildings, annual replacement will cost KSh 7.2 per hectare. None of the new buildings will need to be replaced.

Road Construction

Road construction is assumed to occur during the year before planting (year 1). All equipment used for project road construction is expected to be completely used up at the end of the six year construction period. In addition, it is assumed that existing roads and roads constructed during the project have no value other than for forestry (no inherited or salvage values). However, equipment existing at the beginning of the project was used for road construction. Table 4 of Appendix 1 lists that equipment. The inherited value plus 10 percent for physical contingency is KL 141 625, or KSh 71.1 per hectare.

APPENDIX 2

KENYA II PLANTATION PROJECT

Table 1 Replacement Costs of Existing Buildings

Type[1/]	Number[2/]	Unit Replacement Cost (KL)	Total Replacement Cost (KL)
Nurseries	90	1,450	130,500
Fire Towers	44	270	11,880
Division Office	13	1,300	16,900
Station Office	85	600-	51,000
General Stores	116	330	38,280
Fuel Stores	90	110	9,900
Division Garages	13	115	1,495
Station Garages	35	460	16,100
Dispensaries	65	330	21,450
Schools	60	950	57,000
Community Halls	18	330[3/]	5,940
Shops	148	290	42,920
Water Supplies	101	1,650	166,650
DFO/Forester Houses	112	7,250	812,000
Ranger/Clerk/Driver Houses	514	1,450	745,300
Teacher's House	189	1,000[4/]	189,000
Forest Guard House	920	550	506,000
Sub-total			2,822,315
Replacement /Additions to existing staff housing and buildings			200,000
Net			2,622,315
10% Physical contigency			262,231
Total Cost			2,884,546 (KSh 360.6/ha)

1/ All of the types listed in either Table 2 or Table 3 of Appendix 1.

2/ Numbers from Table 3, Appendix 1 (when given). Office and garage numbers split by assuming all 13 divisions had one unit with the remainder located on stations. Numbers for items from Table 8, Appendix 1, assume that each old division/station has same buildings as new division/station.

3/ Community hall cost not given in Table 8, Appendix 1, so assumed equal to store/dispensary cost of KL 330.

4/ Teacher's house cost (not given) is assumed equal to average of junior staff's and forest guards' houses.

APPENDIX 3

ECONOMIC UNIT VALUES OF OUTPUT

The economic price of pulpwood is calculated by determining the cost of importing pulpwood to run the Webuye mill and then shadow pricing it at 130 percent. The price is determined as follows:

	US$
Cost of pulpwood chips from West Coast (North America) f.o.b.	84/BDU 1/
Freight and insurance to Mombasa, Kenya	35/BDU
Cost: C.i.f. Mombasa	119/BDU
Less Chipping costs (including depreciation of equipment)	10/BDU
Equivalent roundwood cost (chip measure)	109/BDU
Equivalent roundwood cost (roundwood measure)	54.5/m^3
Transport to Webuye mill (600 km at KSh 0.178/m^3/km)	10.7/m^3
Handling	1.6/m^3
Cost of pulpwood at mill site	66.7/m^3
Less: Logging costs	13.5/m^3
Transport and handling to mill	2.0/m^3
Economic price of pulpwood stumpage	51.2/m^3
	(KSh 366/m^3)
Economic shadow price of pulpwood stumpage	KSh 475.8/m^3

The economic price of sawlogs which substitute for potential imports is calculated by determining the cost of importing sawnwood to Nairobi. The economic price of sawlogs which are exported is estimated to be KSh 236/m^3. The two prices are averaged (weighted: 75 percent for import substitution output and 25 percent for exported output) and multiplied by 1.30 to reflect the 130 percent shadow pricing of foreign exchange. The average shadow price for sawnwood is then converted to sawlog prices by applying the appropriate conversion factors. These prices are calculated as follows:

	US$/$m^3$
Sawnwood price (lowest grade) in Scandinavia f.o.b.	120
Sea freight to Mombasa	35
Port handling in Mombasa and rail freight to Nairobi	11
Sawnwood import cost in Nairobi	166
Sawmilling cost (less stumpage element plus handling/freight to Nairobi)	55
Economic price (sawnwood measure) of import substitution sawlog stumpage	US$111/$m^3$
	(KSh 793/m^3)
Economic price (sawnwood measure) of export sawlog stumpage	(KSh 236/m^3)
Weighted average economic price (sawnwood measure) of sawlog stumpage	KSh 653.75/m^3
Weighted average shadow price (sawnwood measure) of sawlog stumpage	KSh 849.875/m^3

1/ Bone dry unit of chips. 1 BDU = 2400 lb = 2m^3 roundwood.

APPENDIX 4

SENSITIVITY ANALYSES

1. In Section E of the text the sensitivity of the original ERR to changes in certain assumptions was discussed. The original ERRs are commonly tested by changing an assumption such as the shadow wage rate and then constructing an entirely new cash or value flow table for the new analysis. The new ERR or NPW is then calculated from this new cash flow. This is a complete but somewhat time-consuming method. Much faster computerised methods to test several types of uncertainties are available. One of these was used to carry out the sensitivity analysis of the Kenya II plantation project components. Results are shown in Tables 1 through 6 in this Appendix. These tables show changes in four investment performance measures due to changes in main items of benefits or costs. Changes in benefits/costs are specified in percentage points, in this case 10 percent.

2. If 20 percent had been specified instead, each change in the investment performance measures would have been twice as large. Benefit decreases and cost increases have the effect of lowering the investment performance measures by the amounts listed in these tables, while benefit increases and cost decreases have the effect of raising the measures by the same amounts. For example, if nursery cost increased by 10 percent, in Table 1 the NPW would fall by KSh 42.81 to KSh 16 133.8 and if nursery cost decreased by 10 percent, NPW would rise by KSh 42.81 to KSh 16 219.42. It is evident then, that a change of 10 percent in nursery cost has little effect on NPW, and consequently it is likely to have no effect on the project decision on whether or not to accept the project (assuming that the other costs, the benefits, the discount rate, etc. are as indicated in this analysis).

3. The sensitivity tables also facilitate quick comparisons of the relative sensitivity of the investment performance measures to the various costs and benefits. For example, in Table 1 the largest values in each column are found in row 27, harvest benefit. This means that for a 10 percent change in any cost/benefit all four investment performance measures are most sensitive to a ten percent change in the harvest value. Likewise, the measures are least sensitive to a 10 percent change in the tools salvage value (row 29).

4. Besides indicating the relative sensitivity of investment performance measures to changes in costs and benefits, the tables can be used to directly calculate NPW, NFW, SEV, and SR for changes in one or more benefits or costs. A common test is to check for changes in one or more costs and/or benefits which cause NPW or NFW to become equal to zero at a discount rate equal to the project's alternative rate of return. If such changes were considered likely, the decision whether or not to implement the project could be affected since the profitability of the project would have changed.

5. For example, the figures in Table 1 indicate that a 10 percent benefit decrease or cost increase in any or all benefits and costs would reduce NPW very little. In fact, the complete elimination of the final harvest would reduce NPW by KSh 16 086.6 to KSh 90. In other words, the rate of return on the sawlog sub-project if the only direct benefits were sawlog thinnings would be about 8 percent. Clearly, then, in order for the ERR to drop below 8 percent, very substantial benefit decreases and/or cost increases must occur.

6. The exact changes in prices necessary to reduce NPW to zero can be calculated from Table 1. For example, the average stumpage price reduction which lowers the present value of the thinnings and final harvest benefits by an amount equal to NPW is calculated as follows:

Activity	change in NPW
10 percent thinning value decrease	- 595.55
10 percent harvest value decrease	- 1 608.66
10 percent total removals value decrease (assumed due decrease in stumpage price)	- 2 204.21
NPW	16 176.61

$$NPW/(10\%\ \text{total removals value decrease}) = \frac{16\ 176.61}{-2\ 204.21} = -7.339$$

Therefore, a 73.39 percent (i.e. 7.339 x 10) reduction in the average stumpage price would reduce NPW to zero.

APPENDIX 4

Table 1, Economic Analysis for Sawlogs (8% discount rate)

Original analysis	KSh/ha
- net present worth (NPW)	16 176.61
- net future worth (NFW)	129 219.75
- soil expectation value (SEV)	18 491.50
- soil rent (SR)	1 369.74

SENSITIVITY ANALYSIS

(8.00 PERCENT DISCØUNT RATE)
KEN SHILLING/HECTARE

DUE TØ A 10 PERCENT CHANGE IN	CHANGE IN: NPW	NFW	SEV	SR
1 NURSERY	42.81	342.01	48.94	3.63
2 LAND PREPARE	35.85	286.37	40.98	3.04
3 PLANTING	11.41	91.15	13.04	.97
4 PLANT SURVEY	.81	6.51	.93	.07
5 BEATING UP	1.14	9.11	1.30	.10
6 WEEDING	10.48	83.73	11.98	.89
7 PRUNING	31.56	252.08	36.07	2.67
8 PRE-CØM THIN	3.89	31.04	4.44	.33
9 MARKING ØUT	.32	2.56	.37	.03
10 FIRE PRØTECT	1.09	8.68	1.24	.09
11 ØTH. PRØTECT	.99	7.90	1.13	.08
12 VE/EQUIP PUR	43.60	348.25	49.84	3.69
13 VE/EQUIP RUN	90.40	722.08	103.33	7.65
14 VE/EQU INHER	6.74	53.84	7.70	.57
15 TØØLS/STØRES	6.66	53.16	7.61	.56
16 TØØLS INHER	.52	4.15	.59	.04
17 BUILD CØNSTR	23.77	189.88	27.17	2.01
18 BUILD MAINT	22.74	181.66	26.00	1.93
19 BUILD REPLAC	14.71	117.47	16.81	1.25
20 BUILD INHER	9.47	75.65	10.83	.80
21 RØAD CØNSTRU	51.44	410.94	58.81	4.36
22 RØAD MAINT	7.51	60.01	8.59	.64
23 RØAD INHER	8.56	68.34	9.78	.72
24 STAFF REQUIR	151.54	1210.52	173.23	12.83
25 ØFFICE RUN	11.40	91.03	13.03	.96
26 THINNINGS	595.55	4757.31	680.78	50.43
27 HARVEST	1608.66	12850.11	1838.87	136.21
28 VE/EQU SALVG	.84	6.74	.96	.07
29 TØØLS SALVG	.07	.52	.07	.01
30 BUILD SALVG	1.93	15.42	2.21	.16

APPENDIX 4

Table 2, Economic Analysis for Pulpwood - Eldoret Division
(8% discount rate)

Original analysis	KSh/ha
- net present worth (NPW)	24 660.18
- net future worth (NFW)	91 243.11
- soil expectation value (SEV)	33 793.52
- soil rent (SR)	2 500.22

SENSITIVITY ANALYSIS

(8.00 PERCENT DISCOUNT RATE)
KEN SHILLING/HECTARE

DUE TO A 10 PERCENT CHANGE IN	CHANGE IN: NPW	NFW	SEV	SR
1 NURSERY	42.81	158.42	58.67	4.35
2 LAND PREPARE	35.85	132.65	49.13	3.64
3 PLANTING	11.41	42.22	15.64	1.16
4 PLANT SURVEY	.81	3.01	1.12	.08
5 BEATING UP	1.14	4.22	1.56	.12
6 WEEDING	10.48	38.78	14.36	1.06
7 FIRE PROTECT	.87	3.23	1.19	.09
8 OTH. PROTECT	.79	2.93	1.09	.08
9 VE/EQUIP PUR	29.19	107.99	40.00	2.96
10 VE/EQUIP RUN	60.46	223.71	82.85	6.14
11 VE/EQU INHER	5.78	21.39	7.92	.59
12 TOOLS/STORES	4.49	16.63	6.16	.46
13 TOOLS INHER	.44	1.63	.60	.04
14 BUILD CONSTR	20.38	75.41	27.93	2.07
15 BUILD MAINT	15.21	56.26	20.84	1.54
16 BUILD REPLAC	9.82	36.34	13.46	1.00
17 BUILD INHER	8.12	30.04	11.13	.82
18 ROAD CONSTRU	51.44	190.35	70.50	5.22
19 ROAD MAINT	6.02	22.29	8.25	.61
20 ROAD INHER	8.56	31.66	11.72	.87
21 STAFF REQUIR	101.39	375.14	138.94	10.29
22 OFFICE RUN	7.67	28.39	10.51	.78
23 HARVEST	2893.36	10705.50	3964.97	293.70
24 VE/EQU SALVG	1.56	5.78	2.14	.16
25 TOOLS SALVG	.12	.44	.16	.01
26 BUILD SALVG	4.12	15.26	5.65	.42

APPENDIX 4

Table 3, Economic Analysis for Pulpwood - Turbo
(8% discount rate)

Original analysis	KSh/ha
- net present worth (NPW)	24 189.75
- net future worth (NFW)	89 502.52
- soil expectation value (SEV)	33 148.86
- soil rent (SR)	2 455.47

SENSITIVITY ANALYSIS

(8.00 PERCENT DISC0UNT RATE)
KEN SHILLING/HECTARE

DUE T0 A 10 RERCENT CHANGE IN	CHANGE IN: NPW	NFW	SEV	S_R
1 NURSERY	46.71	172.84	64.01	4.74
2 LAND PREPARE	35.85	132.65	49.13	3.64
3 PLANTING	11.41	42.22	15.64	1.16
4 PLANT SURVEY	.81	3.01	1.12	.08
5 BEATING UP	2.28	8.44	3.13	.23
6 WEEDING	28.61	105.87	39.21	2.90
7 FIRE PR0TECT	.87	3.23	1.19	.09
8 0TH. PR0TECT	.79	2.93	1.09	.08
9 VE/EQUIP PUR	31.90	118.03	43.71	3.24
10 VE/EQUIP RUN	66.15	244.75	90.65	6.71
11 VE/EQU INHER	6.29	23.27	8.62	.64
12 T00LS/ST0RES	4.91	18.15	6.72	.50
13 T00LS INHER	.48	1.78	.66	.05
14 BUILD C0NSTR	22.18	82.07	30.39	2.25
15 BUILD MAINT	16.61	61.44	22.76	1.69
16 BUILD REPLAC	10.74	39.76	14.72	1.09
17 BUILD INHER	8.84	32.71	12.11	.90
18 R0AD C0NSTRU	51.44	190.35	70.50	5.22
19 R0AD MAINT	6.02	22.29	8.25	.61
20 R0AD INHER	8.56	31.66	11.72	.87
21 STAFF REQUIR	110.86	410.18	151.92	11.25
22 0FFICE RUN	8.39	31.03	11.49	.85
23 HARVEST	2893.36	10705.50	3964.97	293.70
24 VE/EQU SALVG	1.70	6.29	2.33	.17
25 T00LS SALVG	.13	.48	.18	.01
26 BUILD SALVG	4.49	16.61	6.15	.46

APPENDIX 4

Table 4, Financial Analysis for Sawlogs (8% discount rate)

Original analysis	KSh/ha
- net present worth (NPW)	1 560.83
- net future worth (NFW)	12 467.98
- soil expectation value (SEV)	1 784.18
- soil rent (SR)	132.16

SENSITIVITY ANALYSIS

(8.00 PERCENT DISCOUNT RATE)
KEN SHILLING/HECTARE

DUE TO A 10 PERCENT CHANGE IN	CHANGE IN: NPV	NFV	SEV	SR
1 NURSERY	55.73	445.19	63.71	4.72
2 LAND PREPARE	47.80	381.83	54.64	4.05
3 PLANTING	15.22	121.56	17.40	1.29
4 PLANT SURVEY	1.09	8.70	1.24	.09
5 BEATING UP	1.53	12.19	1.74	.13
6 WEEDING	13.98	111.68	15.98	1.18
7 PRUNING	42.10	336.26	48.12	3.56
8 PRE-COM THIN	5.18	41.39	5.92	.44
9 MARKING OUT	.42	3.36	.48	.04
10 FIRE PROTECT	1.48	11.84	1.69	.13
11 OTH. PROTECT	1.28	10.26	1.47	.11
12 VE/EQUIP PUR	33.53	267.85	38.33	2.84
13 VE/EQUIP RUN	74.71	596.83	85.41	6.33
14 VE/EQU INHER	5.18	41.38	5.92	.44
15 TOOLS/STORES	5.44	43.45	6.22	.46
16 TOOLS INHER	.42	3.35	.48	.04
17 BUILD CONSTR	23.08	184.36	26.38	1.95
18 BUILD MAINT	22.78	181.95	26.04	1.93
19 BUILD REPLAC	14.27	114.03	16.32	1.21
20 BUILD INHER	9.20	73.49	10.52	.78
21 ROAD CONSTRU	45.39	362.57	51.88	3.84
22 ROAD MAINT	6.33	50.53	7.23	.54
23 ROAD INHER	6.58	52.59	7.53	.56
24 STAFF REQUIR	147.12	1175.16	168.17	12.46
25 OFFICE RUN	11.08	88.51	12.67	.94
26 THINNINGS	116.79	932.94	133.51	9.89
27 HARVEST	315.47	2520.00	360.62	26.71
28 VE/EQU SALVG	.65	5.18	.74	.05
29 TOOLS SALVG	.05	.42	.06	.00
30 BUILD SALVG	1.87	14.97	2.14	.16

APPENDIX 4

Table 5, Financial Analysis for Pulpwood - Eldoret (8% discount rate)

Original analysis	KSh/ha
- net present worth (NPW)	2 456.32
- net future worth (NFW)	9 088.43
- soil expectation value (SEV)	3 366.06
- soil rent (SR)	249.34

SENSITIVITY ANALYSIS

(8.00 PERCENT DISCOUNT RATE)
KEN SHILLING/HECTARE

DUE TO A 10 PERCENT CHANGE IN	CHANGE IN: NPW	NFW	SEV	SR
1 NURSERY	55.73	206.21	76.37	5.66
2 LAND PREPARE	47.80	176.86	65.50	4.85
3 PLANTING	15.22	56.31	20.85	1.54
4 PLANT SURVEY	1.09	4.03	1.49	.11
5 BEATING UP	1.53	5.65	2.09	.15
6 WEEDING	13.98	51.73	19.16	1.42
7 FIRE PRØTECT	1.19	4.40	1.63	.12
8 ØTH. PRØTECT	1.03	3.81	1.41	.10
9 VE/EQUIP PUR	22.47	83.14	30.79	2.28
10 VE/EQUIP RUN	50.00	184.98	68.51	5.07
11 VE/EQU INHER	4.44	16.43	6.08	.45
12 TØØLS/STØRES	3.63	13.44	4.98	.37
13 TØØLS INHER	.36	1.33	.49	.04
14 BUILD CØNSTR	19.79	73.22	27.12	2.01
15 BUILD MAINT	14.76	54.62	20.23	1.50
16 BUILD REPLAC	9.54	35.31	13.08	.97
17 BUILD INHER	7.89	29.19	10.81	.80
18 RØAD CØNSTRU	45.39	167.94	62.20	4.61
19 RØAD MAINT	5.07	18.77	6.95	.51
20 RØAD INHER	6.58	24.36	9.02	.67
21 STAFF REQUIR	98.42	364.15	134.87	9.99
22 ØFFICE RUN	7.45	27.58	10.21	.76
23 HARVEST	182.43	675.00	250.00	18.52
24 VE/EQU SALVG	1.20	4.44	1.64	.12
25 TØØLS SALVG	.10	.36	.13	.01
26 BUILD SALVG	4.00	14.81	5.49	.41

APPENDIX 4

Table 6, Financial Analysis for Pulpwood - Turbo (8% discount rate)

Original analysis	KSh/ha
- net present worth (NPW)	2 938.07
- net future worth (NFW)	10 870.92
- soil expectation value (SEV)	4 026.24
- soil rent (SR)	298.24

SENSITIVITY ANALYSIS

(8.00 PERCENT DISCOUNT RATE)
KEN SHILLING/HECTARE

DUE TO A 10 PERCENT CHANGE IN	CHANGE IN: NPW	NFW	SEV	SR
1 NURSERY	60.80	224.95	83.31	6.17
2 LAND PREPARE	47.80	176.86	65.50	4.85
3 PLANTING	15.22	56.31	20.85	1.54
4 PLANT SURVEY	1.09	4.03	1.49	.11
5 BEATING UP	3.04	11.26	4.17	.31
6 WEEDING	33.91	125.47	46.47	3.44
7 FIRE PROTECT	1.19	4.40	1.63	.12
8 OTH. PROTECT	1.03	3.81	1.41	.10
9 VE/EQUIP PUR	24.56	90.86	33.65	2.49
10 VE/EQUIP RUN	54.65	202.22	74.90	5.55
11 VE/EQU INHER	4.84	17.91	6.63	.49
12 TOOLS/STORES	3.97	14.68	5.44	.40
13 TOOLS INHER	.39	1.44	.53	.04
14 BUILD CONSTR	21.54	79.70	29.52	2.19
15 BUILD MAINT	16.13	59.69	22.11	1.64
16 BUILD REPLAC	10.44	38.65	14.31	1.06
17 BUILD INHER	8.58	31.75	11.76	.87
18 ROAD CONSTRU	45.39	167.94	62.20	4.61
19 ROAD MAINT	5.07	18.77	6.95	.51
20 ROAD INHER	6.58	24.36	9.02	.67
21 STAFF REQUIR	107.65	398.30	147.52	10.93
22 OFFICE RUN	8.13	30.10	11.15	.83
23 HARVEST	182.43	675.00	250.00	18.52
24 VE/EQU SALVG	1.31	4.84	1.79	.13
25 TOOLS SALVG	.11	.39	.14	.01
26 BUILD SALVG	4.36	16.12	5.97	.44

APPENDIX 5

KENYA – SAWLOG AND PULPWOOD PLANTATION PROJECT

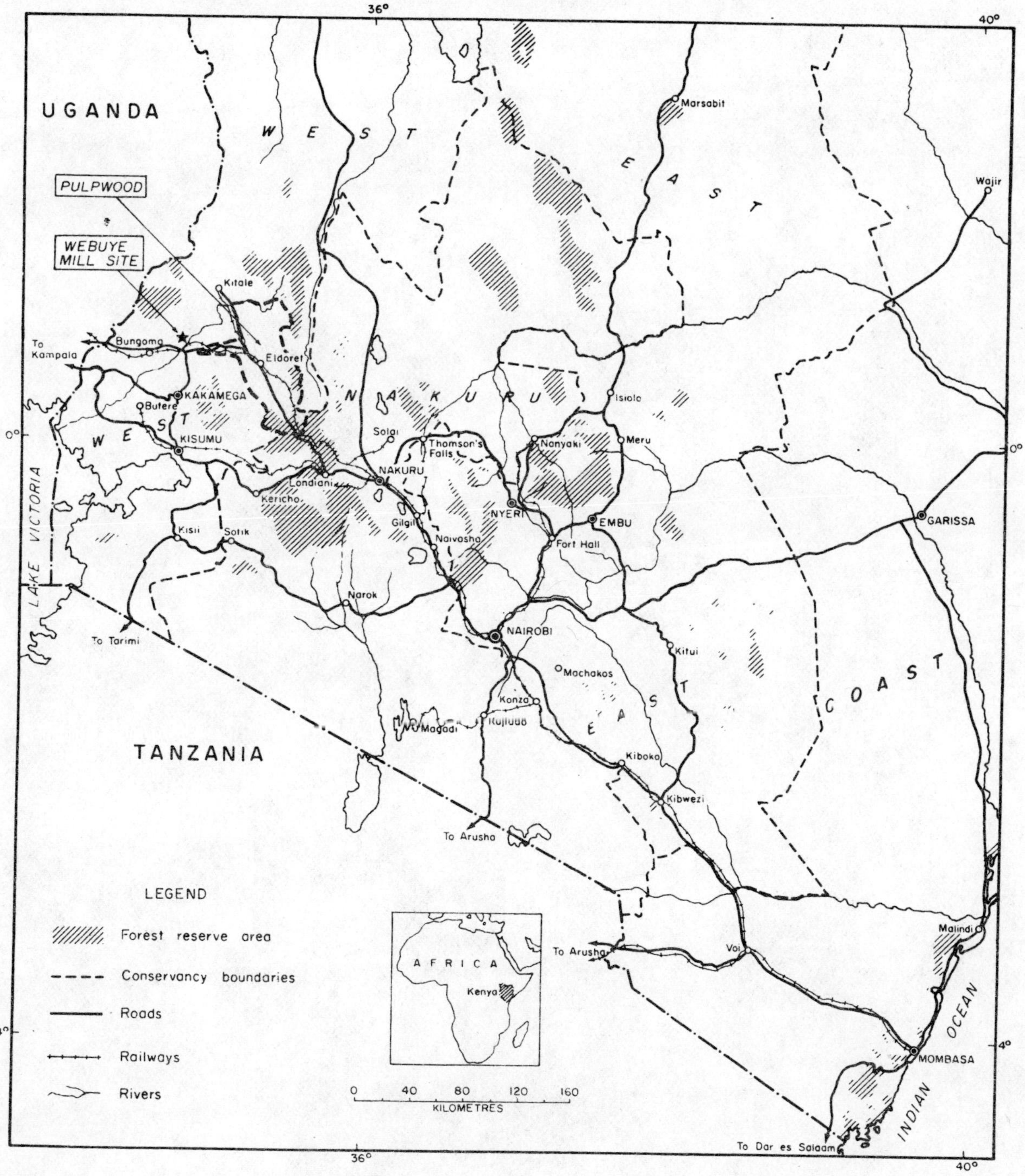

ECONOMIC ANALYSIS OF FORESTRY PROJECTS: CASE STUDIES

FAO FORESTRY PAPER 17, SUPPLEMENT 1

CASE STUDY NO. 6

ZAMBIA INDUSTRIAL FORESTRY PROJECT

FOOD AND AGRICULTURE ORGANIZATION OF THE UNITED NATIONS

Rome, 1979

PREFACE

This is one in a series of case studies of forestry and forest industries projects that has been prepared by the Food and Agriculture Organization of the United Nations (FAO) in order to demonstrate methods of identifying, preparing and appraising projects in the forest sector. The programme of case studies has been made possible by a special allocation to FAO for this purpose from the Swedish International Development Authority (SIDA).

The Project reported on in this case study is a project of the Government of Zambia which has been supported by a loan from the World Bank. The case study is based largely on material contained in an appraisal of the Project carried out by the World Bank (Report No. 1429-ZA, 20 April 1977). Additional information was gathered through a visit to the Project during 1977. The case study therefore describes the Project as it was perceived at that time. This material has been released to FAO by the World Bank and the Government of Zambia exclusively for the purpose of teaching forestry planning methods. It may not be quoted or reproduced in part or in whole without permission.

CURRENCY EQUIVALENTS

Currency Unit = Zambian Kwacha (K)
K1 = US$1.2444
US$1 = K0.8036

ABBREVIATIONS

LNRT	-	Ministry of Land, Natural Resources and Tourism
FD	-	Forest Department
IPD	-	Industrial Plantations Division
KITE	-	Kafubu Industrial Timber Enterprises
ERR	-	Economic rate of return
FRR	-	Financial rate of return
IRR	-	Internal rate of return
f.o.b.	-	Free-on-board
c.i.f.	-	Cost, insurance and freight
p.a.	-	per annum
p.d.	-	per day

TABLE OF CONTENTS

INTRODUCTION

1. This case study is based upon a Project which forms part of Zambia's continuing forestry and forest industry development programme.[1]

2. It is a five-year "time-slice" project, covering planting and maintaining 19 500 ha of industrial forestry plantations during 1978-82, maintaining existing plantations during that period, logging and transporting 120 000 m^3 annually from pre-project plantations, and establishment and operation of a sawmill to produce about 46 000 m^3 of sawnwood annually from pre-project plantations. It is the second such five-year project within the programme, and it is envisaged that it will be followed by further projects. The project costs over the five-year period total approximately US$ 33.7 million.

3. The case study is divided into three parts. Section A gives background information on the country, the forest sector and the industrial forestry programme. Section B provides technical details about the Project, including estimates of input requirements and output levels. Section C contains analyses of Project costs and Project benefits, and a comparison of costs and benefits. This includes an examination of the sensitivity of the economic rate of return (ERR) of the Project, and of some of its component parts, to changes in assumptions about input and output values.

4. This case study provides a basis for discussion of a number of issues that commonly occur in planning and analysing joint forestry and forest industry projects. These centre round the relationship of the different components to each other (e.g., harmonizing the timing and scale of tree growing and processing, the relationship of the species properties to the final product requirements, and valuing the roundwood which constitutes the output of the forestry component of the project and an input into the industries component). The case study also covers other points which often arise in analysing afforestation projects (e.g. the shadow pricing of land, labour and project outputs).[2] It also illustrates issues related to the analysis of "time-slice" projects.

A. BACKGROUND

General

5. Zambia is a land-locked country in central Africa. The total surface area is about 735 000 km^2, and the population is nearly 5 million. Zambia has a sub-tropical climate, with a warm wet season from November to April, a cool dry season from May to September and a hot dry season from September to November. Average annual rainfall decreases from about 1 300 mm in the northwest to about 800 mm in the central region, around the capital, Lusaka.

6. GNP per capita is about $540 (1975). Copper mining dominates the economy, accounting for more than 90 percent of exports and about 25 percent of GDP in 1975. Although manufacturing, services, transport and communications have all been growing rapidly, aggregate growth has been slow due to near stagnation in mining and a low growth rate in agriculture. Agricultural production does not yet adequately reflect the country's huge and usable land resources. The "agriculture, forestry and fishing" group accounts for less than 10 percent of GDP.

1/ Policies and practices described in this case study are not necessarily those actually employed in Zambia either at present or during the period covered by the Project.

2/ For additional coverage of industrial afforestation project issues see also Case Study No. 4: Kenya I Sawlog and Pulpwood Plantation Project, and Case Study No. 5: Kenya II Sawlog and Pulpwood Plantation Project.

Forest Resources and Wood Utilization

7. About 50 percent of the total area of Zambia is classified as forest land, the greater part of which is open woodland ("miombo"). The natural hardwood forests have so far been providing all the country's requirements of fuelwood, most of its poles and some of its sawnwood, particularly mining timber, while most of the sawn timber and wood-based panels and all the paper products are imported. Accessible productive forests are shrinking, and Zambia's needs for industrial wood of all kinds are rising fast and are being met increasingly from imports. In 1976 imports of forest products amounted to US$ 13 million.

8. The Government of Zambia has accorded high priority to the development of forest resources so that in due course it can become self-sufficient in meeting its timber requirements including the products of wood-based industries. It also recognizes that forests are important in providing protection against soil erosion and in maintaining river flows. To this end, the Government has established a National Forest Estate[1] of over 8 million ha (some 9 percent of Zambia's total land area).

9. Forest management is the responsibility of the Forest Department (FD) of the Ministry of Land, Natural Resources and Tourism (LNRT). To solve the problems of long gestation periods and difficult commercial exploitation associated with the indigenous species, the Forest Department about 30 years ago began the planting of fast-growing exotic species (mainly Pine and Eucalyptus). This was followed in the 1960's by the creation of an Industrial Plantations Division (IPD) within the Department to handle the then considerably expanded afforestation programme.

10. Roundwood consumption was estimated in 1975 to be about 5 million m^3 of which some 4.5 million m^3 was fuelwood and charcoal. Other uses of roundwood include poles for building and fencing and other domestic purposes. The main industrial uses of roundwood are as poles for the mining industry (pit-props and smelter and refinery poles), transmission poles (power, telegraph and telephone), building poles and fence posts. The current demand for sawnwood is about 100 000 m^3, about half of which is imported.

11. Primary forest industries in Zambia consist essentially of sawmilling and the conversion of Eucalyptus roundwood to transmission poles and posts. Although IPD started sawmilling of exotic species in 1970, more than 80 percent of all sawmilling in Zambia (i.e. all non-IPD sawmilling) is still based on indigenous timber. Excluding pit sawing and a few small local saws, about a dozen sawmills account for over 90 percent of all sawmilling.

12. Many sawmills in Zambia have in the past operated below capacity, but the situation has improved considerably in recent years, as obviously misplaced sawmills went out of business, and those remaining in business have begun to operate at or near full capacity. The only IPD sawmill operating below capacity - Dola Hill - should reach full capacity production in a couple of years when more mature pine will become available.

13. For many years, most of the transmission poles (electricity, telephone, etc.) used in the country were imported. Increasingly, however, locally-grown Eucalyptus poles are being substituted for the imports.

---

[1] The forest estate or reserves within which exploitation of forest resources is closely supervised by the Forest Department.

14. Zambia's secondary forest industries consist of a blockboard factory, a door manufacturing company, and several joinery and furniture manufacturing enterprises all using mainly imported wood. In the ten years 1964-74, Zambia imported annually on average some 10 000 m^3 of plywood and blockboard, about 800 m^3 of particle board, some 5 200 m^3 of fibreboard, and about 25 000 tons of paper. These imports suggest some scope for import-substitution industries, using the smallwood and sawmill waste in the manufacture of one or more types of wood-based panel with or without a pulp and paper mill.

15. Past studies have indicated that the optimum economic benefit to Zambia would result from a programme which included a pulp and paper mill. This arises mainly because sawmilling alone does not utilise all forest thinnings and results in about 60 percent waste, and wood panel industries are relatively small users of such waste. Moreover, past studies have shown that, by the mid-eighties, domestic demand for kraft and industrial papers would be adequate to justify a pulp mill coming on stream. However, its cost may exceed Zambia's capability to mobilise the necessary finance in the next few years and so far no technical or commercial partner has been firmly identified. Accordingly, the timing of the establishment of a pulp mill is uncertain and investments in afforestation should be justified on the basis of sawmilling and wood-based panel manufacture alone.

The Industrial Forestry Programme

Origin, Concept and Objectives

16. In 1949, the Forest Department started growing Pines on the Copperbelt, at first on an experimental scale near Ndola, and then on an increasingly larger scale around Ndola and west of Kitwe (see map). The Government decided to embark on a long-term industrial forestry programme on the Copperbelt and, in order to implement this programme, the Industrial Plantations Division of the Forest Department was created in 1962. In the sixties, Eucalypts were also tried and from 1967 onwards were planted on a large scale. The main species were *P. kesiya* for Pines and *E. grandis* for Eucalypts. The target was the planting of 40 000 ha by 1993, of which approximately 3/4 would be Pine and 1/4 Eucalyptus. The Government is now considering increasing the target to at least 80 000 ha before the turn of the century. The objectives of this planting programme are to supply:

(a) most of Zambia's needs in sawn lumber, using mainly Pine when Pine sawlogs become available in sufficient quantity;

(b) smallwood and sawmill waste, both of Pine and Eucalyptus, for a pulp and paper mill and for various board mills;

(c) poles (Eucalyptus) of all kinds for smelting, telephone, transmission and other uses.

17. By the year 2000, it is believed that most of these needs can be met from domestic supplies, provided the planting programme is implemented. Regarding sawnwood, however, it is expected that some imports will still be required to supplement domestic production.

18. By 1975 IPD, which had become a virtually independent and entirely self-accounting unit, was employing a total staff of 134, of whom 13 were professional or senior officers, and a total labour force of just over 1 000. Its activities included afforestation, construction and road work, forest exploitation (logging and transport), wood processing and marketing.

Phase I Project

19. The Project, which was managed by IPD, was part of the Zambian Government's Forest Development Programme and provided for the annual planting of 1 000 ha each of Pine (mainly *P. kesiya*) and Eucalyptus (mainly *E. grandis*) over an 8-year period (1969-1976).

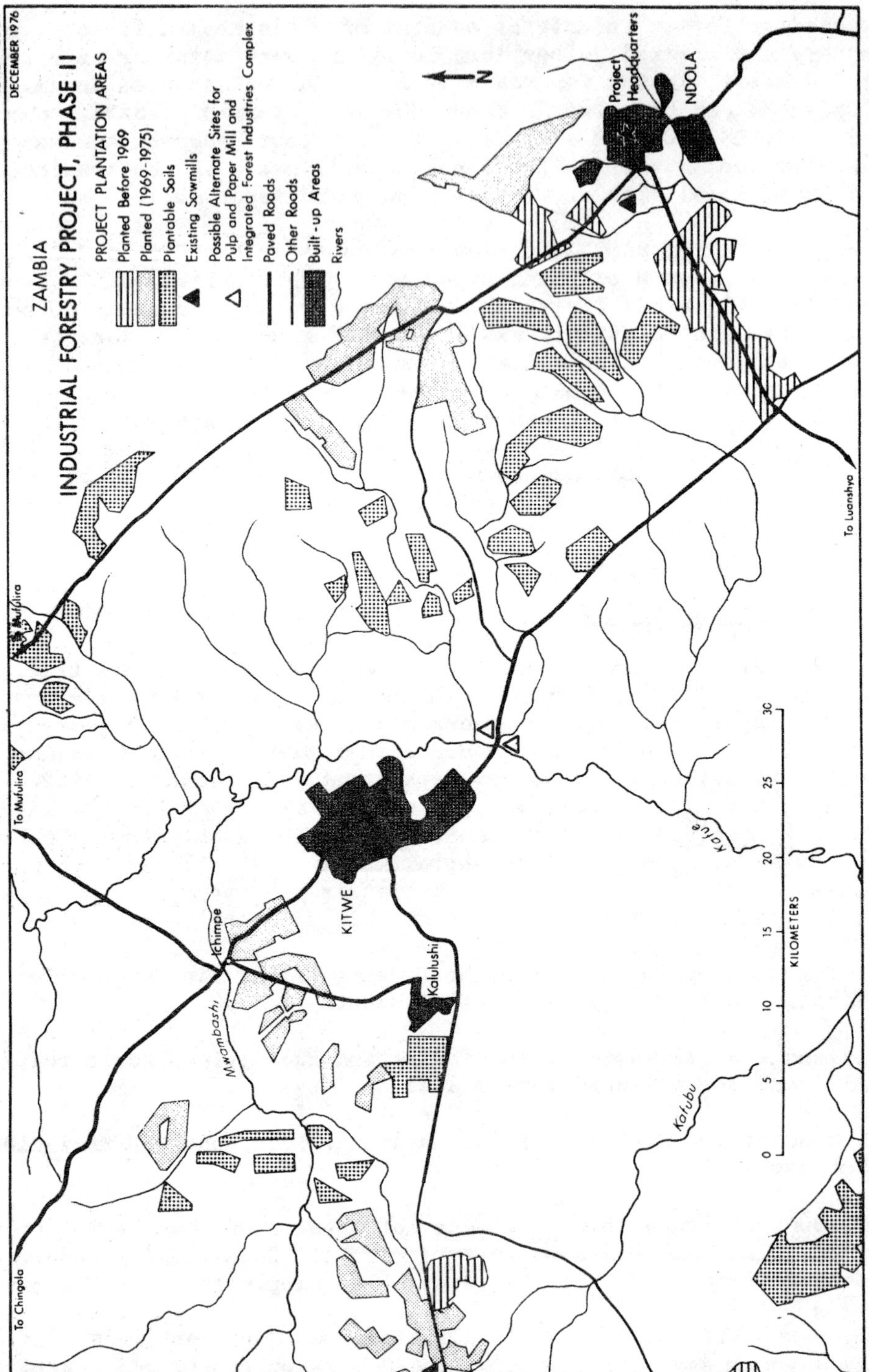
IBRD 12528
DECEMBER 1976
ZAMBIA
INDUSTRIAL FORESTRY PROJECT, PHASE II
PROJECT PLANTATION AREAS
Planted Before 1969
Planted (1969-1975)
Plantable Soils
Existing Sawmills
Possible Alternate Sites for Pulp and Paper Mill and Integrated Forest Industries Complex
Paved Roads
Other Roads
Built-up Areas
Rivers
N
Project Headquarters
NDOLA
To Luanshya
To Mufulira
Mufulira
To Chingola
KITWE
Ichimpe
Kalulushi
Mwambashi
Kafue
Kafubu
KILOMETERS
0
5
10
15
20
25
30
CHATI
Mukutuma
Fibale
Lunga
TANZANIA
MALAWI
MOZAMBIQUE
ZAIRE
ANGOLA
ZAMBIA
Location of Project
RHODESIA
BOTSWANA
NAMIBIA

The Pine would be grown on a 30-year rotation at a mean annual volume increment rate of approximately 17 m^3/ha, while the Eucalyptus would be grown on a 12-year rotation with a mean annual increment of 18 m^3/ha. It was anticipated that over 90 percent of the Pine production and about half of the Eucalyptus production would be used for sawlogs and construction timber, the balance being for poles of various kinds.

20. The total plantation target of 16 000 ha was reached in 7 years instead of 8, but, based on a reassessment of wood requirements, the ratio of pine to eucalyptus was 72:28 instead of 50:50. The pine management regime was also revised, in order to reduce the quantities of smallwood produced early in the rotation. The mean annual increment of P. kesiya is now considered to be around 14 m^3/ha (3 m^3 less), after due allowance for losses, but that of E. grandis is probably around 25 m^3/ha (7 m^3 more). P. kesiya is now grown on a 25 year rotation.

21. Two new species were introduced into the planting programme on a fairly large scale: Pinus oocarpa and Eucalyptus cloeziana. This was partly to achieve diversification to reduce the risks of disease, poor performance, etc. In addition E. cloeziana has better stem form than E. grandis, which should lead to better recovery in sawmilling, and it is better for transmission poles and mine use. Though it has slower initial growth, it tends to catch up with E. grandis later in the rotation. Similarly, P. oocarpa has better form than P. kesiya (small branches; less susceptible to wind sweep), and is also a slow starter which tends to catch up later.

22. Costs have proved to be high, mainly due to the intensive land clearing and preparation necessary in order to achieve high yields. Research is currently focused on reducing these costs, and on offsetting them to a greater degree by using more of the wood from the cleared woodland for charcoal production and sale.

23. The main issue that plagued the first project was that of uncertainties surrounding development of industries to process the project output. As sufficient processing capacity was not available elsewhere in the country, IPD was obliged to expand the scope of the project to include harvesting and sawmilling. Uncertainty about the timing, scale and technical specifications of the expected panel and pulp industries hindered the planning and management of the plantations.

B. THE PROJECT

General Description

24. The Project would, during the five years 1978 through 1982, continue IPD's long-term planting and maintenance programme, expand IPD's logging and sawmilling capacity, and conduct studies designed to lead to more economical and efficient land clearing and charcoal production. The Project would provide for:

(a) about 3 500 ha of new plantations annually, of which about 3 000 ha would be pine and about 500 ha eucalyptus;

(b) replanting annually with eucalyptus of approximately 400 ha of clear-felled plantations;

(c) maintenance of (a) and (b), together with all existing IPD plantations;

(d) doubling of IPD's logging and transportation capacity from about 60 000 m^3 to about 120 000 m^3 roundwood per year;

(e) construction and equipment of a sawmill of about 46 000 m^3 per year sawnwood capacity;

(f) staff training and fellowships to facilitate (a) to (e) above; and

(g) research, experimentation and studies designed to reduce land clearing costs and to improve charcoal production methods.

25. Plantation maintenance costs have been capitalized in accordance with normal plantation practice and included in Project costs for the plantation development period. All operating costs beyond the five year Project period are excluded from Project costs. The Project would be implemented by IFD.

The Project Area

26. The Project area (see map) is situated on the Copperbelt around the towns of Ndola and Kitwe. The new plantations would be extensions of the existing plantations. There are two main blocks: the first is north and west of Ndola and is generally extending towards Kitwe (this is mostly under Pine); the second is located some 20-50 km west of Kitwe, centre around Chati, with a small forest industrial complex (KITE), and is extending mainly north and south (most of the eucalyptus is planted here, though there is also some pine).

27. Forest plantations are only created inside National Forests. There is, therefore, no land tenure problem. Furthermore, there is an abundant supply of other arable land in the area, and consequently no pressure to use National Forest land for cultivation. The plantations are on good soils, which are readily available on the Copperbelt. Over 500 000 ha of land, or almost 20 percent of the total area of the Copperbelt Province, are reserved under the Forests Act, of which less than 20 000 ha would be planted under the Project. Of some 300 000 ha sampled, about 90 000 ha have been found suitable for tree-planting. The land is generally flat, undulating very gently and varying in altitude from about 1 500 m in the north to about 1 200 m in the south. Average annual rainfall in the Project area is about 1 200 - 1 300 mm, falling mainly from late October to early April. Mean temperatures are 15°C in winter and 23°C in summer. Although violent thunderstorms and torrential rains are common during the wettest months, strong winds are rare, and there are only about three nights of frost a year on the average. In short, conditions inside the Project area are excellent for growing pines and eucalyptus.

28. There is no need within the foreseeable future to plant fuelwood on the Copperbelt, as large areas of woodland are being clearfelled annually for afforestation, and enough firewood and charcoal should be obtained from these areas to supply all the needs of the Copperbelt. This region also contains the greatest concentration of population, industrial development and wealth, hence of potential markets. The area is well served by public roads, forest roads and tracks, and a railway line.

29. Once the choice of a site for the proposed pulp and paper mill has been settled, it would be desirable to locate all future sawmills, the pulp and paper mill, and wood-based panel manufacture within the same complex, in order to minimize wood transport costs and fully utilize the infrastructure within the complex. In the meantime, the Project sawmill will be located at Chati because it is nearest the immediately available roundwood source, and there is already sufficient infrastructure - official buildings, roads, water and electricity - as well as labour and scarce managerial talent there which can be utilized by the new sawmill.

Afforestation, Reforestation and Maintenance (See Annex 1)

30. The detailed planting and replanting would be as follows:

	Area (ha)			
		Eucalyptus		
Year	Pine (new)	New Plantations	Replanting	Total
1	3 000	500	200	3 700
2	3 000	500	300	3 800
3	3 000	500	400	3 900
4	3 000	500	500	4 000
5	3 000	500	600	4 100
Total	15 000	2 500	2 000	19 500

31. It is assumed that about 60 percent of the pine would be Pinus kesiya and 40 percent Pinus oocarpa and other species, and that about 2/3 of the Eucalyptus would be E. grandis and 1/3 E. cloeziana and other species. These proportions ensure good average yields, while providing some diversification, as well as sufficient supplies of wood for specific purposes (e.g. E. cloeziana for poles).

32. "Miombo" woodland would be cleared prior to planting. The soil would then be prepared by disc-harrowing. Both pine and eucalyptus would be raised in polythene tubes. The planting distances would be 2m70 x 2m70 (1 370 plants per ha) for pine and 3m60 x 3m60 (770 plants per ha) for eucalyptus. Borate would be applied to eucalyptus plantations at the time of planting, but no fertilizers would be used for pine. Very intensive weeding would be required during the first 3 years for pine, but for 1-2 years only for eucalyptus.

33. Land clearing for afforestation would continue to be done initially by local contractors but would be gradually taken over by IPD during the Project. This transfer has been necessitated by the decline in competition and high prices charged by the remaining private contracting capacity. Road and other construction and maintenance, as well as the sawmill building, would be carried out by competitive bidding advertised locally or by IPD itself if this proves less expensive.

34. Pine would be pruned at 5, 8 and 11 years, thinned at 11, 14 and 19 years, and clearfelled at 25 years. Eucalyptus would be pruned at about 2 years, thinned at 5 and 8 years, and clearfelled at 12 years.

Road Construction and Maintenance

35. The required road density has been estimated at 10 km of access roads (total 175 km), 3 km of "logging" roads (total 53 km) and 31 km (total 543 km) of compartment roads per 1 000 ha of new plantations. Road maintenance is estimated at 10 percent of capital costs per annum for all categories of roads.

Protection

36. So far pines have been remarkably free from pests and diseases. Nursery techniques are so well developed that losses caused by "damping off" fungi are minimal, and there is no problem of mycorrhiza. Scattered dieback does occur, probably caused by some pathogenic fungi, but this is not a problem.

37. Eucalyptus, on the other hand, particularly E. grandis, are liable to attack by both insects and fungi. The Eucalyptus Longhorn Bark Beetles (Phoracantha spp.) may cause serious losses in some Eucalyptus plantations at Chati, to such an extent that certain areas are no longer planted or replanted with Eucalyptus. Mostly pole crops are affected. Various control measures have been tried, but so far with little or no success. Research is continuing in this field. Concern has also been expressed about the incidence of heart-rot in the lower part of the stem of E. grandis, but this disease does not appear to be serious, especially in view of the relatively short timber rotation adopted for this species (12 years).

38. The fire hazard is high on the Copperbelt, particularly in pine plantations 5 to 8 years old. External firebreaks 15 m wide would be constructed and maintained around all plantations. Internal roads would be used as firebreaks, and 20 m strips would be maintained on either side of main roads. The main fire protection measure would be controlled burning of firebreaks and plantations while there is a serious fire hazard. Finally, fire control would involve maintaining fire-fighting vehicles, equipment and fire towers, as well as manning fire crews.

Project Management

39. Provision would be made for all IPD personnel during the plantation development period. The staff would include a Director, a Financial Controller, 3 Conservators, 3 Auditors and Senior Accounts, 11 Engineers, Managers and Senior Forest Officers, and technical, clerical and an accounting supporting staff. Provision is included for salaries and all the usual civil service allowances. Provision would also be made for the cost of all surveys and for the running of the IPD Training Centre. Baseline project administration and overhead expenses are estimated at K 2.4 million. These costs exclude logging, sawmilling, and other operating costs which should be recovered from annual revenues.

Buildings and Housing

40. Additional housing would be provided for IPD staff and workers. Provision would also be included for normal building maintenance. Total baseline costs of house construction and maintenance are estimated at K 2.7 million. (Details in Annex 1).

Equipment and Vehicles

41. The project would require investment in equipment for afforestation (land clearing, soil preparation, transport of labour and materials, and all fire protection measures), road construction and maintenance, construction of buildings, for workshops and garages, and for administration. (Details in Annex 1). Provision would also be made for tools, office and workshop equipment and an adequate supply of essential spare parts. The project workshops would be responsible for the repair and maintenance of project vehicles and equipment.

Logging and Transport

42. All logging and log transport will be carried out by the Logging Section of IPD. Its logging and transportation operations consist of felling and bucking with chainsaws, bowsaws, and axes; skidding with tractors of various sizes; loading with a front end loader or small hydraulic crane mounted on a tractor; hauling by truck or log trailer; and unloading with a large hydraulic crane. In 1975 the Logging Section felled, extracted and transported about 60 000 m^3 of roundwood, mostly in the Chati-Lamba division. As of June 1976 the Logging Section employed some 120 persons, including six supervisory staff.

43. The Project provides for the logging and transport capacity of IPD's Logging Section to be doubled to enable it to supply the project sawmill, through the phased acquisition of 18 small to medium sized tractors, 4 small hydraulic tractor-mounted loaders, 12 logging trucks of 15-ton and 2 of 35-ton capacity, 2 tippers of the dump truck variety, each of 10 m^3 capacity, two 5-ton trucks and a small grader, 2 four-wheel drive transporters and 3 motorcycles.

Sawmill (See Annex 2)

44. Existing IPD sawmill capacity will be fully utilized by 1978. Log supply projections indicate that thereafter additional capacity will be needed. The project therefore includes provision for a new sawmill capable of producing about 23 000 m^3 sawnwood annually on a single shift of 8 hours, or 46 000 m^3 on a double shift. The sawmill will be located at Chati - the present home of Kafubu Industrial Timber Enterprises (KITE) - to saw mainly IPD's Eucalyptus logs in the first instance, and mainly pine starting from about 1982. The Project provides for the cost of a sawmill building with a concrete slab, and all costs of importing, transporting to Chati, and installing the sawmill and ancillary equipment.

Staff Training and Fellowships

45. The Project would provide for the training costs of Zambians required to replace expatriate staff on short-term contracts now heading many of the units and sections in IPD. The local training of mechanics and artisans will be intensified in order to ensure an adequate supply of this class of workers for IPD.

Trials on Land Clearing and Charcoal Production

46. Land-clearing operations as currently practised on the Copperbelt prior to tree planting are extremely costly. These land-clearing practices are necessary in order to get the high yields achieved. However, in a country in which fuelwood and charcoal are the chief source of fuel, there appears to be scope for using the trees pulled down by bulldozers for these purposes; at present most are simply burnt on the spot. With the twin objectives of reducing the net costs of land clearing and of furthering the development of efficient charcoal production for domestic markets, the Project would provide about K 225 000 for equipment, materials, labour, supervision, and 6 man-months of consultancy, to be spent on land clearing and charcoal production trials.

Future Supply and Demand of Wood Raw Materials

47. Allowing for normal losses at various stages of plantation activity, the mean annual increment is expected to be 14 m^3/ha for *Pinus kesiya* and 11 m^3/ha for other pines on a 25-year rotation. For Eucalyptus, it is expected to be 25 m^3/ha for *E. grandis* and 15 m^3/ha for other Eucalyptus species on a 12-year rotation. Sawlogs would represent about 66 percent of the total volume pines, while sawlogs and large poles would represent about 60 percent of the total volume of eucalyptus products. (See Annex 3, Table 1). The remaining volume would represent smallwood which, at the moment, is largely wasted, but which would be the basic raw material for any future pulp and paper mill or wood-based board industries.

48. Under these assumptions, output from the Project should develop as shown in Table 1, and overall IPD wood production is expected to increase as indicated in the table below, where it is compared with projections of national wood demand:

	Volume '000 m^3 Roundwood Equivalent			
	1975	1980	1990	2000
National Demand				
Fuelwood	4 500	5 000	6 000	7 000
Sawnwood	250	325	530	860
Poles	70	90	145	235
Other (paper and wood-based panels)	200	235	415	675
IPD Production				
Sawnwood	80	135	420	465
Poles	25	30	40	45
Smallwood and sawmill waste	100	195	515	700

Table 1

ESTIMATED PROJECT OUTPUT [1]
(1000 m^3)

Year	Pine		Eucalyptus		
	Sawlogs	Smallwood	Sawlogs	Poles	Smallwood
1978	-	-	-	-	-
1979	-	-	-	-	-
1980	-	-	-	-	-
1981	-	-	-	-	-
1982	-	-	-	-	-
1983	-	-	-	-	-
1984	-	-	-	-	-
1985	-	-	18.9	14.0	18.9
1986	-	-	21.6	16.0	21.6
1987	-	-	24.3	18.0	24.3
1988	78.0	108.0	27.0	20.0	27.0
1989	78.0	108.0	29.7	22.0	37.9
1990	78.0	78.0	66.4	24.0	21.6
1991	180.0	192.0	74.7	27.0	24.3
1992	180.0	192.0	83.0	30.0	27.0
1993	102.0	84.0	91.3	33.0	29.7
1994	102.0	84.0			
1995	102.0	84.0			
1996	108.0	60.0			
1997	108.0	60.0			
1998	108.0	60.0			
1999	108.0	60.0			
2000	108.0	60.0			
2001 [2]	107.6 [3]	28.0			
2002 [2]	213.4 [3]	56.0			
2003	326.0 [3]	56.0			
2004	326.0 [3]	56.0			
2005	326.0 [3]	56.0			
2006	326.0 [3]	56.0			

1/ Derived from the average yield figures given in Annex 4, Table 1, and the areas planted annually during 1978–82 (Annex 1)

2/ One third of planting year 1978 felled in year 24 (2001) rather than year 25 (2002)

3/ Loss of 3% of useable volume **assumed** in clearfelled pine sawlog volume due to logging damage

49. The above figures show that IPD's key output is going to be sawnwood, of which IPD produced about 30 percent of the national consumption in 1975. IPD's proportion is likely to go up to some 50 percent in the year 2000, by which time the output from the rapidly diminishing and virtually irreplaceable natural woodlands would be even less than the 170 000 m^3 of sawlogs those woodlands contributed in 1975. Even if the sawlog output from the indigenous forests were to remain constant at 170 000 m^3, it would still be necessary to import annually some 225 000 m^3 of sawlogs by the year 2000, after allowing for the output from this Project.

50. In theory it is possible to import pine sawn timber from Kenya at a landed price of approximately K 2 per m^3 above the K 126 currently charged by IPD. However, imports of sawn timber or sawlogs from surrounding countries where production costs may be lower than in Zambia are, in practice, not as frequent as might be expected, because overland transport has not been easy to arrange. Coniferous sawnwood from Europe and North America - the traditional sources - costs considerably more than IPD's product and caters for the luxury section of the sawnwood market. There are no known alternative sources of Eucalyptus logs or poles. As long as IPD's prices continue to be below the landed prices of wood from other countries, it is likely to continue to sell all the sawnwood and poles it can produce.

51. As regards the proposed pulp and paper mill, production is expected to be entirely for the domestic market which is protected by high transport costs from external sources of supply. The output is also expected to consist of kraft and other commercial grade papers which are different from types of paper planned for manufacture in most of the surrounding countries

C. PROJECT COSTS AND BENEFITS

52. The Project is a time-slice of a long-term planting, replanting and wood processing programme started in 1949 and intended to extend many years after this Project, and to involve the planting of a total of about 80 000 ha of pine and eucalyptus with their associated processing costs.

53. The heterogeneity of the Project's components, while inevitable in a time-slice project of this nature, is such that they are not directly related to each other, and do not, either together or individually (except possibly for the Project sawmill), contain a complete cost or benefit stream. It is therefore not possible to calculate a rate of return on Project components alone without introducing some costs which were either incurred prior to the Project (e.g. wood from the pre-Project plantations for the sawmill), or costs to be incurred 10 or more years hence and therefore not included in the Project (e.g., sawmills to saw wood from the proposed 19 500 ha of plantations), or components which are still somewhat problematic (e.g., the pulp and paper mill to process smallwood and sawmill waste from the same proposed Project plantations). These costs and benefits which, although outside the Project, have to be brought into any meaningful rate of return calculations, have therefore been introduced and combined with the Project costs and benefits proper in rate of return calculations for the Project and necessary associated investments. (The exact non-Project costs thus introduced into the calculations are identified in Table 3.)

54. The primary direct benefits from the Project would consist of pine and eucalyptus sawnwood, eucalyptus transmission poles and building posts, as well as smallwood and sawmill waste for the proposed pulp and paper mill. The plantations are therefore planned on the assumption that a pulp and paper mill would be built within the next few years, and that all the wood from the plantations would be used by one or other of the wood processing industries. The direct benefits have therefore been measured - both in the financial and economic analysis - at the processed product stage and not at the roundwood production stage. Costs consequently include costs of logging, and costs of roundwood transport and processing.

55. Because of the uncertainties about the pulpmill, Project costs and benefits have been calculated both with the mill and without it. In other words, one set of calculations has been developed on the assumption that there would be no pulpmill, in which case the smallwood would consequently have no value.

Project Costs

Financial Costs

56. The Project is estimated to cost K 27.1 million (US$33.7 million), of which K 13.0 million (US$16.2 million) would be in foreign exchange. Project costs are summarized in Table 2. Project base costs, which do not include any duties or taxes, are estimated as of March 1977. A physical contingency of 10 percent has been added to all costs, and starting from April 1977 price contingencies totalling about 15 percent of baseline costs have been added.[1]

57. The derivation of the afforestation costs is shown, item by item, in Annex 1. Sawmill costs, and the associated logging and transport costs, are given in Annex 2, and pulpmill costs in Annex 3.

58. The following stumpage costs of wood to be supplied to the Project sawmill and to the pulpmill from plantations established prior to the beginning of the project have been calculated:

	Rate (K/m^3)			
Category	Pine		Eucalyptus	
	I. All produce saleable	II. Smallwood[2] unsaleable	I. All produce saleable	II. Smallwood[2] unsaleable
Sawlogs	18	22	6	8
Large poles	-	-	10	13
Smallwood	5	-	3	-

59. These stumpage rates have been calculated as the replacement cost of establishment, maintenance and administration compounded through the length of the rotation at a rate of 10 percent. The details of these calculations are shown in Annex 4. The higher of the two levels of sawlog stumpage (i.e. assuming smallwood unsaleable) was used in the Project sawmill calculation[3]. The smallwood value was used as the stumpage value of pulpwood.

60. Total annual financial costs are shown in Table 3.

---

[1] Price contingencies are excluded from the FRR and NPV calculations.

[2] If smallwood could not be sold - i.e., if the pulpmill did not get built, all costs would have to be recovered from sales of sawlog and pole products.

[3] The sawmill would be mainly using sawlogs from plantations from which the smallwood had been thinned out before the pulp mill came on stream.

Table 2

SUMMARY PROJECT COSTS
(K'000)

	1978 Year 1	1979 Year 2	1980 Year 3	1981 Year 4	1982 Year 5	Total Amount	Foreign Exchange Amount	Foreign Exchange %
Plantations 1/	3,679	2,601	3,248	2,738	2,672	14,938	6,248	42
Logging and Transportation 2/	707	281	-	-	-	988	968	98
Sawmilling 3/	2,600	256	513	864	1,212	5,445	3,025	56
Staff Training and Fellowship	20	40	20	-	-	80	64	80
Studies and Trials on Land Clearing and Charcoal Production	66	99	60	-	-	225	99	44
Total Base Costs	7,072	3,277	3,841	3,602	3,884	21,676	10,404	
Physical Contingencies 4/	707	328	385	360	388	2,168	1,040	48
Price Contingencies 5/	1,060	492	576	540	583	3,251	1,560	48
Total Project Costs	8,839	4,097	4,802	4,502	4,855	27,095	13,004	48

1/ Annex 1
2/ Capital Costs only
3/ Annex 2
4/ 10% of Base Costs
5/ 15% of Base Costs

Table 3

FINANCIAL CASH FLOW FOR THE PROJECT (K. MILLION)

Years (1)	Project Base Costs 1/ (2)	Physical Contin-gencies 2/ (3)	Wood Cost for Project Sawmill Using Pre-Project Wood 3/ (4)	Total Project Costs Without Pulp Mill (5)	Cost of Pulp Mill 4/ (6)	Total Costs Includ-ing a Pulp Mill (7)	Revenues from Project Affores-tation (8)	Revenues of Project Sawmill Using Pre-Project Wood (9)	Total Project Revenue Without Pulp Mill (10)	Revenues from Pulp and Paper Mill (11)	Total Project Revenue Including Pulp Mill (12)
1978	7.0	0.7	-	7.7	2.4	10.1	-	-	-	-	-
1979	3.3	0.3	0.4	4.0	15.4	19.4	-	1.0	1.0	-	1.0
1980	3.8	0.4	0.8	5.0	27.3	32.3	-	2.1	2.1	-	2.1
1981	3.6	0.4	1.4	5.4	18.4	23.8	-	3.1	3.1	-	3.1
1982	3.9	0.4	2.0	6.3	12.4	18.7	-	4.5	4.5	11.1	15.6
1983	2.6	0.3	2.0	4.9	9.4	14.3	-	4.5	4.5	14.5	19.0
1984	4.3	0.4	2.0	6.7	9.0	15.7	-	4.5	4.5	16.1	20.6
1985	2.8	0.3	2.0	5.1	8.9	14.0	1.5	4.5	6.0	17.0	23.0
1986	3.3	0.3	2.0	5.6	9.2	14.8	1.7	4.5	6.2	17.9	24.0
1987	3.4	0.3	2.0	5.7	8.8	14.5	1.9	4.5	6.4	18.2	24.6
1988	5.1	0.5	2.0	7.6	7.8	15.4	5.7	4.5	6.2	18.4	24.6
1989	5.8	0.6	2.0	8.4	7.8	16.2	9.3	4.5	13.8	18.6	32.4
1990	4.6	0.4	2.0	7.0	7.8	14.8	8.8	4.5	13.1	18.6	31.7
1991	5.9	0.6	2.0	8.5	7.7	16.2	12.8	4.5	17.3	18.6	35.9
1992	6.1	0.6	2.0	8.7	7.7	16.4	13.3	4.5	17.4	18.6	36.0
1993	4.7	0.5	2.0	7.2	7.9	15.1	9.7	4.5	14.2	18.6	32.8
1994	2.0	0.2	2.0	4.2	7.9	12.1	4.7	4.5	9.2	18.6	28.8
1995	2.0	0.2	-	2.2	7.9	10.1	4.7	-	4.7	18.6	23.3
1996	1.9	0.2	-	2.1	8.1	10.2	5.0	-	5.0	18.6	23.6
1997	1.9	0.2	-	2.1	8.1	10.2	5.0	-	5.0	18.6	23.6
1998	1.9	0.2	-	2.1	8.1	10.2	5.0	-	5.0	18.6	23.6
1999	1.9	0.2	-	2.1	8.1	10.2	5.0	-	5.0	18.6	23.6
2000	1.7	0.1	-	1.8	8.1	9.9	5.0	-	5.0	18.6	23.6
2001	1.9	0.2	-	2.1	8.3	10.4	5.0	-	5.0	18.6	23.6
2002	7.0	0.7	-	7.7	8.1	15.8	10.1	-	10.1	18.6	28.7
2003	6.0	0.6	-	6.6	8.0	14.6	15.1	-	15.1	18.6	33.7
2004	4.5	0.4	-	4.9	8.0	12.9	15.1	-	15.1	18.6	33.7
2005	4.4	0.4	-	4.8	8.0	12.8	15.1	-	15.1	18.6	33.7
2006	4.4	0.4	-	4.8	8.0	12.8	15.1	-	15.1	28.5	45.9

1/ Costs from 1978 to 1982 are from Table 2. Thereafter costs include all operating maintenance and capital replacement expenses of Phase II plantation and project sawmill, as well as logging, pole treatment, and sawmilling costs of all project trees and costs of logging old wood for the single project sawmill.

2/ 10% of base costs.

3/ Based on new recommended IPD stumpages, i.e. K22/m3 for pine and K13/m3 for eucalyptus (see Annex 4).

4/ Pulpwood purchased from the project output is given zero value as the pulpmill's cost would be IPD's revenue and would cancel each other when consolidated as here in one FRR calculation. Pulpwood purchased from pre-project plantations is valued at the recommended IPD stumpage, i.e. K5/m3 for pine and K3/m3 for eucalyptus. (Annex 3).

Economic Costs

61. To arrive at costs at economic values, from the financial costs, the costs of traded inputs have been converted from foreign to domestic values by using a shadow foreign exchange conversion factor of US$1 = K 0.9 instead of the official US$1 = K 0.8, as the latter appears to overvalue Zambian currency by this margin. The proportion of the financial costs which is attributable to traded items, and which hence is shadow priced in this way, is shown in Table 2 (in the "Foreign Exchange" column), and in the detailed cost tables.

62. In addition, unskilled labour has been shadow-priced at 60 percent of actual cost to reflect the relative abundance of this class of labour. The labour component of costs is identified separately (under the heading "labour") in each of the cost annexes.

63. No account has been taken in the financial or economic calculations of the cost of land. Actual payments for land used for the plantations have been few, and would not make any difference to the calculations. Also, the alternative cost of the land either in agricultural or charcoal production terms is equally small. However, this aspect would need to be kept under review if planting continues to expand beyond the scope indicated in this Project.

64. Total annual economic costs are shown in Table 4.

Project Benefits

Financial Revenues

65. As has been noted above, revenues/benefits from the Project occur from sales of the products of the industries which will use, or are expected to use, the Project roundwood. The estimates of revenues from the sawnwood and poles produced from roundwood from the Project plantations are based on the prices prevailing at present in Zambia[1] for produce from the existing industrial plantations, as follows:

	K per m^3
Eucalyptus lumber, green and rough	
Boards longer than 1.5 m (longs)	86
Shorter boards (shorts)	48
Waney edge boards	25
Eucalyptus poles	
from first thinnings	65
from second thinnings	85
Pine lumber, green and rough	
Boards longer than 1.5 m (longs)	126
Shorter boards (shorts)	63

66. It has been assumed that the eucalyptus lumber from the Project sawmill will consist of 65 percent long boards, 25 percent shorts and 10 percent waney edge boards. Similarly, it was assumed that pine lumber from the mill will consist of 75 percent longs and 25 percent shorts.

67. The estimates of quantities of lumber and poles are derived from the estimates of roundwood yields shown in Table 1, and an assumption of a 42 percent yield of sawnwood from sawlog volumes in the course of processing the latter. The resulting annual revenues are shown in column 8 of Table 3.

[1] This would prove to be a very conservative consumption if local sawnwood prices rose in real terms at the rate predicted for world prices on page 17.

Table 4

ECONOMIC VALUE FLOW FOR THE PROJECT

(K million)

Year (1)	Financial labour costs 1/ (2)	Financial foreign exchange costs 1/ (3)	Other project financial costs 2/ (4)	Foreign exchange shadow priced 3/ (5)	Labour shadow priced 4/ (6)	Project economic costs without pulp mill (7)	Pulp mill costs shadow priced 5/ (8)	Project economic costs with pulp mill (9)	Project benefits without pulp mill 6/ (10)	Pulp mill benefits 3/ 5/ (11)	Project benefits with pulp mill (12)
1978	1.3	5.6	0.8	6.3	0.8	7.9	2.7	10.6	-	-	-
1979	1.3	1.8	0.9	2.0	0.8	3.7	17.2	20.9	1.2	-	1.2
1980	1.6	2.2	1.2	2.5	1.0	4.7	30.2	34.9	2.7	-	2.7
1981	2.2	0.8	2.2	0.9	1.3	4.4	20.5	24.9	4.0	-	4.0
1982	2.0	1.0	3.3	1.1	1.2	5.6	12.7	18.3	5.8	12.5	18.3
1983	1.1	2.1	1.7	2.4	0.7	4.8	9.2	14.0	5.8	16.3	22.1
1984	1.7	3.1	1.9	3.5	1.0	6.4	8.7	15.1	5.8	18.1	23.9
1985	1.2	2.2	1.7	2.5	0.7	4.9	8.5	13.4	8.2	19.1	27.3
1986	1.4	2.5	1.7	2.8	0.8	5.3	8.6	13.9	8.4	20.1	28.5
1987	1.4	2.5	1.8	2.8	0.8	5.4	8.6	14.0	8.8	20.5	29.3
1988	2.0	3.6	2.0	4.0	1.2	3.6	7.8	11.4	8.6	20.7	29.3
1989	2.2	4.1	2.1	4.6	1.3	8.0	7.8	15.8	18.9	20.9	39.8
1990	1.8	3.3	1.9	3.7	1.0	6.6	7.8	14.4	18.0	20.9	38.9
1991	2.3	4.2	2.0	4.7	1.4	8.1	7.8	15.9	23.8	20.9	44.7
1992	2.3	4.2	2.0	4.7	1.4	8.1	7.8	15.9	24.0	20.9	44.9
1993	1.9	3.4	1.9	3.8	1.1	7.8	7.8	15.6	19.6	20.9	40.5
1994	0.9	1.7	1.6	1.9	0.5	4.0	7.8	11.8	6.4	20.9	27.3
1995	0.7	1.3	0.2	1.5	0.4	2.1	7.8	9.9	6.4	20.9	27.3
1996	0.6	1.2	0.3	1.4	0.4	=.1	7.8	9.9	6.9	20.9	27.8
1997	0.6	1.2	0.3	1.4	0.4	2.1	7.8	9.9	6.9	20.9	27.8
1998	0.6	1.2	0.3	1.4	0.4	2.1	7.8	9.9	6.9	20.9	27.8
1999	0.6	1.2	0.3	1.4	0.4	2.1	7.8	9.9	6.9	20.9	27.8
2000	0.6	1.0	0.2	1.1	0.4	1.7	7.8	9.5	6.9	20.9	27.8
2001	0.6	1.2	0.3	1.4	0.4	2.1	7.8	9.9	6.8	20.9	27.7
2002	2.4	4.4	0.9	4.9	1.4	7.2	7.8	15.0	13.8	20.9	34.7
2003	2.0	3.8	0.8	4.3	1.2	6.3	7.8	14.1	20.6	20.9	41.5
2004	1.5	2.8	0.6	3.1	0.9	4.6	7.8	12.4	20.6	20.9	41.5
2005	1.5	2.7	0.6	3.0	0.9	4.5	7.8	12.3	20.6	20.9	41.5
2006	1.5	2.7	0.6	3.0	0.9	4.5	7.8	12.3	20.6	31.6	52.2

1/ See Table 2 and detailed cost tables.
2/ Total financial costs from table 3 net of labour and foreign exchange costs.
3/ Shadow priced at US$1.00=K0.9 instead of the official rate of US$1.00=K0.8
4/ At 60% of financial labour costs.
5/ See Annex 3, page 3.
6/ IPD prices assumed to rise as set out in Annex 2, Table 3, footnote 6.

68. Column 9 of the same table shows the additional revenue obtained from sawnwood produced by the Project sawmill using sawlogs obtained from plantations established prior to the Project. These revenues are arrived at from the unit prices shown above and the volumes given in Annex 2.

69. The revenues from the pulpmill (column 11 of Table 3) are derived in Annex 3.

Economic Benefits

70. As the final outputs of the Project will replace imported supplies, in order to arrive at their economic value they need to be valued in terms of world market prices. Kenya seems the most likely source of imports for the kind of pine sawn timber to be produced by the Project. Such timber can be delivered in the Zambian Copperbelt at a total cost of about K 128/m^3 for a 1.5 m length, compared with K 126/m^3 currently charged by IPD. There is no known alternative source of eucalyptus. The parity prices of sawn timber have therefore been assumed to be approximately equal to the current IPD prices.

71. In addition, an allowance has been made for an expected rise in the real value of sawn timber. The World Bank has forecast that the price of sawn timber will rise, at constant 1974 values (US$), as follows:

1978	1979	1980	1985 (and after)
168.1	170.8	174.8	184.1

72. Project sawn timber output prices have therefore been assumed to rise at the same rate, i.e. as follows (1978 = 100):

1978	1979	1980	1985 (and after)
100	102	104	113

73. Paper prices used in the financial analysis were based on world market prices, and therefore no adjustment was needed in this respect in the economic accounts. The value of all importable project outputs - sawnwood, poles and paper - was valued using the shadow price for foreign exchange of US$1 = K 0.9 instead of the official US$1 = K0.8.

74. Annual benefits are shown in Table 4.

Indirect Benefits

75. The Project is expected to be neutral in its environmental impact. The new plantations would help to preserve water catchment areas and prevent soil erosion, but since indigenous forests (which were themselves quite adequate for these purposes) would have to be removed to plant the new trees, there is really no incremental benefit, nor any adverse effects from this source. Similarly, the other components of the Project neither improve nor worsen the environment.

Comparing Costs and Benefits

76. On the basis of the total cost and total revenue/benefit streams shown in Tables 3 and 4, the Project and necessary associated investments would earn an internal financial rate of return (FRR) of 10.5 percent and an internal economic rate of return (ERR) of 14.3 percent.

77. The sawmill to be financed through this Project, together with its associated logging and transport, would process wood from trees planted under previous projects. This group of investments alone would earn a FRR of 15.6 percent and an ERR of 32 percent (Annex 2).

78. The pulp and paper mill, processing sawmill waste and smallwood first from pre-Project plantations and later from Project plantations as well, would earn an FRR of 10.4 percent and an ERR of 11.7 percent[1]. (Annex 3).

Sensitivity Analysis

79. As was noted earlier, consideration has to be given to the possible effects on the Project if, due to lack of funds, lack of technical expertise, or any other reasons, the proposed pulp and paper mill does not materialize. Cash flow statements of the project with and without a pulp and paper mill are shown in Tables 3 and 4. Without the pulpmill, the Project and its necessary associated future investments would earn an FRR of 10.0 percent and an ERR of 20.0 percent, based on sawmilling and pole manufacture alone, but on a much reduced investment and level of activities. The reduced economic impact is shown by the fact that the discounted net value[2] with the pulpmill is K 61 million, and without the pulpmill K 26 million.

80. In addition, the ERR of the Project with a pulp and paper mill has been tested for sensitivity to changes in the cost and benefit streams with the following results:

	ERR
Basic Model	14.3
Costs increased 10%	12.5
Costs increased 20%	10.6
Benefits reduced 10%	12.1

81. It would seem that the ERR is not sensitive to small changes in costs or benefits. A substantial reduction in benefits could result from fires, disease or inadequate production. However, this risk is common to all forestry projects. IPD's fire record has been normal in the past, and the conventional fire preventive measures have been built into the Project.

[1] To calculate the return on the pulp and paper mill as a separate entity, the pulpwood raw material was costed at the cost-price stumpage values calculated in Annex 4 in the financial calculation, and was given a zero stumpage value in the economic calculation because it would have no use other than in the pulpmill (Annex 3).

[2] Discounted at 8.5%, which is the interest rate to be paid on most of the money invested in the project.

ANNEX 1

AFFORESTATION AND REFORESTATION

A. Establishment of Plantations

1. Land Clearing. Afforestation nearly always takes place in "miombo" woodland (see Annex 1) with more or less open canopy and trees 15-20 m high. The land must therefore be completely cleared before planting.

2. The land clearing method which is used on the Copperbelt is very thorough and expensive, but IPD claims that it should not be changed as this would reduce growth and yields. The work has so far been carried out mainly under contract (IPD has recently started undertaking some clearing with direct labour and some land is also cleared by charcoal burners), and the cost per hectare was approximately K200 in 1976. The following operations are included:

(a) Knockdown. A heavy chain, weighing about 50 kg per link and some 200 m long, is pulled by two 300 HP bulldozers moving forward at an average distance of 30 m apart. The chain can easily pull down all small and most medium-sized trees in its path. One or two 200 HP bulldozers follow some 60 m behind the others and help to pull down larger-sized trees.

(b) Windrowing. All the uprooted trees are piled into windrows some 80 m apart by means of bulldozers (200 or 300 HP). When the wood is sufficiently dry it is all burnt in the rows. (It is hoped that during the phase II project the wood will increasingly be converted into charcoal.)

(c) Stumping. Any stumps remaining in the ground after the passage of the bulldozers are removed by hand and piled on to the windrows, and the holes are then filled by hand.

(d) Ploughing. The whole area is ploughed after the stumps have been removed.

3. All these operations have until recently been carried out under contract, except for some of the ploughing which has been done directly by IPD. In order to reduce costs and be less dependent on contractors, IPD would like gradually to take over all the land clearing operations, phased over the Project period. However, in view of the heavy investment involved and of the period of training required, it is assumed that the average cost per ha will be more or less the same throughout the Project period, regardless of who performs the work. Thus, assuming an average total area of 3 500 ha cleared annually at an average cost of K200/ha, the total (base line - 1976) costs are estimated at K 700 000 p.a. or K 3.5 million over the whole Project period.

Equipment

4. A complete unit consists of two 300 HP and at least one 200 HP bulldozers with two sets of chains. IPD already has one such unit, but another unit, with a standby 300 HP dozer, would be purchased during year 1 of the Project, and another replacement unit in year 3 of the Project.

5. The estimated costs are:

Capital Costs

Equipment	Unit Price K'000	Number	Year 1 K'000	Number	Year 3 K'000
300 HP bulldozer, with all attachments	157	3	471	2	314
200 HP bulldozer, with all attachments	114	1	114	1	114
50-60 HP tractor	10	5	50	5	50
Ancillary equipment 1/			70		40
Ploughs, winches, chains, etc.			40		32
TOTAL			745		550

1/ Low bed trailers, tanker, welding sets, etc.

Operating Costs (for one unit during year 1, and 2 units from year 2 to year 5).
N.B. 2000 hours per unit p.a.

		Cost per unit p.a. (K'000)
Materials, spares, maintenance and repairs		45
Fuel - 300 HP: 50 litre/hr; 200 HP: 36 litre/hr; Tractors: 3.5 litre/hr; Ancillary equipment	K0.15/litre	55
Lubricants	K0.81/litre	10
TOTAL		110

Total Operating Costs (K'000)

Year	1	2	3	4	5
	110	220	220	220	220

Salaries and Wages (2 units from year 2)

Category	Number	Annual Salary/wages K	Cost per unit p.a. K
Supervisor	1	3 650	3 650
Dozer drivers	6	1 620	9 720
Tractor drivers	5	1 416	7 080
Labour (incl. Transport)	6	492	2 950
TOTAL			23 400
Ancillary equipment drivers	3	1 416	4 248 (not doubled with second unit)

Total Salaries and Wages (K'000)

Year	1	2	3	4	5
	28	51	51	51	51

Recapitulation of Land Clearing Costs (K'000)

Year	1	2	3	4	5	Total
(a) by IPD						
Equipment	745	-	550	-	-	1 295
Fuel and Lubricants	110	220	220	220	220	990
Salaries & Wages (incl. Transport)	28	51	51	51	51	232
TOTAL	883	271	821	271	271	2 517
(b) by contract	400	300	200	83	-	983
TOTAL (a) + (b)	1 283	571	1 021	354	271	3 500

Soil Preparation

6. Ploughing has been included in the land clearing costs. There only remains pre-planting discing. The unit cost per operation is K7.30/ha (1976), of which labour represents about 25 percent or K1.80/ha. No special equipment need be purchased, apart from disc harrows. (The tractors used in ploughing can also be utilized for discing.) Normally only one discing operation is required, unless planting takes place after too long an interval, when a second discing may be necessary.

Year	1	2	3	4	5	Total
Area (ha)	3 700	3 800	3 900	4 000	4 100	19 500
Costs (K'000)						
Equipment and fuel	20	21	22	23	24	110
Labour (incl. transport)	7	7	7	7	7	35
TOTAL	27	28	29	30	31	145

Nurseries and Seedling Production

7. No new nurseries will be required during the Project, the existing nurseries being conveniently located and adequate. Nursery techniques are well developed and satisfactory. Pine is spot-sown in tubes (1-3 seeds per tube, according to the tested germinating power of the seed stock), while Eucalyptus is transplanted from seedbeds into tubes. The tubes are made of black polythene, 5 cm wide when flat and 12 cm long. Sterilized soil is used to fill the tubes. No special mycorrhizal problem exists as regards Pine. Sowing normally takes place in June-July and the seedlings are ready for planting out as from November-December. Fertilizers are applied diluted in water, one month after germinating (or transplanting) and regularly thereafter, as required. The average task for the filling of tubes is 1 175 p.d. Allowing 20-25 percent for various losses, it is advisable to allow 1 700 Pine or 1 000 Eucalyptus plants per ha, giving a weighted average of 1 500 plants per ha for both Pine and Eucalyptus. The weighted average cost is K10 per thousand seedlings, giving a weighted average cost of K15 per ha, of which approximately one-third or K5 is for materials, and two-thirds or K10 for labour.

Year	1	2	3	4	5	Total
Area (ha)	3 700	3 800	3 900	4 000	4 100	19 500
Costs (K'000)						
Materials	19	19	20	20	21	99
Labour (incl. transport)	37	38	39	40	41	195
TOTAL	56	57	59	60	62	294

Planting and Fertilizing

8. The present planting distances of 2m 70 x 2m 70 (1 370 plants/ha) for Pine and 3m 60 x 3m 60 (770 plants/ha) for Eucalyptus will be maintained. The average cost (all labour) is K15/ha for Pine and K10/ha for Eucalyptus, giving a weighted average of K13/ha, which includes 10 percent for the replacement of failures.

9. No fertilizers are used for Pine, but borate and NPK are applied in Eucalyptus plantations at the time of planting. The cost of this treatment is K20/ha (1976), of which K10 is for the borate and NPK, and K5 is for labour and transport. The planting programme and costs may be summarized as follows:

Year	1	2	3	4	5	Total
Area (ha)						
Pine	3 000	3 000	3 000	3 000	3 000	15 000
Eucalyptus	700	800	900	1 000	1 100	4 500
TOTAL	3 700	3 800	3 900	4 000	4 100	19 500
Costs (K'000)						
(a) Planting						
Labour (incl. transport)	48	49	51	52	53	253
(b) Fertilizing						
Borate & NPK	7	8	9	10	11	45
Labour (incl. transport)	7	8	9	10	11	45
TOTAL	14	16	18	20	22	90
TOTAL (a) + (b)	62	65	69	76	75	343

Weeding

10. Very intensive weeding is required. Most of the weeding is done mechanically, in lines, using a tractor and rotorvator, but hand weeding is sometimes carried out simultaneously around the trees. The following operations are done on the average:

(a) Pine	Operation	Average Cost (K/ha)
1st year	6 mechanical, 4 manual	47
2nd year	4 mechanical, 2 manual	45
3rd year	2 mechanical	23
	Total weeding costs (3 yrs.)	115

(b) Eucalyptus	Operation	Average Cost (K/ha)
1st year	6 mechanical, 4 manual	47
2nd year	0-2 mechanical	7
	Total weeding costs (2 yrs.)	54

Average weighted cost/ha p.a. for Pine and Eucalyptus: K100 (assuming areas at the start and end of the Project balance each other out for the 2-3 years of weeding). This cost includes equipment (33 percent), fuel (17 percent) and labour (50 percent).

Year	1	2	3	4	5	Total
Area (ha)	3 700	3 800	3 900	4 000	4 100	19 500
Costs (K'000)						
Equipment (tractors, etc.)	122	125	128	132	135	642
Fuel & lubricants	63	65	67	68	70	333
Labour (incl.transport)	185	190	195	200	205	975
TOTAL	370	380	390	400	410	1 950

Recapitulation - Total Establishment Costs (K'000)

Operation	Year 1	2	3	4	5	Total
Land clearing	1 283	571	1 021	354	271	3 500
Soil preparation	27	28	29	30	31	145
Nursery production	56	57	59	60	62	294
Planting and fertilizing	62	65	69	72	75	343
Weeding	370	380	390	400	410	1 290
TOTAL	1 798	1 101	1 568	916	849	6 232

Recapitulation - Total Establishment Costs by Categories of Expenditure (K'000)

Category	Year 1	2	3	4	5	Total
Equipment	877	135	688	142	145	1 987
Materials - Local	26	27	29	30	32	144
Fuel and Lubricants	183	296	299	301	304	1 383
Labour (inc. transport)	312	343	352	360	368	1 735
Contract	400	300	200	83	-	983
TOTAL	1 798	1 101	1 568	916	849	6 232

B. Maintenance of Plantations (mainly pre-Project)

11. **Pruning**

(a) Pine

Operation	Unit Cost (1976) K/ha	Average area (ha) p.a.	Total cost p.a. (K'000)
1st pruning (5 yrs.) (1 000 stems/ha)	10	2 000	20
2nd pruning (8 yrs.) (260 stems/ha)	5	1 500	8
3rd pruning (11 yrs.) (260 stems/ha)	10	1 000	10
		Total cost p.a. Pine	38

(b) Eucalyptus

Pruning (at about 2 yrs.)	5	750	4
Average total cost of pruning p.a. Pine + Eucalyptus			42

12. Thinning. Except in the case of first thinnings for Pine, the only cost involved is for the marking of thinnings as it is assumed that thinnings will be saleable and the felling and logging costs will be borne by logging operations. The first Eucalyptus thinning (at 5 years) is a line thinning and involves no marking costs.

	Pine				Eucalyptus	
	Thin. I (11 yrs.)	Thin.II (14 yrs.)	Thin.III (19 yrs.)	Total P.	Thin. II (8 yrs.)	Total P + E
Cost K/ha	10	3	2	-	1	
Average area (ha) p.a.	1 500	1 000	250	2 750	800	3 550
Total Cost (K'000)	15	3	1	19	1	20

Recapitulation Total Cost of Pruning and Thinning (all labour)

Operation	Year 1	2	3	4	5	Total
Pruning	42	42	42	42	42	210
Thinning	20	20	20	20	20	100
TOTAL (Labour) (including transport)	62	62	62	62	62	310

C. Roads

13. All road construction and maintenance is included here, even of so-called "logging" roads, which are also used for other purposes. However, external firebreaks, which are sometimes used as roads, are dealt with separately under Fire Control.

I. Road Construction (new roads)

Category:	Access Roads (Class B)	"Logging" Roads (Class C)	Compartment Roads (Class D)	Total all roads
Density (km/1 000 ha)	10	3	31	
Average cost 1/ (K/km)	3 125	2 000	280	
Total cost (K/1 000 ha)	31 250	6 000	8.680	
Total cost 2/ (K'000) p.a.	109	21	30	160
Total cost (K'000) - 5 yrs	545	105	150	800
Total length (km) - 5 yrs	175	52.5	542.5	

1/ Mid-1976 costs have been increased by 12 percent to allow for the 20 percent devaluation of the Kwacha in 1976, which affects all imports (55 percent of total road costs).

2/ An average of 3 500 ha of new plantations p.a.

II. Road Maintenance

14. Roads to be maintained annually are as follows:

Category	Access Roads (Class B)	"Logging" Roads (Class C)	Compartment Roads (Class D)	Total all roads
Length (km) of existing roads (1975)	80	90	100	-
Length (km) of new Project roads (total) divided by 2 1/	90	25	270	-
Total length (km) (average)	170	115	370	-
Average cost p.a. (K/km) 2/	312.5	200	28	-
Total cost p.a. (K'000)	53	23	10.4	86.4
Total cost - 5 yrs (K'000)	265	115	52	432

1/ To obtain yearly average over the whole 5-year period

2/ 10 percent of capital costs

15. The total base-line Project costs for road construction and maintenance are K 1 232 000, of which approximately 60 percent (K 470 000) is for equipment and materials, and 40 percent (K 492 000) for labour (including transport).

16. **Detailed Project costs are given below:**

Equipment (machinery)	Number	Unit Cost (K'000)	Total Cost (K'000)
150 HP motor graders	2 (1 yr. 1 & 1 yr. 4)	80	160
5-ton tipper trucks	4 (2 yr. 1 & 2 yr. 4)	33	132
		Total equipment (machinery)	292
Spares			40
Fuel and lubricants			288
Workshop equipment			40
Materials - imported			40
local			40
Total non-labour			740
Total labour (including transport)			492
Total road costs (construction & maintenance)			1 232

Year	1	2	3	4	5	Total
Machinery and equipment	146	-	-	146	-	292
Spares	20	5	5	5	5	40
Fuel & lubricants	44	52	58	64	70	288
Workshop equipment	20	5	5	5	5	40
Material						
imported	8	8	8	8	8	40
local	8	8	8	8	8	40
Labour (including transport	80	88	98	108	118	492
TOTAL	326	166	182	344	214	1 232

D. Fire Protection

17. The fire hazard is high in forest plantations on the Copperbelt and losses may sometimes be appreciable despite normal precautions. For example, some 200 ha of Pine were completely lost and another 400 ha of Pine were more or less severely scorched in 1975. Fire damage is worst in Pine stands 5 to 8 years old. External firebreaks 15 m wide are constructed and maintained around all new plantations. Internally, roads are used as firebreaks. 20 m wide breaks are usually maintained on either side of main roads. But essentially fire protection consists of controlled burning of firebreaks and of all plantations while there is a serious fire hazard. Finally, fire control includes maintaining fire-fighting vehicles, equipment and fire towers, as well as fire crews.

18. Fire protection is considered under the following headings:

(i) Construction of external firebreaks (new)

The cost of construction of new external firebreaks is the same as the cost of land clearing, as the same operation is involved. This cost is assumed to be about K200/ha. As the width is 15 m, the average area of external firebreaks is $1\frac{1}{2}$ ha per km and the average cost is K300/km (1976). The average density is 23 km/1000 ha of new plantations, which gives a total of 80.5 km p.a. (3 500 ha of new plantations p.a.). The average annual cost is, therefore, K24 000, divided as follows:

Cost (K'000)

Year	1	2	3	4	5	Total
(a) by IPD						
Equipment	8	8	8	8	8	40
Fuel & lub., etc.	7	7	7	7	7	35
Labour (inc. transport)	1	1	1	1	1	5
(b) by contract	8	8	8	8	8	40
TOTAL	24	24	24	24	24	120

(ii) Maintenance of firebreaks

19. All firebreaks are maintained by regular annual strip-ploughing or discing and controlled burning. The average cost is K25/km. The average area of plantations during the project is assumed to be 26 000 + 18 000 = 34 000 ha, and the average length of firebreaks to be maintained (at 23 km/1 000 ha) is assumed to be 34 x 23 = 782 km. The total average cost of firebreak maintenance is K20 000 p.a., made up of:

Equipment	(25%)	-	K 5 000
Fuel & lub., etc.	(50%)	-	K 10 000
Labour (inc. transport)	(25%)	-	K 5 000

Cost (K'000)

Year	1	2	3	4	5	Total
Equipment	2	3	5	7	8	25
Fuel & lub., etc.	8	9	10	11	12	50
Labour (incl. transport)	4	5	5	5	6	25
Total	14	17	20	23	26	100

(iii) Controlled burning

20. Assuming half the total area of plantations is control-burnt annually, the average area involved will be $\frac{34\,000}{2}$ = 17 000 ha. At an average cost of K 0.5/ha, the annual average cost will be K 9 000, nearly all of it labour.

Cost (K'000)

Year	1	2	3	4	5	Total
Labour (incl. transport)	7	8	9	10	11	45

(iv) Fire Control

21. This includes all fire-fighting equipment and services, and the manning of fire towers and standby crews during periods of high fire hazard.

(a) Fire tenders

There are two types of fire tenders: the light one which consists of a fast, 4WD, 3-ton load-carrying capacity cross-country vehicle, and the heavy one which is a proper fire engine. The requirements are one light tender for 3 500 ha of plantations and one heavy tender for 7 000 ha of plantations. The average total area of plantations during the Project is 34 000 ha (see the two previous headings). The total requirements, including replacements, will therefore be 10 light tenders and 5 heavy tenders.

	Light tenders	Heavy tenders	Total (K'000)
Total requirements (number)	10	5	
No. available in working order	5	2	
Additional no. required (new)	5	3	
Average unit cost (K'000) 1/	30	40	
Total cost (K'000)	150	120	270
Running costs (average)			
(Fuel, lub., etc. (p.a. K10 000)			50
(Labour (p.a. K10 000)			50
Total cost over 5 years			370

1/ Allowing a 20 percent increase over mid-1976 costs (1976 devaluation).

Year	1	2	3	4	5	Total
Equipment	100	-	100	-	70	270
Fuel, lub., etc.	8	9	10	11	12	50
Labour (incl. transport)	8	9	10	11	12	50
Total	116	18	120	22	94	370

(b) Fire towers

22. Requirements - 1 large tower (30 m high) for 1 800 ha.
1 small tower (8 m high) for 4 000 ha.

	Large tower	Small Tower	Total
No. required p.a.	2	1	3
Unit cost (K'000)	7	2	
of which (materials	4	1	5
(labour	3	1	4
Total cost p.a.	14	2	16
of which (materials	8	1	9
(labour	6	1	7

Year	1	2	3	4	5	Total
Materials (incl. transport)	9	9	9	9	9	45
Labour (incl. transport)	7	7	7	7	7	35
Total	16	16	16	16	16	80

23. (c) Fire fighting equipment[1]

	60 W. radios	25 W. radios	10 W. radios	Other Equipment
No. required (5 years)	6	50	50	
Unit cost (K'000)	2.5	1	0.8	
Total cost (K'000) - 5 yrs.	15	50	40	20
Grand total cost (K'000) - yrs.				125

Year	1	2	3	4	5	Total
Equipment	75	-	-	50	-	125

[1] Allowing a 20% increase over mid-1976 costs (1976 devaluation).

(d) Fire crews and fire-fighting

Average cost - all labour - (K'000)
p.a. - 20
5 years - 100

Year	1	2	3	4	5	Total
Labour	16	18	20	22	24	100

24. Recapitulation fire control costs (iv) (a-d) (K'000)

Year	1	2	3	4	5	Total
Equipment	175	-	100	50	70	395
Materials	9	9	9	9	9	45
Fuel, lub., etc.	8	9	10	11	12	50
Labour (incl. transport)	31	34	37	40	43	185
TOTAL	223	52	156	110	134	675

25. Recapitulation - all fire protection costs (i) to (iv) (K'000)

Year	1	2	3	4	5	Total
Equipment	185	11	113	65	86	460
Materials local	9	9	9	9	9	45
Fuel, lub., etc.	23	25	27	29	31	135
Labour (incl. transport)	43	48	52	56	61	260
By contract	8	8	8	8	8	40
TOTAL	268	101	209	167	195	940

26. **Recapitulation - all afforestation costs (A to D)**
Direct costs (K'000)

Year	1	2	3	4	5	Total
1. Establishment of plantations	1 798	1 101	1 568	916	849	6 232
2. Maintenance of plantations	62	62	62	62	62	310
3. Roads - construction and maintenance	326	166	182	344	214	1 232
4. Fire protection	268	101	209	167	195	940
TOTAL	2 454	1 430	2 021	1 489	1 320	8 714

27. Recapitulation - all afforestation costs (Direct costs) (A to D)
By categories of expenditure (K'000)

Category Year	1	2	3	4	5	Total
Equipment	1 256	164	819	371	249	2 859
Materials - local	43	44	46	42	49	229
Fuel and lubricants	250	373	384	394	405	1 806
Labour (incl. transport)	497	544	564	586	609	2 797
Contract	408	308	208	91	8	1 023
TOTAL	2 454	1 430	2 021	1 489	1 320	8 714

E. Non-Afforestation (Indirect) Costs (Project Management)

Administration and Overheads

28. The average IPD establishment (staff) costs during the Project would be as follows:

Posts	No.	Unit cost p.a.[1/] (K'000)	Total cost p.a. (K'000)
Director	1	8.7	8.7
Financial Controller	1	8.7	8.7
Conservators	5	8.2	41.0
Auditors & Senior Accountants	3	6.8	20.4
Engineers, Managers, Senior Forest Officers	11	6.4	70.4
Senior Mechanics & Clerks of Works	6	5.8	34.8
Electrical Officers & Building Officers	5	4.9	24.5
Foresters	10	4.9	49.0
Accountants	4	4.6	18.4
Executive Officers, Surveyors, Transport Officers	13	3.2	41.6
Assistant Accountants	5	3.0	15.0
Stenographers	3	2.7	8.1
Typists	8	2.0	16.0
Senior Officers	10	1.9	19.0
Clerical Officers	15	1.7	25.5
Junior Clerical Officers	8	1.3	10.4
		Total	411.5
+ 12½% housing allowance			51.5
Total cost (salaries and all allowances) p.a.			463
Office expenses (average) p.a.			27
Average Total Cost of Administration and Overheads p.a.			490

1/ Including salary and allowances (excluding housing allowance).

29. **Recapitulation - administration and overhead costs**

Year	1	2	3	4	5	Total
Salaries and allowances	463	463	463	463	463	2 315
Office expenses (materials and services)	27	27	27	27	27	135
TOTAL	490	490	490	490	490	2 450

F. Housing (additional)

30. The only new buildings required during the project are additional housing for the staff.

Type	Unit cost (K'000) (at 1976 prices)	Year 1 Qty.	Year 1 T. Cost (K'000)	Year 2 Qty.	Year 2 T. Cost (K'000)	Year 3 Qty.	Year 3 T. Cost (K'000)	Year 4 Qty.	Year 4 T. Cost (K'000)	Year 5 Qty.	Year 5 T. Cost (K'000)	Total Qty.	Total T. Cost (K'000)
Low density	16	3	48	2	32	2	32	2	32	3	48	12	192
Medium density	12	4	48	2	24	2	24	3	36	4	48	15	180
High density	4	60	240	60	240	70	280	70	280	90	360	350	1400
Ablution blocks	6	10	60	6	36	7	42	7	42	9	54	39	234
Electricity and Water Services			50		50		50		50		50		250
TOTAL			446		382		428		440		560		2256

31. It is assumed that on the average labour and materials will account for about half the total costs each, and that the foreign exchange component will be about 25 percent of total costs.

Recapitulation - housing costs (K'000)

Year	1	2	3	4	5	Total
Labour (inc. transport)	223	191	214	220	280	1 128
Materials						
local	112	96	107	110	140	565
foreign	111	95	107	110	140	563
TOTAL	446	382	428	440	560	2 256

G. Building Maintenance

32. The average cost of maintaining all Project buildings, old and new, is estimated at K100 000 p.a., of which about half is for labour and half for materials (50 percent foreign).

Building maintenance costs are, therefore, estimated thus:

Costs (K'000)

Year	1	2	3	4	5	Total
Labour (incl. transport)	40	45	50	55	60	250
Materials						
local	20	23	25	28	30	126
foreign	20	22	25	27	30	124
TOTAL	80	90	100	110	120	500

H. Administrative Vehicles & Labour Transport

33. At the end of the Phase I project, the following administrative vehicles were in working order:

Trucks and vans	37
Personnel vehicles	22
Motorcycles	8

34. It is assumed that almost half of these vehicles will be used in afforestation (the other half in logging, forest industries, etc.). Provision is made for the following replacements during the Project period:

Category	Number		Average Unit Cost (1976) (K'000)	Total Cost (5 years) (K'000)
1. 8-ton flat trucks	10)		19.4	194
2. 5-ton trucks or vans	10)	2 p.a.	16	160
3. Personnel vehicles	10)		9	90
4. Motor cycles	10)		2	20
			TOTAL	464

35. The average annual running costs of the existing administrative vehicles and for the transport of labour to the work sites (for afforestation operations only) are:

Item	Number	Unit Cost p.a. (K'000)	Total Cost p.a. (K'000)
Labour - drivers	25	1 416	35 400
skilled workers	10	1 000	10 000
unskilled workers	20	492	9 840
Total labour	55		55 240
Fuel and lubricants (imported)			50 000
Spares (imported)			20 000
TOTAL			125 240

Cost of administrative vehicles, including labour transport

Item Year	1	2	3	4	5	Total
Vehicles & equipment (new)	100	100	100	100	64	404
Running cost:						
Labour	55	55	55	55	55	275
Fuel and lubricants	50	50	50	50	50	250
Spares	20	20	20	20	20	100
TOTAL	225	225	225	225	189	1 089

The cost of labour transport for all afforestation operations has been included in the cost of labour in each case, and it is assumed that it represents about 10 percent of total labour costs. As the average annual labour costs for afforestation come to K559 400, labour transport represents approximately K56 000 of it. This must be deducted from the totals given above in order to avoid double counting. The actual cost of administrative vehicles and transport under non-afforestation (indirect) costs is therefore K1 089 000 less K280 000 (K55 000 x 5) = K809 000. Labour transport thus represents 25 percent of the total cost of administrative vehicles + labour transport, and by subtracting 25 percent we obtain:

Revised cost of administrative vehicles (excluding labour transport)

Item	Year 1	2	3	4	5	Total
Vehicles & equipment	75	75	75	75	48	348
Labour	41	41	41	41	41	205
Fuel & lubricants	38	38	38	38	38	190
Spares	15	15	15	15	15	75
TOTAL	169	169	169	169	142	818

I. Surveys and Training

36. The average annual cost of surveys is K30 000, made up of labour (K15 000) and materials (K15 000).

IPD has its own training centre, and average annual costs of training are estimated at K10 000 (for materials).

The total costs of surveys and training come to:

Costs (K'000)

Year	1	2	3	4	5	Total
Surveys - labour	15	15	15	15	15	75
materials	15	15	15	15	15	75
Training - materials	10	10	10	10	10	50
TOTAL	40	40	40	40	40	200

Recapitulation - Non-afforestation: (Indirect)Costs (E-I)
(Project Management)

Costs (K'000)

Year	1	2	3	4	5	Total
Salaries & allowances	463	463	463	463	463	2 315
Vehicles & equipment	75	75	75	75	38	348
Fuel & lubricants	38	38	38	38	38	190
Materials - foreign	161	147	162	167	200	837
local	169	156	169	175	207	876
Labour	319	292	320	331	396	1 658
TOTAL	1 225	1 171	1 227	1 249	1 352	6 224

Plantations

Direct and Indirect Costs (K'000)

	1978 Year 1	1979 Year 2	1980 Year 3	1981 Year 4	1982 Year 5	Total 5 Years	Foreign Exchange Amount	Foreign Exchange %
Equipment and vehicles	1 331	239	894	446	297	3 207	3 046	95
Materials: foreign	161	147	162	167	200	837	795	95
local	212	200	215	222	256	1 105	-	-
Fuel and lubricant	288	411	422	432	443	1 996	1 896	95
Labour: afforestation	497	541	564	586	609	2 797	-	-
other	319	292	320	331	396	1 658	-	-
Contract	408	308	208	91	8	1 023	511	50
Salaries & allowances	463	463	463	463	463	2 315	-	30
TOTAL	3 679	2 601	3 248	2 738	2 672	14 938	6 248	46

ANNEX 2

PROJECT SAWMILL

Introduction

1. The conditions in Zambia provide ample justification for the construction of a new sawmill for the production of pine and eucalyptus lumber. The present capacity of existing sawmills at Chati and Dola Hill are not sufficient to handle the increasing volume of pine and eucalyptus sawlogs from IPD after 1978. Log yield projections indicate that an additional sawmill would be required by then.

2. A sawmill would have an estimated capacity of 250 m^3(r) log input per 8-hour shift, corresponding to a lumber output of 113 m^3(s) per 8-hour shift, assuming the average lumber yield to be 45 percent.

3. It has been assumed that the domestic demand for pine and eucalyptus lumber in Zambia will be sufficient to keep the proposed new sawmill at Chati operating on one shift per day in 1980, producing 20 000 m^3(s) per year, and 2 shifts per day in 1982, producing 46 000 m^3(s) in that year.

4. The most economical site for the Project sawmill is at Chati, where there is already sufficient land, labour, water, management and other supporting infrastructure. A future sawmill (probably 1982) would be built at the site of the proposed pulp and paper mill.

5. Proper utilization of sawmill waste should be of particular importance to the sawmills in Zambia which are processing pine and eucalyptus sawlogs from IPD plantations. Since the lumber yield is about 45 percent, the main product of these mills in terms of volume is not lumber but sawmill waste. The production of chips from sawmill waste for production of pulp or panels would require that a suitable chipper be installed at the end of the main waste conveyor, together with chip-conveyor and chip-storage bins for loading of trucks.

Capital Requirement for Sawmill 1/

	Kwacha
Sawmill equipment, crated for export, fob Chattanooga, Tenn., USA (estimated gross weight 180 tons) (estimated total volume 490 m^3)	625 000
Estimated land and sea freight	100 000
Sawmill equipment, cif Chati	725 000

1/ In addition, logging and transport equipment to supply the sawmill is required, at a capital cost of K 990 000.

Installation Cost	175 000
Sawmill Equipment Installed	900 000
Contingencies (including chipping plant)	75 000
Fees paid to Engineering and Contracting firms	75 000
Site preparation (levelling and surfacing)	250 000
Building, 2 storey, on concrete slab foundation	300 000
Moving Equipment	250 000
Machine shop equipment	100 000
Extra spare parts and stores	150 000
Interest during construction	100 000
Total Plant Cost	K 2 200 000
Working Capital	400 000
Total Capital Requirement	K 2 600 000

Manufacturing Cost[1] at Full Capacity (two-shift operation)

Wood Cost		K per year
Eucalyptus sawlog stumpage 29 950 m^3 (r) x K 8		239 200
Pine sawlog stumpage 72 500 m^3 (r) x K 22		1 595 000
Total Wood Cost		1 834 200
Logging and log transport cost 102 400 m^3 (r) x K 6		614 400
Electric energy 40 000 m^3 (s) x 75 kWh x 0.22		60 000
Operating and Maintenance Supplies		120 000
Petroleum products and tires		60 000
Direct Labour:		
20 skilled at K 1 700	34 000	
30 semi-skilled at K 1 000	30 000	
60 unskilled at K 600	36 000	
Total Direct Labour		100 000

[1] Excluding taxes, interest payments and depreciation allowances.

Administration and Mill Supervision	120 000
Allowances and Bonuses:	
45% of wages for direct labour	46 000
35% of salaries	42 000
Miscellaneous Overhead Costs	50 000
Total Conversion Costs	1 212 400
Total Production Costs	3 046 000
Physical Contingencies	304 700
Total Manufacturing Costs	3 351 300

Projected Sales

6. Estimates of sales revenue are based on the following prices and product mix:

Eucalyptus lumber, green and rough	K per m^3 (s)
Boards longer than 1.5 m	86
Shorts	48
Waney edge boards	25
Pine lumber, green and rough	
Boards longer than 1.5 m	126
Shorts	63

7. It has been assumed that the eucalyptus lumber from this mill will consist of 65% long boards, 25% shorts and 10% waney edge boards. Similarly, it was assumed that pine lumber from the mill would consist of 75% longs and 25% shorts.

Projected Sales Volume and Revenue

Year	Sales Volume			Sales Revenue		
	Pine	Eucalyptus ('000 m3)	Total	Pine	Eucalyptus (K million)	Total
1979	6.5	3.7	10.2	0.72	0.26	0.98
1980	13.1	7.3	20.4	1.44	0.52	1.96
1981	19.9	13.5	33.4	2.19	0.95	3.14
1982-93	32.6	13.5	46.1	3.60	0.95	4.55

Financial Rate of Return

8. Projections of financial cash flow[1] are given, in constant 1977 prices, in Table 2 of this Annex. These projections value sawlogs at the replacement cost unit values calculated in Annex 4 (plus a 10% allowance for physical contingencies). The build-up to full capacity is at the rate indicated in the section above on sales, and costs have been pro-rated accordingly. The financial rate of return computed from these projections is 15.6%.

Economic Rate of Return

9. In order to arrive at projections of economic costs and benefits (Table 3 of this Annex) the following adjustments to the financial accounts have been made:

(1) Imported inputs and outputs which will substitute for imported supplies have been valued at US$1.00 = K 0.9 instead of the official US$1.00 = K 0.8.

(2) The real value of sawnwood outputs has been assumed to rise at the rate set out on p. 17.

(3) Local labour was shadow priced at 60% of the financial cost.

10. On this basis, the economic rate of return has been established to be 32.0%.

[1] Exlcuding transfer costs and financial transactions, such as taxes, interest payment and allowances for depreciation.

ANNEX 2
TABLE 1

Project Sawmill

	1978	1979	1980	1981	1982	Foreign Exchange
		(K million)				%
Revenue	-	0.98	1.96	3.14	4.55	100
Capital Costs						
Logging	0.71	0.28	-	-	-	98
Sawmill	-	-	-	-	-	65
Moving equipment	0.25					
Other equipment	1.70					
Other	.25					
Working capital	.40	-	-	-	-	-
Total	2.60	0.28	-	-	-	
Manufacturing Costs						
Wood		0.39	0.77	1.21	1.83	-
Logging and transport[1]		0.14	0.27	0.44	0.61	50
Conversion costs						33
Supplies		0.05	0.10	0.17	0.24	
Labour		0.03	0.06	0.10	0.15	
Other		0.04	0.08	0.15	0.21	
Total		0.65	1.30	2.07	3.04	
Physical contingencies[2]	0.33	0.09	0.13	0.21	0.30	
Total Costs	3.64	1.02	1.43	2.28	3.34	

[1] of which 40% accounted for by labour costs.

[2] at 10% of base costs.

ANNEX 2
TABLE 2

FINANCIAL RATE OF RETURN FOR THE PROJECT SAWMILL (1978–1993)

(K million)

	Wood costs 1/	Other costs 2/	Total costs	Total revenues
1978	-	3.6	3.6	-
1979	0.4	0.6	1.0	1.0
1980	0.8	0.6	1.4	2.0
1981	1.3	1.0	2.3	3.1
1982	2.0	1.3	3.3	4.5
1983	2.0	2.4	4.4	4.5
1984	2.0	1.7	3.7	4.5
1985	2.0	1.3	3.3	4.5
1986	2.0	1.3	3.3	4.5
1987	2.0	2.9	4.9	4.5
1988	2.0	2.4	4.4	4.5
1989	2.0	2.0	4.0	4.5
1990	2.0	1.3	3.3	4.5
1991	2.0	1.3	3.3	4.5
1992	2.0	1.3	3.3	4.5
1993	2.0	1.3	3.3	5.1

1/ Stumpage costs only; using the average unit replacement costs of K22/m^3 for pine and K13/m^3 for eucalyptus (See Annex 4)

2/ Capital costs (including costs of logging equipment acquired by the project to supply the mill (Table1)) and operating costs (including logging and transport operating costs at K6/m^3); excludes depreciation but includes equipment replacement (assuming a life of 5 years for logging equipment and moving sawmill equipment and 10 years for other equipment and structures).

ANNEX 2
TABLE 3

ECONOMIC RATE OF RETURN FOR THE PROJECT SAWMILL (1978-1993)

Year	Wood costs1/	Foreign exchange costs2/	Foreign exchange shadow priced3/	Labour Costs 3/	Labour shadow priced 4/	Other costs5/	Total economic costs	Total benefits6/ 3/
1978	-	2.1	2.4	-	-	1.5	3.9	-
1979	0.4	0.4	0.45	0.1	0.05	0.2	1.1	1.1
1980	0.8	0.15	0.15	0.2	0.1	0.25	1.3	2.4
1981	1.4	0.3	0.3	0.25	0.15	0.45	2.3	3.7
1982	2.0	0.5	0.55	0.4	0.25	0.4	3.2	5.5
1983	2.0	1.4	1.6	0.4	0.25	0.6	4.3	5.6
1984	2.0	0.8	0.9	0.4	0.25	0.5	3.7	5.6
1985	2.0	0.5	0.55	0.4	0.25	0.4	3.2	5.8
1986	2.0	0.5	0.55	0.4	0.25	0.4	3.2	5.8
1987	2.0	1.5	1.7	0.4	0.25	1.0	5.0	5.8
1988	2.0	1.4	1.6	0.4	0.25	0.6	4.5	5.8
1989	2.0	1.1	1.25	0.4	0.25	0.5	4.0	5.8
1990	2.0	0.5	0.55	0.4	0.25	0.4	3.2	5.8
1991	2.0	0.5	0.55	0.4	0.25	0.4	3.2	5.8
1992	2.0	0.5	0.55	0.4	0.25	0.4	3.2	5.8
1993	2.0	0.5	0.55	0.4	0.25	0.4	3.2	6.4

1/ From Annex 2, Table 2

2/ From Annex 2, Table 1

3/ At U.S.$ 1.00 = K 0.9 instead of the official rate of U.S.$ 1.00 = K 0.8

4/ At 60% of actual labour costs

5/ "Other costs" in Annex 2, Table 2, net of foreign exchange and labour costs

6/ Assuming IPD prices rise as follows (1978=100): 1979 = 102, 1980 = 104, 1981 = 106, 1982 = 108, 1983 = 110, 1984 = 112, 1985 and after = 113 (see page 17)

ANNEX 3

PRELIMINARY EVALUATION OF PROPOSED PULP AND PAPER MILL

Introduction

1. The establishment of a pulp and paper mill to utilize part of the wood generated by the industrial plantations has been under consideration by the Government for several years. Conceptual planning for this project is reasonably well advanced. This Annex presents a brief description of the major features of the development as presently conceived, and reviews financial and economic implications.

2. Final exact configuration of the project has not been determined but there now appears to be a concensus with respect to the following features:

(i) the mill would produce linerboard, corrugated medium, sack paper, bag paper and wrapping paper using the kraft pulping process;

(ii) the mill would have a total production capacity of the order of 40 000 tons annually; and

(iii) the mill would start up in about 1982.

3. It has been assumed that fibre furnishes would be 100% pine pulp for all grades except corrugating medium where eucalyptus pulp together with waste paper would be used. The mill would be located on the bank of the Kafue river, close to Kitwe. Earlier studies showed this river to have an adequate flow of water for the mill's requirements. The selected location for the mill eliminates the requirement for any substantial investment in infrastructure. However, some costs may be incurred in extending the railway line into the site, improving the access from the main Kitwe-Ndola road, extending power lines to the site, and providing housing for key staff. For these purposes, a lump sum estimate of K 3 million has been included in the capital cost estimates.

Projected Sales Volume and Revenue

Year	Sales Volume			Sales Revenue		
	Exports	Domestic	Total	Exports	Domestic	Total
	(Thousand tons)			(US$ Million)		
1982	-	24.0	24.0	-	11.1	11.1
1983	3.4	28.6	32.0	1.2	13.3	14.5
1984	5.7	30.3	36.0	2.0	14.1	16.1
1985	5.9	32.1	38.0	2.1	14.9	17.0
1986	6.0	34.0	40.0	2.1	15.8	17.9
1987	3.9	36.1	40.0	1.4	16.8	18.2
1988	1.7	38.3	40.0	0.6	17.8	18.4
1989-2001	-	40.0	40.0	-	18.6	18.6

Project Costs

4. The total project cost, in 1977 prices, is estimated to be K 70.0 million (US$88 million).

5. Projected manufacturing costs per ton of paper produced, excluding depreciation, is summarized in the following table. These costs are based on operation at the full design production rate.

Manufacturing Cost at Full Capacity

Item	Amount (K/ton)	Amount (US$/ton)	Amount (%)
Wood	45.3	56.6	23.5
Chemicals	16.6	20.8	8.6
Other Operating Supplies	24.0	30.0	12.4
Fuel and Power	41.4	51.7	21.4
Labour	19.8	24.8	10.3
Administration & Overhead	45.9	57.4	23.8
Total	193.0	241.3	100.0

6. The wood costs are based on the projected delivered price of the various types of wood used, developed as follows:

Estimated Delivered Wood Cost to Pulp Mill

Type of Wood	Volume (m^3/A)	Price[1] (K/m^3)	Delivery K/m^3)	Total (K/m^3)	Total ($US\$/m^3$)
Pine smallwood	135 000	5.0	6.0	11.0	13.7
Eucalyptus smallwood	20 000	3.0	6.0	9.0	11.3
Sawmill waste	40 000	3.5	-	3.5	4.4
Total	195 000			9.3	11.6

Financial Projections

7. Financial projections for rate of return calculations have been prepared in constant 1977 prices and are attached to this Annex as Table 1. These projections incorporate realistic assumptions concerning the rate of build-up to full production capacity, and for the cost of an expatriate start-up and training crew. The financial rate of return after taxes computed from these projections is 10.4%.

[1] The "price" represents stumpage for the smallwood (see Annex 4), and a transfer price based on fuel equivalent for the sawmill waste. An assumption implicit in this is that a new sawmill will be constructed adjacent to the proposed pulp mill.

Economic Rate of Return

8. In order to get an estimate of the economic internal rate of return of the project, the following adjustments to financial costs and benefits have been made:

(i) Foreign exchange costs[1] and benefits were estimated using a shadow exchange rate of US$ = K 0.9 instead of the official US$1 = K 0.8; and

(ii) Local labour[2] was shadow priced at 60% of the financial cost.

(iii) The pulpwood raw material was given a stumpage value[3] of zero as it has no alternative use.

9. On this basis, the economic rate of return from the pulp and paper mill project has been determined to be 11.7%.

1/ 100% of plant and expatriate assistance costs, 60% of infrastructure and conversion costs and 20% of wood costs.

2/ 13% of conversion costs and 17% of wood costs.

3/ or ex-sawmill value in the case of sawmill waste.

ANNEX 4
TABLE 1

CALCULATIONS OF STUMPAGE RATES - DIRECT COSTS

Basic data for per hectare model calculations

1. Costs

Direct costs (K)		Pine	Eucalyptus
Year 1	Establishment		
	Land clearing	200	200 (18 for replanting)
	Soil preparation	7	7
	Nursery (plant product)	17	10
	Planting	15	10
	Fertilizers	-	20
	Weeding (first year)	47	47
Total establishment (direct) costs		286	294 (112 for replanting)
Year 2	Weeding	45	11 (including a light pruning)
Year 3	Weeding	23	-
Year 5	Pruning I	10	- (see year 2)
Year 8	Pruning II	5	-
Year 11	Pruning III	10	-
Year 11	Thinning I	4(10)1/	-(10)1/ (line thinning at year 5)
Year 14	Marking of Thinning II	3	1 (at year 8, not 14)
Year 19	Marking of Thinning III	2	-
Clearfelling at	year	25	12

1/ If smallwood is unsaleable.

ANNEX 4
TABLE 2

Stumpage rates - indirect costs may be divided into two categories:

(a) New (i.e. referring to new plantations only: 17.5 ha over whole 5-year Project period).

Item		Total Project cost (K'000)	Total average area involved (ha '000)	Average cost per ha p.a. during Project (K)
1.	Road construction	800	17.5	9
2.	Fire protection (new firebreaks and fire towers)	200	17.5	2
3.	Housing (additional)	2 256	17.5	26
			TOTAL	37

(b) Recurrent (i.e. referring to whole area of plantations: 23 000 + 3 500 = 26 500 ha before Project and 44 000 ha at end of Project, giving an average area of 35 000 ha over the whole Project period).

Item		Total Project cost (K'000)	Total average area involved (ha'000)	Average cost per ha p.a. during Project (K)
1.	Road maintenance	432	35	3
2.	Fire protection (recurrent)	740	35	4
3.	Administration	2 450	35	14
4.	Building maintenance	500	35	3
5.	Administrative vehicles and transport	818	35	5
6.	Surveys and training	200	35	1
			TOTAL	30

Indirect costs would be charged at the rate of K37 (for new plantations only) and K30 (for recurrent expenses for all plantations), i.e. a total of K67 per annum during the first five years (Project period), and at the rate of only K30 (recurrent) per annum thereafter.

ANNEX 4
TABLE 3

2. Yields (m^3/ha)

The following yields have been assumed:

(a) Pine

Operation	Age	P. kesiya Sawlogs	P. kesiya Smallwood	P. kesiya Total	P. oocarpa and others Sawlogs	P. oocarpa and others Smallwood	P. oocarpa and others Total	(60% P.K., 40% P.O.) Weighted average for Pine (all species) Sawlogs	Weighted average Smallwood	Weighted average Total
Thin I	11	30	40	70	20	30	50	26	36	62
Thin II	14	40	30	70	25	25	50	34	28	62
Thin III	19	40	20	60	30	20	50	36	20	56
Clearfell	25	120	30	150	100	25	125	112	28	140
TOTAL		230	120	350	175	100	275	208	112	320
MAI		(9.2)	(4.8)	14	(7)	(4)	11	(8.3)	(4.5)	12.8

(b) Eucalyptus

Operation	Age	E. grandis Sawlogs	E. grandis L.poles	E. grandis Smallw.	E. grandis Total	E. cloeziana and others Sawlogs	E. cloeziana and others L. poles	E. cloeziana and others Smallw.	E. cloeziana and others Total	(2/3 E.g., 1/3 E.c.) weighted average for Eucalyptus (all species) Sawlogs	Weighted average L.poles	Weighted average Smallw.	Weighted average Total
Thin I	5	-	-	60	60	-	-	20	20	-	-	47	47
Thin II	8	30	20	30	80	20	20	20	60	27	20	27	74
Clearfell	12	100	30	30	160	50	30	20	100	83	30	27	140
TOTAL		130	50	120	300	70	50	60	180	110	50	101	261
MAI		(10.8)	(4.2)	(10)	25	(5.8)	(4.2)	(9.0)	15	(9.0)	(4.2)	(8.6)	21.8

ANNEX 4
TABLE 4

3. Stumpage rates

On the basis of the assumptions given above for costs and yields, the following stumpage rates have been calculated in order to give an IRR of 10 percent.

Rate (K/m^3)

Category	Pine		Eucalyptus	
	I. All produce saleable	II. Smallwood unsaleable	I. All produce saleable	II. Smallwood unsaleable
Sawlogs	18	22	6	8
Large poles	-	-	10	13
Smallwood	5	-	3	-

ANNEX 4
TABLE 5

STUMPAGE RATES - PINE, SAWLOGS AND SMALLWOOD (PER HECTARE - K)

Year		Direct Costs-K	Indirect Costs-K	Total Costs-K	Volume (m^3/ha) Sawlogs	Volume (m^3/ha) Smallwood	Revenue Sawlogs @K18/m^3	Revenue Smallwood @K5/m^3	Total Benefits	Net Benefits	FIRR 10% -	FIRR 10% +
1	Establishment	286	67	353						(353)	353	
2	Weeding	45	67	112						(112)	101	
3		23	67	90						(90)	74	
4			67	67						(67)	50	
5	Pruning I	10	67	77						(77)	53	
6			30	30						(30)	19	
7			30	30						(30)	17	
8	Pruning II	5	30	35						(35)	18	
9			30	30						(30)	14	
10			30	30						(30)	13	
11	Thinning I and Pruning III	14	30	44	26	36	468	180	648	604		232
12			30	30						(30)	10	
13			30	30						(30)	10	
14	Thinning II	3	30	33	34	28	612	140	752	719		208
15			30	30						(30)	8	
16			30	30						(30)	7	
17			30	30						(30)	6	
18			30	30						(30)	6	
19	Thinning III	2	30	32	36	20	648	100	748	716		129
20			30	30						(30)	5	
21			30	30						(30)	4	
22			30	30						(30)	4	
23			30	3C						(30)	4	
24			30	3C						(30)	3	
25	Clearfelling		30	3C	112	28	2 016	140	2 156	2 126		215
										TOTALS:	779	784

<u>Conclusions</u>: The stumpage rates for Pine should be about K18/m^3 for sawlogs and about K5/m^3 for smallwood assuming both are saleable.

ANNEX 4
TABLE 6

STUMPAGE RATES - EUCALYPTUS (2/3 E.G., 1/3 E.C. & OTHERS) - SAWLOGS, LARGE POLES AND SMALLWOOD

(per hectare - K)

Year		Direct Costs	Indirect Costs	Total Costs	Volume (m^3/ha) Sawlogs	L. poles	Smallwood	Revenue Sawlogs @K6/m^3	L. poles @K10/m^3	Smallwood @K3/km^3	Total Revenue	Net Benefits	FIRR 10% -	+
1	Establishment	304	67	371								(371)	371	
2	Weeding & Pruning	11	67	78								(78)	71	
3			67	67								(67)	55	
4			67	67								(67)	50	
5	Thinning I	-	67	67	-	-	47	-	-	141	141	74		51
6			30	30								(30)	19	
7			30	30								(30)	17	
8	Thinning II	1	30	31	27	20	27	162	200	81	443	412		211
9			30	30								(30)	14	
10			30	30								(30)	13	
11			30	30								(30)	12	
12	Clearfelling	-	30	30	83	30	27	498	300	81	879	849		297
13	Reestablishment	112	30	142								(142)	45	
14	Weeding & Pruning	11	30	41								(41)	12	
15			30	30								(30)	8	
16			30	30								(30)	7	
17	Thinning I	-	30	30	-	-	47	-	-	141	141	111		24
18			30	30								(30)	6	
19			30	30								(30)	5	
20	Thinning II	1	30	31	27	20	27	162	200	81	443	412		67
21			30	30								(30)	4	
22			30	30								(30)	4	
23			30	30								(30)	3	
24	Clearfelling	-	30	30	83	30	27	498	300	81	879	849		86
												TOTALS:	716	736

Conclusion: The stumpage rates for Eucalyptus should be about K6/m^3 for sawlogs, K10/m^3 for large poles and K3/m^3 for smallwood assuming all are saleable.

ANNEX 4
TABLE 7

STUMPAGE RATE - PINE (60% P.K., 40% P.O. & OTHERS) - SAWLOG UTILIZATION ONLY

(Per hectare - K)

Year	Operation	Direct Costs	Indirect Costs	Total Costs	Volume of sawlogs m^3/ha	Revenue at $K22/m^3$	Net Benefits	FIRR 10% -	FIRR 10% +
1	Establishment	286	67	353			(353)	353	
2	Weeding	45	67	112			(112)	102	
3		23	67	90			(90)	79	
4			67	67			(67)	50	
5	Pruning I	10	67	77			(77)	53	
6			30	30			(30)	19	
7			30	30			(30)	17	
8	Pruning II	5	30	35			(35)	18	
9			30	30			(30)	14	
10			30	30			(30)	13	
11	Thinning I & Pruning III	20	30	50	26	572	522		201
12			30	30			(30)	10	
13			30	30			(30)	10	
14	Thinning II	3	30	33	34	748	715		207
15			30	30			(30)	8	
16			30	30			(30)	7	
17			30	30			(30)	6	
18			30	30			(30)	6	
19	Thinning III	2	30	32	36	792	760		137
20			30	30			(30)	5	
21			30	30			(30)	4	
22			30	30			(30)	4	
23			30	30			(30)	4	
24			30	30			(30)	3	
25	Clearfelling		30	30	112	2464	2434		246
							TOTALS	785	791

<u>Conclusion:</u> The stumpage rate for pine sawlogs should be about $K22/m^3$ to produce an IRR of 10%, assuming smallwood is not saleable, but thinnings must be done as scheduled.

ANNEX 4
TABLE 8

STUMPAGE RATE - EUCALYPTUS (2/3 E.G., 1/3 E.C. & OTHERS) - SAWLOG AND POLE UTILIZATION ONLY

(Per hectare - K)

Year		Direct Costs	Indirect Costs	Total Costs	Volume (m^3/ha) Sawlogs	Large Poles	Revenue Sawlogs at K8/m^3	Large poles at K13/m^3	Total Revenue	Net Benefits	FIRR 10% -	+
1	Establishment	304	67	371						(371)	371	
2	Weeding & Pruning	11	67	78						(78)	71	
3			67	67						(67)	55	
4			67	67						(67)	59	
5	Thinning I	10	67	77						(77)	46	
6			30	30						(30)	19	
7			30	30						(30)	17	
8	Thinning II	1	30	31	27	20	216	260	476	445		228
9			30	30						(30)	14	
10			30	30						(30)	13	
11			30	30						(30)	12	
12	Clearfelling		30	30	83	30	664	390	1054	1024		358
13	Reestablishment		30	142						(142)	45	
14	Weeding & Pruning	11	30	41						(41)	12	
15			30	30						(30)	8	
16			30	30						(30)	7	
17	Thinning I	10	30	40						(40)	9	
18			30	30						(30)	6	
19			30	30						(30)	5	
20	Thinning II	1	30	31	27	20	216	260	476	445		73
21			30	30						(30)	4	
22			30	30						(30)	4	
23			30	30						(30)	3	
24	Clearfelling		30	30	83	30	664	390	1054	1024		114
										TOTAL	771	773

Conclusion: The stumpage rate for eucalyptus should be about K8/m^3 for sawlogs and K13/m^3 for large poles, to produce an IRR of 10%, assuming smallwood is not saleable, but thinnings must be done as scheduled.

FAO FORESTRY PAPER 17 Sup. 2

economic analysis of forestry projects: readings

policy and planning service
forestry department

FOOD AND AGRICULTURE ORGANIZATION OF THE UNITED NATIONS
Rome, 1980

First printing 1980
Second printing 1983

The designations employed and the presentation of material in this publication do not imply the expression of any opinion whatsoever on the part of the Food and Agriculture Organization of the United Nations concerning the legal status of any country, territory, city or area or of its authorities, or concerning the delimitation of its frontiers or boundaries.

M-30

ISBN 92-5-100955-4

FOREWARD

This volume of papers forms part of a series of publications that have been produced by the Planning and Investment Studies Unit of FAO's Forestry Department in order to make available information on analysis and planning in the forestry sector for teaching and reference use. The series, which has the title "Economic Analysis of Forestry Projects", consists of a guide to analysis 1/ and a volume of case studies 2/ in addition to the present volume of readings.

This volume of readings has been edited for FAO by H.M. Gregersen, Professor of Forestry and Agricultural and Applied Economics at the University of Minnesota. The coverage of the four papers is reviewed in his Editor's Introduction. All of them have been specially commissioned for this FAO series from the authors, each of whom is writing here in his personal capacity. The work on the two papers by authors at the University of Minnesota was made possible through a special budgetary contribution to FAO for this purpose from the Swedish International Development Authority (SIDA).

1/ Economic Analysis of Forestry Projects. FAO Forestry Paper No. 17, FAO, Rome, 1979.

2/ Economic Analysis of Forestry Projects: Case Studies. FAO Forestry Paper No. 17, Supp. 1, FAO, Rome, 1979.

TABLE OF CONTENTS

EDITOR'S INTRODUCTION

The present publication is provided as a complement to FAO's recently published guide, Economic Analysis of Forestry Projects[1], (henceforth referred to as EAFP), which deals mainly with the steps involved in organising and carrying out an analysis of the economic efficiency associated with a forestry project. As mentioned in EAFP, economic efficiency is only one of the factors with which decision-makers are concerned in their quest to make "better" decisions concerning forestry projects. In addition, they may be concerned explicitly and quite separately with income distribution impacts, effects of projects on local areas as opposed to the nation as a whole, balance of payments or trade effects, employment implications, and so forth.

Both the McGaughey and Schuster papers in this volume deal with the question of income distribution and employment impacts of forestry projects — topics of increasing concern to decision-makers in developed as well as developing countries. While much of the recent dicussion in the project analysis and evaluation literature has been concerned with the means to quantitatively incorporate income redistribution impacts directly with efficiency impacts in an integrated "social economic" analysis framework, both McGaughey and Schuster treat income redistribution impacts using a partial analysis approach such as suggested in EAFP. The idea of developing "weights" for costs and benefits associated with different income groups and introducing such in an economic efficiency analysis is appealing in theory and concept, but difficult to implement in practice, due to lack of agreement on specific weights to be assigned different groups. While labour shadow prices implicitly recognise the lower social cost (or higher social benefit) associated with using unemployed labour, McGaughey and Schuster both provide additional separate measures of employment and income distribution impacts which can be calculated and used in making additional comparisons between forestry projects.

McGaughey treats improved income distribution and increased employment as complementary goals in rural development projects. He discusses a number of different specific measures which can be calculated and used for various purposes in assessing project impacts on employment, regional income and income distribution. However, he concludes that: "What is clear is that no income distribution indicator can be universally recommended, since such indicators will have to be adapted in each country to the local data conditions and limitations which project analysts face."

Schuster discusses several different dimensions associated with income redistribution and employment impacts. He distinguishes between regional income redistribution and "individual welfare" impacts of projects. The impact indicators recommended for consideration in terms of individual welfare were selected on the basis of "their presumed relevancy and their feasibility to be measured". Based on this criterion, he chose project impact on 1) unemployment rates, 2) average wage rates, and 3) income distribution among the population of the region affected by the project. He further discusses measures of project effect on economic equilibrium and stability in local communities or regions, recognising that these two factors affect community welfare in a broader sense.

---

[1] FAO Forestry Paper No. 17, FAO, Rome, 1979.

McGaughey deals at some length with various measures of project impact on foreign exchange flows (or balance of payments impacts), recognising that this is a factor of considerable importance to many decision-makers. Schuster, recognising the importance of regional distribution impacts of projects, discusses several useful measures of project impact on local government. He specifically looks at intergovernmental payments (revenue sharing, in-kind payments, and other types of payments) and discusses types of local government costs often associated with projects which are not considered in economic efficiency analyses. McGaughey also discusses appropriate measures of regional economic growth which can be calculated as part of a broader project analysis.

While Schuster in general sticks with impact measures related to single objectives, such as employment, income redistribution, regional stability, etc., McGuaghey goes on to discuss several multiple objective evaluation procedures which can be applied in practice. These include the Delphi method, scoring models and various means for combining single goal indicators. McGaughey further indicates the usefulness of sensitivity analysis in looking at project impacts and in providing information for decision-makers in cases where data are scarce or unreliable.[y]

EAFP provides a general approach to economic analysis of all types of forestry and forest industry projects. Peculiarities or unique aspects associated with specific types of projects, such as pulp and paper, or watershed and wildlife, are not discussed. The paper by Gregersen and Brooks in this volume deals with some of the specific problems and conditions encountered when analysing watershed projects.

The paper by Houghtaling and Gregersen provides a more detailed treatment of compounding and discounting procedures than is found in EAFP. In addition, derivations of various useful compounding and discounting formulas are provided together with comments on how to treat some common time related problems encountered in project analysis - e.g. how to treat inflation and how to deal with situations where multiple discount and/or compound rates have to be used for a given analysis.

Finally, it should be pointed out that EAFP and the present volume are limited to discussions concerning economic efficiency, income distribution, local area economic impacts, balance of payments and employment impacts - all items that in the broadest sense can be associated with various economic and social aspects of welfare. It is explicitly recognized that further work on treatment of impacts from an environmental point of view is needed. However, this is beyond the intended scope of the present effort.

---

[y] See also EAFP, Chapter 10.

INVESTMENT CRITERIA FOR AGRICULTURAL AND RURAL DEVELOPMENT PROJECTS:

The Measurement of Their Multiple Economic and Social Consequences

by

Stephen E. McGaughey [1]

---

[1] S.E. McGaughey is Chief of the Agricultural Economics Section in the Economic and Social Development Department of the Inter-American Development Bank.

1. INTRODUCTION

In recent years extensive improvements have been made in project evaluation methods in order to incorporate considerations beyond the traditional concern of project analysts with economic efficiency or profitability. Most of these improvements have been in the realm of more theoretically complete project evaluation systems. The most notable procedures include those developed by Little and Mirrlees (70) and Dasgupta, Marglin and Sen (117). Much of the later work has been an improvement and a refinement of this earlier seminal thinking.

The recommended methods are severely limited in many circumstances because (1) they are not designed to incorporate a large number of objectives; (2) a large data base is needed to complete the evaluation; (3) not much success has been achieved in simplifying the system for easy application by project analysts; and (4) little recognition has been given to the decision making process in developing countries which often proceeds in circumstances in which economic considerations may be of marginal importance.

The present paper is contributed in the view that while the investment decision making process is little known to economists, engineers and other project analysts, policy makers should welcome additional knowledge of the multiple economic and social consequences of the many projects that must be appraised in formulating an investment programme. While increasingly individual projects will be evaluated using social cost-benefit analysis, the procedures herein recommended are posited as an intermediate step between essentially what occurs now (little or no formal appraisal of multiple objectives) and the widespread use of social benefit-cost analysis (which may be possible some years from now). In this paper multiple benefits and costs are first introduced via single objective evaluation criteria permitting the ranking of projects for the separate goals. A large number of project performance indicators are reviewed, compared and evaluated for use under different situations.

Alternative procedures for ascertaining, estimating and combining multiple objectives for project ranking are discussed, such as scoring models and a more direct weighting scheme proposed by McGaughey and Thorbecke (75). Finally, procedures for completing relevant sensitivity analysis are developed including simple graphical techniques and a more detailed elasticity analysis.

All of the proposed procedures should be of primary interest to national and sector planners who need tools for defining sector-wide investment programmes. Among other subsectors, the present procedures are formulated for application in devising water resource programmes, sector forestry plans and regional rural development programmes.

2. THE NEED FOR MEASURABLE INVESTMENT CRITERIA

While economic and social project evaluation methodologies, as they are now proposed, may one day be widely employed by international leading agencies and national planning authorities, there are few prospects that this will occur on a large scale soon, either in countries or by international agencies, especially if it is required that aspects such as income redistribution effects are to be accounted for in the evaluations. 1/

Less developed countries need evaluation techniques that can be uniformly applied throughout all subsectors at a low cost with minimal data demands, producing results which are reasonably understandable to policy makers. 2/ In most countries there is a distinct propensity to apply narrow financial and commercial criteria to project selection. An advantage in these criteria is that most policy makers understand the concept of the financial and commercial criteria to project selection. An advantage in these criteria is that most policy makers understand the concept of the financial rate of return. The economic evaluation is of prime interest to agencies with a broad national or sector-wide viewpoint. The national and sector planning authorities or the planning office of the development finance corporation all have an interest in allocating resources to high priority sectors based upon general economic and social goals stated in development plans; but, project evaluations are not of great interest to policy makers in their day-to-day activities because they are primarily concerned with the allocation of financial (budgetary) resources.

A second reason why the most complicated socio-economic project evaluation methods have not been quickly accepted by public authorities is that they place heavy demands on scarce technical talents. Thus, while many project practitioners without advanced economic training may understand the use of financial criteria, it is doubtful that they will understand the evaluation techniques proposed by Squire-van der Tak (120), Little-Mirrlees (70), or UNIDO (117). In developing countries, trained specialists are mainly in demand as managers or administrators. They are assigned to the execution of programmes and projects rather than to their ex-ante evaluation. While a social cost-benefit analysis may be needed for projects or programmes to be financed by external lending agencies, the assignment of local specialists to these projects often occurs only to ensure financing by the agencies, by presenting the required economic and financial evaluations.

Economists tend to take an egocentric stance in designing investment criteria, overemphisizing the importance of economic considerations in the final project selection. It is clear that a broad spectrum of objectives - economic, social, political and historical - are all combined in the minds of policy makers who allocate public investment budgets on a project-by-project basis or as a part of a medium- to long-term investment programme. While financial investment criteria will continue to be the centerpiece of the analysis, appraisals which occasionally take into account shadow price adjustments will be undertaken for large investment projects destined for external finance. Hence, the new evaluation techniques will likely be applied on a piecemeal, project-by-project basis for some years to come and their full integration into national, sectoral or regional planning is not likely to occur for one or two decades.

1/ It may be argued that international lending agencies have no right to set weights on consumption or income benefits received by groups within a particular society. (See D. Lal (62). The weights logically would have to be set by national authorities and consistently applied to all projects, either for external assistance or those to be financed exclusively by local resources. To the extent that evaluation techniques are not used similarly by all parties, their value would seem to be severely limited.

2/ See R. Chambers (19) who forcefully argues that "simple is optimal".

What is needed now is a perception of a sequence of improvements in prevailing evaluation procedures. Project evaluation techniques should meet the following tests: they should be measurable with available data; they should be applicable to a wide variety of project investment categories; they should be consistent with rural development or agricultural sector programme analysis; they should provide rankings of individual investment projects; they should be understandable to most policy makers who review the sector or project plans; and, finally, they should take into account multiple economic and social objectives and, thus they should be amenable to a multidisciplinary approach to project analysis. With the present short supply of trained economists and the low quality of the basic statistical information, the Little-Mirrlees/UNIDO social cost-benefit analysis does not seem to meet many of these tests.

First, concerning measurability, it is clear that most current appraisals are conducted on the basis of expert judgements (educated guesses) of the values of many project technical and economic parameters. This will continue to occur in years to come because the prospects of obtaining an independent, reliable data base for the massive application of cost-benefit analysis do not seem good.[1] Also, that which is possible for a single project is not always possible for a large number of projects. A set of decision criteria is required, either partial indicators or social cost-benefit ratios, which can be applied on the basis of expert judgements and little information on the regions where the projects are to be located. As cost, price, production and market information improves, social cost-benefit analysis may be viable in a larger number of circumstances.

Second, the decision criteria for rural development projects should be suitable to a wide variety of projects as well as economic and social sectors. Rural development projects generally contain a complicated mixture of sectors, subsectors, investments and annual expenditures;[2] decision criteria that would permit an improvement in their design will have to be compatible with these diverse subsectors and sectors.

Third, it is necessary that the decision rules used for rural development projects be integrated into the regional and sector, agricultural and industrial planning processes. This means that the decision rules would permit policy makers to rank the rural development projects along with other national development alternatives. The sector and regional planning authorities should be able to incorporate the projects into the budgetary process and to understand their economic and social impacts to the greatest degree possible. Short of rather complicated general equilibrium or multi-level planning models (see the models developed for Mexico (86) and the Ivory Coast (33)), it is difficult to imagine that multiple objective social cost-benefit analysis will be integrated into sector and regional planning in the foreseeable future.

---

[1] For example, physical yield data, farm costs of production, the effects of soil conservation schemes among others may not ever be amenable to estimation except on a pilot or experimental basis.

[2] Two recent examples of complex rural development programmes and projects include the Mexican PIDER and the Colombian DRI.

Fourth, the proposed decision criteria must be used to establish the optimum project mix within regions or within regional development programmes, determining the distribution of investment funds among directly productive and productive support activities and social infrastructure. When there is abundant information and trained specialists, programming models may be used for this purpose. The knowledge provided through social cost-benefit analysis of multiple objective effects will be extremely limited in most circumstances.

Fifth, the decision criteria should be useful to a multidisciplinary team of project designers and planners, allowing for the participation of all disciplines - economic, social, financial and technical - in the design and organization of the project. An evaluation tecnique that is the sole purview of a limited group of technicians, such as economists, will not likely receive the support needed to be adapted to a large number of project appraisals.

Six, an essential characteristic of the decision criteria is that they are understandable to policy makers who make the final investment and expenditure decisions. While much time and effort has been given to the simplification of evaluation techniques, they are still a mystery to most policy makers. Concepts such as shadow prices, conversion factors, the accounting rate of interest, and the consumption rate of interest are not familiar to most of them.

Finally, it is essential that decision criteria be applied with the understanding that a gradual improvement in evaluation techniques will be made; techniques which are useful during the next decade will be replaced, on a larger scale, by more sophisticated social cost-benefit analyses. A possible sequence of improvements might be the following: begin with current methods (which are, primarily, financial evaluations with added elements of an economic evaluation); gradually improve the economic evaluation for single projects, particularly, large scale projects which place major demands on the public budget; introduce simple indicators for the planning and evaluation of projects for multiple objectives and multiple sectors; adopt simple programming techniques such as goal programming to project and sector analysis; apply the complete social cost-benefit analysis to more sectors and to more projects; and, finally, undertake full scale social cost-benefit analysis and general equilibrium multi-level programming by sectors and regions. There are evaluation techniques suitable to each of these stages. However, the pre-conditions needed to apply them all are not extant. Decision criteria are needed which apply to rural development projects having multiple objectives.

The current challenge in most developing countries is to apply appraisals without great budgetary cost and using presently available data. The economist should take a more modest stance. Rather than pushing for the application of optimal investment criteria, he could provide a greater service by developing techniques which can be used without large changes in today's public agency staffs and budgets.

3. ALTERNATIVE ISSUES IN CHOOSING INVESTMENT CRITERIA

3.1 The Main Issues

The purpose of the present section is to examine the use to which investment criteria will be assigned in the evaluation of rural development projects. This use derives from the issues which have been mentioned relating to the ultimate pruposes of rural development projects. The form that the analysis takes will depend upon the following principal contrasting elements: (1) financial versus economic analysis; (2) the measurement of multiple objectives and their weights versus single objectives; (3) constraints on the analysis, particularly institutional constraints versus unconstrained cases; (4) design versus project ranking, and (5) project versus regional programme planning.

3.2 Financial versus Economic Analysis [1]

In traditional cost-benefit analysis practitioners tended to rigidly separate financial and economic analyses. The former, among other things, is concerned with the commercial profitability of a project, while the latter is germane to improving the allocation of resources. As a consequence of incorporating income distribution and equity considerations into social cost-benefit analysis, a linkage between financial and economic analysis is made. Thus, the financial analysis, which treats the distribution of payments and repayments, and sources and applications of funds, is the major tool for tracing the incidence of benefits and costs of a particular project; financial analysis is now more important than a mere indicator of the capacity of the project beneficiaries to repay loans. This is especially crucial to the design of rural development projects.

3.3 Multiple Objectives

One of the perplexing problems in economic and financial analysis is how to systematically introduce multiple goals into project design. This problem has been broadly examined by Norton, Bassoco, and Silos (86), Sfeir-Younis and Bromley (106), Papandreau and Zohar (89), and McGaughey and Thorbecke (75, 76). Other specific applications of multiple objective analysis are discussed later in relation to specific investment criteria.

Multiple objectives may be introduced in several ways. One way is to specify explicitly the separate partial impacts of projects on one or another economic and social objective. The purpose of this approach is to emphasise the individual effects of a project without combining them into a single measure. As long as decision making is limited to a few projects, regions and objectives, mathematical programming models may be an appropriate technique for introducing multiple objectives and selecting priorities, but the number of circumstances in which the models can be applied seems rather limited, especially if the main purpose is to design a comprehensive (national) rural development programme.

---

[1] This issue is treated in Economic Analysis of Forestry Projects, FAO Forestry Paper No. 17.

In addition to the problem of how to combine multiple economic objectives, regions and projects in a consistent fashion, a further complication is how to set welfare or preference weights on these objectives. This is, if a policy-maker can identify, for example, four objectives, it is clear that he might not place equal importance on all of the objectives. It is more likely that greater weight might be placed on one or two principal objectives, assigning less significance (or weight) to "secondary" objectives.

In later sections an effort is made to design simple systems to incorporate multiple objective weights into the evaluation and design of rural development projects. Four economic objectives are considered: (1) an increase in production and income in rural areas; (2) an increase in remunerative employment and an improvement in the income distribution; (3) the attainment of an improved regional balance in growth and development; and (4) an increase in foreign exchange earnings.

An increase in income, consumption and value added - one of the most important objectives of economic analysis - is referred to as the economic efficiency objective. If projects are chosen solely on the basis of the efficiency goal, there is no guarantee that beneficiaries of the investment programme will represent the lowest income groups. Indeed, if one were to select projects only on the basis of the efficiency goal then commercial agricultural producers with access to financial resources and other assets often would be the main beneficiaries. Usually, rural development projects are designed with a particular beneficiary group in mind, so the "second-best" efficiency objective is to maximize the net benefits of the beneficiary group within a certain project area. However, planners may be disappointed to find that the opportunity costs associated with investments for increasing the benefits to smallholders and other poor rural inhabitants may be high.

The goal of increasing the amount of remunerative employment from the project investments is closely related to the income redistribution effect. This goal is so much akin to the income redistribution goal that several authors, including Marglin (80), have argued that the employment objective is a surrogate for the income redistribution goal. They emphaisze that employment is not desired for its own sake, but rather as a vehicle for increasing income and welfare.

Various dimensions of the problem will have to be considered to introduce employment indicators. A distinction must be made between permanent and seasonal agricultural and non-agricultural, high-paid and low-paid employment, and the preferences of policy-makers for employment creation at different points in time. These differences are not always directly dissaggregated in social cost-benefit analyses since equity considerations are introduced in the form of welfare weights on the benefits and costs of the low income beneficiaries. Therefore, one advantage of using partial indicators of project performance is that these qualitative features of employment can be explicitly enumerated.

Rather than taking employment as a substitute measure of income redistribution effects, an alternative is to take the direct effects of a project on the prevailing regional income distribution. Procedures for treating income distribution consequances have been fully discussed by, among others, Balassa (5), Byerlee (16), Harberger (39), Kalter and Stevens (61), Little and Mirrlees (70), UNIDO (117).

The promotion of regional economic balance is another important economic objective, i.e. promoting backward regions relative to regions which have attained acceptable levels of per capita income. This difference is more than the mere distinction between promoting development in rural as opposed to urban areas. There may be urban areas in backward regions which have an important role to play in the income growth of the poorest income groups in surrounding rural areas. Three major policy questions dominate the issue of regional economic balance. First, there is the desire on the part of national authorities to achieve greater diversification to improve the terms of trade between agriculture and industry, from one region to another. Secondly, the government may want to change the regional location of an activity, particularly rural development activities; direct or indirect government influence can be used to achieve a desired locational mix of rural development activities. Thirdly, allocations might be made among regions to maximize the opportunities for further long-term growth.

Improving the balance of payments is one of the most widely propounded economic objectives in less developed countries. It is also a goal which is extremely difficult to interpret at the project level expecially when considering single projects. The objective might be stated in terms of maximizing the net foreign exchange benefits of a project, increasing the exports of specific items or substituting the imports of others. Such objectives are introduced in project evaluation via the introduction of a foreign exchange shadow price.

3.4 Institutional and Related Concerns

Project appraisal must proceed under a gamut of constraints which limit the scope and ultimate effectiveness of a project both for the executing agencies and the beneficiaries. To the extent possible, project appraisal methodologies should help identify and quantify these constraints, to determine the consequences of eliminating the most important ones. The task becomes more complex for rural development projects in which there are constraints which, in effect, cut accross many subsectors in several regions. Likewise, there are constraints which are national in scope not originating in a particular region or sector which may alter the ambiance in which the project functions. These constraints can be divided into (1) financial, (2) human and manpower, (3) institutional and administrative, and (4) social categories. This categorization is not a complete one.

Financial constraints refer to limits on the availability of funds for project fixed and working capital and operating expenditures. The national treasury limits the amount of funds which go to the various ministries while sector (e.g. education, transportation, health) allocations limit what might be available to a particular rural development project or region. Without a national political decision there is little that can be done to alter these national financial constraints. At the regional and local level, project authorities may have greater access to funds such that, for example, crop and livestock lending at the local level can be redirected to the beneficiaries of an indicated rural development project. These possibilities will depend upon the circumstances prevailing in each country where rural development projects are being executed. It is important in project evaluation that the type of funds constraining project execution be determined. But little can be done by project authorities to change fundamental constraints although the purpose of designing rural development projects per se is to coordinate the assignment of public funds for specific projects within specific regions where such coordination did not occur before.

Financial constraints are accounted for in project appraisal through the conventional flow-of-funds analysis and by an analysis of the incidence of expenditures on beneficiary groups within the economy. Thus, certain beneficiaries may be required to repay relatively more than others so that the allocation of project funds will have differential effects on the beneficiaries' income distribution.

A second fundamental project constraint relates to the availability of trained specialists and other manpower categories for project execution. When such human resources are not available, a training programme must be undertaken with the cost generally attributed to the project. Start-up and execution time for the project will likely be lengthened. Or if the project draws trained manpower away from other activities, then extraordinary funds may have to be assigned to pay these specialists at wage rates that are sufficiently attractive for them to make the changes. Rural development projects involve redirecting the use of existing public sector specialists to the new projects as well as requiring additional numbers of specialists because of the overall increase in project activity and financing. Such manpower constraints will have to be incorporated into the project cost estimates and specific indicators of project labour use will have to be obtained in order to compare rural development project needs.

Institutional problems are the most pervasive and abstruse project constraints as well as the most difficult to alleviate. They take so many forms that the category - institutional constraints - is probably too extensive. Rather sub-categories including structural features, personnel management, financial management, technical management, and overall administration procedures have to be considered to fully define the nature of the institutional problem.

Structural problems refer to whether the institutional framework is compatible with the project design and objectives (obviously, project design and objectives have to be adjusted to the institutional structure as well, but here we are concerned with the obverse of this proposition). If a project involves the active participation of local beneficiaries in its design and execution, it is likely that the decentralisation of public services will have to be undertaken to be consistent with this kind of project organisation. Likewise, if a particular basic service is supplied by several competing offices, then it might be necessary to consolidate their functions under a single responsible agency. The decentralisation of government activities can in itself add to costs and this will have to be balanced with the expected benefits.

Personnel management which improves the utilisation of the skills of all agency and project employees is needed for successful project implementation. Personnel standards are often different among the several agencies involved in the execution of a rural development programme and they may have to compete for staff because of different salary levels among the agencies or because of staff dissatisfaction from similarly qualified professionals being paid different salaries for the same work requirements.

Financial management and administration is crucial to the smooth functioning of a project to ensure that materials and operating supplies are provided at the location and at the rate which is concomitant with staff execution capacity. Information on the length of the production cycles and operating funds demands are the basis for the estimates of the requirements for working capital and other financial resources. Indicators of the efficiency of the internal use of the financial funds and the viability of funds management should be obtained.

The technical staff of an institution needs extra guidance as new projects, such as rural development projects, are initiated. The staff may be unaccustomed to work with the new beneficiary groups and also may not be familiar with sharing project design and execution control. This situation requires special skill, tack and understanding on the part of the technical managers. When project staff have to be geographically shifted or when their functions radically change because of the introduction of a new project, then these changes should be quantified.

The administrative procedures which bear on whether a project is successful or not are not easily identified. Whether internal administrative coordination among major subdivisions takes place has a direct bearing on the coordination of the separate projects components. It is clear that one cannot quantify such relationships and only with a long historical perspective on the performance of the agency in a region can one see where the improvements must be made.

Institutional weakness tends to be used as a catchall category referring to a large number of problems endemic to most newly evolving agencies, including social, political and economic problems. While it may be easy to prescribe internal administrative improvements, they will only take place gradually over a long period in which staff is trained and new relationships established. Project evaluators must look to the internal capacity of agencies before introducing complex procedures which take up the time and resources of these agencies.

Social constraints refer to those broad classes of problems deriving from historical, cultural, behavioral and anthropological features of a society which affect the anticipated outcome of the project. Production systems, operating procedures, incentive and control mechanisms must all be adjusted to the presence of such fundamental constraints. While quantitative project evaluations provide little direct information on these features, it is essential that sociological studies be undertaken before the basic project benefit and cost flows can be constructed, because such flows subsume the economic and social system. The need to understand the social dynamics of an investment project is especially important for rural development projects in which contact with the rural poor is being made on a scale previously unknown to public agencies.

3.5 Designing and Ranking

While the recommended way to evaluate and select a project is to apply a single criterion to its simultaneous designing and ranking, in practice, this often does not occur. Separate agencies are responsible for the two activities - designing and ranking - and a considerable period of time, even years, may divide the period in which these choices are made. The designing of a project concerns the choice of its size (e.g. height of a dam, length of a canal, number of beneficiaries, volume of waste treatment, volume of potable water) and location within a region or district, as well as the timing of investments and the choice of technology. The ranking of a project involves a comparison between one project and another within a particular set of economic, political and social constraints so as to make a "go, no-go" decision to determine the time sequence in which the project might be executed compared to other projects in the list.

Project design is primarily a micro-oriented activity which relates technical, economic and social feasibility to determine a project's eventual characteristics. Local beneficiaries may participate in the design process in rural development projects or the basic project parameters may be handed down by a regional or national planning group. Simple criteria are needed which allow for the comparison of project benefits and costs at an early stage, well before detailed information of a project's cost and benefit flows are known. Thus, the final features of a project are often set very early during the long process of design, selection and execution.

The second problem is to rank a project within a comprehensive basket of project alternatives. Such ranking is often done at the national level and less frequently at a regional level by multisectoral planning groups, finance ministry or development corporation. While national development plans have evolved in many countries, there are few cases in which social cost-benefit analysis has been employed to rank a large number of projects. This occurs because, while for a single project there may be adequate detailed data to apply the complete social cost-benefit analysis, there may be limited data for several other projects. The most complex methods are effectively neutralized by only one important project within the set which has less than a complete data base. Therefore, simplified investment criteria are needed to compare projects even at a more advanced state of design and preparation. Likewise, a large number of projects are often not available to rank at any moment in time since projects are considered by planning authorities as they are identified and studied.

3.6 Project and Regional Development Planning

Investment criteria should be constructed as aids for regional and local economic and social planning. Present day criteria are especially accommodated to national planning needs in countries where local and regional planning is still in an incipient stage. But because of growing interest in rural development projects, which increase local participation in social and economic planning, simplified and measurable investment criteria are also a requisite for local and regional planners. While certain objectives such as income distribution improvements and economic growth have their equivalents at the regional level, methodological problems may be more severe at this level because of the limited data base as well as the diverse viewpoints of local planners. Usually, objectives such as increasing foreign exchange earnings are of secondary interest to local planners who are mainly concerned with many factors which affect local beneficiary groups. Hence, while if a preference weighting scheme reflects national goals, it is probable that the preference weights of local policy makers might be substantially different, even in the rather unlikely case in which these two groups of policy makers are concerned with the same objectives. National planners may be able to impose their preferences on the control of a large share of the regional and local givernment's revenues.

It is difficult to balance the two desirable features of simplicity and comprehensiveness in the same evaluation techniques and satisfy the needs of local rural development planners. At the local level techniques should capture the essence of the local social, economic and political aspirations, providing clear indicators of project performance that are relevant to local problems. Local notions of equity, income growth, preferred beneficiary groups and desirable product lines and employment activities should all be measurable by the available investment criteria. Furthermore, the concept of national economic parameters (33) (e.g. shadow exchange rates and interest rates) may have to be reviewed in light of the increasing importance of regional and local planning activities.

The separate viewpoints of national, regional and local public authorities should be explicitly introduced into project analysis because of their different perceptions of project benefits and costs. A benefit to a national planner (e.g. greater national employment) may be viewed as irrelevant to a regional project planner (because the employment does not fall within his region) or viewed as a cost (if revenues in the regional plan are used to pay for the employment generating activities). In the same example, elsewhere, a local municipality may have to incur local health and other social infrastructure expenditures because of the new residents attracted by the new investment opportunities. Hence, what is a cost (benefit) to one group may be a benefit (cost) to another. Project planners will, therefore, have to use great care in defining project net benefit flows in terms of specific points of view.

4. SINGLE OBJECTIVE MEASURES [1] [2]

4.1 The Objective

The principal economic objectives associated with national and sectoral level planning generally include (1) the income or efficiency objective, (2) the employment and income distribution objective, (3) the foreign exchange earnings objective and (4) the regional growth objective. Other objectives such as price stability, properly pursued by monetary and fiscal policies, are not considered in project appraisal. While a project may contribute to price instability, there is no way this effect can be measured using present economic techniques. Rural development projects are specifically atuned to promoting national income growth and redistribution as well as correcting strong regional imbalances in per capita income and social services. The goal of increasing foreign exchange earnings may enter the choice of projects activities even though a rural development project may not have strong, positive balance of payments effects.

---

[1] This and later sections are substantially based on approaches presented in McGaughey and Thorbecke (75), as well as Papandreou and Zohar (89).

[2] Project evaluation criteria have been thoroughly considered by project practitioners. This topic is covered in a variety of books, manuals and articles on the subject including economic and financial writers, especially in FAO's Economic Analysis of Forestry Projects and books by Gittinger (32), Little-Mirrlees (70), UNIDO (117), and Roemer and Stern (96). The three criterion are: (1) the discounted present value criterion; (2) the benefit-cost ratio (a version of the present value criterion); and (3) the internal rate of return. They permit a comparison of the net benefit flows of alternative projects (or designs) by incorporating the comparison of net benefits received at different points of time.

It is recommended here that a version of the present value criterion such as the benefit-cost ratio be used in ranking rural or agricultural development projects. This seems most appropriate for the design of rural development projects, especially if there is a large number of alternatives that have to be ranked, a severe budget constraint, project components are from greatly varying economic subsectors, and if many of the component parts are likely to be fairly divisible.

In the following sections each of the distinct economic and social objectives will be considered and individual indicators will be constructed which measure the effects of individual projects. The single objective criterion for each will be developed on the assumption that practitioners may confront a wide variety of circumstances in which data are severely limited at the early stage of project design. For later refined feasibility analysis it can be expected that the economic and social data will be more abundant and, consequently, more complex investment criteria may be applied to measure simultaneously the impacts of the projects on the several social and economic objectives.

4.2 The Income (Efficiency) Objective

This objective is to maximize the income or consumption contribution of the project to the economy. The objective may be measured on the basis of present value criterion using shadow prices. The total benefits of a project are the real income flows received by the beneficiaries. Thus, a rural development project has directly productive activities which result in increases in net farm incomes, increases in incomes to those providing market or agricultural credit services and indirect increases in incomes from rural education, health or other social services. It is exceedingly difficult to estimate the long-term income effects of a rural education programme. If these net income flows can be estimated, then they belong as a part of the total benefit flows of the project.

First, determine whether the project is feasible given an interest rate for computing the present discounted value of the net income benefits of the project. Secondly, follow the standard procedures outlined in the project evaluation manuals. If the project has a net positive present value or if its internal rate of return exceeds the opportunity costs of funds reflecting the capital constraint, then the project can be considered feasible in terms of the income objective. It is recommended here that some value criteria should be applied on a trial and error basis. For the purpose of ranking project alternatives it is recommended that the net benefit-cost ratio be applied to rural development projects. Depending upon the circumstances of available data or the detail of the required project design, four alternative forms of the benefit cost-ratio can be used to rank projects in terms of the national income objective. These are:[1]

$$\frac{X}{I} = \frac{PV_i\,(X_t)}{PV_i\,(I_t)} \qquad (1)$$

$$\frac{V}{I} = \frac{PV_i\,(X_t - O_t)}{PV_i\,(I_t)} \qquad (2)$$

$$\frac{B}{C} = \frac{PV_i\,(X_t - C_t)}{PV_i\,(I_t)} \qquad (3)$$

$$SMP = \frac{PV_i\,(V_t - D)}{PV_i\,(I_t)} + s\,\frac{PV_i\,(F_t)}{PV_i\,(I_t)} \quad \text{and} \qquad (4a)$$

$$PV_1\,(F_t) = PV_i\,(B_1 + B_2 + B_3) \qquad (4b)$$

---

[1] $PV_i(X)$ means the sum of the present discounted values of gross production X, each year t, at interest rate i.

Terms are defined:

X_t = gross production, year t;

I_t = fixed investments, year t;

C_t = operating costs, year t;

V_t = domestic value added, year t;

D_t = domestic operating costs, year t;

O_t = purchased inputs other than labour and land, year t;

F_t = net foreign exchange effects, year t; and

s = the ratio of the difference between the shadow and official exchange rate to the official exchange rate.

The ratio shown in equation (1) is the output-capital ratio - the value of the gross production in each year X_t and the fixed investment outlays in each year t, I_t, discounted at rate i. This ratio is appropriate only when little project information is available. It may even occur that project practitioners will have to employ an undiscounted version of this ratio in the preliminary selection of project designs. While one can agree that this ratio is not justified under most circumstances because it excludes project costs, it may occur that due to lack of farm budgets and other basic cost data, analysts would be forced to rely upon this simple output-capital ratio.

A second indicator of the income objective is the ratio of project value added to investment expenditures. This relationship is shown in equation (2) as the ratio of the present discounted value of gross production less purchases to the present value of the project investments. This ratio measures the direct contribution per dollar of investment expenditure in the project. Shadow prices should be included if they are available.

It can be expected that the shadow prices will not easily be obtained. If the national or sector planning authorities have supplied shadow prices to the project analysts then, of course, they can proceed to undertake a more complete cost-benefit analysis as shown in equation (3), if all project cost and benefit components are known. This is the well known ratio of the present value of net benefits of a project to the present value of the capital expenditures including adjustments for the shadow prices of foreign exchange and labour, as well as the social opportunity cost of capital. The ratio may be used to obtain a ranking of investment projects according to the income objective. 1/

1/ The preferred form of the benefit-cost ratio is the "net" ratio; in the numerator are the net variable benefits accruing to beneficiaries during the project's operation and the denominator contains all fixed cost entering the budget constraint including all initial investments.

A form that may be used to analyse the income or efficiency contribution of a project is the Social Marginal Productivity criterion (SMP) introduced nearly twenty years ago by Chenery (20). The criterion, while similar to the benefit-cost ratio, separates the balance of payments effect of the project from the domestic value added-investment ratio. The SMP is defined as the ratio of the domestic value added to the capital expenditure plus the balance of payments effect per dollar of investment expenditure adjusted by an appropriate shadow exchange rate. The balance of payments effect (F) is made up of three elements which include, B_1, the indirect and direct effects of the investment expenditures, B_2, the direct effects of the project's operation and B_3, the indirect operating impact of the project.

4.3 Employment and Income Distribution Objectives

The objectives of increased employment and an improved income distribution may be treated separately or as complementary goals. While it is possible to conceive of circumstances in which the two objectives are at odds, - e.g. that greater employment may produce a more unequal income distribution or an improved income distribution may result from less employment - it is assumed here that employment and the income distribution are complementary goals. Thus, the creation of greater employment in rural areas would lead to an improved income distribution and in circumstances in which the precise regional income distribution is not known, it is possible to use an employment performance indicator as a substitute for the income distribution indicator. Since rural development projects are designed to benefit the low income sector then greater employment and income in this sector will, paripassu, produce an improved regional income distribution. The employment or income distribution indicator should be a ratio of employment or income distribution improvements to the initial capital expenditures or other relevant financial constraints.

Initiating the discussion with the employment objective (a surrogate of the income distribution objective), several considerations must be kept in mind in applying employment creation indicators for the purposes of ranking rural development projects. First, consideration should be given to separating employment into that produced during the initial stage in which the infrastructure and other facilities are being installed; this employment can be thought of as temporary compared to the employment generated during the project's operating phase. This does not mean that public works employment is undesirable, rather that as the public works are completed within a region, the volume of this type of employment declines rapidly. It is possible that unskilled, semi-skilled and skilled workers employed on the construction may be able to obtain other employment once the irrigation projects and marketing facilities, research facilities, and educational programmes are underway. But it also may occur that construction workers will not be able to obtain permanent employment, except for a few agricultural labourers who participated during the construction of the infrastructure.

Table 1. Selected Project Indicators [1]

Project	X/I	V/I	B/C	SMP	L/I	L^o/I	L^o/I	F/I	s(B/I)
A	3.7	2.4	4.7	2.8	0.46	0.24	0.23	1.6	0.38
B	3.2	1.2	2.7	1.4	0.57	0.33	0.24	1.2	0.23
C	2.8	0.8	2.1	1.1	0.52	0.36	0.16	1.3	0.26
D	2.5	0.4	1.4	0.3	0.47	0.34	0.13	0.4	0.15

Source: McGaughey and Thorbecke (75), Tables 2 and 3.

[1] See text for definitions. Irrigation projects names are: A = Antaura; B = Chococo; C = Olmos; D = Moquegua.

A second consideration concerns the distinction that must be drawn between the added employment of various categories of workers, whether skilled, semi-skilled or unskilled. Since employment is a substitute for the income redistribution indicator, the employment ratios should be limited to unskilled or semi-skilled workers or others who have been beneficiaries of education or training programmes as part of the integrated rural or agricultural development programme. Since the incidence of unemployment or underemployment is smaller for the more highly paid skilled workers, additional employment of this type is not as urgent as for the unskilled or semi-skilled labourers. Some workers will be making the transition from semi-skilled to skilled work categories such as apprentices, and they should be identified, perhaps, by their wage rates which will be less than the rate for the skilled workers.

Therefore, it is necessary to estimate the number of man/days employed during the operating and construction phases of a project and these man/days must be added or valued at a wage rate for the purpose of ranking projects. This rate may not necessarily be the shadow wage rate referred to in traditional social cost-benefit analysis. For the purposes of ranking projects it is essential that the wage rates for different categories of unskilled and semi-skilled workers be comparative, otherwise, differences in employment generated will derive from spurious differences among non-comparable wage categories in different regions.

Five indicators may be considered to establish the employment impact of a project. These are shown in the following equations (see Table 1):

$$\left(\frac{L}{I}\right) = \frac{PV(L_t^o) + PV_i(L_t^c)}{PV\ (I_t)} \quad (5)$$

$$\frac{L^c}{I} = \frac{PV_i\ (L^c)}{PV_i\ (I_t)} \quad (7)$$

$$\left(\frac{L^o}{I}\right) = \frac{PV_i\ (L^o)}{PV_i\ (I_t)} \quad (6)$$

$$\frac{L}{x} = \frac{PV_i\ (L_t^o + L_t^c)}{PV_i\ (X_t)} \quad (8)$$

Where:

L^o = employment of skilled and semi-skilled (poor) workers during the operating phase.

L^c = employment during the construction phase.

I_t = the total fixed investment outlay.

X_t = the gross output of the project.

Another appropriate measure substituting for output might be project value added of other indicators such as net project income. The employment generated by the project should be discounted at an interest rate which reflects the urgency or the impatience (time preference) of public authorities in creating employment at different points in time. It may be that the interest rate applied to the employment ratio will be higher (lower) than the interest rate applied to the discounting of the consumption flows of the project depending upon the rate of time preference for these benefits, although Lal (63) argues that they should be equal.

It is necessary to separate the employment created during the operating phase from the employment generated during the construction phase since they will involve different technological choices affecting the labour intensity of the project. It is also clear that the direct labour-capital ratio does not take into account the indirect employment generated in a region from a rural development project; in order to estimate the indirect employment impacts of a project, input-output tables and employment multipliers may be used to trace first, second, third, and fourth round employment effects within a region. If the input-output table is available the labour-capital ratio of equation (5) could be expanded to include a third element in the numerator - the present value of the net indirect employment created. The labour-capital ratio is simply interpreted: a project with a larger labour-capital ratio is more desirable to public authorities than one with a smaller ratio. Accordingly, when comparing projects that produce similar outputs, with similar capital expenditures, then a labour-output ratio as shown in equation (8) would be appropriate for a comparison of project investments.

In a study of irrigation projects McGaughey and Thorbecke (75) found that the bulk of employment created by the projects occured during the operating phase and the employment created during the construction phase was around one-third of the employment obtained during the operating phase. It is not difficult to imagine that, through experience, project planners will be able to establish a rather definite idea as to what the employment-capital ratios are likely to be for different types of projects using alternative technologies, making it possible during the design phase, to eliminate projects which are grossly inferior in their generation of employment opportunities for unskilled, poor labourers.

$$PRI = \frac{(N^o + N^c)}{(rI + C)} \qquad (9)$$

The poverty redressal index (PRI) of equation (9), proposed by Lal (63), is an annual employment/project cost ratio showing the number of poor people employed by a project during the construction (N^c) and operating (N^o) phases per dollar of project costs, consisting of the equivalent annual investment and depreciation costs (rI) plus the equal annual variable operating costs (C). The initial investment, I, is converted into an annaul equivalent by the factor r, which reflects the social rate of return to investments and the annual capital maintenance and replacement costs; the initial employment contribution is also transformed into an annual equivalent by the time preference rate i, for present and future consumption. The larger the ratio PRI the more people benefit from a project, hence a project which provides a larger number of employment positions for the same or larger wage and investment bill is preferred to a project which provides few employment slots at a smaller wage and investment bill.

If there are data on the size distribution of income by regional and national level, it may be possible to adopt more direct measures of the effect of investments on the distribution of income. The most direct approach is that used by Kalter and Stevens (61) who trace the distribution of project costs and benefits to each individual income class within the region and show the net redistribution benefits of each income class, as the difference between the direct benefit that is received from the project and the direct and indirect outlays that the different income classes are required to make in reimbursing the project costs, directly and through taxes. Other measures of income distribution can be used for the appropriate project rankings and comparisons. Szal and Robinson (112) have surveyed the available income distribution measures. Of those reviewed, the ones that seem most appropriate for the evaluation of rural development projects include the following: (1) the Gini coefficient; (2) the (so-called) Population in Relative Poverty index defined as the percentage of the population having less than one-half of the median income; (3) the Maximum Equilization Percentage defined as the share of aggregate income that needs to be redistributed to the lowest income groups in order for the distribution of income to be perfectly equal or, alternatively, the proportion of total income that must be transferred from those above either the mean income or the poverty income in order for those below the poverty level to have their income increased to a level equal to the poverty cut-off level; (4) inequality indexes which relate the mean income of those who have less than the national mean income and the mean income of those who have more than the national mean income; and (5) indicators of regional beneficiary income expressed as a share of the total regional income or the region's income accruing to the rural poor compared to a national average income. If there is detailed knowledge of the size distribution of family, regional and national income before the initiation of a rural development project, then it would be possible to indicate the extent to which any of these proposed indicators change consequant to the project investments.[1]

---

[1] It must be clearly admitted that detailed income distribution information is usually not available although this situation may be gradually improving in some countries.

Each of the individual income inequality measures will be considered in turn. First, the Gini coefficient falls between zero when the income distribution is perfectly equal and a value of one when the distribution is perfectly unequal. 1/ Therefore, an appropriate measure of the impact of the project on regional or national income distribution is the proportional reduction in the Gini coefficient per dollar of investment in the rural development project. Projects which effect a larger proportional reduction in the Gini coefficient would be considered more desirable than projects that make an incrementally smaller reduction in the Gini coefficient per dollar of investment outlay. Unfortunately, a weakness of this measure is that the Gini coefficient can be shown (see Szal and Robinson/ 112 \to be equal for several possible income distributions, allowing a shift in the distribution without changing the numerical value of the Gini coefficient itself. Likewise, Paglin (88) has shown the inadequacy of the traditional Lorenz curve in reflecting a reasonable concept of "equality".

Regarding the Population in Relative Poverty index (PRP) an appropriate project indicator could be the reduction in the percentage of the population which has less than one-half of the national median income per dollar of project outlay. For the Maximum Equalization Percentage, if the share of aggregate income needed to be redistributed to make the income distribution perfectly equal (or to provide the lowest income with an income equal to the average), is reduced by the project, then it can be argued that a net improvement in the income distribution has occurred. Consequently, the appropriate indicator would be the reduction in the share of aggregate income redistributed per dollar of expenditure on the rural development program or project.

Another alternative is to utilise the inequality indexes which relate the average income of individuals above the national mean and below the national mean income. Three measures, $\bar{Y}_p$ the average income of individuals below the national mean $(\bar{Y})$, and $\bar{Y}_s$ the average income of individuals above the national mean, can be related in the following fashion:

$$u = \frac{\bar{Y} - \bar{Y}_p}{\bar{Y}} \quad \text{and} \tag{10}$$

$$w = \frac{\bar{Y}_s - \bar{Y}_p}{\bar{Y}_s} \tag{11}$$

$$PID = \frac{w_o - w_t}{w_o} \Big/ (\bar{I}) \tag{12}$$

1/ See Szal and Robinson (112) for a definition of the Gini coefficient.

Equation (10) shows a relationship between the national mean and the mean income of people who are below the national mean; this index ranges from u = 0 when the national mean and the below-national mean incomes are equal, to a value of u = 1 when there is perfect inequality. Similarly, one can relate the average income of those above the national mean ($\bar{Y}_s$) to those below the national mean by index w which ranges from a value of w = 0 for perfect equality to w = 1 when there is perfect inequality. It is suggested that an approach might be to compare the indexes in a region before and after the development project and a ranking found by showing the percentage reduction in the values of the coefficients u and w per dollar of total outlay on the respective project or the per beneficiary dollar expenditure in the respective regions.

Table 2 shows the inequality indexes of equations 11 and 12 for two regions A and B in the year zero before the project begins and the year five after the first stage of the project terminates. Region A has the lowest per capita income of 100 compared to region B with a per capita income of 200 at the beginning of the period. Over the five-year period the average income in region A increased 20 percent while the average income of the "above-average" income groups (s) increased by nearly 7 percent and the income of the below-average groups (p) increased by 60 percent. In region B the income of all groups is higher than region A. The inequality indexes declined, correspondingly, by 11 and 10 percent in region A and 20.0 and 15 percent in region B indicating, thereby, that the increase in incomes and the reduction in inequality was more substantial in region B than in region A.

This illustrates a conflict in policy making that may occur in selecting rural development regional priorities in situations such as between A and B in the illustration. Region A is clearly worse off than region B, having a lower per capita income for all income groups in years before and after the project investments. However, the incomes in region B, especially of the lower income groups, rose much more than in region A. Therefore, if the resource costs of increasing the incomes in region A and B are the same, the index of performance, a ratio between the reduction in the inequality measures per dollar of expenditure, would have given preference to investment in region B.

Thus, policy makers may have to decide on a less proportional reduction in inequality by initially concentrating activities in the lowest income regions. This is a "merit want" of the kind referred to by Marglin (80) which derives from underlying political priorities that go beyond the simple application of economic investment criteria. A part of this conflict may be resolved by using a general indicator of project performance as shown in equation (12). This inequality index is defined as a ratio between the percentage reduction in inequality as measured by the inequality index before and after the project investments, to the per capita investment outlays ($\bar{I}$) during the project's life. This Project Income Distribution index (PID) is posited as a direct measure of project performance that corrects for regional population sizes and public expenditures that might be more cost-effective in one region compared to another. For example, assuming that region A (Table 2) has a population of 500 000 and region B a population of 250 000 and each project has a total cost (I) of US$50 million; the per capita outlays to raise the regional incomes from year 0 to year 5 are, respectively, US$100 in region A and US$200 in region B.

Table 2. Project Inequality Indices and Regional Ranking

Region or Project Area		Year 0	Year 5	Percent
A	$\bar{Y}$	100	120	20
	$\bar{Y}_s$	150	160	7
	$\bar{Y}_p$	25	40	60
	u	0.75	0.67	-11
	w	0.83	0.75	-10
	PID*	-	-	100
B	$\bar{Y}$	200	250	25
	$\bar{Y}_s$	385	390	1
	$\bar{Y}_p$	50	100	100
	u	0.75	0.60	-20
	w	0.87	0.74	-15
	PID*	-	-	75

Notation: See discussion in text of equation (10), (11) and (12)

* converted to an index with a maximum of 100.

The reduction in the inequality index w in the two regions provides a PID of 100 in region A and 75 in region B. This shows that a project located in region A which gives rise to increases in income within the region is preferred to region B where fewer higher income beneficiaries are found. The PID index is an example of the kind that might be constructed by national and regional planners to produce a project ranking on the basis of their impact upon income inequality within the beneficiary regions. It is evident that income distribution data often are not available for region or project areas, making it necessary to recur to direct measures of the cost-effectiveness of projects in increasing incomes of lower income groups, including the employment-investment ratios referred to in the previous section.

Finally, a statistic for regional income comparisons is the ratio of average regional income to average national income or regional income inequalities to national income inequalities. Table 3 shows two such simple indicators: namely, the ratio of the average regional income ($\bar{Y}_R$) to the average national income ($\bar{Y}$) or the mean income of the low income regional population ($\bar{Y}_p$) compared to $\bar{Y}$. In the example, region A has the lowest per capita income for either regional income indicator; region C has the highest average income while regions B and D are ranked distinctly by the two indicators. Region B has the lowest average regional income compared to D, while region D has the lowest average income of the poorest segment of the population. It is clear from this that, if planning offices are to make choices among regions in rural development projects, considerable energy will have to be given to the construction of regional income surveys which permit comparisons among the average, sub-average, and supra-average regional income groups. Undoubtedly, over the coming years, the role of regional economic planning will rise in importance.

Table 3. Regional Income Indicators

Region (R)	$\bar{Y}_R$	$\bar{Y}_p$	$\bar{Y}_R / \bar{Y}$	$\bar{Y}_p / \bar{Y}$
A	100	25	0.53	0.13
B	200	50	1.05	0.26
C	350	60	1.84	0.32
D	260	40	1.37	0.21
Total ($\bar{Y}$)	190	-	1.00	-

Notation: $\bar{Y}$ = average national income.

$\bar{Y}_R$ = average regional income.

$\bar{Y}_p$ = average income of the regional population having less than the regional average income.

Several indicators have been posited as vehicles to compare the national or regional income distributional consequences of alternative investment projects. All of these indicators can only be applied with difficulty because of the severe weakness of all data on income distribution and the lack of local and regional income and product accounts. Once the income distribution (whether national or regional) is known it is also essential that the income levels of the direct beneficiaries be obtained and this may require costly local socio-economic surveys. Initially, the poverty redressal index (PRI) (equation (9)) could be a simple and direct measure of the number of beneficiaries arising from a national or local investment programme. Another applicable indicator could be the proportional change in the Gini coefficient per unit of project investments. Likewise, the PID can be obtained if the income distribution is known. This latter has the convenience that the cost-effectiveness of reducing income inequalities is better taken into account.

It is clear that no self-evident income distribution indicator can be universally recommended since they will have to be adapted in each country to the local data conditions and limitations which affect project analysts. Countries with a good data base will be able to introduce detailed regional income measures such as those discussed by Kalter and Stevens (61). Indicators which use rather fixed or arbitrary income categories (e.g. one-half the median income) to measure the distributional changes will be of little practical use in project evaluations unless a national agreement on the cut-off line for these categories can be obtained.

4.4 Foreign Exchange Objectives 1/

An often espoused objective of economic development is to increase the supply of foreign exchange. It is natural that this objective would be introduced into the micro-economic project evaluation by measuring the foreign exchange effects of individual projects. Hence, a project is justified on the basis that it increases foreign exchange more than another project. Some project manuals suggest that foreign exchange is not a legitimate, separate goal of project evaluation because it is accounted for in the shadow pricing of factors of production in the social cost-benefit analysis. Therefore, it is argued, that if all primary inputs are shadow priced using border prices reflecting international scarcities, the comparative advantage of project production activities is reflected in the net present value criterion. This procedure is appropriate when planning authorities are ascertaining the economic feasibility of an individual project without comparing it with a large number of project alternatives. However, if project designers want to rank investment possibilities, it is appropriate to calculate an indicator of the "partial" effects of a project on the foreign exchange earnings or savings of the country. In addition, external lending agencies are particularly interested in estimating such effects since they primarily supply the foreign exchange costs of agricultural, rural development or other investment projects in which they participate.

Two approaches are herein suggested for computing the foreign exchange consequences of a project. The first involves taking a view that economic and financial project data are severely limited. This approach is to make a conservative assumption regarding the traded share of the project's output. This seems particularly appropriate for rural development projects which initially tend to produce output of domestically consumed food or staple products; it is unlikely that the project will produce for direct export or import substitution but rather will benefit local consumers by making net additions to their already deficient diets.

Two crude ratios may be used to determine the partial foreign exchange impact of a rural development project. The first is a ratio of the net direct foreign exchange earnings per unit of project investment defined as the ratio:

$$F/I = PV_i(F_t) \Big/ PV_i(I_t) \qquad (13)$$

1/ This objective is closely related to the national income objective - foreign exchange problems are really constraints on national income growth. Governments often treat the foreign exchange goal as a separate objective.

An alternative to this simple foreign exchange-capital ratio is the ratio of the direct and indirect foreign exchange earnings to the capital outlays, the second part of the social marginal productivity ratio (see Chenery (20)) discussed earlier:

$$s(B/I) = s\left[PV_i\ (B_1 + B_2 + B_3)\right] \Big/ PV_i\ (I_t) \qquad (14)$$

The only difference between the two ratios is that in the second the indirect effects are derived from the multiplier of the project outlays on imports and exports. In either case, the ratios are meant as only the simplest of all approaches and may not be employed to determine the social profitability of the projects in the same way as the social cost-benefit analysis.

A second approach to establishing the foreign exchange effects of a project is to calculate the domestic resource cost (DRC) per unit of foreign exchange earned or saved by the project. This ratio, discussed by Bruno (15), is a restatement of the net present value or the internal rate of return criterion which, instead of using the social profitability per unit of investment costs, uses the ratio of the domestic primary resource cost per unit of foreign exchange earnings and savings. Thus the DRC (or d_h) is defined as:

$$d_h = (V_j + P_k) \Big/ (X_s - M_k) \qquad (15)$$

where V_j = the total domestic value added per unit of output, at shadow prices, each year.

P_k = shadow priced, nontraded commodities per unit of output each year.

X_s = foreign exchange earnings per unit of output, each year.

M_k = import requirements per unit of output, each year.

As shown, the DRC criterion may be applied to a project on an annual basis assuming that the project cost and benefit flows are uniform throughout the project's life. An alternative would be to obtain the present value of the ratio of the domestic value added to the present value of foreign exchange earned or saved per unit of output, and convert the irregular annual flows into uniform annual flows by applying the capital recovery factor to the uneven annual flows.

The domestic resource cost d_n is compared to a shadow foreign exchange rate d_o such that if d_h is less than d_o, the project is feasible. This can be shown to be equivalent to the net present value criterion such that the $PV_i \geq 0$ when the $d_h \leq d_o$. Likewise, when $PV_i < 0$, then $d_h > d_o$, indicating that the domestic resource cost per unit of foreign exchange earned or saved exceeds the opportunity costs expressed in the shadow foreign exchange rate.

The domestic resource cost criterion is not an efficient ranking device, just as the present value criterion cannot be used for ranking projects (see Bruno (15)). Since the implementation of each set of projects to be compared may cause the relative prices in the economy to change, it is not always possible to rank all projects, excluded from an initial feasible group, by their domestic resource costs. However, if the projects are small and they do not seriously alter the supply and demand conditions in the economy, the DRC may be taken as a close approximation for ranking project alternatives, particularly, discriminating among projects which may be producing either exceedingly large volumes of foreign exchange per unit of domestic resource cost or projects which appear to be extremely poor suppliers of foreign exchange. Projects in the middle ground between the extremes may not be ranked with the same degree of confidence.

Another criterion which has been used extensively, as reported by Balassa (3, 4), is based on the "effects method" of project evaluation. This method computes the benefits of a project as the increase in the domestic value added "... taken to equal changes in domestic incomes (wages, profits, rent and government revenue) associated with the project's implementation ..." (Balassa (3)). These project benefits are shown to be equal to the increase in foreign exchange stemming from the project computed in terms of the domestic currency. The costs of the project are variously identified as being "... three possible alternatives: identifying cost with the domestic cost of investment in the project, with the value of imports embodied in the investment, or with the loss in budgetary revenue." (Balassa (3)). Balassa demonstrates that the effects method of project evaluation is inappropriate for the determination of the feasibility of a project in the sense that the social cost-benefit analysis determines feasibility. The method excludes the non-capital domestic resource cost of executing a project and, therefore, implicitly assigns a zero shadow price to labour and non-traded domestic commodities, including those embodied in domestic investments. It is shown that with the inclusion of the missing factors, the effects method can be transformed into the domestic resource cost or the internal rate of return criterion and the feasibility of a project is obtained according to the usual social cost-benefit criterion.

4.5 Regional Economic Growth

An important objective of rural development programmes is to improve the balance of economic growth and per capita incomes among regions, placing special emphasis on low income regions. Direct project effects have been referred to in previous sections, where the increase in net farm income, employment or foreign exchange is used to determine project feasibility and ranking, although not necessarily from the viewpoint of regional authorities. Therefore, it is necessary to consider the indirect or secondary impacts of project investment on cash incomes within the region. The regional secondary effects of a project will differ from secondary effects attributed to a project at the national level. The major difference lies in the extent of the secondary income expansion. At the national level the net benefits of the multiplier effect may be quite small since the cost of displaced output reduces the magnitude of gross secondary benefits and since resources being used in one project are drawn from alternative uses throughout the economy. At the regional level, the cost of resources displaced from other regions is not a part of the project's secondary costs. The gross regional project benefit is regional value added, i.e., gross project benefits less the purchase of inputs from outside the region. A multiplier effect at the regional level is fully operative for cash incomes at three levels: wages and profits spent within the region; the regionally spent portion of value added of input-supplying industries; and the value added of output processing firms. For the region one would have to estimate the portion of extra income spent and the total amount of new added expenditure generated by the project within the region itself.

For each year one would obtain for the region:

$$R = \frac{b}{1 - b} (Y + W + X), \qquad (16)$$

where b is the fraction of annual project cash income spent within the region, Y the net project cash income retained within the region (net of all loan repayments and transfers of income to other regions), W the value of annual cash wage payments, and Z the value of net income and wage payments of input-supplying and output-processing activities within the region. If 75 percent of the regional income is spent ($b = .75$), then the regional multi plier effect is $3.0\ (Y + W + Z)$. It would be the maximum secondary expansion assuming no additional induced secondary investments arising from the project. In each case the value of b should be carefully justified and a high degree of regional interdependence established before secondary income expansions are included in computing project benefits. Furthermore, it is cash income that is relevant in each case so that the marketable proportion of gross and net benefits will have to be estimated for each rural development project.

A second approach in determining the local effects of a development project is to determine the direct consequences on the regional income distribution. Following Kalter and Stevens (61), it might be possible to estimate the income consequences on separate regions in a rural development programme. On the one hand, the regional benefits are defined. They should be a measure of regional net incomes stemming from the project investments both inside and outside of the region, identifying those net benefits which accrue to individuals within the specified region. A project is said to change the income distribution if "... the distribution of project net benefits is non-proportional to the income distribution projected to occur without the project in question." (Kalter and Stevens (61); p.207). The income distribution effects will be the difference between the net benefits accruing to the regions and the payments made by the regions in order to receive the project benefits. These latter payments include the tax payments made to finance the project's construction including income, sales, import or export taxes. These taxes payments are assumed to be made in the same proportion in which income, sales and other taxes were paid by the region in the past. To the extent that the region pays more or less than its usual burden of taxes, a reimbursement adjustment must be included which is the difference between what is actually reimbursed by the region and the "proportional reimbursement" defined as the reimbursement that would have been made using the traditional tax payment share of the region.

To the extent that debt repayments incurred in order to undertake the project can be separated from the regional net benefits, then it might be advisable to include them as a separate item in the calculation of the regional net distribution benefits.

The final component of the regional net benefits is the operating maintenance and replacement costs which the region must make during the project's operating phase. To the extent that these operating, maintenance and replacement costs are paid by other regions then, of course, the net regional benefits are greater. This also holds for the other costs of the project. The component parts of the net regional benefits to region 1 are shown as follows:

$$B^1 = PV(B_1) - \left[Z_1 + \{PV(R_1) - K_1\ PV\ (\sum^{r} R_r)\} + PV(O_1) \right] \qquad (17)$$

where $PV(B_1)$ = the present value of the net benefits to region 1.

Z_1 = the initial payments made through taxes by project beneficiaries to finance the project construction.

$PV(R_1)$ = the present value of the actual repayments (excepting taxes) made by region 1.

$PV(\Sigma R_r)$ = the present value of total repayments made by all regions for the development programme.

K_1 = regions 1's proportion of the total initial tax paid by all regions.

$PV(O_1)$ = the present value of the operating, maintenance and replacement costs incurred by region 1.

It is obvious that this technique of estimating the regional distributional impacts of rural development programmes requires data on the distribution of public sector tax revenues among regions and beneficiary groups. However, if the project cost and benefit flows are known and if the beneficiary group is well defined, it is possible that the volume of taxes paid by the beneficiaries will be small and the principal components of the project repayments will be the direct reimbursements of the initial construction cost and the operating costs of the project, i.e. the value of K_1 in the previous equation will be very small.

The spending gererated by an investment may also lead to additions to national and regional employment. The direct employment generation of a project has been previously discussed. Initially, the unskilled labour required per unit of output of input-supplying and output-processing industries may be obtained on the basis of standard labour requirements for these industries. But it should be emphasized that to the extent that employment in other activities within the region is displaced, it must be treated as a negative contribution within the overall, regional employment estimates.

5. SIMPLE MULTIPLE OBJECTIVE EVALUATION PROCEDURES

5.1 Introduction

In an earlier section (3.3) the difficulty of introducing multiple objective analysis into project evaluation is raised as a central issue in designing measurable investment criteria for agricultural project evaluation. Four major objectives are considered including the efficiency goal, the employment-income distribution objective, the foreign exchange objective and the goal of more equitable regional (urban-rural) development.

In the preceeding section, single objective evaluation criteria are suggested for circumstances in which decision makers are concerned only with the partial effects of a project. It is obvious that attention must be given to combining these partial effects in a way that the goals of policy makers are introduced. The purpose of this section is to explore various ways in which these partial effects might be combined or weighted so that projects are ranked for the various combined criteria. From the onset, it is important to clarify for those who expect a theoretically pure and satisfying way of identifying and combining policy makers preferences that they likely will be dissatisfied. Rather the proposed procedures are thought of as supplementary information to an informal decision-making process that presently goes on with little or no formal project analysis. A small improvement in existing planning procedures is proposed as a first step toward more complicated methods that might be employed at some future time when detailed project data are known.

A further issue concerns whether the objectives are expressed in general or specific terms. Thus, a specific goal might be to "increase the production of a product x by y amounts", while a general objective might be to "increase the production of x". Experience in the use of scoring models for planning shows that it is easier to obtain initial acceptance of the goals when they are stated in broad terms, while the introduction of specific statements of precise targets is made at a later stage after the scoring system has evolved.

Once policy makers have accepted a list of objectives $(O_1,\ldots O_m)$, the succeeding step is to establish the weights on each objective.

Two approaches are recommended for the preliminary introduction of multiple objective considerations into project evaluation. Both of the approaches fit situations in which little is known about the miltiple partial effects of the projects. While data may be available for a simple benefit-cost analysis for a few projects, a number of project alternatives may not have been studied to the detail that allows for a full-fledged social cost-benefit analysis.

The first approach is one in which expert opinions are elicited through a systematic interview technique called the Delphi method. The results of the interview technique are used to produce scores on the effects of each project on each objective. A scoring model is combined with the Delphi technique to obtain combined project scores and project rankings.

The second approach involves the combination of the partial effects quantified in previous sections with alternative weighting factors which approximate policy makers preferences for the separate economic goals. This will allow for the ranking of projects using the combined project performance measures.

While it is important to introduce a wide range of multiple objectives it is also important that the efficiency objective not be made subordinate to all other objectives in the group of objectives under consideration. Likewise, the efficiency objective may be introduced via the goal of maximizing the foreign exchange contribution of a project. This is done by the inclusion of foreign exchange shadow prices. But using the simple scoring model and the Delphi method the objectives cannot be promoted by the proper (shadow) pricing of factors or production but must be assigned a separate performance score.

5.2 The Delphi Method

Stemming from the lack of historical experience and economic and social data on the execution of rural development programmes, subjective judgements are needed in order to undertake even a simple financial cash-flow analysis. Furthermore, a variety of heroic assumptions principally based upon expert or observer evaluation may have to be made to complete a detailed social cost-benefit analysis. Therefore, recognizing that reliance will have to be given to these subjective judgements, it is far better that they be obtained in a systematic and coherent fashion, rather than in a disorganized and spontaneous way. Personal judgements of technicians, experts and policy makers may be requested on a variety of topics including, for example, the objectives of economic and social policy, the choice of beneficiary groups, the selection of regional priorities, the design of individual projects and sub-projects, and the values of project coefficients such as yields per hectare, health visits per nurse, and investment costs per kilometre.

The Delphi method, which has gained respect in recent years, is a method by which the expert opinions of individuals, panels or committees are systematically combined to produce information on the feasibility, desirability and other features of various policy options. Linstone and Turoff (68) have thoroughly surveyed the application of this method to government, industry, social, economic and political questions. The Delphi technique is a systematc interviewing technique to obtain an expression of goals, opinions, and judgements of individual participants. The results of the judgements are weighted and combined in some appropriate fashion to obtain a performance value for the various project or policy options under review. The final result may be to rank projects or determine their feasibility or combine sub-projects or activities.

The method, which has several variants, is applied in the following specific stages: determination of the principal issues to be considered, an analysis of the options open to achieve these objectives or to resolve the issues, setting the initial position of each expert, exploring the areas of disagreement among various experts, evaluating the causes for disagreement and, finally, determining the options open to resolve the issues. In the policy Delphi (Turoff (116)) participants are asked to numerically rate the options or instruments according to the criteria of desirability, feasibility, importance and confidence. A rating scale is set for each of these alternative criteria so that the participants are able to rank the policies or projects according to whether they are, for example, very desirable, desirable, somewhat undesirable or very undesirable and as definitely feasible, possibly feasible, possibly infeasible, definitely infeasible. For example, participants may be asked to give their opinion as to whether a project will benefit the lowest income groups within the target regions and the alternatives may include (1) very probable, for a (subjective) probability greater than 0.75; probable for a probability greater than 0.5 but less than 0.75; improbable for a value of less than 0.5 and greater than 0.25 and very improbable for a probability of less than or equal to 0.25.

In the preparation and design of rural development projects the Delphi interview method may be used for two purposes: The identification of the principal social and economic objectives of planners and policy makers and estimating the probable consequences of a project investment upon the individual objectives that have been established. Furthermore, within the literature on the Delphi method some attention has been given to the question of determining the weights on the distinctive objectives in order to combine them with the probable outcomes of the projects for each of the relative objectives. 1/

1/ See the report by Gum, Roefs and Kimbell (35) in which an effort is made to quantify preference weights on societal goals and sub-goals for water resource investments.

There are no special rules for the application of the Delphi method. However, there are several common difficulties referred to by Linstone (68). These include (1) a tendency by participants to heavily discount the future, making judgements oriented to immediate goals; (2) a tendency of the participants to attempt to predict consequences far into the future, although they may have serious uncertainties regarding the prediction; (3) a tendency to try to simplify statements of objectives or results in order to keep conflict to minimum; (4) problems in executing the Delphi technique including the selection of participants, designing questionnaires, construction of feed-back and review mechanisms; (5) the so-called optimism-pessimism bias which is referred to as "... a bias toward over-pessimism in long-range forecasts and overoptimism in short-range forecasts." Linstone (68, p.84); (6) a tendency to oversell the value of Delphi techniques; and (7) a possibility that participants will systematically deceive or design their answers to increase their own immediate gain from the exercise.

The Delphi interview techniques is appropriate for rural development projects in which project beneficiaries participate in designing and structuring the programme. Various interest groups such as small farmers, landless workers, urban dwellers, regional planners and politicians, and government bureaucrats can be asked to provide assessments of whether alternative sub-projects or programmes would be included in rural development programme. The respondents could be asked to rank objectives of rural development programmes and within these objectives a variety of subgoals including alternative levels of income, employment and social services for the region. The alternatives may be ranked using a scale of zero to 100 points, asking each participant to allocate the 100 points among the several objectives, so that the sum is always equal to one hundred. An alternative approach is to ask the participants to rank the objectives according to a fixed scale of one through five or one through six and then averaging these results across all participants or the interest groups. The Delphi technique is closely tied to the construction of scoring models which is the quantitative representation of the results of the Delphi interviews. The design and construction of scoring models is now discussed.

5.3 Scoring Models

In this section a project evaluation system is elaborated for the assignments of scores to each project by informed technicians. Scoring models have been discussed especially in conjunction with the evaluation of industrial research and development programmes.[1] A substantial gain in clarity and in organization is achieved by the scoring technique because of the explicit enumeration of preferences for alternative planning goals. Likewise, project evaluators are encouraged to explicitly express their judgements on the potential contribution of each project to each planning goal.

Scoring individual projects involves the following steps: (1) the selection and enumeration of the social and economic objectives of the investment programme; (2) the determination of a weighting scheme that reflects policy makers assessments of the relative importance of each of the objectives; (3) the choice of a numerical scale for measuring the value of the contribution of a project to each objective in (1) above; and (4) the determination of a ranking mechanism (function) that combines the policy makers preferences and the project scores in obtaining a total score for each project. As a final step (5), all possible projects under consideration are ranked according to their individual scores.

---

[1] See Moore and Baker (83) and Dean and Nishry (25).

A project scoring model has the following properties and elements as shown in the accompanying Table 4.[1] Decision makers identify m objectives $O_1, O_2, \ldots, O_m$. In some (to be specified) fashion, m weights $W_1, W_2, \ldots, W_m$ are obtained which express the relative importance of each of the objectives during the planning phase. Thus, for example, objectives might be to maximize income, to obtain an improved distribution of income or to increase foreign exchange earnings. In the system (i) preferences are expressed, (ii) projects are scored and (iii) projects are ranked.

The initial problem is to outline a list of objectives for the economic sector that is under study. It is unnecessary that all decision makers and technicians agree on the exact number, form or specification of the objectives. Some agreement might be made, to limit the number of objectives to a manageable total that can later be weighted by the policy-makers. The objectives should be independent, in the sense that each can be assigned a separate weight and is conceived as a separate preference by policy-makers.

Table 4. Elements of a Scoring Model

Objectives Weights	O_1 W_1	O_2 .... O_m W_2 .... W_m	Total Score (S_i)
Project			
1	P_{11}	P_{12} .... P_{1m}	$S_1 = \sum_j^m W_j P_{1j}$
2	P_{21}	P_{22} .... P_{2m}	$S_2 = \sum_j^m W_j P_{2j}$
⋮	⋮	⋮ ⋮	
m	P_{n1}	P_{n2} .... P_{nm}	$S_n = \sum_j^m W_j P_{nj}$

[1] See especially Dean and Nishry (25).

The choice of weights is the most crucial step in constructing the system since, depending upon the relative importance attached to each goal, the structure and composition of the investment programme will be greatly affected. Furthermore, the underlying structure and composition of the weights is a sensitive political decision, reflecting not only the importance of each goal but the ultimate political structure of the decision making process of the society. Thus, to cite some examples, a simple democratic process might require the selection of the weights, giving equal importance to the preferences expressed by each member of the group. A purely representative political process might attach equal importance to the views expressed by each representative. A purely centralized system would, perhaps, set the weights according to the views of a small group of economic and social planners. Evidently, there is a continum of possible ways to select the individual participants, and, hence, the choice will be made by the appropriate planning authorities and by a mechanism chosen by each society.

Assuming the objectives have been identified, two procedures are now elaborated. The first is based on the assumption that all policy makers interviewed are of equal importance in expressing their views and that the objectives are ranked by a pre-set scale based on the number of objectives under consideration. Once the interview group is determined the preference weights are derived.

The first weighting scheme might be called the "mean preference" alternative. Each government official examines a list of objectives and ranks the objectives one through the total number of objectives. Suppose there are six objectives and three officials. In terms of the scale, the objectives are ranked as in the first three columns of Table 5.

Table 5. Preference Weights Estimates

Objective	Individual Rankings			Mean Ranking	W_j
	1	2	3		
1	6	2	2	3.3	31.7
2	5	4	1	3.3	31.7
3	2	5	3	3.3	31.7
4	3	6	4	4.3	41.4
5	4	3	5	4.0	38.5
6	1	1	6	2.6	25.0
				Total	200.0

Individual 1 ranks objective 1 in first position (six points); objective 2, second (five points); objective 3, fifth (two points); etc. The highest ranked objective receives the largest value. Then, as shown, the values are averaged for the three individuals. Each individual is assigned equal weight in the scheme, and the results can be described as a "compromise" weighting. The final weights can be adjusted to provide any particular total sum of the weights W_j. Following the notation of Table 4, the score (S_i) of any project i is equal to

$$S_i = \sum_j^m W_j P_{ij} = 31.7 P_{i1} + 31.7 P_{i2} + 31.7 P_{i3} + 41.4 P_{i4} + 38.5 P_{i5} + 25.0 P_{i6} \quad (18)$$

Assuming that the projects are scored on a scale of one through five, a "perfect" project (a project receiving a score of five for each objective P_{ij}) would receive a total score of 1 000.

Two variations on this system of estimating weights are now considered in order to highlight an obvious weakness of the above method. In Table 5 the decision makers are restricted to a strict ordering according to a pre-established scale. Not only are ties among two or more objectives prohibited, but the relative "distance" among the objectives is fixed. Hence, individual 1 of Table 5 cannot pick objectives of equal value, even though the average ranking yields equality among the first three goals.

To extend the range of choices for policy makers, ties are admitted. One may proceed as shown in the following example:

Table 6. Preference Weights

Objectives	Individual Ranking			Mean Ranking
	1	2	3	
1	6	6	2	4.7
2	6	5	4	5.0
3	5	4	5	4.7
4	4	3	5	4.0
5	3	2	3	2.7
6	3	2	3	2.7

whereby, individual 1 expresses equal preferences for objectives 1 and 2 and for objectives 5 and 6.

An alternative way of computing the weights could be as follows:

Table 7. Mean Preference Weights

Objectives	Individual Ranking						Mean Preference
	1	2	3	1	2	3	
1	6	6	1	5.5	6.0	1.0	4.2
2	6	5	3	5.5	5.0	3.0	4.5
3	5	4	2	3.5	4.0	4.5	4.0
4	5	3	1	3.5	2.5	6.0	4.0
5	4	3	2	2.0	2.5	4.5	3.0
6	3	2	4	1.0	1.0	2.0	1.3

In Table 7, if two objectives fall into the first two slots, but are of equal value (6 and 6), an average of the corresponding position numbers (6 and 5) is used, i.e. 5.5 for individual one. Likewise, objective 5, for individual 1, is in position 2 of the rankings and, therefore, receives a weight of 2. The main result of this method is to permit a wider spread among the objectives.

The final step is to determine the contribution of each project. This is based upon a scale of score values set by the project evaluation team. Assume that a scale of one to five is chosen—a project receiving a score of five for objective one makes the maximum contribution and a score of one is the minimum contribution.

Consider the separate example of Table 8 below:

Table 8. Project Scores and Rankings

Project \ Objectives W_j	O_1 18	O_2 74	O_3 54	O_4 54	Total = 200
1	5	4	2	4	710
2	1	1	5	4	670
3	3	3	1	4	546

There are four objectives with weights, respectively, of 18, 75, 54 and 54. There are three projects each with a score for each objective. Project one receives scores of 5, 4, 2 and 4, respectively, for each objective and a total score of 710 is obtained,

$$S_1 = 18(5) + 74(4) + 54(2) + 54(4) = 710 \qquad (19)$$

There are two serious inherent problems in the application of scoring techniques. The first is in the choice of the scoring scale and the second is in the choice of the scoring and evaluation group. The scale can be treated as a range of project effects that can be correlated with quantitative values. For example, a project output-capital ratio can be transformed into a numerical scale. Employing a point's scale of five to one, a project output-investment ratio of (say) .35 - .30 might receive five points, of .30 - .25, four points, etc. Considering the difficulty in determining these values the scale should be limited to about five intervals.

The choice of a scoring and evaluation (Delphi) group is crucial in the application of the technique to rural development projects. Several criteria should be considered. First, the members should be reasonably familiar with the projects, activities or programmes to be ranked. At least one member of the committee should be chosen from the entity directly responsible for making the feasibility studies. To maintain a degree of objectivity, people should be included from outside of the group expected to benefit from the project investments. The number of committee members should be no more than five, otherwise it is difficult to obtain agreement of the group on the score of each project or activity.

5.4 Combining Single Goal Indicators

5.4.1 Introduction

The present section analyses the alternative methodologies for combining multiple objectives through simple measures of project performance to ascertain project feasibility ranking. The weighted combination of individual objective measures will be effectuated in a variety of ways including: (1) a combination of the benefit-cost ratio with the scoring technique; (2) combining equity and distributional considerations in the benefit-cost ratio; and (3) a numerical ranking scheme which allows for the simple weighting of project performance for a number of single objective measures. The criteria at this stage do not require the use of a complex programming technique such as goal programming or multi-level programming (33). A rather simple procedure is required to provide more information to policy makers who must make choices among different projects. Clearly, these are not optimization techniques in the sense of consistent mathematical programme models, but they are techniques which can be applied and which will supply more information to policy makers than they would have received if they wait (often without result) for the economy-wide application of more sophisticated programming models.

5.4.2 Benefit-Cost Analysis and Non-quantificable Effects

A first approach to combining several economic objectives is to use the benefit-cost or cost-effectiveness ratio and combine it, in some fashion, with the scoring model discussed in previous sections. This approach, suggested by Acar (1), is an appropriate technique for this purpose, using the benefit-cost ratio as a principal criterion and weighting the ratio by the additional effects that a project may have on social objectives. Thus, assume first of all that a project has been evaluated using a standard economic cost-benefit ratio. This ratio could be used to obtain a ranking of the project alternatives, but these may change if the project has non-efficiency consequences that have not been accounted for in the benefit-cost analysis. The solution is a benefit-cost ratio adjusted by a factor I_s as follows:

$$(B/C)' = (B/C)\ (I_s) \tag{20}$$

where B/C = the economic benefit-cost ratio,

I_s = an "index of suitability" and

$$1 - a \leq I_s \leq 1 + a \tag{21}$$

The index of suitability, which modifies the economic benefit-cost calculations, is based upon the policy maker's judgement as to how much he would allow for shifts away from the efficiency criteria (by factor a) such that the suitability index can be no smaller than (1 - a) and no larger than (1 + a). An example demonstrates the value of this particular relationship. In Table 9 three projects are shown with cost-benefit ratios ranging from 2.0 to 1.5. Under this criterion it would be appropriate to choose project (3) with the highest priority, project (1) next, and then project (2) with the lowest priority. However, assume that on the basis of a scoring model, non-efficiency effects of the three projects have been scored and these scores range from 53 to 203. These scores are now transformed into the range (1 - a), (1 + a). If it is assumed that policy makers will not permit a greater variation than twenty percent away from the efficiency level, i.e. a = 0.20, then the scores ranging from 55 to 223 are transformed into the segment 0.80 to 1.20. This transformation (following the explanatory footnote of Table 9) gives a modified benefit-cost ratio for each of the three projects. This new ratio changes the rankings of the three projects because the benefit-cost ratio of the highest ranked project is reduced by twenty percent because it had the lowest score, and the lowest ranked project with benefit-cost ratio of 1.58 has its numerical value increased by twenty percent because it had the highest score in the non-efficency category. In the example, project (1) remains in its relative position after the modified benefit-cost calculations are made. While this procedure contains an element of arbitrariness because of the <u>ad hoc</u> selection of the coefficients the approach does provide a systematic means to combine the benefit-cost ratio with other social and economic objectives.

Table 9. Benefit-Cost Ratio and the Suitability Index

Project	B/C	Score	I_s (a = 0.20)	(B/C)'
1	1.65	170	1.07	1.77
2	1.50	223	1.20	1.80
3	2.00	55	0.80	1.60

$$I_s = 1 + u \frac{(2\,a)}{(v_M - v_m)}$$, where

u = plus or minus deviation of each score from the mid-range score v'

v' = the mid-range score, the average of the highest and the lowest score

a = the policy standard for range of the scores, as in equation (21)

v_M = the highest score

v_m = the lowest score

For example, it might be possible to obtain scores for the social development components of the rural development projects such as health, education and nutrition programmes; they would be combined using the suitability index of the proposed criterion. Thus, if a project has a high benefit-cost ratio because it yields large increases in the net incomes of benefiting farmers from its production and production support components and if the social development component of the project has a low degree of cost-effectiveness, this can be introduced via a score. This procedure would appear to be superior to one of merely neglecting the social development components of a rural development project.

Likewise, it would be possible to extend the scoring analysis to include factors such as the indirect impacts of the investment projects as well as the social consequences of individual project components. Thus, a project which is within a very poor region may have its benefit-cost ratio adjusted by an index of suitability which is inversely proportional to the level of regional per capita income. As recommended, this index can easily be constrained within limits desired by the national or regional planners so as to avoid overwhelming the efficiency effects of the project with the non-efficiency elements.

These procedures are related to a suggestion by Professor Harberger (39), who proposes that the standard efficiency benefit-cost ratio may be combined with information on the effects of a project on the poorest income groups. Assume for the sake of discussion that project's costs are separated into two categories: the total benefits and costs of a project (irrespective of who might receive them) and the benefits and costs that go to the poorest segment of the rural population. The former will be called the benefits, B, and the costs, C, and the latter will be designated as the benefits, B_p, to the poorest groups and the costs, C_p, to the poorest groups.

The distributional benefits of a project should not be allowed to make up more than a certain share of the total project net benefits perhaps no more than 20 percent. Any project with net distributional benefits that exceeded (say) 15 percent of its net efficiency benefits would have its distributional benefits constrained to that 15 percent differential so that the project would not have an unfair advantage in any comparison made with an "efficient" project which had no distributional benefits at all. Thus,

$$(B_p - C_p) / (B - C) \leq a \qquad (22)$$

5.4.3 Weighting Multiple Objectives

Short of undertaking a complete programming exercise of the type suggested by Candler and Boehlje (17) or Sfeir-Younis and Bromley (106), a compromise solution is to use the separate rankings obtained from the partial indicators of project performance and weight the rankings to obtain an aggregate project ordering under alternative (preference) weights for each of the economic and social objectives. For each of the rural development objectives, an indicator is chosen from the alternatives discussed in previous sections and is used to rank each of the projects according to each of the individual objectives.

There are two types of ranking that may be used for this purpose - the ordinal or the cardinal rankings. The former, obtained by using a pre-set scale, is the ranked position of each of the projects for each of the objectives. The income objective is represented by the benefit cost-ratio, the employment objective by the employment-investment ratio and the balance of payments objective by a foreign exchange-investment ratio or domestic resource cost criterion.

An example of such a ranking is shown in Table 10 where five projects and three objectives are represented and the two systems of ranking are compared. The ordinal ranking shows that the project with the highest value for a given objective receives a ranking of five, the next a ranking of four and the lowest receives a ranking of one. This is an ordinal ranking because the determination is whether a project's position is higher than the other and the ranked "distance" between the two projects is of no concern under this scheme.

Table 10. Ordinal (Y^P, E^P, F^P) and Cardinal ($\bar{Y}^P$, $\bar{E}^P$, $\bar{F}^P$) Project Rankings by Objective

Project	Benefit-cost ratio (BCR)	Employment capital ratio (ECR)	Foreign exchange-capital ratio (FCR)	Rankings: Income		Rankings: Employment		Rankings: Foreign Exchange	
				Y^P	$\bar{Y}^P$	E^P	$\bar{E}^P$	F^P	$\bar{F}^P$
A	1.8	0.8	1.2	4	1.14	5	1.86	2	0.61
B	2.3	0.5	0.6	5	1.46	4	1.16	1	0.30
C	1.2	0.35	2.7	2	0.76	3	0.81	4	1.36
D	1.1	0.2	1.8	1	0.70	1	0.47	3	0.91
E	1.5	0.3	3.6	3	0.95	2	0.70	5	1.82
Average	1.56	0.43	1.98						

5 = highest ranking

1 = lowest ranking

An alternative procedure is to use a cardinal ranking which takes into account the "distance" between the projects for each of the investment criteria so for the income objective Y, project A is ranked the highest with a numerical value of 1.46, project B is ranked next with a value of 1.14, and so forth, until project D is ranked with a value of 0.7. These cardinal values are obtained by taking the ratio for each criterion of the individual value for the project to the average for each criterion such that, for example, for project B, 1.46 = (2.3)/(1.56) for the BCR criterion. Thus, the ordinal ranking for each of the objectives is obtained by the following relationships:

$$\overline{Y}^{p} = BCR^{p} \Big/ \frac{\sum^{p} BCR^{p}}{n} \qquad (23)$$

$$\overline{E}^{p} = ECR^{p} \Big/ \frac{\sum^{p} ECR^{p}}{n} \qquad (24)$$

$$\overline{F}^{p} = FCR^{p} \Big/ \frac{\sum FCR^{p}}{n} \qquad (25)$$

for each project p and n projects.

The purpose of the exercise is to combine the three objectives - income, employment and foreign exchange - into a single ranking such that the ordinal ranking is defined as:

$$R^{p} = y\, Y^{p} + e\, E^{p} + f\, F^{p} \qquad (26)$$

where p is any project and R is its respective ranking for the weights y, e and f such that y + e + f = 1. Likewise, the <u>cardinal</u> ranking is defined as:

$$\overline{R}^{p} = y\, \overline{Y}^{p} + e\, \overline{E}^{p} + f\, \overline{F}^{p} \qquad (27)$$

For each set of weights y, e and f there will be a distinct ranking R_j of the projects. For example, in Table 11 two alternative weights are shown for the two ordinal rankings (R_1, R_2) and the two cardinal rankings ($\overline{R}_1$, $\overline{R}_2$). Under the relationship R_2, the ranking across each of the individual objectives is

$$E = 3.75 = (0.25)\,(3) + (0.25)\,(2) + (0.50)\,(5), \text{ and} \qquad (28)$$

the value of project E = 3.75 is a weighted average of project scores under the ordinal ranking. Similarly, for the cardinal ranking $\overline{R}_2$ for project E with the same weights

$$E = 1.33 = (0.25)\,(0.95) + (0.25)\,(0.70) + (0.5)\,(1.82) \qquad (29)$$

Table 11. Ordinal (R_1, R_2) and Cardinal ($\bar{R}_1$, $\bar{R}_2$) Combined Rankings

R_1 (.33, .33, .33) y	R_2 (0.25, 0.25, 0.50)	$\bar{R}_1$ (0.33, 0.33, 0.33)	$\bar{R}_2$ (0.25, 0.25, 0.5)
A = 3.67	E = 3.75	A = 1.20	E = 1.33
B = 3.33	A = 3.25	E = 1.16	A = 1.07
E = 3.33	B = 2.75	C = 0.98	C = 1.07
C = 3.00	C = 2.25	B = 0.97	B = 0.81
D = 1.67	D = 2.00	D = 0.69	D = 0.76

y The project ranking factors (y, e, f) such that y + e + f = 1.

It is important to emphasize that the weights y, e and f are not social welfare preference weights and that the assignment of value 0.50 to the balance of payments objective does not mean that it is twice as important as either the income or the employment objective which each have weights of 0.25. Rather it means simply that the ranking of the project has twice the value as the ranking of the other two objectives, i.e. they are merely relative weights. It is an additional step to make these weights y, e and f consistent with preference weights as discussed in McGaughey and Thornbecke (76). It is enough to state at this point that it is possible to transform the weighting equation into a relationship that would represent the preference weights of the policy maker, but it is necessary to obtain equivalent preferences values for each objective from the policy maker.

It is useful to note that the proposed procedures will not always produce the same orderings for the ordinal and the cardinal rankings. For example, under comparisons R_1 and $\bar{R}_1$, for the ordinal ranking there is a small numerical gap between the highest and the next highest projects A and B, and for the cardinal ranking there is a substantial gap between projects A and B. Likewise, under the cardinal ranking project B does very poorly relative to the other projects and this poor performance is more reflected in the value that it receives for the cardinal ranking, B = 0.38. Projects B and C have different positions within the ranking since project B has a high ordinal ranking of 2.75, while it has a very low cardinal ranking of 0.81 due to the relative separation between the projects is more nearly reflected by the ranking $\bar{R}_2$.

The proposed ranking mechanism is designed to be of utility to project designers and planners when there is a serious underlying lack of information about the effects of rural development projects in different regions.

The objective of the recommended evaluation procedure is to contribute more information to the existing decision making process by identifying projects which are likely to perform well under many circumstances and those which might be expected to do poorly under a wide range of weights that policy makers might place on these objectives. No single decision rule or procedure - however cosistent and theoretically proper - can be the sole or even main criteria for making investment decisions.

It is unlikely that policy makers will be able to express the "weights" that they place on economic and social objectives. This means that the economic planner will have to follow a procedure that uses a wide range of possible weights giving preponderant importance to one and then another objective or by giving similar importance to two or more objectives. The number of such possibilities are large and those that seem the most relevant ones will have to be worked out by planners. For each set of possible weights a ranking of the project possibilities will be obtained, and for several rankings (and set of weights) planners may begin a process of choice and elimination. There often will be projects that are ranked high for many of the weighted rankings - these projects can be chosen with greater confidence as those contributing to a wide variety of economic and social objectives; likewise, projects that perform poorly (ranked low) for many alternative weights can be eliminated with a greater degree of confidence. Project D (Table 11) ranks poorly and project E does well under a number of presented circumstances. Of course, as new projects are discovered and designed these should be added and the procedure repeated. Likewise, projects which are discarded at one time should constantly be reviewed as a part of the procedure because they might rank higher at a later date. Therefore, the essence of this system is to continuously evaluate and re-evaluate a constantly changing batch of project alternatives.

The procedure can be applied to circumstances in which there are rural development projects containing both productive and social elements, as in the rural development programmes of Mexico and Colombia. Thus, if rural development projects can be appraised under criteria such as the cost-benefit ratio, employment-capital ratio, and domestic resource cost ratio, one can proceed to apply alternative ranking weights and point out the projects that seem to perform especially well.

A second alternative would be to use the procedure to weight the internal rural development project elements. Suppose that there are projects A and B, each with two major categories of investments - production development and social infrastructure. If the productive activities can be appraised using the benefit-cost ratio or the employment-capital ratio and the social activities can be evaluated using a form of the benefit-cost ratio or, lacking data, a cost-effectiveness ratio, then weights can be set on the importance of the productive and social inputs of the project. If planners assume that the productive support benefits are of primary importance and the social services are of secondary importance, they might attach a relative weight to these two elements of (say) .75 for production effects and .25 for social development effects. This would limit the extent to which any one of the separate services of the project, productive or social, would overwhelm the ranking of the project.

One can define a ranking:

$$R^{P} = x X + s S \tag{30a}$$

$$= x\ (BCR) + s\ (CER) \quad \text{and} \tag{30b}$$

$$x + s = 1$$

to obtain a weighted average of its production and social development effects wherein the production development effects (X) are represented by the benefit-cost ratio (BCR) and the social development impact (S) is represented by the cost-effectiveness ratio (CER). Following the example of Table 12, there are two projects A and B each with production and social development components X and S, measured by the benefit-cost ratio and the cost-effectiveness ratio. Project A has a benefit-cost ratio of 2.8 and project B a benefit-cost ratio of 1.9. Project A has a cost-effectiveness ratio (say) for an education subprogramme of thirteen students per US$100 expended in the programme and project B seventeen students trained per US$100 expended in the programme. The cardinal ranking is obtained as in early examples as the ratio between the specific indicator and its average such that the production effect of project A is 1.19 = (2.8) / (2.35). Finally, a weighted ranking is obtained under the assumption that the production effects are weighted by the factor .75 and the social development effects by the factor .25. While project A has a higher production effect than B and project B a higher social development effect than A, project A is preferred. In fact, not until the social development weight is given a value of s = .6 and the production component a value of x = .4 are the projects equal in their overall rankings. The implications of such calculations in these comparisons provide policy makers with considerable additional information on the impacts of their decisions as well as the importance that one or another component might play in an investment.

Table 12. Ranking of Projects with Production and Social Sub-projects

Project	Production Component X = BCR	Social Component S = CER	Cardinal Ranking $\bar{X}^{P}$	$\bar{S}^{P}$	Weighted Ranking 1/ $\bar{R}_1$ (0.75 , 0.25)
A	2.80	13	1.19	0.87	1.11
B	1.90	17	0.81	1.13	0.89
Average	2.35	15			

1/ x = 0.75; s = 0.25

6. SENSITIVITY ANALYSIS

6.1 Present Procedures

Following the conduct of the initial social cost-benefit analysis a common technique is to incorporate effects of risk or uncertainty by preparing a detailed sensitivity analysis. This sensitivity analysis treats the benefit-cost ratio, the present value criterion or the internal rate of return criterion under the assumption that there are systematic differences in the important parameters making up the original calculations.

There is considerable literature on alternative ways of incorporating risk and uncertainty into project evaluation, notably, Pouliquen (91) and Reutlinger (94). These approaches emphasize estimates based upon either historical experience or technical perception of the probability that the benefits, costs, prices, outputs and other technical parameters will achieve the values that have been estimated under the initial certainty approach. Thus, for example, it is suggested that one can obtain an estimate of the probability distribution of project investment costs. This can be combined with appropriate probability distributions for inputs and outputs of the projects and they all can be combined into a probabilistic present value or benefit-cost criterion.

It is not the purpose of this section to present a detailed review of alternative approaches to treating risk and uncertainty in project evaluation. While most project evaluation manuals pay lip service to the importance of introducing risk and uncertainty, there are no widely used simple methodologies for introducing probability as an element of the project cost and benefit calculations. While specific projects may have been evaluated using such techniques, there is a considerable element of personal subjective judgement in determining the probability distributions for each of the individual cost and benefit flows. If a probability distribution is not known for every variable making up the net benefit stream, then the risk analysis is nearly meaningless. Therefore, probability analysis cannot easily be introduced at a practical level of evaluation of rural development projects which are composed of many sub-projects each with different technical, economic and social consequences. While one can conceive that probability estimates might be obtained for a simple agricultural project which produces, say, five or six commodities with a relatively well-known technology, it would not seem possible to do the same for a project that has ten or fifteen sub-projects each with a large variety of services and products. Therefore, in the foreseeable future it will be necessary to proceed using rather simple sensitivity analysis in which technicians make arbitrary adjustments in the important variables of the cost-benefit flows and check the response of the measures of project worth to these systematic variations in relation to the chosen criteria.

The current practice of international lending agencies as well as national planning groups is to approach sensitivity analysis in a rather pragmatic, conservative and somewhat arbitrary way. The sensitivity analysis usually proceeds along the following lines: first, the cost and benefit flows of the project are obtained using the "best estimates" of the individual technician in the field based upon recent historical experience; secondly, on the cost side, systematic increases are made in the investment and operating costs by ten, twenty and thirty percent over the project's life; thirdly, on the benefit side, variations are made to reduce the product prices, yields or similar variables by five, ten, fifteen, and twenty percent over the life of the project; fourthly, the internal rate of return of the project is recomputed for the alternative cost and benefit flows constructed from the percentage variations in the values indicated previously, showing thereby, variations in the internal rate of return or benefit-cost ratio of the projects; fifthly, the internal rates of return are compared to some cut-off rate and if all or most of the internal rates of return, even under the more conservative cost and benefit variations, are acceptable the project is deemed viable.

Table 13 shows the simple sensitivity analysis used by external lending agencies in computing internal rates of return for candidate projects. In this hypothetical case a project has an initial outlay of US$100 in year 0 and a uniform net benefit of US$25 annually over ten years and an internal rate of return of 21.4 percent without any adjustment in the benefits or costs. The rate of return of the project is recalculated for operating and investment costs which are expected to increase by five, ten, fifteen and twenty percent, respectively, and annual project benefits are expected to diminish by factors of five, ten, fifteen and twenty percent. In the hypothetical example, the original rate of return is a respectable 21 percent, but it diminishes rather rapidly as adjustments are made in the benefits and costs of the project. Thus, if the project economist had underestimated investment and operating costs by five percent and overestimated project benefits by five percent, the internal rate of return would be fully seven percentage points less. It is rather easy to anticipate that project rates of return will be highly sensitive to variations in the costs and benefits.

Therefore, it is advisable that project evaluators undertake a more detailed cost and benefit sensitivity analysis than that just referred to. The project benefit and cost flows should be disaggregated into their most important components, including the technical coefficients and the prices - both shadow and market - used to obtain the real input and output flows. Hence, it might occur that a project's profitability is insensitive to some of these variables such as the price of labour or the price of foreign exchange and more sensitive to other variables such as the rate of interest, the price of domestic inputs or the price of skilled labour.

Table 13. Internal rate of return for percentage variations in costs and benefits. (Initial investment I = 100; annual benefits B = 60; annual costs C = 35; time period, 10 years.)

% ΔB \ % ΔTC	0	+5	+10	+15	+20
0	21.4	17.9	14.5	11.2	8
-5	17.7	14.1	10.8	7.5	4.3
-10	13.7	10.2	6.8	3.2	0
-15	9.6	6.0	-	-	-
-20	5.1	1.3	-	-	-

% ΔTC = percentage increase in total costs (investment plus annual costs).

% ΔB = percentage decrease in benefits.

(-) = a negative internal rate of return.

6.2 Graphical Analysis

Two approaches are recommended in obtaining estimates of the sensitivity of measures of project worth to variations in the prices and technical parameters of a project. The first is to take a rather simple graphical approach as shown in McGaughey and Thorbecke (75), using alternative values of the shadow exchange rate, the shadow wage rate and the project discount rate and examining the variation in the benefit-cost ratio to changes in these variables. A hypothetical example is shown in Figure 1, which depicts the benefit-cost ratio for two interest rates, 10 percent and 15 percent; shadow exchange rate adjustments between one times the domestic price of foreign exchange to twice the domestic price of foreign exchange; and two wage rates W_1 and W_2. This shows that the benefit-cost ratio at an interest rate of 10 percent increases as the exchange rate is adjusted upward from 1.0 to 1.5 to twice its official value. Likewise, if a lower shadow wage rate is used, the benefit-cost function shifts upward from the solid to the dashed line as depicted. Correspondingly, as the interest rate is increased from 10 to 15 percent the benefit-cost curve drops substantially. At this higher rate of interest, 15 percent, the shadow wage rate and exchange rate adjustments are introduced and as the shadow exchange rate increases, the benefit-cost ratio increases and as the shadow wage rate is reduced, the benefit-cost function shifts upward but to a lesser extent, displaying a "flatter" curve than at an interest rate of 10 percent. Of course, a similar graphical presentation [1] may be made for the net present value criterion, as well as the internal rate of return criterion for similar variations in the wage rate, the foreign exchange rate or any other variable of concern to the project analyst.

---

[1] The graphical relationship is likely to be nonlinear.

Figure 1. Sensitivity Analysis, Benefit-Cost ratio

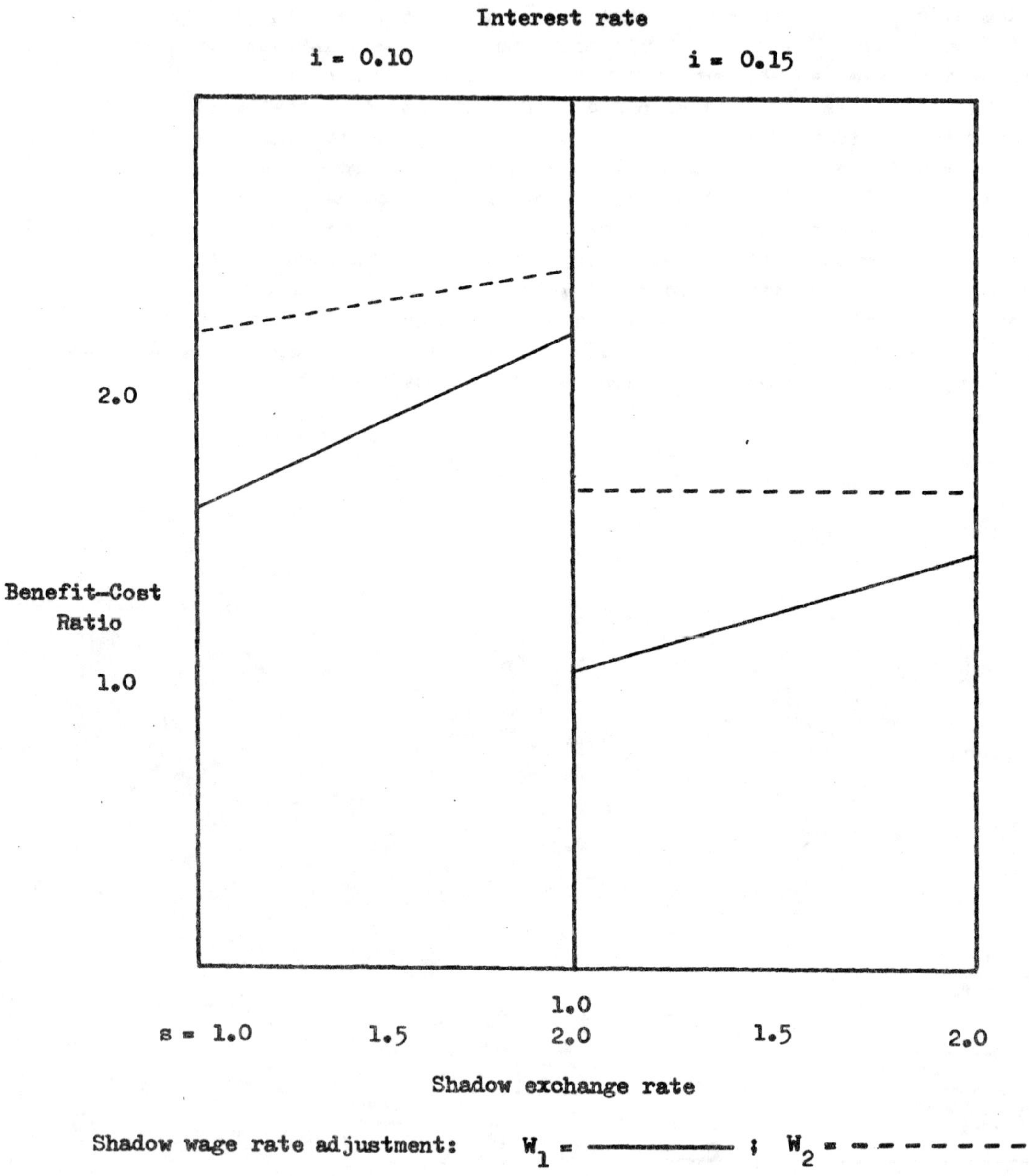

6.3 Shadow Price Elasticity

A second approach is to undertake a quantitative analysis of the sensitivity of an investment indicator, two changes in the shadow price of labour, the shadow price of foreign exchange and the rate of discount. The present example will utilize the benefit-cost ratio (BCR) but the present value criterion can be as easily used. The BCR of any single project may display differing degrees of variation to changes in each of the above mentioned shadow prices. Furthermore, project priorities may be altered substantially by changes in input and output prices. Therefore, in choosing an investment programme it is essential to examine the shadow price sensitivity of each project, because if the benefit-cost ratios vary rather uniformly for all projects, priorities among projects will not change, and if it is found that project priorities and individual project feasibility are insensitive to, say, the wage rate, a smaller proportion of data gathering resources and human talent can be used in selecting this rate.

The benefit-cost ratio is defined

$$BCR = Pv_i\,(B_t)\;/\;PV_i\,(C_t),\; t = 0, 1, \ldots\ldots, n \qquad (31)$$

where $B_t = X_t - C_{xt}$ = the project benefits in year t measured in market prices,

X_t = gross output, year t

C_{xt} = variable production operating costs, year t

$C_t = I_t + O_t$ = total project costs in year t valued at market prices,

I_t = investment costs,

O_t = investment operating and maintenance costs,

n = the final year of the project's useful economic life, and

i = the rate of interest.

The BCR of equation (31) is expressed in market prices, so that shadow price adjustments are made by multiplying the corresponding benefits or costs by the ratio of the shadow price of the input (output) to the market price of the input (output). Hence

$$B_t = (fXF_t - XD_t) - (xCF_{xt} + bCB_{xt} + CDN_{xt}) \qquad (32)$$

where f is the ratio of the shadow exchange rate to the official exchange rate and b is the ratio of the shadow wage rate to the market wage rate.

Also,

$$X_t = fXF_t + XD_t \tag{33}$$

XF_t = the market value of (traded) exports and import substitutes of the project, year t,

XD_t = the value of domestic (nontraded) production

$$C_{xt} = fCF_{xt} + bCB_{xt} + CDN_{xt}, \tag{34}$$

where, CF_{xt} = the traded component of the variable input costs

CB_{xt} = the labour component of the variable input costs, and

CDN_{xt} = the balance (domestic non-labour nontraded component) of the variable input costs.

Furthermore,

$$C_t = fIF_t + bIB_t + IDN_t + O_t \tag{35}$$

IF_t = the foreign exchange (traded) component of public investment, year t

IB_t = the labour component of public investment, and

IDN_t = the domestic non-labour nontraded investment costs.

The values of the coefficients f and b are assumed constant throughout the project's life. To initiate the sensitivity analysis of the BCR, limits are placed on the values of the coefficients f and b, and the rate of interest, i, so that $(f_1 \leq f \leq f_2)$, $(b_1 \leq b \leq b_2)$ and $(i_1 \leq i \leq i_2)$. It can be assumed that f = 1.0, i.e., the lower bound for the foreign exchange adjustment is the official exchange rate, and $b_1 = 0$ and $b_2 = 1$ since labour is likely to be overvalued at the market wage rate and, at the very limit, its marginal productivity is equal to zero. A wide range of values for i may have to be introduced. The next step is to systematically alter the values of f, b and i and to check the (shadow price) sensitivity of the BCR of each project and the project priorities.

To obtain the desired measure of project sensitivity the concept of the shadow price elasticity of the BCR is introduced. (Again, the concept can easily be extended to the next present value criterion.) The present value of a project's benefits and costs can be viewed as a function of the values of f, b and i. Thus,

$$BCR_{ifb} = PV_{ifb}(B_t) / PV_{ifb}(C_t), \; t = 1, \ldots, T \tag{36}$$

is the ratio of the present value of the benefits of equations (32) to the present value of the costs of equation (35), for a given interest rate, i, shadow exchange rate adjustment f and shadow wage rate adjustment b.

The elasticity of the BCR to a change in the foreign exchange rate is

$$E_f = \frac{\partial \, BCR}{\partial \, f} \cdot \frac{f}{BCR}$$

for a given i and b, and since

$$\frac{\partial \, BCR}{\partial \, f} = \frac{PV_i \, (XF_t - CF_{xt})}{PV_i \, (C_t)} - (BCR) \, \frac{PV_i \, (IF_t)}{PV_i \, (C_t)} \qquad (37)$$

it follows that

$$E_f = f \left[\frac{PV_i \, (XF_t - CF_t)}{PV_i \, (B_t)} - \frac{PV \, (IF_t)}{PV \, (C_t)} \right] \qquad (38)$$

Thus, the elasticity of the BCR to changes in the foreign exchange costs and benefits of the project depends upon the value of the coefficient f and the difference between (1) the ratio of the present value of foreign exchange benefits to the present value of total benefits and (2) the ratio of the present value of foreign exchange costs to the present value of total costs.

The elasticity of the BCR to changes in the shadow wage costs of the project is defined similarly, such that

$$E_b = \frac{\partial \, BCR}{\partial \, b} \cdot \frac{b}{BCR} \text{ with f and i fixed.} \qquad (39)$$

It can be shown that

$$E_b = - b \left[\frac{PV_i \, (CB_{xt})}{PV_i \, (B_t)} + \frac{PV_i \, (IB_t)}{PV_i \, (C_t)} \right] \qquad (40)$$

such that the elasticity with respect to the shadow wage is determined by the level of b, (1) the ratio of the present value of variable labour costs to the present value of benefits and (2) the ratio of the present value of the labour component of the public investment costs to the present value of total costs.

Thirdly, the elasticity of the BCR to changes in the interest rate i is defined as

$$E_i = \frac{\partial \, BCR}{\partial \, b} \cdot \frac{i}{BCR} \text{ for fixed f and b} \qquad (41)$$

Specifically, it can be shown that

$$E_i = -i \left[\frac{\sum_{t=0}^{T} tB_t d^{-(t+1)}}{PV_i\ (B_t)} - \frac{\sum_{t=0}^{T} tC_t d^{-(t+1)}}{PV_i\ (C_t)} \right] \qquad (42)$$

where $d = (1 + i)$

The selection of the interest rate is crucial for the feasibility of the capital intensive projects and the feasibility of labour intensive projects is, of course, sensitive to changes in the shadow wage.

Table 14. Shadow Price Elasticity of the BCR to Changes (Δ i) in i.

Projects 1/	f b	1.0 1.0	2.5 0.25
	Δ i		E_i
A	5-10%	-1.12	-1.06
	10-15%	-1.63	-1.44
B	5-10%	- .97	- .93
	10-15%	-1.52	-1.26
C	5-10%	-1.17	-1.11
	10-15%	-1.84	-1.54
D	5-10%	-0.84	-0.80
	10-15%	-1.15	-1.09
E	5-10%	-0.95	-0.90
	10-15%	-1.20	-1.13

1/ Based upon eleven irrigation projects cited in McGaughey and Thorbecke (75) such that

A = a large capital intensive project
B = a medium-sized capital intensive project
C = medium-sized moderately capital intensive projects
D = small moderately capital intensive projects
E = small relatively labour intensive projects

Table 15. Shadow Price Elasticity of the BCR to Changes (Δf) in f, the Shadow Exchange Rate Adjustment

Projects [y]	Δ f	1.5–1.0	2.0–1.5	2.5–2.0
	b	0.75	0.5	0.25
	i		E_s	
A	10	.66	.51	.40
	15	.74	.55	.42
B	10	.06	.02	-.09
	15	.15	.03	-.05
C	10	.17	.08	.01
	15	.27	.15	.05
D	10	.51	.46	.41
	15	.51	.46	.40
E	10	.69	.66	.62
	15	.71	.67	.63

[y] See footnote Table 14.

Table 16. Shadow Price Elasticity of the BCR to Changes (Δb) in b, the Shadow Wage Rate Adjustment.

Projects [y]	f	1.5	2.0	2.5
	Δ b	0.75–1.0	0.5–0.75	0.25–0.5
	i		E_b	
A	10	-0.43	-0.25	-0.13
	15	-0.46	-0.27	-0.14
B	10	-0.45	-0.26	-0.13
	15	-0.51	-0.29	-0.14
C	10	-0.49	-0.28	-0.14
	15	-0.57	-0.31	-0.15
D	10	-0.53	-0.33	-0.18
	15	-0.56	-0.35	-0.19
E	10	-0.49	-0.33	-0.19
	15	-0.50	-0.34	-0.20

[y] See footnote Table 14.

As seen in Tables 14, 15 and 16 a project's BCR is not uniformly sensitive to the choice of shadow prices. The elasticity with respect to i, E_i, is greater than (negative) one for most values of f and b; the BCR is inversely related to i. Furthermore, the higher the rate of interest the greater is the sensitivity of the BCR to changes in i. Shadow wage and exchange rate adjustments alter the elasticity but less than shifts to higher rates of discount. The foreign exchange rate adjustments is uniformly inelastic except for a very high rate of interest at the market wage rate. Subsequent adjustments in the wage and exchange rates reduce the elasticity. Finally, the BCR is positively related to an increasing exchange rate. The elasticity of the BCR to reductions in the shadow wage is likewise inelastic but highly variable ranging from $E_b = -.57$ to $E_b - .13$. Reductions in the shadow wage rate increase the BCR. [1]

For the projects in question it can be concluded that (1) the BCR of any single project is highly dependent upon the combined choice of the discount rate, the shadow wage rate and the shadow exchange rate; (2) the greater is the adjustment in the latter two rates the lower is the sensitivity of the BCR to each; and (3) as the interest rate is increased its effect over the BCR is increased. The selection of the interest rate is crucial for the feasibility of the capital intensive projects and the feasibility of the small labour intensive projects is, of course, sensitive to changes in the shadow wage.

It is recognized that a major difficulty in project appraisal and ranking is the choice of shadow prices to translate private investment choice into social benefit-cost analysis. To this end, the evaluation of the sensitivity of an investment project's feasibility and ranking may yield information which would permit planners to estimate with less urgency the social opportunity cost of a project's input (output). It is suggested that the shadow price elasticity of the benefit-cost ratio provides the desired measure of sensitivity.

In particular, it may be found that the benefit-cost ratio - or any other investment criterion adopted - is relatively insensitive to changes in any one or more of the prices. Among the projects referred to it was found that the BCR displayed varying degrees of sensitivity to the shadow prices. The BCR's of the small labour intensive projects were more sensitive to the shadow wage rate - although with an elasticity less than unitary - than the medium-sized projects. The latter projects, as well as the one large capital intensive project were more sensitive to the choice of the interest rate than the small projects. All projects displayed similar (low) elasticities to the shadow exchange rate, a somewhat unexpected result.

The methodology is general enough to be applicable to any alternative investment criteria - the present value of net benefits, the internal rate of return - and to all classes of investment projects. Both the project's feasibility and ranking among a set of projects can be treated for any assigned limits of the shadow prices.

---

[1] Elasticities are not computed as absolute values so that a negative elasticity reflects an inverse relationship between the direction of the shadow price adjustment and the BCR. In general, this is necessary since it is possible that the sign of the elasticity may vary for a change in the assumed level of the remaining shadow prices which are held constant while computing the own-elasticity.

BIBLIOGRAPHY

1. Acar, William. 1976. An integrated approach to multidimensional evaluation and cost-effectiveness analysis. Public Finance, 31(1): 58-72.

2. Anker, Desmond L.W. 1973. Rural development problems and strategies. International Labor Review, 108(6): 461-484.

3. Balassa, Bela. 1976. The "effects method" of project evaluation. Oxford Bulletin of Economics and Statistics, 38: 219-231.

4. Balassa, Bela. 1977a. The "effects method" evaluation once again. Oxford Bulletin of Economics and Statistics, 38: 345-353.

5. Balassa, Bela. 1977b. The income distribution parameter in project appraisal. Economic Progress, Private Values and Public Policy, Amsterdam: North-Holland, pp. 217-232.

6. Balogh, Thomas. 1978. Failures in the strategy against poverty. World Development, 6(1): 11-22.

7. Baumol, William J. & Bradford, David F. 1970. Optimal departures from marginal cost pricing. American Economic Review, pp. 265-283.

8. Benor, Daniel & Harrison, James Q. 1977. Agricultural extension: the training and visit system. Washington, D.C.: The World Bank.

9. Biggs, Huntley H. & Tinnermeier, R.L. (eds.) 1974. Small farm agricultural development problems. Colorado State University: Fort Collines, Colorado.

10. Blair, Harry W. 1978. Rural development, class structure and bureaucracy in Bangladesh. World Development, 6(1): 65-82.

11. Blitzer, Charles R., Clark, Peter B. & Taylor, Lance (eds.) 1975. Economy-wide models and development planning. Wo rld Bank and Oxford University Press.

12. Brown, M.L. 1974. Preparing farm budgets for agricultural project analysis. IBRD - Economic Development Institute.

13. Bruce, Colin. 1976. Social cost-benefit analysis: a guide for country and project economists to the derivation and application of economic and social accounting prices. IBRD, World Bank Staff Working Paper No. 239.

14. Bruno, Michael. 1967. The optimal selection of export-promoting and import substituting project in planning the external sector: techniques, problems and policies. First Interregional Seminar on Development Planning, (Ankara, Turkey, September 1965) United Nations: New York, No. 67, II. B5.

15. Bruno, Michael. 1972. Domestic resource costs and effective protection: clarification and synthesis. Journal of Political Economy, 80(1): 16-33.

16. Byerlee, D. Indirect employment and income distribution effects of agricultural
1973 development strategies: a simulation approach applied to Nigeria. African Rural Employment Paper No. 9, Michigan State University: East Lansing, Michigan.

17. Candler, Wilford & Boehlje, Michael. Use of linear programming in capital budgeting
1971 with multiple goals. American Journal of Agricultural Economics, 53(2): 325-330.

18. Carnemark, Curt, Biderman, Jaime & Bovet, David. The economic analysis of rural road
1976 projects. World Bank Staff Working Paper No. 241, 61 p.

19. Chambers, Robert. Project selection for poverty-focused rural development: simple is
1978 optimal. World Development, 6(2): 209-219.

20. Chenery, Hollis B. The application of investment criteria. Quarterly Journal of
1953 Economics, 67: 76-96.

21. Chervel, Marc. The rationale of the effects method: a reply to Bela Balassa. Oxford
1977 Bulletin of Economics and Statistics, 39: 333-344.

22. Crosson, Pierre R., Cummings, Ronald G. & Federick, Kenneth D. (eds.) Selected water
1978 management issues in Latin American agriculture. Baltimore and London: The John Hopkins University Press.

23. Dalkey, Norman C. Toward a theory of group estimation in Harold A. Linstone and Murray
1975 Turoff (eds.). The Delphi Method: Techniques and Application. London: Addison-Wesley Publishing Co. pp. 236-261.

24. Dasgupta, P. & Stiglitz, J.E. Benefit-cost analysis and trade policies. Journal of
1974 Political Economy.

25. Dean, B.V. & Nishry, M.J. Scoring and profitability models for evaluating and selecting
1965 engineering projects. JORSA, 13(4): 550-570.

26. De Vries, A.B. The export experience of developing countries. IBRD, World Bank Staff
1967 Occasional Paper No. 3. Baltimore: The John Hopkins Press.

27. De Weille, J. Quantification of road user savings. World Bank Staff Occasional Paper
1966 No. 2. The John Hopkins Press.

28. Dobb, Maurice. An essay on economic growth and planning. Modern Reader Paperbacks,
1969 New York.

29. Freeman, A. Myrick, III & Haveman, Robert H. Benefit-cost analysis and multiple
1970 objectives: current issues in water resources planning. Water Resources Research, 6(6): 1533-1539.

30. Freeman, A. Myrick, III. Income distribution and planning for public investment.
1967 American Economic Review, 62(3): 495-508.

31. Feldman, Paul. Efficiency distribution and the role of government in a market economy.
1971 Journal of Political Economy, 79(3): 508-526.

32. Gittinger, J. Price. Economic analysis of agricultural projects. Baltimore: The John
1972 Hopkins University Press.

33. Goreux, Louis M. Interdependence in planning: multilevel programming studies of the
1977 Ivory Coast. Published for the World Bank by The John Hopkins University Press, Baltimore.

34. Guerrero, Pablo, Howard, Elenor, Lal, Deepak & Powers, Terry. Pilot study on national
1977 accounting parameters: their estimation and use in Chile, Costa Rica and Jamaica. Vol. 1: General Overview; Vol. 2: Country and Project Case Studies; Washington, D.C., Inter-American Development Bank - Economic and Social Development Department.

35. Gum, R.L., Roefs, T.G. & Kimball, D.B. Quantifying societal goals: development of a
1976 weighting methodology. Water Resources Research 12(4).

36. Hansen, J.R. A guide to the guidelines: the unido method of economic project evaluation.
1974 IBRD, Bank Staff Working Paper No. 166, Washington, D.C.

37. Harberger, Arnold C. Basic needs versus distributional weights in social cost-benefit
1978 analysis. mimeo, University of Chicago.

38. Harberger, Arnold C. Project evaluation: collected papers. The University of Chicago
1976a Press.

39. Harberger, Arnold C. On the use of distributional weights in social cost-benefit
1976b analysis. Paper presented at a conference on Research in Taxation sponsored by the National Bureau of Economic Research. Standford, California.

40. Havemore, R.H. & Margolis J. Public expenditures and policy analysis. University of
1970 Pennsylvania. Chicago: Markhom Publishing Co.

41. Hawkins, C.J. & Pearce. D.W. Capital investment appraisal. University of Southampton,
1971 Department of Economics. The MacMillan Press.

42. Hayden, F. Gregory. Toward a social welfare construct for social indicators. The
1977 American Journal of Economics and Sociology, 6(2): 129-146.

43. Heller, Peter S. Issues in the costing of public sector outputs: the public medical
1975 services of Malaysia. World Bank Staff Paper No. 207, 85 p.

44. Hinrichs, Harley H. & Graeme M. Taylor. Program budgeting and cost-benefit analysis:
1969 cases, texts and readings. Pacific Palisades, California: Goodyear Publishing Co.

45. Inter-American Development Bank. Guidelines for the preparation of agrarian reform
1974a projects. (mimeo), Division of Analysis of Agricultural Development Projects. 13 p.

46. Inter-American Development Bank. Guidelines for the preparation of agricultural
1974b diversification projects. (mimeo) Division of Agricultural Development Projects, 13 p.

47. Inter-American Development Bank. Guidelines for the preparation of agricultural
1974c marketing projects. (mimeo) Division of Analysis of Agricultural Projects, 12 p.

48. Inter-American Development Bank. Guidelines for the preparation of agricultural
1974d research and extension projects. (mimeo) Division of Analysis of Agricultural Projects, 11 p.

49. Inter-American Development Bank. Guidelines for the preparation of community develop-
1974e ment projects. (mimeo) Division of Analysis of Agricultural projects, 11 p.

50. Inter-American Development Bank. Guidelines for the preparation of projects for
1974f financing agricultural cooperatives. (mimeo) Division of Analysis of Development Projects. 9 p.

51. Inter-American Development Bank. Guidelines for the preparation of integrated
1974g agricultural projects. (mimeo) Division of Analysis of Agricultural Development Projects. 11 p.

52. Inter-American Development Bank. Guidelines for the preparation of forestry development
1974h projects. (mimeo) Division of Analysis of Agricultural Development Projects, 14 p.

53. Inter-American Development Bank. Guidelines for the preparation of land settlement
1974i projects. (mimeo) Division of Analysis of Agricultural Development Projects, 16 p.

54. International Bank for Reconstruction and Development. Appraisal methodology for
1970 feeder road projects. Economics Department Working Paper No. 70, Washington, D.C.

55. International Bank for Reconstruction and Development. The export experience of
1967 developing countries. World Bank Staff Occasional Paper No. 3, Baltimore: The John Hopkins Press.

56. Israel, Arturo. Appraisal methodology for feeder road projects. World Bank Economics
1970 Department Working Paper No. 70, 38 p. and two annexes.

57. Johnson, Harry G. The theory of effective protection and preferences. Economica,
1969 36(142): 119-138.

58. Johnston, B. & Kilby, D. Rural urban linkages. (mimeo) Rural Development Projects
1974 Course, IBRD, Economic Development Institute.

59. Johnston, Bruce F. & Meyer, Anthony J. Nutrition, health and population in strategies
1977 for rural development. Economic Development and Cultural Change, 26(1): 1-23.

60. Kalter, R.J. & Stevens, T.H. Resource investments, impact distribution and evaluation
1971 concepts. American Journal of Agricultural Economics, 53(2).

61. Lal, Deepak. Distribution and development: a review article. World Development,
1976 4(9): 725-738.

62. Lal, Deepak. Employment, income distribution and a poverty redressal index. World
1972 Bank Economic Staff Working Paper No. 129, 14 p.

63. Lal, Deepak. Methods of project analysis: a review. IBRD, World Bank Staff Occasional
1974 Paper No. 16, Baltimore: The John Hopkins University Press.

64. Lal, Deepak. Wells and welfare. Development Centre Studies, Series on Cost-Benefit
1972 Analysis, Case Study No. 1. Paris: OECD.

65. Layard, R. (ed.) Cost-benefit analysis. England: Penguin Books, Ltd.
1972

66. Linn, Johannes F. Economic and social analysis of projects: a case study of Ivory
1977 Coast. World Bank Staff Working Paper No. 253, 68 p.

67. Linstone, H. & Turoff, M. (eds.) The delphi method: techniques and applications.
1975 Reading Massachusetts: Addison-Wesley Publishing Co.

68. Little, I.M.D. & Tipping, D.G. Cost-benefit analysis of the Kulai oil palm estate,
1974 West Malaysia. OECD, Development Centre Studies, Case Study No. 3,
Paris.

69. Little, I.M.D. & Mirrlees, J.A. Project appraisal and planning for developing countries.
1974 New York: Basic Books Inc.

70. Lee, Sang M. Goal programming for decision analysis. England: Auerback Publishers
1972 Inc., 387 p.

71. MacArthur, J.D. & Galal A. Amin. Cost-benefit analysis and income distribution in
1978 development countries: a symposium. World Development, Special Issue,
6(2): 240.

72. McColl, G.D. & Throsby, C.D. Multiple objective benefit - cost analysis and regional
1972 development. Economic Record, 48: 201-219.

73. McGaughey, Stephen E. La evaluación económica y financiera de proyectos de inversión.
1972 Lima: Ministerio de Agricultura - Universidad de Iowa, USAID,
Publication No. 35.

74. McGaughey, Stephen E. & Thorbecke, Erik. Project selection and macroeconomic objectives:
1972 a methodology applied to Peruvian irrigation projects. American Journal
of Agricultural Economics, 54(1).

75. McGaughey, Stephen E. & Thorbecke, Eric. Project selection and macroeconomic objectives:
1973 reply. American Journal of Agricultural Economics, 55(1): 107-109.

76. Major, David C. Benefit-cost ratios for projects in multiple objective investment
1969 programs. Water Resources Research, 5(6): 1174-1178.

77. Mangahas, M., Meyers, W.H. & Barker, R. Labour absorption in Philippine agriculture.
1972 Paris: OECD Development Centre.

78. Manning R.C. & Ela, O.M. An economic evaluation of irrigation rehabilitation projects
1971 in Mexico. Washington, D.C.: IRDR)

79. Marglin, Stephen A. Public investment criteria: benefit-cost analysis for planned
1967 economic growth. Cambridge, Mass.: The M.I.T. Press.

80. Mehmet, Ozay. Benefit-cost analysis for employment creation. International Labor
1971 Review. 104(1-2): 37-50.

81. Meyer, Richard L. & Larson, Donald W. Rural non-farm employment: the recent East Asian
1978 experience. Journal of Economic Development, 3(1).

82. Moore, John R., Jr. & Norman R. Baker. Computational analysis of scoring models for
1969 R and D project selection. Management Science, 14(4): 212-232.

83. Mosher, A.T. Projects of integrated rural development. Agricultural Development
1972 Council.

84. Mumy, Gene E. & Steve H. Hanke. Public investment criteria for underpriced public
1975 products. American Economic Review, pp. 712-720.

85. Norton, Roger D., Luz María Bassoco & José S. Silos. Appraisal of irrigation projects
1974 and related policies and investment. World Bank Staff Working Paper
No. 184, 27 p.

86. Offedal, Olav T. & Levinson, James F. Equity and income effects of nutrition and health
1977 care, in Charles R. Frank, Jr. and Richard C. Webb (eds.). Income
Distribution and Growth in the Less Developed Countires. The Brookings
Institution: Washington,D.C. pp. 381-433.

87. Paglin, M. The measurement and trend of inequality: a basic revision. American
1975 Economic Review, 65(4).

88. Papandreou, Andreas G. & Zohar, Uri. The impact approach to project selection, Vol. II,
New York: Praeger Publishers.

89. Pearce, D.W. Cost-benefit analysis. London: The McMillan Press.
1971

90. Pouliquen, L.Y. Risk analysis in project appraisal. World Bank Staff Occasional
1970 Paper No. 11. Baltimore: The John Hopkins Press.

91. Rahman, M.A. Regional allocation of investment. Quarter Journal of Economics.
1963 11(1): 26-39.

92. Ray, Anadarup. Cost recovery policies for public sector projects. World Bank Staff
1975 Working Paper No. 206.

93. Rutlinger, S. Techniques for project appraisal under uncertainty. World Bank Staff
1970 Occasional Paper No. 10. Baltimore: The John Hopkins Press.

94. Roemer, Michael. Planning by "revealed performance": an improvement upon the traditional
1976 method. World Development 4(9): 775-783.

95. Roemer, Michael & Joseph J. Stern. The appraisal of development projects: a practical
1975 guide to project analysis with case studies and solutions. New York, Praeger Publishers.

96. Roncón M. Ernesto. Los DRI como alternativa al desarrollo rural. Seminario sobre
1977 proyectos agrícolas, MACA-IICA- BID, La Paz, Bolivia.

97. Rondinelli, Dennis A. (ed.). Planning development projects. Stroudsburg, Pennsylvania,
1977 Dowden, Hutchinson & Ross, Inc.

98. Ronmasset, J.A. Risk and uncertainty in agricultural development. New York: The
1977 Agricultural Development Council.

99. Ruttan, Vernon, W. Integrated rural development programs: a skeptical perspective.
1975 International Development Review, 17(4).

100. Saunder, Robert J., Warford, Jeremy J. & Mann, Patrick C. Alternative concepts of
1977 marginal cost for public utility pricing: problems of application in the water supply sector. World Bank Staff Working Paper No. 259.

101. Scheibe, M., M. Skutsh & J. Schofer. Experiments in delphi methodology, Harold A.
1975 Linstone and Marry Turoff (eds.). The Delphi Method: Techniques and Applications, London: Addison-Wesley Publishing Co., pp. 262-287.

102. Schwab, Bernard & Lusztig, Peter. A comparative analysis of the net present value and
1969 the benefit-cost ratio as measures of the economic desirability of investment. Journal of Finance, pp. 507-516.

103. Schwartz, Hugh & Berney, Richard (eds.). Social and economic dimensions of project
1977 evaluation. Washington, D.C.: Inter-American Development Bank.

104. Seton, F. Shadow wages in the Chilean economy. Paris: OECD, Development Centre
1972 Studies.

105. Sfeir-Yunis, Alfredo & Bromley, Daniel W. Decision-making in developing countries:
1977 multiple objective formulation and evaluation methods. New York: Praeger Publishers.

106. Sjaastad, Larry A. & Wise Carver, Daniel L. The social cost of public investment.
1977 Journal of Political Economy, 85(3): 513-547.

107. Squire, Lyn & Van der Tak, Herman G. Economic analysis of projects. Baltimore: The
1975 John Hopkins University Press.

108. Stern, Joseph J. The employment impact of industrial investment: a preliminary report.
1977 World Bank Staff Working Paper No. 255.

109. Stern N.H. An appraisal of tea production on small holdings in Kenya. Paris: OECD Development Center Studies.

110. Su-Chin, Y. Social rate of returns for project evaluation: an estimate for Yugoslavia. World Bank Staff Working Paper No. 205.

111. Szal, Richard & Robinson, Sherman. Measuring income inequality, appendix I in Charles
1977 R. Frank, Jr. and Richard C. Webb (eds.). Income Distribution and Growth in the Less Developed Countries. Washington D.C.: The Brookings Institution.

112. Thomas, Vinod. The measurement of spatial differences in poverty: the case of Peru.
1978 World Bank Staff Working Paper No. 273.

113. Trinkl, Frank H. Allocations among programs having counteractive outcomes. Policy
1972a Sciences, 3(2): 163-176.

114. Trinkl, Frank H. Resolving opposed judgements in resource allocation decisions.
1972b Policy Sciences, 3(4): 421-434.

115. Turoff, Murray. The policy delphi, in Harold A. Linstone and Murray Turoff (eds.).
1975 The Delphi Method: Techniques and Applications, London: Addison-Wesley Publishing Co.

116. United Nations Industrial Development Organization. Guidelines for project evaluation.
1972 New York: United Nations.

117. U.S. Department of Agriculture. Local secondary effects of watershed projects: a
1964 case study of Roger Mills County Oklahoma. ERS-178.

118. U.S. Water Resources Council. Principles for planning water and land resources.
1970 Washington, D.C.

119. Van der Tak, H.G. & Ray, A. The economic benefits of road transport projects. World
1971 Bank Staff Occasional Paper No. 13. Baltimore: The John Hopkins Press.

120. Van der Tak, H.G. The economic choice between hydroelectric and thermal power develop-
1966 ment. World Bank Staff Occasional Paper No. 1. Baltimore: The John Hopkins Press.

ECONOMIC IMPACT ANALYSIS OF FORESTRY PROJECTS:

A Guide to Evaluation of Distributional Consequences

by

Ervin G. Schuster [1]

---

[1] E. G. Schuster is Associate Professor of Forestry at the University of Montana.

1. INTRODUCTION

This paper is designed to serve as a general guide to designing evaluations of the distributional consequences of forestry projects. A model for distributional evaluations will be presented together with some tentative approaches to analysis. Each analyst will need to adapt these recommendations to specific circumstances, since it is impossible to prescribe an analysis model that will be applicable to all situations. For one thing, there is no reason to believe that all decision-makers have the same set of distributional goals and objectives. For another thing, expertise and data will vary with the situation.

As one reads the literature on distributional analysis of forestry projects, it becomes apparent, very quickly, that a set of issues keeps reoccurring. These issues deal with the analytic context within which distributional evaluations take place. They deal neither with why to do the analysis nor with how to do the analysis. Rather, they answer the question: What concepts should the analyst consider when planning the analysis? It is worthwhile to briefly but explicitly discuss some of these —— the concept of situational analysis first.

1.1 Situational Analysis

It seems obvious, almost trivial, to assert that the type of distributional analysis conducted should be dictated by the specifics of the situation. What constitutes the situation? Nothing other than the unique nature of the "decision problem" —— the decision-maker, the objectives, the alternatives, and the environment within which the decision will be made. The types of variables and the measurements taken on these variables should be determined for each individual decision problem. To the extent that different decision problems share many common aspects —— decision-maker objectives in particular —— the analysis of distributional consequences can be appropriately similar. To the extent that differences are found, analyses should also differ. In order to be relevant to a decision-maker, the analysis should be tailor-made to the specific situation.

While seemingly obvious, the importance of this concept has apparently escaped many analysts. It basically means that there exists no single form of distributional analysis that will have universal applicability. Why? Because the decision problems are not universally identical. Can we expect situational differences to exist relative to distributional analyses? Yes. Differences will exist whenever one encounters differences in economic growth goals and economic development goals from one decision-maker to another. There is no *a priori* reason to believe that distributional goals for one part of a country will be the same as those for other parts, let alone one nation compared to another. For a distributional analysis to be responsive to the decision-maker's needs, it must be geared to the specific situation.

1.2 Impacts, Adjustments and Geographical Scope

One major difference between efficiency analyses and analyses of distributional consequences is the geographical scope of the analysis. Efficiency analyses do not typically make conscious reference to a geographical area of applicability, dealing exclusively with the existence of costs and revenues. Distributional analyses normally make explicit reference to geographical boundaries. It is important that both the analyst and decision-maker be aware of the implications of geographical limitations.

Whenever an analyst delineates a specific area of land as appropriate for distributional assessments, this necessarily implies that areas outside the borders are analytically irrelevant. A dichotomy exists between the area of concern and the "rest of the world". The area of concern is often termed the "impact area". This may consist of a local community in Mexico, a multi-county area in Montana, a province in Canada, or an entire country. Typically done at an early stage in the analysis, this geographical delineation is critically important because it necessarily controls all subsequent analysis results. Analysts should probably give much more attention to the issue of boundaries. If boundaries are inappropriately drawn, analysis results will also be inappropriate. This is so because many of the tools used in analysis take on meaning in terms of the specific geographical area defined.

Beyond the question of geographical scope lies the issue of "impact". The language used in many distributional assessments would refer to a forestry project as having an impact on employment. But some analysts are extremely hesitant to accept the notion that a forestry project will result in the increase of Y jobs or a decrease of Y wages - the employment and wage impacts. Why is this? The concern is that a redistribution of employment and wages, rather than absolute gains and losses, will actually occur. Considering a specified area only, one argument states that workers released from one activity will simply shift to another line of work. This also can be reversed. No net gain nor loss results, simply shift to another line of work. Given this type of rationale, Waggener (1970) and the Consulting Services Corporation (ca. 1969) prefer the use of the term "adjustment burdens" rather than "impacts". Whether the consequences of forestry projects are reflected in absolute change or simply shifts is a matter lending itself to further analysis.

A final point should be made. Consider the argument stating that if employment opportunities are created that cannot be filled by workers in the geographical area, labour resources will migrate from another geographical area to fill the void. The question is as before: Is there a net gain in local economic activity if one area gains at the expense of another? A decision-maker charged with responsibility of stabilizing all local areas should seriously consider the implications of the dichotomy between "impact area" and "rest of world".

1.3 Data Base Importance

It is difficult to overemphasize the importance of data base in analysis of the distributional consequences of forestry projects. While almost any economic analysis requires data, distributional analyses are particularly sensitive to the existing data base - especially availability of secondary data. This is so because most of the analytical tools are quantitative in nature and generally rely on secondary data. What are secondary data? These data normally consist of numerical information routinely collected and disseminated by public agencies, such as the U.S. Bureau of the Census. Analysts typically rely on secondary sources of data. It is simply too expensive to collect original or primary data. Accordingly, many of the analytical tools discussed later require data from secondary sources for their implementation. In absence of these data, ability of an analyst to conduct an analysis of distributional consequences is seriously impaired -- to the point of infeasibility.

A dilemma is now developing. On the one hand, distributional analyses ought to be geared to the specific needs of a decision-maker. On the other hand, secondary data are required for analytical tools. This guarantees a problem. The analyst may not be able to evaluate certain distributional consequences that are important to the decision-maker. This may be due to either the lack of an analytical tool or lack of needed data. Consequently, analyses are often conducted only in those areas for which both tools and data exist. Except for token treatment, other important consequences are omitted from the analysis. Most distributional analyses are partial analyses, at best.

But the data base is important in other ways. Consider the ralationship between available secondary data and definition of the impact area. Many analytical tools require data that describe characteristics of industries within a particular geographical boundary. Some system must usually be devised by which firms are organized into industries to facilitate record keeping. In the United States, the Standard Industrial Classification (SIC) system is the official taxonomic structure and numerical code used by most organizations collecting industrial data (OMB, 1972). For example, SIC 242 is commonly understood as referring to "sawmills and planning mills".

One difficulty with heavy reliance on secondary data organized under any classification system is that the system may not be appropriate. For example, the U.S. system has a category for establishments engaged in production of sheep and goats (SIC 0214). The problem faced by the analyst concerned with these establishments from the standpoint of the range livestock (as opposed to feedlot) is that the distinction simply cannot be made - appropriate codes do not exist. But a more serious problem is that data based on some classification systems can systematically bias analyses. For example, it is relatively easy to make measurements on the timber-using industry under the United States' Standard Industrial Classification system. The system has directly applicable classifications. But what if an economic activity gain in the timber-using industry is associated with a loss in the outdoor recreation industry? No code exists for this latter industry. The problem is that establishments constituting the outdoor recreation industry are subsumed in several (currently unspecified) industrial sectors and therefore cannot be accurately identified. Important decision-making information may be thereby lost.

Another difficulty with heavy reliance on secondary data is the potential for "information disclosure". This may be a problem in some countries and not in others. When present, data collecting agencies are legally prohibited from releasing data on a particular class of industry whereby doing so, some characteristics of a specific firm is revealed (for example, U.S. Code, 1939). Procedures are then established to preclude such revelations. What do disclosure requirements have to do with distributional analyses? This: in order to provide data and yet avoid disclosing information on individual firms, one common remedy is to expand the relevant geographical boundaries. There is a much better chance that data can be released on a specific industry for a multi-county area or a state, as opposed to a single county. The aggregations then dictate boundaries of the local area. There has been noticeable tendency for distributional assessments to focus on the multi-county and state levels. Another way of avoiding disclosure problems is by combining industries, instead of geographical areas. For example, sawmills may be combined with logging, furniture and paper into a wood products industry category. While detailed geographical information may be available, detailed industry information is not. It is, therefore, entirely possible for the data base, and not the decision-maker, to specify both the geographical boundaries and industrial categories used in analysis.

One final aspect of the data base should be discussed. This is the problem where some data available to analysts - particularly income and employment data - are based on "covered employment". This term as used in the United States refers to workers covered by some type of unemployment insurance, either local or national. Non-covered workers and industries are not included in these data. In the case of the United States, omissions include self-employed loggers and others; farmers are almost totally excluded. An associated problem is that definitions change. Definitions as to both industrial classes and "covered" employment can change with time. This is particularly troublesome when interpreting time series data. What appears to be an increase in wages over-time may simply be the result of a change in data collection definitions resulting in more workers being counted; the wages always existed - now they are being recorded. While it is difficult to be aware of changing definitions, it is even more difficult to make appropriate adjustments in analyses.

There are undoubtedly other important data base considerations. Those discussed are illustrative, not exhaustive. The analyst should understand that the results of analysis are strongly dictated by the data used. Where inappropriate or incomplete, the analysis will follow accordingly. The analyst should make every effort to comprehend the data base and its implications for the analysis.

The remainder of the discussion is divided into seven major sections. The next section (2) discusses various evaluation models and recommends a general model. Sections 3 through 7 discuss methods and techniques that can be used to evaluate each component of the recommended model. Examples of techniques and methods will be drawn from a variety of sources available to the author in the United States. This seems preferable to use of comprehensive case studies, since most actual case studies are not particularly comprehensive. The discussion of methods and techniques is not intended to be exhaustive, but rather to provide an overview of the empirical state of the art. Several methods, such as input-output analysis, are themselves the topic of books (see Miernyk, 1965). Detailed treatment is simply beyond the scope of this paper. Several topics are not discussed at all, most notably the goodness or badness of equity implications and interporal (or intergenerational) considerations (see Okum, 1975). The topics are of great theoretical, but not empirical, importance. The model recommended is general enough so that it can be adapted to the specific data limitations existing in many countries. The final section provides some concluding remarks.

2. DISTRIBUTIONAL CONSEQUENCES MODELS

Evaluating the distributional consequences of alternative forestry projects is a relatively recent addition to forestry planning processes. Admittedly, much evidence exists that forestry programmes have evolved partly in response to concerns with community stability and other non-efficiency aspects of community welfare (see Waggener 1966 and 1977). But evaluations in this area generally have not been very formal nor explicit. Marty (1975) has recently called for explicit, comprehensive assessments of forestry projects, including distributional consequences. Leven (1970) has recommended a generalized framework for distributional analyses. The following considers selected analysis models directly applicable to forestry.

2.1 Review of Literature [1]

Bentley (1968) was interested in analysis of policy alternatives with regard to U.S. Forest Service timber sales. Relevant policy alternatives included short-term sales, long-term sales, purchaser quota systems and others. These alternatives were evaluated in terms of four aspects:

(i) Economic efficiency - in terms of maximization of net returns to the treasury from timber sales;

(ii) Economic progress - regarding long-term growth and development;

(iii) Economic stability - in terms of employment instability and other social problems;

(iv) Equity - referring to barriers that distort the distribution of economic opportunity for reasons other than long-term efficiency.

The latter three items relate to the distributional consequences of the policy alternatives. The alternatives were evaluated and compared: a) to a standard and rated as below, equal to, or above the standard, and b) to each other and an ordinal ranking was assigned each alternative relative to the four areas of evaluation. Bentley's model was designed to evaluate timber sale policy, not to serve as a general model of distributional consequences.

Zinn (1972) did attempt to develop a more general model. The objective was to develop a means for analyzing the contributions of forestry to a region. Contributions included both economic and social dimensions. The forestry sector was defined to include a specific set of important forest-based and forest-oriented activities - those concerned with management, production and distribution of wood and recreational resources. Over time this model was refined to entail (Zinn, 1976):

(i) The amount of economic activity generated in the sector measured by employment, annual payroll and value added in production in the sector;

(ii) The productivity of the sector - value added per man/hour of production labour;

(iii) The sector's direct effect on individual welfare in terms of wage rate, vacation credits and pay, annual incomes, social security benefits, health insurance, working conditions and others;

(iv) The sector's effect on regional economic stability - turnover of enterprises, permanency of employment, etc.;

(v) The sector's effect on the geographic distribution of economic activity over the region;

(vi) The sector's interactions with the public sector;

(vii) The sector's effect on the distribution of income in the region;

(viii) Characteristics and conditions of entrepreneurship in the sector including legal structure of enterprises and the quality or performance of entrepreneurship in the sector;

(ix) The sector's generation of external costs and environmental impacts;

(x) The secondary economic effects of the sector.

---

[1] As mentioned, this review is limited to United States literature.

While the model was originally intended to deal with the present contributions of forestry to a region, it could serve as the basis for an analysis of distributional consequences of forestry projects. It is comprehensive.

Convery(1973) attempted to develop a model for measuring local economic impacts of forest management practices by the U.S. Forest Service in the southern part of the United States. This model was to be: a) inexpensive, b) easy to implement, c) universally applicable, and d) intellectually defensible. Convery eventually focused on the following effects:

(i) Income - where value added defined as value of gross output less cost of material inputs was selected as a first measure of income generated;

(ii) Employment - where man-years was used as the measure of employment quantity and quality of employment was assayed by examining wage rates, seasonality of employment, and physical work conditions;

(iii) Tax revenue - where the U.S. Forest Service revenue sharing programme with local governments as "payments in kind" were measured for each management alternative.

To further enhance application of this model, a computer programme was written to accomplish the necessary calculations (Field and Convery, 1976). This programme, BENEFIT, provides forest planners with information after they enter appropriate data at local computer terminals.

2.2 The Approach of This Paper

Recently, Schuster (1976b) developed a model to evaluate the distributional consequences of alternative timber harvest schedules. This model calls for evaluation of distributional consequences in the following areas: a) economic activity, b) individual welfare, c) area equilibrium, and d) local government. The model was based on availability of methodology, availability of data, and a desire to promote more comprehensive evaluations. The framework is applicable in evaluations of a wide range of forestry projects, not just scheduling of timber harvest. As such, it is being used in a more general way by the U.S. Forest Service (1977). This model will be presented later.

All of the models discussed deal with major categories of distributional consequence - stability, employment and individual welfare. These represent the areas of distributional consequences with which the analyst will work. But these areas cannot be measured directly. For example, we cannot directly measure "equity". Each area must be defined in terms of something that can be measured. And then the analyst must select a specific unit of measurement. Thus, before going on, we should adopt some common terms, not only to better communicate, but also to indicate the several decisions an analyst must make when planning an evaluation.

A comprehensive analysis of a forestry project has several parts. For example, one part may deal with environmental consequences, another with distributional consequences and a third with efficiency. Call this the "type" of consequence. What aspects of each "type" should be evaluated? The answer will be specified in terms of the "classes" of consequences within the "type" that will be evaluated. In terms of earlier discussion, stability and individual welfare would each be a class within the distributional consequences "type". The analyst must then define each class of consequences in terms of one or more "indicators". Consider the class of distributional consequences dealing with individual welfare. There exists several different indicators of welfare change - wage rates, home ownership, and disposable income are but a few. Since consequences associated with any given class may be diverse, it seems appropriate to use several indicators. This is particularly important since the indicators need not give consistent signals. For example, employment levels and value-added levels each may serve to indicate one aspect of the level of economic activity. It is perfectly reasonable to speculate that in some cases value-added may increase while employment level declines. Is economic activity going up or down? Is it good or bad? Answers require major value judgments best left to the decision-maker, not the analyst (Davis and Bentley, 1967). The analyst would be well advised to provide both types of information to the decision-maker. One step remains: each indicator must be measured or quantified. Some variable must actually be specified. For example, in the case of employment level, the analyst may choose to use "annual average employment of the 'covered' labour force over sixteen years of age", or some other measurement variable.

Figure 1 shows the hierarchical nature of the elements in any model to evaluate the distributional consequences of forestry projects. It is less important for the reader to accept the terms used in the hierarchy - types, classes, indicators - than to accept the notion that a hierarchy exists. This is important because the relationship between the analyst and the decision-maker varies along the hierarchy. What is the basis for identifying "types" of analysis of forestry projects? The decision-maker's goal structure defines the "types". What is the basis to determine "classes" of distributional consequences? The decision-maker's goal structure also defines them. What is the basis for identifying "indicators" of economic activity? The theoretical and empirical expertise of the analyst mixes with the needs of the decision-maker to specify appropriate indicators. What is the basis for identifying measurement variables? Enter the analyst. This area is almost exclusively the domain of the analyst. But importantly, the relationship between the analyst and the decision-maker is seen to vary along the hierarchy. This approach to analysis will help ensure maximum usefulness of the analysis in the decision process. Other approaches based on providing decision-makers with information they "ought to know" (determined exclusively by the analyst) will likely be found inappropriate and/or irrelevant. The analyst must function within the information needs of the decision-maker.

The remainder of this paper will proceed on the basis of a rather generalized model of distributional analysis. Table 1 shows this model in terms of classes and indicators. The ones shown may be applicable to many situations, inapplicable to others. But the main advantage of this format lies in its utility as a framework within which methods and techniques can be discussed. In applying this model, the first step involves definition of the economic setting and baseline data to be used in evaluating each class of consequence.

Figure 1. Hierarchy of elements in a distributional consequence model

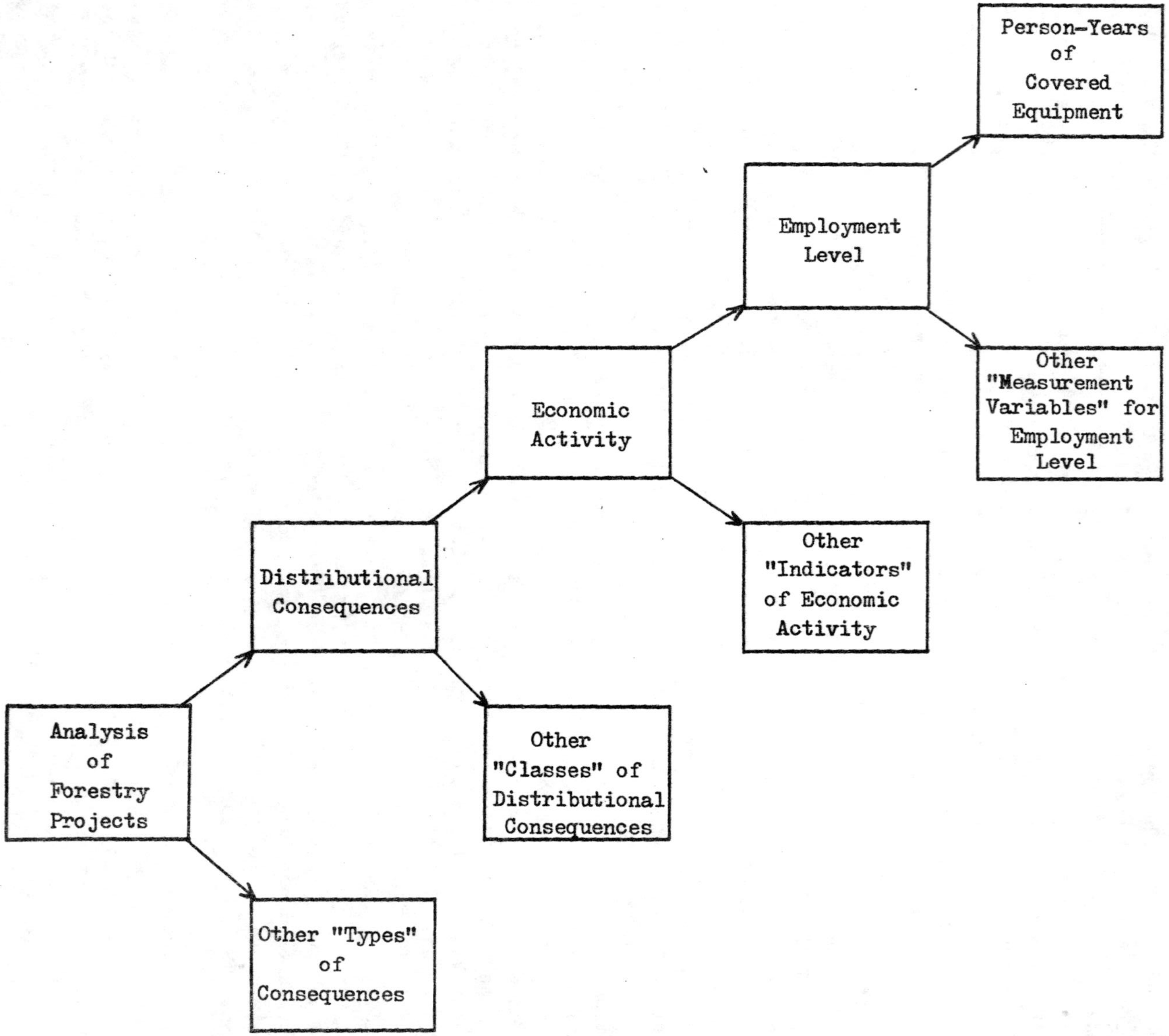

Table 1. Classes and Indicators of Distributional Consequences for Use in Evaluating Alternative Forestry Projects

Class of Consequence	Indicator of Area
Economic activity	Employment Payroll/Wages/Earnings Value added Sales
Individual welfare	Unemployment rate Average wage rates Income redistribution
Area equilibrium	Economic diversity Community adjustments
Local government	Intergovernmental payments Cost to local government

3. THE ECONOMIC SETTING - BASELINE DATA

One of the most important areas of information to provide a forestry decision-maker is a description of the existing socio-economic environment in the geographical area of concern. This could be considered the first step in any analysis of distributional consequences of forestry projects. The goal of this effort should be to provide enough of the right kinds of information to give the decision-maker a "feel" for the geographical area of concern. The existing environment provides the setting within which the forestry project will function. Without this knowledge, it would be difficult to gauge the desirability of any proposed project which will change the environment.

But description of the economic setting is also essential to the analyst, for two other reasons. First, since the distributional analysis should always be geared to economic change resulting from the proposed project, the status quo must be identified. The analyst must assess the status of an economy "with" and "without" the effects of the proposed project. Knowledge of the current situation will help the analyst make the "without" project assessment. The recommended "with" and "without" analysis may be contrasted to a "before" and "after" analysis. They are not the same, the difference mainly being that economies are constantly undergoing change, quite apart from forestry projects. A "before" situation does not reflect this change. Assessment of an economy "without" the forestry project is recognition that the effect of the project is the difference between two states of change, one with and one without the forestry project (Bell, 1976). Second, knowledge of the existing economic setting is essential to interpreting or understanding the implications of a distributional analysis. Too many distributional analyses are sterile, presented as facts and figures only. Suppose an analysis determined that a forestry project would result in employment of one hundred additional persons in the wood products industry. The long-term implications of additional employment in an area of chronic unemployment may be quite different than in a more "healthy" economy.

A question immediately arises: what constitutes the geographical boundaries of the project area? The answer can be reasonably simple or very complicated. On the simple side, just ask the decision-maker. Two types of responses are likely. First, the answer may be given in terms of some combination of political units (counties, districts, states, provinces, or regions). For some reason - possibly because of political pressure, possibly because of a previous study that delineated impact areas, possibly because of organization policy - the decision-maker may be able to specify the area of concern. From the analyst's view, this response is the easiest to handle. Simply carry out the analysis for the specified area. If the answer does not correspond to political boundaries - for example, watershed boundaries - the analyst will likely encounter data problems. Data generally available for analysis are compiled on the basis of political boundaries. Desirable or not, practical necessities of analysis will likely compel a political boundary delineation of the project area.

Alternatively, the decision-maker might respond: "Do the analysis for the area that is affected by the forestry project." This means the analyst must do some preliminary work before actual analysis can start. The analyst must somehow select a delineation for the area. Two general approaches may be used: a) use existing, delineated boundaries, or b) develop other relevant boundaries. Depending on availability of existing data, the analyst may well find that someone has already identified aggregations of political units that are homogeneous with respect to economic structure. For example, the U.S. Department of Commerce has delineated Standard Metropolitan Statistical Areas (SMSAs) and Bureau of Economic Analysis (BEA) Economic Areas; the U.S. Water Resources Council has delineated Water Resources Regions and Subregions. The analyst should select one such region delineation system based on its correspondence with the forestry project being evaluated and availability of subsequent analysis data. Once a regional system is selected, the analyst simply identifies that region containing the forestry project and then uses it in analysis.

While use of an existing system to delineate the area of concern has the distinct advantage of expediency, it has a disadvantage. The delineation may not be exactly appropriate - too big or too little in terms of the project being analyzed. The analyst may have to delineate the boundaries. How? On the one hand, the analyst could conduct a rigorous, special study to define the boundaries (for example, see Fox and Kumar, 1966; and Bouderville, 1966). But since this may well require more resources and time than available for most project analyses, sophisticated approaches to identifying the region will not be discussed further here. Rather, the analyst will probably adopt a less rigorous approach. Any number of information sources may be used. Officials or colleagues may be consulted. Availability of data may be assessed in relation to the project being considered. For example, if a project that will eventually result in a timber harvest change is being considered, the analyst may decide to restrict consideration to those political units (e.g. counties or provinces) containing the sawmills that will process the harvested timber.

By one way or another, the analyst must determine the boundaries of the area to be studied. It is this area then for which a description of the economic setting or environment will be developed.

What economic aspects of the area should be described? There is no general answer, only a recommendation. The analyst should describe those economic aspects that are: a) important to the decision-maker, b) necessary for the "without" part of the distributional analysis, and c) useful to the analyst in interpreting other results of the analysis. Nevertheless, some generally applicable items can be identified. The following list of socio-economic characteristics was adapted from that used as the basis to develop an overview of the population and economic activity in several planning areas of the U.S. Forest Service's Northern Region (USFS, 1975):

Population

- levels over time
- distribution by age, race and sex
- change components including births and deaths
- density and crowding
- migration patterns
- future projections

Employment

- levels by industry
- distribution by occupation
- opportunities
- labour force participation and unemployment

Income

- per capita and family
- distribution
- cost of living
- welfare statistics

Others

- housing
- family stability
- freedom from crime
- educational achievements
- health of people

The listing above is illustrative only. The analyst will have to add and subtract from the list depending on the circumstance being evaluated. Obviously, not all of the analyst's effort could be devoted to describing the existing situation. How much effort should be devoted? Again, there is no general answer; the analyst's judgement is required. Presumably, this judgement will be based on the relationship between the cost of providing additional information and the value of that additional information in decision-making. Before considering the next topic, Table 2 presents an example outline of a rather compresive description of an economic setting. Once an adequate description of the economic setting has been developed, the remainder of the distributional analysis can be pursued.

Table 2. Example: Outline of comprehensive description of economic setting

BEAVERHEAD NATIONAL FOREST
SOCIO-ECONOMIC OVERVIEW FOR UNIFIED PLANNING

Contents

Table 2. (Cont.d)

Source: USFS, 1974.

4. IMPACTS ON ECONOMIC ACTIVITY

Consider now the first class of consequences shown in Table 2. Distributional consequences of forestry projects can be measured and reflected in many ways. But the dominant way - and typically the only way - is to show these consequences in terms of changes in the level of economic activity. Note again that the analyst should be concerned with measuring change resulting from (or attributable to) the project. The goal of this effort should be to measure aggregate, net change in economic activity. The work "aggregate" is used because a forestry project may require inputs from supporting economic sectors; concern is with all sectors. The word "net" is used because any particular forestry project may have negative effects on some other forestry or non-forestry sectors. The most obvious example of this is where forestry projects compete for a fixed budget; expansion of one aspect (e.g. recreation) can only be done at the expense of restricting another (e.g. timber). In this event, the analyst should identify the total net project effects as the sum of two parts, one positive and the other negative.

Measurement of changes in economic activity requires the analyst to have some knowledge of the underlying relationship between inputs and outputs - the production function. Figure 2 shows that all stages in a production activity require use of economic resources. The typical economic efficiency analysis evaluates the relationship between the value of inputs and outputs over time.[1] This phase of the distributional analysis measures the levels of economic activity associated with the production process.

What indicators of economic activity should the analyst consider? The possibilities are many. But in practice only a few are used and, of these, only one or two may actually be measured. The listing below shows the four commonly evaluated indicators of economic activity.

(i) Employment levels - Changes in the level of total employment in various industrial sectors should be determined on a full-time basis. That is, employment should be couched in terms of "person-years". Part-time and seasonal employment must be converted to an annual equivalent. The analyst must be aware of the distinction between "total" and "covered" employment. If the distinction is great, the analysis results should be modified as appropriate.

(ii) Payroll and wage levels - This generally refers to income of workers in the local economy. Again, aggregate and sector data should be developed. The analyst should be aware of definitional problems. Depending on the data source used, earnings, income, payroll and wages will likely be different and have different clientele-based meanings. An important difference may involve wages and earnings. While "wages" often refers to payments to employees, the term "earnings" additionally includes payments to proprietors.

(iii) Value added - On a local basis, this indicator of economic activity is analogous to gross national product. It measures value or worth added to a product by industrial activities. It is the total of all differences between sales receipts and costs of intermediate goods. As a measure of total income generated, its biggest limitation for local area analysis is that some of this income (profits and interest payments) may go to economic units outside the local area (Convery, 1973).

---

[1] See Economic Analysis of Forestry Projects (EAFP), FAO Forestry Paper No. 17.

(iv) Sales levels - This amounts to the value of gross output or aggregate retail sales. It is part of the data needed to make value-added calculations. One major difficulty with this indicator is that it typically only registers market transactions.

Which one or combination of these indicators should be evaluated? The empirical literature certainly does not help answer the question. All, either singly or in combinations, can be found. Experience, however, suggests that decision-makers can more easily comprehend changes in employment and wage levels, than value-added or sales. The analyst should seriously consider use of more than one indicator. This is because changes in economic activity need not be consistent between indicators. A forestry project may be associated with an increase in wages and a decrease in employment. Both should be provided to the decision-maker.

Figure 2. Hypothetical Production Process

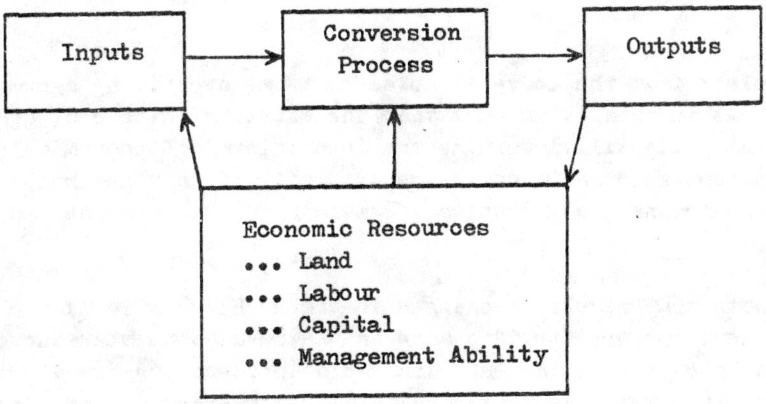

4.1 Primary Impacts

Once the analyst selects an indicator(s) of economic activity to evaluate, the next step is to link the forestry project to a change in that economic activity. This task will require the analyst to use either existing relationships or to develop needed relationships. Skill, ingenuity, and imagination may be required. The analyst must first relate the forestry project to changes in commodity outputs. The listing below shows some potential output measures. In all cases, the output measure used should be: a) related to a traditionally accepted unit of output measurement and b) capable of being translated into one of the indicators of economic activity change discussed earlier.

(i) Timber - timber harvest change measured in board feet, cubic feet, cubits, or cubic metres;

(ii) Water - water yield change measured in acre-feet or cubic metres per hectare;

(iii) Recreation - recreation activity change measured in terms of the opportunity to generate occasions, visitor days, or recreation days;

(iv) Range - forage or range livestock production change measured in animal unit months or grazing or animals produced;

(v) Wildlife - wildlife population change measured in terms of population level or population density.

Two illustrations of approaches to this problem may be useful. First, Ferguson and Phillips (1975) estimated the effect on total sales associated with changing a recreation area to a national park. Data indicated that the 1 562 000 visitor days of recreation in 1975 were associated with US$58.6 million of expenditures by recreationists. They then developed a series of projections for change in the level of recreation activity associated with establishment of a national park. After accounting for normal growth, they identified the levels of total sales in the region associated with each projection level. Second, Johnson (1972) determined that there were approximately six wood products workers per million board feet of timber harvested in Montana. After adjustments were made, it was determined that a change of 100 million board feet in timber harvest would result in a change of 500 jobs.

It should be clear from the above examples that one overriding assumption is usually implied: linearity. It is usually assumed that the existing pattern of commodity output relative to economic activity will determine the future level of economic activity. Technically speaking, constant returns to scale are assumed. If this assumption is realistic with regard to the forestry project, results are meaningful. Otherwise, adjustments must be made.

Forestry projects that result in changes in timber harvest level are so common that further elaboration seems appropriate. A change in the level of timber harvest will have a first-order effect on those industries that consume or process roundwood. These industries are referred to alternatively as the timber-using industry, the wood-products industry, or the forest-products industry. Whatever the term used, no standard definition or description exists. Analysts appear to generally adopt one of two approaches: a) the narrower approach which includes only processors of industrial roundwood, and b) the broader approach which includes secondary processors, such as furniture plants. For the purpose of most distributional analyses, the narrower approach is more practical since secondary processors do not generally use forestry outputs directly and will eventually be included as part of "aggregate impacts".

In general, measures for primary impact indicators regarding timber harvest take the form of relationships involving employees or dollars of wages per million board feet of timber cut in a particular area. In the example mentioned earlier, it was estimated that about six wood-products workers per million board feet or timber harvested were employed in Montana (Johnson, 1972). These measures are sometimes called consumption ratios. They can be made more refined or sophisticated by taking changes over time into consideration, and by taking the geographical pattern of timber-origin and processor-destination into consideration.

The first two refinements can be illustrated by work in the Pacific Northwest (Wall and Oswald, 1975). Figure 3 shows estimates of employment consumption ratios based on time-series data. The equation resulting from fitting these data to some curve form is used to estimate future ratios. For example, if an analyst desired the appropriate ratio for target date 1975, that date would be entered into "X" and the equation shown solved for 1.55 employees per million board feet; note: "X" = (target date) - (1950). As a second refinement, the authors developed equations for several industrial classes; the multiplier above is applicable only to the logging industry. For these types of ratios to be useful, the analyst must be able to estimate the change in timber volume reaching, for example, western Washington. Timber source is an open question.

Figure 3 - Average annual employment in logging (SIC 2411) per million board feet of wood harvested in Washington by state area, 1950-1970

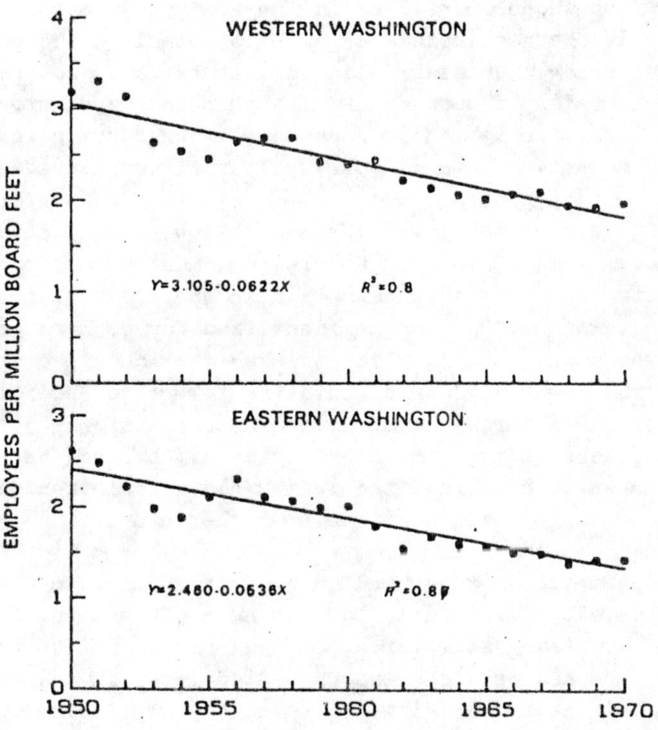

Source: Wall and Oswald, 1975

How does one construct consumption ratios? The following is the procedure outlined by Wall and Oswald (1975):

> To develop ratios for use in predicting employment impacts of changing material supplies, we need a historical base of employment and raw material consumption data. Employment data are generally available from state employment security departments. Raw material consumption data are often difficult to come by. We were able to get raw material consumption data for logging and pulping. However, for lumber and plywood manufacture, we had to develop consumption estimates based on historic production levels. We used recent production-consumption ratios to develop the estimates of consumption for the historic time series. This process assumes static utilization of raw material and possibly could result in an understatement of raw material use and, consequently, an overstatement of employment requirements in the early years of the time series data. The effect of this possible bias is not known, but we feel it is nominal and does not substantially affect the relationships developed in this paper.
>
> Since the technique of employment prediction through these ratios is generally used to determine impacts of changes in resource supply, it should be used for those industries whose operations in the area of concern will be affected by resource supply changes in that area. These include the primary resource oriented manufacturing processes: logging, lumber manufacture, plywood manufacture, and conversion of wood fiber into pulp and paper stock. The ratios should not be used for typically secondary manufacturing industries, whose location is due to market, or other non-raw material considerations.
>
> The factors that determine the level of labour input in the production process tend to change over time in most industries and most areas; consequently, the ratio of employment to raw material consumption in the production process also tends to change over time. If a trend can be identified for such ratios - that is, if we can identify the direction and rate at which they change over time - then we can extrapolate the identified trend into the future to predict employment required per unit of raw material consumed in the production process. This approach, of course, assumes that the aggregative effects of the changes in the underlying factors which have resulted in the historical trend in employment - consumption ratio(s) will continue into the future.
>
> The employment-consumption ratios used in projecting employment levels should, if possible, be developed from data for the area of concern. For instance, if projections are being made for a county, use of ratios developed for large areas, such as the half-state areas presented later in this report, could be quite inappropriate and misleading. In case of insufficient data in a small area, ratios from a broader area can be used if it can be determined that they are similar to those for the smaller area.

Some of the curves, if extended for many years, will indicate an employment-consumption ratio approaching or reaching zero employees per unit of wood consumed. This is unrealistic. In such cases the curves should be extrapolated for only short periods into the future. When extrapolations for longer time periods are required, the user must temper the curves based on the best available evidence and expert opinion of what the employment-wood consumption relationships will be like in the future.

Finally, any one of several curve forms may adequately fit the historical trend data. The potential user is then faced with making a selection. It is often expedient to select the simplest form that adequately defines the trend. But in this as well as other aspects of selection and use of ratio trends for projection of employment in the forest products industries, considerable judgement is required for the user.

An alternative set of procedures has been outlined by Bell (1977). These approaches follow, one being quick and inexpensive and the other more costly and time consuming:

Using Published Data

Step 1. From the equations presented in Wall and Oswald (1975) for Oregon, we get the following ratios for employees per million board feet.

Industry	Ratio
Logging	1.46
Sawmills & planning mills	3.49
Veneer and plywood	6.44

Step 2. From ... Oregon or local forest data, we get the distribution of the national forest harvest.

Industry	Volume (mm bd.ft.)	Percent of Total
Sawmill, planning mills etc.	322 914	77
Veneer and plywood	96 089	(23)
Total	419 003	100

Step 3. Assuming changes in harvest will be distributed to the mills in the same proportion as the present harvest, we derive a combined employees per million board-foot ratio:

$$\underset{\text{(logging)}}{1.46} + \underset{\text{(sawmills)}}{0.77 \times 3.49} + \underset{\text{(veneer)}}{0.23 \times 6.44} = \underset{\text{(ratio)}}{5.63}$$

In some cases, employment for pulpmills may also be added if they absolutely depend on the local residues from the veneer and sawmills for wood input. Generally, they have alternative wood supplies so that pulp production and thus employment will probably not change with changes in local harvest.

Using Original Data

Step 1. From a survey of individual mills, determine the average annual consumption of wood and average annual employment. Divide the number of employees by the amount of wood used to get the average employment wood-consumption ratio for each mill such as is shown in the fictitious example below.

Firm	Ratio
Average of eight local loggers	1.5
Silesia Milling Co.	3.0
Sidney Scrag Mill	2.1
Porter Veneer and Plywood, Inc.	8.6

A better procedure if the data can be obtained is to calculate the marginal rather than the average employment-wood consumption ratio. This could be obtained by observing the effect on employment of year-to-year changes in mill inputs, assuming no change in technology. Or it could be arrived at by asking the mill operator how employment would change with a given change in his wood supply.

Step 2. Estimate the proportion of increase or decrease in harvest going to each mill. This could be done on the basis of past harvest or it could reflect mill bidding power or the type of raw material involved. For the example, we will assume a decrease in the number of large logs will prevail.

Firm	Past harvest (percent)	Anticipated harvest (percent)
Silesia Milling Co.	30	40
Sidney Scrag Mill	20	20
Porter Veneer and Plywood, Inc.	50	40

Step 3. Apply the same principles as in the previous step 3.

$$\underset{\text{(logging)}}{1.5} + \underset{\text{(sawmills)}}{(0.4 \times 3.0 + 0.2 \times 2.1)} + \underset{\text{(veneer)}}{0.4 \times 8.6} = \underset{\text{(ratio)}}{6.6}$$

The third modification takes the relationship between the origin and destination of timber into account. Table 3 is based on official planning regions in Idaho (Schuster, Koss and Godfrey, 1975). All timber-using industry employment is aggregated. The focus of concern for the analyst is the location of timber harvest. For example, each million board feet of timber harvest in Region 1 was associated with 5.53 people employed in Region I, 0.31 in Region II, and no employ in the other regions - that is, timber was not delivered from Region I to Regions III-VI.

While several difficulties exist with use of consumption ratios, they do represent about the best measures available to quantify primary economic activity impacts regarding timber. The indicator used need not be employment. Any indicator - wages, employment, or sales - for which data are available will work. If the results are interpreted as relative indications, rather than absolute truths, misuse will be minimized.

Table 3. Timber harvest and employment in Idaho's timber-using industry, by planning region, 1972

Region of Timber Harvest	1972 Timber Harvest	Idaho State Planning Region of Timber Destination						Total 1972 Direct Employment
		Region I	Region II	Region III	Region IV	Region V	Region VI	
Region I	610.42	5.53	0.31	0.00	0.00	0.00	0.00	3,565
Region II	811.67	0.51	6.53	0.29 (0.23)[a]	0.00	0.00	0.00	5,950
Region III	207.61	0.00	0.07	15.26 (11.95)	0.11 (0.06)	0.00	0.00	3,206 (2,508)
Region IV	13.43	0.00	0.00	0.05 (0.04)	17.08 (9.90)	0.00	0.00	230 (133)
Region V	18.83	0.00	0.00	0.00	0.00	1.81 (0.74)	0.00	34 (14)
Region VI	96.45	0.00	0.00	0.75 (0.59)	0.00	0.00	5.76 (5.50)	628 (587)
Imports[b]	93.72	5.30	0.53	0.65 (0.51)	0.00	0.00	0.00	607 (594)
Totals	1,852.13							14,219[c] (13,302)

a. Counties with employment but receiving no timber were excluded from calculations; multipliers, where different, are shown in parentheses: these multipliers are similar to county multipliers.

b. Imports were estimated from unpublished data obtained in the Idaho Forest Industry Study.

c. Due to rounding errors, this total does not agree with the actual totals of 14,223 and (13,305).

Source: Schuster, Koss and Godfrey, 1975.

This section will be concluded with an example (Table 4) that deals with calculation of primary wage and employment impacts of a 10 million board feet decrease in timber harvest in southwestern Oregon. Primary economic impacts are reflected in the logging, plywood and sawmill industries.

Primary economic impacts should be calculated for every distinct aspect of the forestry project and then combined into a total of primary impacts. Accomplishing this, the question remains: What are the aggregate impacts of the forestry project on economic activity?

Table 4. Example: Calculation of primary employment and wage impacts

Let 10 million board feet now be subtracted from the allowable cut of that portion of the Umpqua in Douglas County. The new level of allowable cut is 347.0 million board feet annually. What are the probable economic impacts or adjustment burdens regarding Douglas County? There will be impacts. The following will discuss these under the four major impact classes described earlier.

Economic Activity. Given the state of economic activity in Douglas County, how will a change in allowable cut be reflected? The first consideration involves the relationship between allowable cut and actual cut. It seems clear from Table 25 that an allowable cut reduction would be directly translated into lower actual cuts. Of the five indicators recommended, two will not be evaluated. Since costs and gains regarding other outputs are assumed to be constant, they will not be evaluated. Additionally, data are not currently available to adequately evaluate changes in value-added. To show the effect of timing on impact evaluations, target years 1970 and 1976 will be used where useful.

How will the decrease of 10 million board feet of timber harvest be distributed among the timber-using industries of Douglas County? Sales between the Forest Service and timber-using industries (Darr and Fight, 1974) suggest the relationship below. If the harvest change were allocated on the basis of percent

Industry	Sales	Percent	Harvest (mmbf)
Logging	US$ 425,152	4.1	- 0.41
Plywood	8,409,481	80.3	- 8.03
Sawmills	1,637,734	15.6	- 3.56
	10,472,367	100.0	-10.00

sales, the 10 million board feet would be distributed as above.

One way to determine primary impacts on employment and wages is to follow the methodology outlined by Wall and Oswald (1975). The following shows the employment-consumption ratios for the industries of concern. These ratios are

Industry	Equation		Consumption ratio (Y) 1970	1976
Logging	Y = 1,748 - 0.012	x	1.4900	1.4132
Plywood	Y = 14,6011 - 0.4969	x	4.6631	1.6817
Sawmills	Y = 7,832 - 0.2135	x	3.5620	2.2810

expressed in terms of employment per million board feet. While the linear form expressing ratios was used, log forms are also available. When the calculated ratios are applied to timber volumes, the following employment and wage levels result. The 1970 wage changes were based on prevailing industry wage rates (USDC, 1971). The 1976 wage levels used rates that assumed the same proportionate increase occurring between 1970 and 1973 would occur to 1976.

Industry	Employment 1970	Employment 1976	Wages 1970	Wages 1976
Logging	- 0.61	- 0.58	- 4,410	- 5,349
Plywood	- 37.44	- 13.50	- 298,472	- 176,985
Sawmills	- 5.56	- 3.56	- 43,268	- 41,367
	- 43.61	- 17.64	- 346,150	- 223,702

Source: Schuster, 1976b.

4.2 Aggregate Impacts - Economic Base Analysis

The consumption ratios just discussed speak only to levels of employment or wages associated with primary consumers of timber. Will the effect on economic activity stop with these primary processors? Probably not. What about other economic activity associated with timber processing - the equipment dealers, food bought by woods workers and other support activities? Income generated in the primary timber-using industry is spent and re-spent. Consequently, additional rounds of economic activity can be envisioned; these can be lumped together and called "secondary" effects. Aggregate change in economic activity is the total of primary and secondary effects. Investigations of aggregate impacts on economic activity normally employ either economic base or interindustry (input-output) analysis. This section deals with economic base analysis, the following with input-output analysis.

Regional scientists have attempted to explain growth of an area in terms of its "economic base". Fundamental to this concept is the fact that, since local areas do not print their own currency, money flows into the area from outside. This normally occurs when markets outside the area demand and purchase goods and services produced internally. Firms that export most of their output and thereby create an inflow of money to the area are classified as "basic". These industries are "basic" to growth and development of an area. They are often termed "export" because they serve markets outside the area. Figure 4 shows that the income flowing to an area is the result of selling goods and services to "export" markets. This income in turn partly goes to paying other local industries for necessary goods and services. Sometimes these other local industries are lumped into a class called "non-basic" and sometimes they are split into "non-basic" and "support" industries. The output of "support" industries are inputs needed in the productive process of the basic industries. For example, a fertilizer industry may provide a necessary input to a timber growing industry that exports its output. Local "support" industries thereby substitute for outside area suppliers. The "non-basic" industries (food stores, barber shops) then provide the goods and services needed to maintain other aspects of the basic industries. In all likelihood, most industries lie along a continuum between completely basic (all income received from outside the area) to completely non-basic (all income received from inside the area).

We should emphasize here that the source of income is important. It is not necessary for output to be exported, but rather for income to be imported. In most cases, these concepts will not be in conflict. In the case of recreation, however, the distinction is important, for recreation commodities are often produced and consumed locally. This industry is "basic" then to the extent that recreationists from outside the region come to the region and spend money earned from outside the region. In this circumstance, the income is effectively imported but the output is not exported.

Figure 4. Role of basic industries in an area's economy.

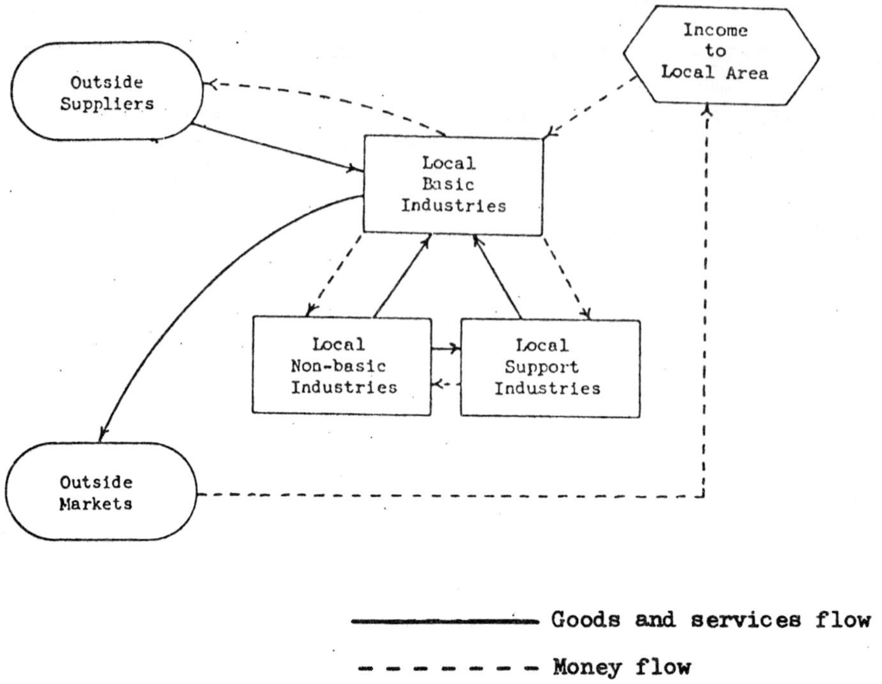

Growth of an area's basic industry is associated with changes in income receipts from outside areas. Changes in other activities (non-basic) are merely the result of overall change in the basic industries. The multiple effect on economic activity is measured by the "economic base multiplier" (Barkley and Allison, 1968). The essential concept underlying any economic base multiplier is that increases in any exogenous variable (e.g. basic industry income or employment) are magnified by other transactions that occur within the area. The area's income or employment is increased by a multiplied amount because of the re-spending that occurs within the area as a result of income received from sales of goods and services produced in the area, but sold outside the area. Multiplier analysis then is relevant and appropriate for analyzing impacts if, and only if, the industry being altered is part of the area's economic base.

Calculation of an economic base multiplier (EBM) is relatively straightforward.[1] The economic base multiplier is applicable to any indicator of economic activity (employment, wages, sales, etc.) for which an appropriate data base exists. The EBM is simply the ratio of total economic activity to economic activity in the economic base. Consider an income multiplier. If the proportion of total income re-spent in an area is "r", the total impact of a change in export income in the area is:

$$\Delta Y = (1 + r + r^2 + r^3 + r^4 + \dots + r^n)\,\Delta Y_B = \left(\frac{1}{1 - r}\right)\Delta Y_B$$

[1] For an excellent discussion of multiplier interpretation, see Coppedge and Youmans (1970).

where:

r = the proportion of each dollar that is re-spent within the region by residents ($0 < r < 1$)

ΔY_B = change in basic industry income

ΔY = change in area's income

n = spending round

Let "r" be replaced by (Y_N/Y) where Y_N is non-basic industry income ($Y_N = Y - Y_B$). The economic base multiplier then becomes:

$$EBM = \frac{1}{(1 - Y_N/Y)}$$

$$= \frac{1}{1 - (Y_N/(Y_N + Y_B))}$$

$$= \frac{Y}{(Y_N + Y_B) - Y_N}$$

$$EBM = Y/Y_B$$

The multiplier is applied:

$$\Delta Y = EBM (\Delta Y_B)$$

If the economic multiplier were calculated to be 1.65, then an income change of \$1 000 in the basic industry would be associated with a \$1 650 income change in the total area (\$1 000 x 1.65). Of the \$1 650 change, \$1 000 is associated with the basic industry and \$650 with the non-basic industries.

The fundamental problem in economic base analysis is determining which industries constitute the economic base. Various approaches have been articulated by Hoover (1971), Convery (1973) and Tiebout (1962):

(i) Definition - Examine each sector individually, and determine whether it is basic or non-basic. For very small economies this approach can to some extent be justified because most of the sectors will either export almost all of their output or fulfill a clearly local supportive role. For a small, rural area using this method, agriculture, mining and manufacturing would be classified as basic, while all of the rest (including construction) would be categorized as non-basic. However, even at this level there are some sectors which are both basic and non-basic; as the economy gets larger these joint sectors are more likely to be found.

(ii) Location Quotients - To separate the basic from the non-basic employment in an industry, the use of a location quotient has been advocated. The employment identified with the portion of the coefficient greater than one is classified as "export". Implicit in this approach is the assumption that local patterns of use and habits of consumption are the same as average national ones. Locations quotients will be discussed shortly.

(iii) Minimum Requirements - To overcome this latter problem, Ullman *et al* (1971) developed what they call a minimum requirements approach. They take all of the cities in a particular size class, and determine the proportion of the total work force in each city employed in each sector of the economy. The lowest percentage for each sector from all the cities is designated as the "minimum requirements" percentage for that sector for cities in this size class. By regressing these minimum requirements (Y) against population (X), parameters (a) and (b) can be determined for:

$$Y = a + b \log X$$

The best estimate of total minimum employment requirement (non-basic employment) is given by this equation.

(iv) Direct Measurement - Given sufficient funds, export industries can be measured directly. Methods available fall into two groups: a) direct measurement of commodity and dollar flows by means of various data sources and b) surveys through interviews and questionnaires of consumers and firms.

Use of location quotients has received much attention in forestry circles, and warrants further discussion. Location quotients (sometimes called the coefficient of localization) are quantitative indexes based on a comparison of economic activity in the industrial sectors of an area to a larger, regional or national system. The listing below shows the hypothetical percentage distribution of employment in industries for a local area and a larger area.

Industry Class	% Employment in Local Area	% Employment in Larger System	Location Quotient
X	.60	.20	3.0
Y	.30	.40	.75
Z	.10	.40	.25
Total	1.00	1.00	

Location quotients are calculated by dividing an area's percentage for an industry by the larger system's percentage for that industry. A location quotient in excess of 1.0 is normally accepted to identify an export or basic industry. The percentage distribution of any economic activity (wages, employment, etc.) can be used to calculate location quotients. Once calculated, the analyst may simply define basic industries to be those which have values in excess of 1.0 for the location quotient. The logic in this approach is that once specialization exists, production predominately serves export markets.

The analyst may use the data above to go to the next step. In the case of employment, estimate "excess employment". Upon reflection, the location quotient will be found to reflect an area's economic specialization. The overall degree of specialization may be determined by subtracting the area's percentage distribution of employment, by industry, from that of the larger system. Where an area's percentage exceeds the larger system's percentage, it is termed "excess"; employment in that industry is in "excess" of national or regional requirements. In this event, only that portion of economic activity associated with the "excess" is included in the economic base. The listing below shows that excess employment levels are calculated from the same data base as location quotients.

Industry Class	Employment in Local Area	% Employment in Local Area	% Employment in Larger System	% Excess	Excess Employment
X	600	.60	.20	.40	400
Y	300	.30	.40	-.10	-
Z	100	.10	.40	-.30	-
Total	1 000	1.00	1.00	.40	

In the case of the location quotient, the percentages are divided while to calculate excess, they are subtracted. This latter approach argues that both internal and external markets are served on a basis proportional to the degree of excess.

Opportunities to critique and extend the several methods available to estimate size of the economic base abound. Others have undertaken this effort (see Isserman, 1977; Mathur and Rosen, 1974; and Tiebout, 1962).

Where does one obtain the data necessary to conduct an economic base analysis? The answer depends on the approach adopted. If an analyst had detailed knowledge of an area's economy, the analyst could simply define, identify or otherwise list the basic industries. However, this method is risky since it relies on the analyst's impressions of reality; these may not be correct. Alternatively, a specialized study of firms in an area could be conducted to determine the amount of export income associated with each firm. But even with these approaches, the analyst will probably have to research secondary data sources to determine levels of total economic activity. These secondary data sources are the primary source of information for the location quotient, excess employment, and the minimum requirements approaches to defining the economic base.

Once the analyst determines a numerical value for the economic base multiplier, accurate estimates of primary impacts become very important. Estimates of primary impacts provide the driving force behind economic base multipliers - that is, if these primary impact industries are part of the economic base. To determine aggregate change, the analyst multiplies the primary change by the multiplier. Calculation of aggregate change is shown in Table 5 that follows, for a complex forestry project. While the economic base approach is reasonably common and inexpensive, a methodologically dissimilar approach to aggregate impacts exists - input-output analyses.

Table 5. Example: Calculation of aggregate impacts with economic base multipliers

THE IMPACT AREA

Using the criteria outlined earlier, the local impact area was identified for the Hiwassee Unit; Bradley, Polk, Monroe and McMinn Counties were included (see fig. 1). Detailed ten year plans (1970-1980) dealing with population, the economy, land use, transport systems and housing are available for Bradley and Polk counties (Tennessee State Planning Commission, 1971a, 1971b), while plans are in preparation for the counties of McMinn and Monroe.1/

The area is one with a record of modest but consistent economic growth centered mainly around Cleveland, Tennessee, the "growth point" of the area. Growth in the manufacturing and mining expanding from 14548 in 1960 to 22291 in 1970 (Table A). Median family income in 1969 was $ 7101, 4.8 percent below the figure for the state, while 18.25% of all families were classified as being below the poverty level, compared with 18.2 percent for the state (Bureau of the Census: General Social and Economic Characteristics).

Table 1. Employment in the Hiwassee Sub-Region1/, 1960 and 1970

Year	Agric., For. & Fisheries	Basic Manuf. & Mining	Total No.	Non-basic Other	Total	Total basic
1960	3974	14548	19522	17659	36181	1.95
1970	2141	22291	24432	22409	46841	1.91

1/ Counties Bradley, McMinn, Monroe and Polk.

Sources: Census of Population 1960: General Social and Economic Characteristics Tennessee: Census of Population 1970 ibid.

IMPACTS MEASURED

The outputs of the Hiwassee unit are listed in Table B. Basic employment generated by each of these alternatives must first be determined, and then the multiplier can be applied to determine non-basic and total employment. Payroll and value added data per employee can be derived and pplied to the employment estimates.

Table 2. Annual Outputs from the Hiwassee Unit under each Management Alternative

Mgmt. Alt.	Wood (MBF) Saw-logs	Pulp-wood	Total1/ value(s)	Recreation (Visitor Days)2/ Developed	Dispersed Motor-ized	Dispersed Non Motor.	Hunting
Present	2786	1984	97587	7500	14214	10386	6150
1	2925	2084	102459	7875	15635	10178	6460
2	2692	1917	94293	8250	15635	10905	6765
3	1388	988	48616	20010	16346	11425	6150
4	2773	1976	97135	8250	14924	10905	6460

1/ Using the average price received over the past 3 years, 1970, 71, 72; $32.25/M.B.F. for sawlogs and $3.9/M.B.F. for pulpwood.

2/ Visitor day means 12 hours of recreation use by one person.

Source: U.S.F.S. Southern Regional Office (1971b). Unit plan for Management of the Hiwassee Unit, Cherokee National Forest, p. 27.

Table 5 (cont.d)

NON-BASIC AND TOTAL IMPACTS

By applying the multiplier (1.91, from Table 1) total and non-basic employment can be derived (Table 12 and Figure 2a). Payroll per employee in the non-basic sector is derived as the state-wide weighted average of annual employee payroll in the selected services, construction, wholesale trade and retain trade sectors, and works out at $4453 for 1967 1/. Non-basic and total payroll can now be displayed for each alternative (Table 12 and Fig. 2b).

Table 12. Basic Non-Basic and Total Employment (E) and Payroll (P) Generated by each Management Alternative, Hiwassee Unit.

Alternative	Basic			Non-Basic		Total	
	E	P	V.A.	E	P	E	P
	No.	$	$	No.	$	No.	$
Present	91.6	453349	736254	83.4	371380	175.0	824729
1	96.4	477474	784241	87.7	390528	184.1	868001
2	90.8	477935	715820	82.6	367818	173.4	815753
3	63.3	303841	367780	57.6	256493	120.9	560334
4	91.4	451849	727426	83.2	370490	174.6	822339

Source: Table 11.

Table 11. Total Basic Employment (E), Payroll (P) and Value Added (V.A.) 1/ Generated by each Management Alternative, Hiwassee Unit

Alternative	Management		Wood					
			Harvesting			Processing		
	E	P	EE	P	V.A.	E	P	V.A.
	No.	$	No.	$	$	No.	$	$
Present	10	70,000	16.3	59413	113268	50	271956	622986
Alt. 1	10	70,000	17.1	62330	118828	53	289747	665413
Alt. 2	10	70,000	15.7	57226	109100	48	263604	606720
Alt. 3	10	70,000	8.1	29524	56287	25	135978	311493
Alt. 4	10	70,000	16.2	59049	112573	49	267780	614853

Alternative	Recreation		Total		
	E	P	E		
	No.	$	No.	$	$
Present	15.3	51980	91.6	453349	736254
Alt. 1	16.3	55397	96.4	477474	784241
Alt. 2	16.8	57105	90.8	447935	715820
Alt. 3	20.2	68339	63.3	303841	367780
Alt. 4	16.2	55020	91.4	451849	727426

1/ With the limitations noted in the text.

Sources: Tables 4, 5, 10

Source: Convery, 1973.

4.3 Aggregate Impacts - Input-Output Analysis

Input-output is one of the most powerful descriptive and analytical tools available to the analyst evaluating distributional consequences of a forestry project. Also called "interindustry analysis", it belongs to the general family of "activity analyses", which also includes linear programming. The nature and degree of interdependence among producing and consuming sectors of an economy is basic to input-output analysis. Analysis of the linkages between sectors in an economy is not new in economics, Quesnay and Walras undertaking such efforts over a century ago. Yet modern efforts dealing with input-output analysis can be traced back only a few decades to the work of Wassily W. Leontief and development of high speed electronic computers.

Input-output analysis can be used in two major ways to evaluate forestry projects. First, it can determine the level of output in all industries of an economy that will be just sufficient to satisfy total demand of forestry products. Second, it can be used to assess the changes in output of various industries in an economy associated with a change in supply of inputs from the forestry sector. These efforts can be accomplished by means of a set of multipliers, somewhat similar to the economic base multipliers previously discussed. This section will briefly review input-output analysis and its application to forestry projects.

Input-output analysis is basically an accounting system that describes dollar or volume flows of commodities between all sectors of an economy. Each sector not only produces goods and services consumed by other sectors but is also a consumer itself, purchasing goods and services to be used in its production process. Data used to describe these linkages are often based on empirical, statistical estimates - typically reflecting one point in time. As such, input-output tables are only as good as the data upon which they are based. Although often used to predict future economic structures, input-output is inherently a "static" model representing a "snapshot" of an economy at a point in time.

The Input-Output Table - A complete input-output analysis basically develops and uses three types of tables. The first table constructed can be called the Input-Output Table - alternatively this can be called the "dollar flow" table or "transactions" table. This table shows the flow of goods and services (measured either in volume or value) among industrial sectors of the economy. Figure 5 shows that each industry appears twice, once as a producer of output and once as a consumer of input. The elements of each row show the amount of a given industry's output that was used by every other sector to produce their own output, and how much was bought by the final consumer. The rows show the distribution of the output, the "market mix", for each industry, summing to total output. The elements of each column show the source of each industry's input of raw materials, semi-finished products and services bought from various supplier industries. Columns show the pattern of input purchases made by a given industry. With some exceptions (the primary input row, the final demand column and the total output column), for every row there is a corresponding column. These rows and columns amount to an itemization of input and output of each designated producing and consuming industry.

Figure 5. Sample Input-Output Table

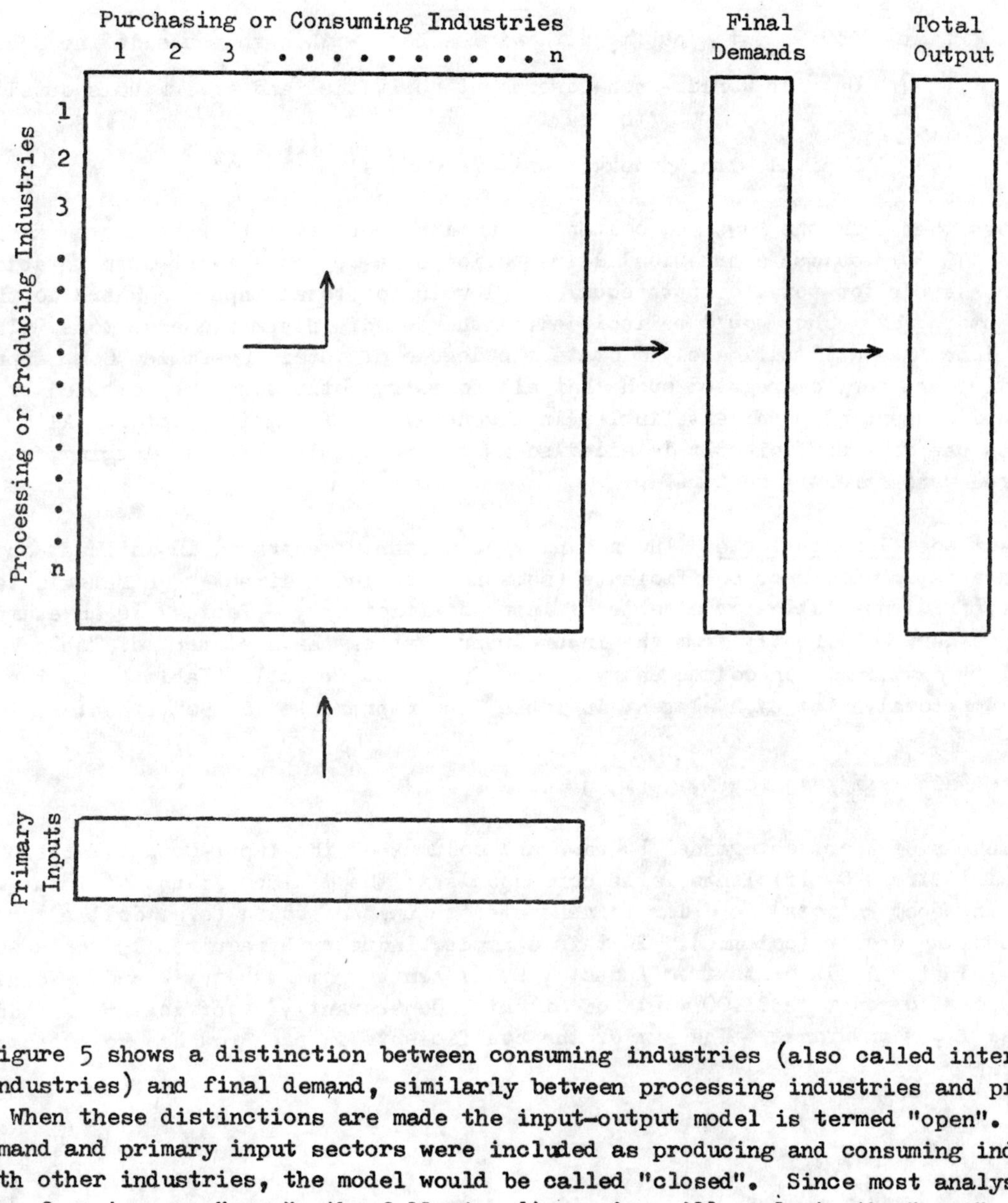

Figure 5 shows a distinction between consuming industries (also called intermediate demand industries) and final demand, similarly between processing industries and primary inputs. When these distinctions are made the input-output model is termed "open". If the final demand and primary input sectors were included as producing and consuming industries along with other industries, the model would be called "closed". Since most analyses concerning forestry are "open", the following discussion will apply to the "open" model.

What sectors are included in an input-output table? Hoover (1971) has identified five major, fundamental sectors:

(i) Intermediate - private business activities within the region;

(ii) Households - individuals and families residing or employed in the region;

(iii) Government - public authorities both within and outside the region;

(iv) Outside World - non-government activities and individuals outside the region;

(v) Capital - the stock of private capital.

These sectors then form the rows and columns of an input-output table. Table 6 (adapted from Hoover, 1971) provides a numerical illustration of a region's input-output table. As can be seen, levels for total outputs equal the levels for total inputs. Where do the forestry sectors fit? They would be included in the set of intermediate sectors. The underlying data base will ultimately dictate the degree of detail possible Some input-output studies are very aggregated such that all forestry activities are combined or even subsumed (and consequently unidentifiable) in a general manufacturing sector. Alternatively, some studies use data sufficiently detailed so as to distinguish between logging, sawmills and other relevant forestry sectors.

Direct Coefficients Table - The second type of table generated in an input-output analysis is a table of direct coefficients (sometimes called a direct requirements table, technical coefficients table, or a table of input coefficients) - Table 7 is an example. This table is derived directly from the Input-Output Table. Each element of Table 7 is constructed by dividing each column entry of the Input-Output Table (Table 6) by the corresponding column total. Let each element in Table 7 be represented as "a", then:

$$aij = Xij/X{\cdot}j \text{ for } i,\ j\ (i,\ j = 1\ .\ .\ .\ n)$$

where the levels of X are determined by rows and columns of the Input-Output Table. Each element of the Direct Coefficients Table corresponds to the minimum amount of input from each producing sector (rows) in order to generate one unit of value (e.g. dollar) or output in the purchasing sector (columns). In this example, Industry D requires 19 cents worth of input from Industry A, 12 cents from Industry B, 27 cents from Industry C and 19 cents from itself in order to generate $1.00 worth of output. Consequently, coefficients in each column equal 1.0 when summed. The sum of the coefficients in each row has no economic meaning.

Table 6. Sample input-output table

	PURCHASING SECTORS								
	Intermediate Sector, by Industry				Final Demand Sectors				
	A	B	C	D	Households (Consumer Goods Sales in Region)	Government (Sales to Government)	Outside (Exports)	Capital (Gross private investment including additions to Inventories	Output Totals
Intermediate processing sector, by industry: A	300	400	100	500	1600	500	200	700	4300
B	50	200	1000	300	100	200	100	900	2850
C	1000	200	100	700	100	300	200	500	3100
D	0	800	200	500	700	0	0	400	2600
Primary input Supply sectors									
Households (Labour Services)	1900	300	1000	400					
Government (Public Services)	200	100	200	100					
Outside (Imports)	200	300	300	0					
Capital (Capital consumption and withdrawals from inventories)	650	550	200	100					
Input Totals	4300	2850	3100	2600					

Table 7. Sample Direct Coefficients Table

Processing Sector	Purchasing Sector			
	A	B	C	D
A	.07	.14	.03	.19
B	.01	.07	.32	.12
C	.23	.07	.03	.27
D	.00	.28	.07	.19
Primary Inputs	.69	.44	.55	.23
Total	1.00	1.00	1.00	1.00

A Direct Coefficients Table can be immediately useful to the analyst of forestry projects. Assume the listing below represents the direct coefficients of the sawmill sector. Suppose the sawmill sector increased its sales of lumber outside the region by one million dollars. In order to do this, it must increase purchases from the construction sector by $600 (.0006 x $1 000 000), from the logging sector by $291 700, from the transportation sector by $26 000, and so forth. The direct effect on other industry sectors can be thereby determined. These direct effects get translated into employment and wages in these support industries. Do the effects of increased lumber sales now stop with these industries? No. In order to produce $291 700 of additional output, the logging sector must increase use of its supply inputs. This generates another round of purchasing, which in turn generates another round. The summation of all subsequent rounds of economic activity generated by increased lumber sales is called "indirect effects".

Producing Sectors	Purchasing Sectors	
	Sawmills	Others
Construction	.0006	.4488
Logging	.2917	.0000
Sawmills	.0923	.0900
Transportation	.0260	.4212
Primary Inputs	.5894	.0400
Total	1.0000	1.0000

Direct and Indirect Coefficients Table - The aggregate or total effect (direct and indirect) of a change in final demand for an industry sector is shown in a table of direct and indirect coefficients - alternatively called total requirements table, direct and indirect requirements table, total direct and indirect effects table, inverse coefficients table and the Leontief inverse table.

Given the information provided in a Direct Coefficients Table, the analyst could conceivably calculate the effect of each additional round of spending discussed above, and then total all such calculations. Fortunately, this labourous task can be accomplished much more simply by use of matrix algebra. Development and use of this technique is generally attributed to Wassily W. Leontief. Information contained in the original Input-Output Table and the Direct Coefficients Table are used to implement this procedure. Let "X" represent the vector of total outputs and "Y" represent the vector of final demands (summed) for the processing sectors shown in an Input-Output Table. Let "A" represent the matrix of coefficients for each processing sector shown in a Direct Coefficient Table. Total output can then be expressed in matrix notation as:

$$X = AX + Y$$

The listing below shows these matrixes using the data provided in Tables 6 and 7:

$$\begin{bmatrix} 4\ 300 \\ 2\ 850 \\ 3\ 100 \\ 2\ 600 \end{bmatrix} = \begin{bmatrix} .07 & .14 & .03 & .19 \\ .01 & .07 & .32 & .12 \\ .23 & .07 & .03 & .27 \\ .00 & .28 & .07 & .19 \end{bmatrix} \cdot \begin{bmatrix} 4\ 300 \\ 2\ 850 \\ 3\ 100 \\ 2\ 600 \end{bmatrix} + \begin{bmatrix} 3\ 000 \\ 1\ 300 \\ 1\ 100 \\ 1\ 100 \end{bmatrix}$$

In matrix notation, the vector of final demand (Y) can be expressed:

$$(I-A)X = Y$$

where I is an identity matrix. This system of numbers is shown in the listing below:

$$\left[\begin{bmatrix} 1\ 000 \\ 0\ 100 \\ 0\ 010 \\ 0\ 001 \end{bmatrix} - \begin{bmatrix} .07 & .14 & .03 & .19 \\ .01 & .07 & .32 & .12 \\ .23 & .07 & .03 & .27 \\ .00 & .28 & .07 & .19 \end{bmatrix} \right] \cdot \begin{bmatrix} 4\ 300 \\ 2\ 850 \\ 3\ 100 \\ 2\ 600 \end{bmatrix} = \begin{bmatrix} 3\ 000 \\ 1\ 300 \\ 1\ 100 \\ 1\ 100 \end{bmatrix}$$

In the language of input-output analysis, the expression (I-A) is called the Leontief Matrix.

Elements of the Direct and Indirect Coefficients Table are defined by the elements of the Leontief Inverse Matrix, $(I-A)^{-1}$. This matrix is used to solve for total output:

$$(I-A)^{-1}Y = X$$

Table 8 shows the results of this matrix inversion process for the data described and as shown in Tables 6 and 7. Table entries, excluding totals, correspond to the Leontief Inverse Matrix. Each entry in the table represents the total output, direct and indirect, required from each of the processing sectors necessary for a one dollar increase in final demand of the purchasing sector. The column total in the table represents the combined effect. This total is called either the "total output multiplier", or the "final demand multiplier", or the "business multiplier".

Table 8. Sample Direct and Indirect Coefficients Table

Purchasing Sector	Purchasing Sector			
	A	B	C	D
A	1.12	.29	.16	.36
B	.13	1.23	.44	.35
C	.30	.28	1.17	.50
D	.07	.45	.25	1.40
Total	1.62	2.25	2.02	2.61

Using our earlier example of sawmills and lumber sales, the listing below illustrates the direct and indirect coefficient for the sawmill sector.

Producing Sector	Purchasing Sector	
	Sawmills	Others
Construction	.0012	1.095
Logging	.3587	.000
Sawmills	1.1198	.250
Transportation	.0367	.928
Total	1.5164	2.273

If sales of lumber outside the region were to increase by one million dollars, changes in the value of output in producing sectors associated with this would eventually amount ot $1 200 in the construction sector, $358 700 in the logging sector, $1 119 800 in the sawmill sector, and $36 700 in the transportation sector. Summing all, the total increase in output value would be $1 516 400. One useful way of visualizing this change (Goldman and Nakazawa, 1974) is to disaggregate the total change into components:

Total Multiplier Effect	=	Initial Economic Change	+	Initial Economic Effect	+	Secondary Economic Effect
∴ ($1 516 400)	=	($1 000 000)	+	($410 600)	+	($105 800)

Where: a) the initial change is due to the increased lumber sales; b) the initial economic effect represents the proportion of additional internal inputs required by the sawmill sector to satisfy the increased sales level (sum of direct coefficients between sawmill sector and producing sectors shown earlier (.4106 = 1.0 - .5894) times the change in sales level); and c) all subsequent adjustments in the economy needed to satisfy increases in the internal purchases by the sawmill sector.

Wage and Employment Impacts: In many instances, change may be more meaningfully reflected in terms of employment, wages or earnings. The analyst may either modify original matrix calculations or adopt a procedure that uses final demand multipliers (FDM) and some additional information. A reasonably straightforward procedure has been recently outlined (BEA, 1977).

Consider the change in wages or earnings first. Given an Input-Output Table (such as Table 6), the analyst can quickly identify gross output (GO) for a specific industry. Next, either wages or earnings for the industry being analyzed must be determined. Since wage or payroll data are more commonly available from secondary data sources than are earnings data, the following focuses on wages. Simply calculate the wage/gross output ratio for the industry in question ($E_j = W_j/GO_j$). However, since the total gross output change applies to all industries, E_j cannot be applied to the total change. This expression must be modified before change in total wages can be determined. The quantity "e_j" must be calculated as follows:

$$e_j = (1/FDM_j)(E_j) + (1 - 1/FDM_j)(E\cdot)$$

where FDM_j is the final demand multiplier for industry j, E_j is as before, and $E\cdot$ is a total wage/gross output ratio for the nation or region. The change in total wages can then be calculated:

$$\Delta \text{ Total Wages} = (\Delta TGO)e_j = \Delta FD_j(FDM_j)(E_j)$$

where ΔFD_j is the change in final demand for the industry in question. In this case, the wage multiplier is:

$$WM_j = FDM_j(E_j)$$

The wage multiplier could then be used as was the FDM to assess the consequences of a change in final demand. If earnings are used instead of wages, these data simply replace wage data in all calculations.

Using the final demand multiplier for the sawmill sector (j) developed earlier, the following illustrates a wage multiplier calculation:

if ... $FDM_j = 1.5164$

if ... E_j = wages/gross output = .239

if ... $E\cdot = .3$

then ... $e_j = (1/1.5164).239 + (1 - 1/1.5164).3$
$= .260$

then ... $WM_j = 1.5164\ (.260)$
$= .394$

if ... ΔFD_j = \$1 000 000

then ... ΔWages = $(\Delta FD_j)(WM_j)$
= 1 000 000 (.394)
= \$394 000

Determination of employment change only requires knowledge of the employment level and the level of total wages for the region. These, again, can be obtained from secondary data sources. Calculate the ratio of total employment/total wages (TE/TW). The change in employment is then calculated:

Δ Employment = Δ Wages (TE/TW)

For example, using the past illustration:

if ... TE/TW = .0001

then ... Δ Employment = \$394 000 (.0001)
= 39.4

Forward and Backward Linkages: The most common way of using input-output analysis is to evaluate changes in any economy that result from increased sales in final demand. For example, the analyst might increase final demand for output of the sawmill sector and then tract the inpacts this would have on other sectors as well as the economy in total. Because this approach essentially works backward from final demand in a specific sector back into the economy, these economic impacts are called "backward-linked". It is particularly useful in analyzing situations including product market or demand forces. However, another situation can exist: supply modification. Waggener (1972) illustrates this point:

> If we assume that the level of timber harvesting is reduced within a particular community, our traditional analysis will trace the consequences backward through the equipment supplier, the logging contractor, the sale of gasoline to the logging truck drive, and the related indirect consequences ... But is this the end? There is certainly reason to believe that it is not. As you know, those logs went someplace - perhaps to the mill on the edge of town ... The change in timber supply in this case spills "forward" into the primary producing sector ..., this forward impact leads to substantially larger impacts than would the backward effect considered above.

Economic impacts based on supply considerations are termed "forward-linked". Recognizing the distinction between demand and supply constraints, Darr and Fight (1974) calculated a backward-linked multiplier of 1.88 and a forward-linked multiplier of 7.53 for Forest Service timber sale activities in Douglas County, Oregon.

There are two general approaches that can be used to determine forward-linked effects. The first involves manipulation of the Input-Output Table (Hoover, 1971). Earlier, a table of direct coefficients was calculated where each of the elements was defined:

$$aij = Xij/X{\cdot}j$$

These could be termed "input coefficients" and were illustrated in Table 7. A similar table must be constructed for the forward linkages by calculating a set of coefficients:

$$a^*ij = Xij/Xi{\cdot}$$

where each element of the Input-Output Table is divided by the row total. Using the data provided earlier, Table 9 shows these forward or supply linked coefficients. This table shows the distribution of output for each of the processing sectors, the distribution of sales. Sales provided by the processing sectors provide the locally supplied inputs to the purchasing sectors. Excluding the total and the coefficients found in the final demand column, the elements of the Table of Forward-Linked Coefficients form a matrix (A*) similar to the matrix described earlier (A). As was done before, the elements of the forward-linked multipliers are constructed:

$$(I - A^*)^{-1}$$

These elements are shown in Table 10. Again, as before, the horizontal sum of all coefficients is interpreted as the supply multiplier for a processing sector.

Table 9. Sample Table of Forward-Linked Coefficients

Processing Sector	Purchasing Sector				Final Demand	Total
	A	B	C	D		
A	.07	.09	.02	.12	.70	1.00
B	.02	.07	.35	.11	.45	1.00
C	.32	.06	.03	.23	.36	1.00
D	.00	.31	.08	.19	.42	1.00

Table 10. Sample Table of Forward-Linked Multipliers

Processing Sector	Purchasing Sector				Total
	A	B	C	D	
A	1.12	.24	.11	.22	1.69
B	.02	1.53	.48	.32	2.35
C	.41	.32	1.18	.42	2.32
D	.12	.61	.11	1.40	2.24

An alternative approach does not involve manipulation of any tables in the backward-linked approach to input-output (CSC, ca 1969; BEA, 1977). The first and fundamental step is to determine the relationship between a change in output of the supply industry (processing industry) and the associated change in final demand in that industry. This relationship can be expressed:

$$\Delta FDj = \Delta Xi(1/Aij)$$

where ΔFDj is the change in final demand in industry; associated with a change in output of industry i, ΔXi is the output change of industry i, and Aij is as before. Given this level of new export demand, the change in total gross output can be calculated as before:

$$\Delta TGO = \Delta FDj(FDMj)$$

Please note: if the change in supply (ΔXi) is expected to result in changes in final demand for several sectors, this process should be extended to these sectors.

Indirect Versus Induced Effects: In the beginning of this discussion of input-output analysis, a distinction was made between "open" and "closed" models. The distinction involved treatment of the final demand sector. If this sector were present, the model was termed "open". In reality, the real issue is the degree of openness. The basic problem involved identification of sectors that are part of the producing economy (endogenous sectors) as opposed to those outside the economy (export or exogenous sectors). These decisions will affect the entire analysis and specifically the multipliers generated.

The nature of any multipliers developed in an input-output analysis depends on how the analyst constructs the flow of transactions between sectors - the linkages. Treatment of the household sector is pivotal, since it can be handled in different ways. One way is to treat households only as primary suppliers of inputs and as demanders of final products - not part of the producing economy. In this case, increases in economic activity are due solely to successive rounds of interindustry transactions called "indirect" or "linked effects" (CSC, ca 1969). Multipliers calculated under this convention are termed Type I. Alternatively, if it seems more reasonable that increased economic activity leading to increased household income will result in increased household expenditures within the region, households should then be treated as part of the producing economy. Increases in economic activity, in this case, are not only due to interindustry transactions of private business (indirect effects), but also to households respending income - called "induced effects". Multipliers calculated on the basis of indirect and induced effects are termed Type II. Since Type II additionally includes induced effects, these multipliers must be larger than Type I counterparts. When Type II multipliers are developed, the household sector is simply included in all matrix calculations which would be expanded to the extent of one row and one column.

Uses of Input-Output Analyses: Most input-output studies available for use in forestry emphasize large geographical areas, such as a state, province or region of a country. This geographical orientation may be inappropriately large. The analyst can either use existing studies, ignore input-output as a tool, or build a more appropriate model. Several input-output studies in forestry have been specifically designed for a county or group of counties such as in New Mexico (Drake et al, 1973), Idaho (Herbst, 1972), Minnesota (Hughes, 1970), California (Fowler, 1974). But because input-output is a very expensive form of study, many analysts rely on existing data to develop models. All other things being equal, the larger the geographical scope, the more expensive the study, and hence the greater tendency to use existing data. To illustrate, a multi-county input-output study in Idaho (Herbst, 1972) relied on an overall Idaho study (Rafsnider and Kunin, 1971a) which in turn was based on an earlier national study (USDC, 1969). On the other hand, a Minnesota study (Hughes, 1970) and a multi-county area study in Indiana (Reimer, 1969) involved collection of primary data. In local economic impact analysis the most useful type of input-output study would probably encompass a one- to three-county area for which data were specifically collected.

While use of existing data helps to avoid the expense of collecting primary data, another problem surfaces: the technical coefficients found in an input-output study pertaining to a large geographical area may be inappropriate for internal areas. The assumptions necessary to apply broad studies to small areas may be prohibitive. Generally, the assumption that the same pattern of linkages that exists for the larger area applies to the smaller one is dubious. Coefficients from the large area must, therefore, be transformed or otherwise made applicable to the smaller area. The problems and prospects of making needed transformations have been discussed elsewhere (Youmans and Stoevener, ca 1973; Rafsnider and Kunin, 1971a).

Input-output studies in forestry have been developed for two distinct purposes: a) to primarily develop a set of multipliers and b) to analyze the net consequences of a specified real change in an economy. These need not be mutually exclusive. The problem faced by most analysts is that existing studies may not be perfectly applicable to the situation being analyzed. What can be done? The data in an Input-Output Table certainly cannot be used because they are applicable only to a different specific situation. But the linkages between industry sectors shown in the other tables may be useful, to the extent that economies are similar. If two economies are similar, their coefficients would also be similar. Multipliers may also be similar. For example, two studies found an income multiplier of about 1.7 for the sawmill and planning mill sector of Idaho (Rafsnider and Kunin, 1971b) and the South (Kaiser and Dutrow, 1971). The difficulty faced by analysts attempting to use previous studies is that detail and computational capabilities available to the original authors may not be available to the analysts. Published multipliers may be the only usable information.

Discussion of input-output analysis will be concluded with a final remark and an example (Table 11) of applying backward linkages to a forestry project. The comment first. Most input-output analyses begin with an initial change in supply or final demand. They end with a measurement of aggregate change either in terms of total gross output, wages or earnings, or employment. The analysis that stops at this point is ignoring a major advantage of input-output analysis. For not only should the analyst be concerned with primary and aggregate impacts, but also with the distribution of these impacts as well. The analysis should disaggregate the aggregate impacts back to the industry sectors from which they came. If employment is to increase by 100, identify the industry sectors which will comprise this change - 50 in the forest products industry, 30 in the construction industry, and so forth. Data available in various tables will allow this to be done. And finally, even if the analyst does not have the computing capability to determine the direct coefficient or to invert a matrix, the original Input-Output Table at least describes a pattern of industry linkage. At minimum, this information can be used to identify and describe sectors likely to be affected by a forestry project.

Table 11. Example: Calculation of aggregate impacts with backward linked input-output multipliers

Backward Linkage Effects

$ 10,000 change in value of output of the Forest Products Sector (11) exported from the region

ΔTGO = change in total gross output of all industries due to change in demand for exported output of industry j

ΔTE = change in total earnings in the region due to change in demand for exported output of industry j

ΔD_{E_j} = change in demand for exported output of industry j

M_j = regional multiplier for industry j

e_j = factor for converting a change in gross output to a change in earnings

a_{h_j} = household coefficient for industry j, representing sales of households (labour) to industry j

E. = national earnings/gross output ratio = .3008

Given: ΔD_{E_j} = 10,000 From Figure 2: M_j = 1.556

E. = .3008 a_{h_j} = .0890

Change in total gross output: $\Delta TGO = \Delta D_{E_j}(M_j)$

$= 10{,}000(1.556)$

$= 15{,}560$

Change in earnings: Step 1 – $e_j = (1/M_j)(a_{hj}) + (1-1/M_j)(E.)$

$= (1/1.556)(.0890) + (1-1/1.556)(.3008)$

$= .0572 + .1075$

$= .1647$

Step 2 – $\Delta TE = \Delta TGO(e_j)$

$= 15{,}560(.1647)$

$= 2562.48$

Change in total employment:

f = regional employment/earnings ratio

ΔTM = change in total employment in the region due to change in demand for the exported output of industry j

Step 1 – f $= \dfrac{\text{total employment in the region}}{\text{total earnings in the region}}$

$= 520{,}800/3{,}744{,}900{,}000$

$= .00014$

Step 2 – $\Delta TM = \Delta TE(f)$

$= 2662.48(.00014)$

$= .36$

Source: Hall, 1977

5. IMPACTS ON INDIVIDUAL WELFARE

The second major component of the distributional consequence model being developed involves assessments of likely adjustments in the economic welfare of individuals living in the local area. Basically, these assessments involve measurements on the "quality of economic life" - as opposed to the "quantity of economic life". The concept of economic welfare has and continues to successfully evade economic theoreticians. The following, therefore, is not to be interpreted as a comprehensive and cohesive treatment of individual welfare. Rather, the impact indicators recommended were selected on the basis of their presumed relevancy and feasibility to be measured. Much of the information needed to measure these indicators comes from the earlier discussion of aggregate changes in economic activity. Three indicators are recommended. Adjustments in unemployment rates will be considered first.

5.1 Unemployment Rates

Unemployment rate refers to the percentage of the total labour force that is unemployed, as determined by some standard or definition. A distinction is normally made between unemployed individuals actively seeking employment and those not actively seeking employment. Unemployment rates usually refer only to those actively seeking employment. Note also, the magnitude of this rate is a function of the definition of the labour force. This definition often refers to the concept of "covered workers" discussed earlier and some age limitation - for example, individuals older than sixteen years of age. This indicator should be viewed as an aggregate measure of individual welfare, reflecting the ability of people to find and secure gainful employment.

The analyst should interpret changes in unemployment rates in a national or regional, as well as local, context. Maintaining proper perspective is important. Consider an expected unemployment rate increase for a local area. A depressed national or regional economy may stimulate further increases in this rate while national or regional expansion may serve to mitigate or offset any negative effects. What about the trend toward increasing the number of females and multiple job holders in the labour force? The analyst concerned with "heads of household" might temper unemployment rate assessments. The point is that both the composition and context of unemployment are important and need to be defined in each case.

It is difficult to assess, *a priori*, whether a forestry project will give rise to a net increase or decrease in unemployment in a locality. For example, it is not altogether clear that an increase in timber harvest will result in a net decrease in unemployment. One aspect of this involves the issue of industrial capacity. Briefly, if the forest products industry has excess capacity - under-utilization of existing labour and capital resources -, more timber to be processed may not give rise to any change in employment and hence have no effect on unemployment rates. Another aspect of this issue involves the notion that industries may be thought of as being complementary or substitute, relative to labour market activity. A positive (or negative) impact on the forest-based industry would have a parallel impact on a complementary industry. The impact would be reverse in the case of substitute industries; a loss for one is a gain for the other. Examples of substitute industries might include sawmills versus plywood mills while complementary industries might be sawmills and transportation. The net effect on unemployment rates is a function of the mix of these industries in a local area. This point is illustrated in the listing below (Rafsnider and Kunin, 1971b). Changes in employment levels result from changes in timber harvest. The exact amount of change and the specific industrial sectors indicated are not important. What is important is that while a decrease in timber harvest activity will have a negative impact on the Forestry and Fishery Products sector, tending to increase unemployment rates, it has a positive effect on the other sectors. Similarly, while the net impact of the harvest increase was +5.95 person-years, it could have been negative, depending on linkages between industries.

Sector	Forestry Decrease	Forestry Increase
Forest and Fishery Products	-1.78961	7.15122
Agricultural, Forestry and Fishery Services	.08546	- .34042
Forest, Greenhouse and Nursery Products	.01673	- .06472
Meat, Animal and Misc. Livestock Products	.20049	- .79595
Total	-1.48690	5.95013

Any assessment of the change in unemployment rate must begin with measurement of employment impacts. Assuming the analyst has information on the labour force and unemployment levels in the local area, the net change in the area's employment could then be used to estimate the new unemployment rate. This procedure, of course, requires the assumption that new employment opportunities would be filled by the existing unemployed labour force. There are problems with this assumption, including the possibility that "full employment" already exists.

The opposite situation, that of an employment decrease, poses a different problem. What assumption should be made relative to where displaced workers will eventually go? Consider the case where discouraging unemployment rates were already projected for an area - Douglas County, Oregon, in this example (Schuster, 1976b). How would a changed level of employment affect these rates? The answer depends on what happens to the displaced workers. For example, assume that a forestry project would decrease employment by 71 employees in 1970 and 32 in 1976. The listing below shows that if the displaced workers cannot find employment in Douglas County, two choices exist: remain unemployed or leave the county. As shown, in either situation, little change in unemployment rate could be expected. The modest consequences in 1976 result from the high unemployment rate already projected for Douglas County and from overall technological advance - fewer timber industry workers per unit of timber processed.

Characteristic	1970		1976	
	Stay	Leave	Stay	Leave
Labour force (without)	28 860	28 860	29 670	29 670
Unemployed	2 290	2 290	5 230	5 230
... Rate	7.9%	7.9%	17.6%	17.6%
Unemployed change	71	71	32	32
Labour force (with)	28 860	28 789	29670	29 638
Unemployed	2 361	2 290	5 262	5 230
... Rate	8.2%	7.9%	17.1%	17.6%
Change in rate	3.8%	0.0%	0.6%	0.0%

Analysis of the effects of forestry projects on unemployment rates usually stops at this point - with calculations of an expected unemployment rate. But should it? Again, the loss of employment opportunity is particularly troublesome. Put in its simplest form, the question is: What happens to forest industry workers when they are permanently displaced from a current job through production cutbacks or plant/operation closure? Schuster (1976a) has argued that what happens to displaced workers can be conceptually linked to adaptability and mobility. Adaptability is the key. The ability of displaced workers to find employment is but one aspect or measure of their overall adaption to a new environment - this can be termed employment adaptability. What determines the employment adaptability of an individual worker? Probably three things: the worker's relevancy, flexibility and circumstance.

> Relevancy refers to the technical or machine aspects of the worker in terms of job performance. The worker is a part in a machine, a factor in a production process. Focus is on the quality and interchangeability of this part, its age and efficiency.
>
> Flexibility refers to the sociopsychological makeup of the worker that gives rise to an inherent propensity toward employment mobility. Mobility can take the form of both geographical and occupational or industrial mobility.
>
> Circumstance refers to the overall socioeconomic environment within which the worker must function. The worker's economic independence, family status and expectations regarding the future are important aspects of the decision context.

Clearly, this model is incomplete and lacks specificity. For example, it is obvious that an interaction term is needed - as circumstance changes, propensity toward mobility may follow. But it is fruitful to separate these elements; conceptually each can have an independent effect on adaptability.

Very little empirical evidence can be presented to indicate the adaptability of forest industry workers to job displacement - a substantial empirical literature does not exist. Much work needs to be done. However, one of the best single sources of data on this point is the work being done by Stevens (1976) on displaced workers. The studied workers that were displaced by mill closure were generally either reemployed or not looking for work. Figure 6 shows the pattern of employment adaptation for workers associated with three mill closures. At the time of data collection, about 81 percent of those looking for work were rehired; 89 percent stayed in the community; and 35 percent of the reemployed left the forest industry. These adjustments in employment took time. Workers at several mills studied average 10-15 weeks of unemployment. At one mill almost half of the workers were immediately reemployed by other mills. Please note, these conclusions are not being presented as universally applicable results. Rather, they are intended to illustrate that initial calculations on unemployment rates need to be further analyzed. Otherwise, the analyst may leave the decision-maker with a false impression of unemployment impacts.

Figure 6. Adaption to closures by sample of 72 workers, Oregon.

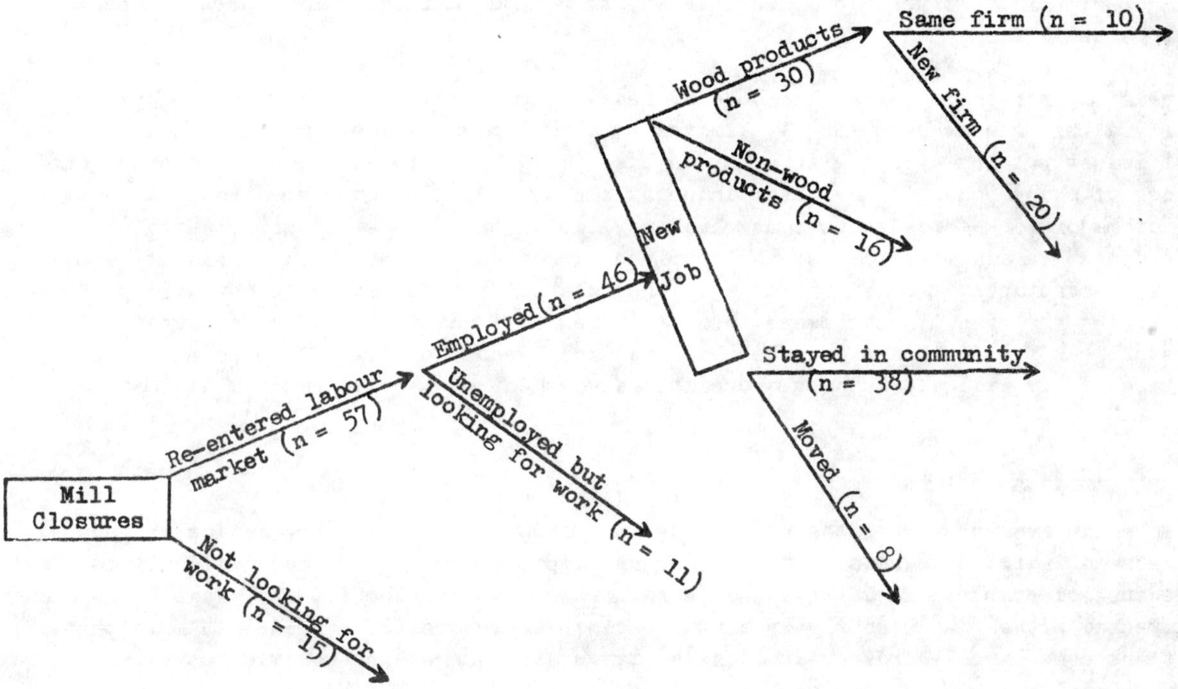

Source: Stevens, 1976.

A concluding comment on unemployment rates seems appropriate. Unemployment rate assessments are based on data aggregates. But are the data too aggregated? Steven's work would suggest so, through the concept of the dual labour force. In the process of conducting a state-wide survey of Oregon forest industry workers, Stevens (1976) identified a labour force almost two-thirds larger than that estimated in census data (about 68 000). Termed a "dual labour force", it consisted of about 65 000 "core" workers and 45 000 "peripheral" workers. Their characteristics are totally different. Core workers constituted the permanent, non-mobile labour force - older, less educated, averaging over thirteen years employment with the forest industry. By contrast, the peripheral labour force is highly mobile and workers averaged only slightly more than two years of employment. Gallaway (1967) makes substantially the same point regarding hired agricultural workers; after the forty-year age class, out-migration stops and, in fact, a low level of in-migration was found.

The upshot of the dual labour force is that:

a) The burden of labour force reductions will be borne by the peripheral labour force which will likely face reduced job prospects because of the history of changing jobs.

b) The core labour force is relatively secure and the seniority of this group will enhance reemployment prospects in the forest industry.

One reasonable conclusion is that stratification of the forest industry labour force into "core" and "peripheral" workers is a sensible way to reduce variation and increase impact assessment capabilities.

A related problem concerns the scope of research presently available on employment adaptability. This research is largely limited to normal fluctuations in the economy of the forest industry. The existing information base would, therefore, allow speculation on consequences of modest changes in forest industry employment. But many are troubled by the prospects of major work force reductions which may lie in the future and would certainly affect the "core" in addition to just the "peripheral" labour force. While seniority would make them more adaptable, age, education and other characteristics would tend toward unadaptability. If the employment impact were to cover a region - say the Pacific Northwest - the picture gets bleak for these are workers characterized by non-mobility. The point is that the data base simply will not support a credible assessment of massive employment impacts.

5.2 Average Wage Rates

Changes in average wage rates prevailing in a local area is another indicator recommended to assess individual welfare impacts. These impacts become somewhat difficult to measure because of problems in determining overall wage rates in the future. Wage rates can be expressed either on an hourly or annual basis. As before, data needed to make these determinations come largely from evaluations of aggregate changes in economic activity.

Determination of wage change has already been discussed. But remember, any particular change in aggregate wages need not be associated with an analogous change in wage rates. The average wage rate that will prevail in a local area after a project is implemented will depend on: a) the way employment gets restructured in industries and b) the prevailing wage rates in those industries. Assume, for example, that a management activity change elicits no net change in employment, but internal adjustments occur. If these adjustments are in favour of high-paying industries, the area's average wage rate will likely go up. If the adjustments are in the opposite direction, the converse will be true. The analyst should make an effort to identify these shifts. If a quantitative assessment of shifts is impossible, a qualitative judgement of probable consequences should be made.

The differential effect on wage rates resulting from a shift of employment from one industry to another was recently illuminated by Polzin and Schweitzer (1975) - although for quite a different purpose. They were evaluating the economic importance of the tourism industry versus the wood products industry in Montana. Their data indicated that while annual wages and salaries in the wood products industry averaged about $8 300 in 1971, wages and salaries in the tourism industry generally averaged less than half that amount. Given that relationship, any employment shift from the high-paying industry to the lower-paying industry will result in a lowering of average wage rates.

Ability of the analyst to accurately measure a change in average wage rate is almost exclusively a function of the data base used. Two situations may exist. On the one hand, the analyst may base assessments on aggregate change and aggregate data. Assume that total wages in an area were $101 160 000 without initiation of a forestry project, and $100 649 122 with the forestry project. The listing below shows that if the "without" average wage rate were used to determine the "with" employment level, no change in average wage rate can be identified.

Characteristic	Without	With
Payroll	$101 160 000	$100 649 122
Employment	16 045	15 963
Average Wage Rate	$6 305	$6 305

The reason for this is that only aggregate changes were involved. That is, workers left the employed labour force at "average" wage rates. Unless the change in employment is associated with differentials in wage rates, no change in average wage rate will be detected. This result will occur when either economic base or input-output analysis uses proportional or linear relationships to assess wage and employment impacts.

However, changes in wage rates can be detected if the analyst can segregate wage or employment changes into industrial categories. Consider the data in the listing below. The "with" column of wages was calculated on the basis of the "without" wage rates together with the "with" distribution of employment. Obviously, any employment shift toward the higher paying industries is shown, resulting in a four percent average wage rate increase. Does the analyst need to use the "without" wage rates as a basis to determine the "with"? No. The "with" rates may be projected on the basis of time series data. Indeed, to use the "without" wage rates requires the major assumption of stable wage relationships.

Industry	Without	With		
	Wage Rate	Employment	Wages (million)	Wage Rate
Forest Industry	$ 8 000	40 000	$ 320	$ 8 000
Other #1	6 000	35 000	210	6 000
Other #2	10 000	25 000	250	10 000
Composite	$ 7 500	100 000	$ 700	$ 7 800

The analyst should make every effort to ensure that average wage rate determinations are as realistic as possible. All other things equal, changes in these rates will affect the amount of purchasing power available to individuals in the area. The discussion above was couched in terms of average annual wages. As before, earnings could be used instead of wages. Similarly, other measures could be used. For example, the analyst may rather convert income levels to a <u>per capita</u> basis or income per household. Final decisions should consider the information needs of the decision-maker.

5.3 Income Re-distribution

Change in income distribution is the final indicator of individual welfare that will be considered. There are several ways this issue could be addressed (Waggener, 1976a). One is to describe the types of workers and individuals in terms of social characteristics and industrial affiliations. There tend to be "winners" and "losers" as a result of a new project; income is being re-distributed in favour of some and away from others. It may be valuable for a decision-maker to know if an already disadvantaged segment of the local area will be aided or further disadvantaged. These types of evaluations may have to be qualitative and subjective. Research is currently underway to improve knowledge in these areas (Stevens, 1975; YoungDay, 1975).

Input-output analyses can also be used to identify distributional effects in a manner less analytical than discussed earlier. Sectors of the economy most likely to be involved in a change in aggregate economic activity can be identified. Any of the tables resulting from an input-output study could be used, depending on the analyst's purpose. Consider the case of the Input-Output Table; it shows dollar flows from producing to consuming sectors, and vice versa. The listing below shows the pattern of sales and expenditures for sawmills in Sullivan County, Pennsylvania (Gamble, 1967). Changes in the sawmill industry would have initial or first-round repercussions for those sectors shown, probably related to the magnitude of the percentages. If knowledge of subsequent repercussions were desired, the table of inverse coefficients would be more useful. Using input-output analysis in this manner is particularly helpful when the analyst has tables available but lacks the facilities (e.g. computer capability) to evaluate the specific project change in question.

Sector	% Purchases	% Sales
Agriculture	0.8	0.4
Sawmills	4.7	4.7
Construction	0.0	0.7
Gas stations	1.6	0.0
Non-profit personal services	3.3	0.0
State A	7.6	0.0
Labour	21.7	0.0
Proprietary income	24.5	0.0
Other internal	3.3	0.1
External to county	32.5	94.0
	100.0	100.0

On the other hand, an operational input-output analysis can be used to actually measure certain types of distributional effects. The Consulting Services Corporation (ca 1969) conducted several evaluations for the Public Land Law Review Commission. One case study involved predicting the impact on employment that would result from a 50 percent increase in carrying capacity of range lands on the upper main stem of the Colorado River by 1980:

	Employment Change	% Output Change
Range livestock	1 692	32.01
Dairy	-51	-32.69
Food & field crops	-121	-30.75
Other retail	97	0.61
Rentals & finance	32	1.33
Household	116	-
Other	151	-
Total	1 916	1.25

While persons in the range livestock and household sectors would be the prime beneficiaries of this policy change, those in the dairy together with food and field crops sectors would be disadvantaged. But again, to make these distributional assessments the analyst must have computational capability that is often not available.

Input-output analysis can be used in a slightly different way. Darr and Fight (1973) computed an index of dependency for each sector relative to all forest-oriented sectors. This index was based on both direct and indirect effects. The listing below shows the five most significant non-government sectors expected to be made relatively worse-off and relatively better-off. While no sector is expected to be made absolutely better-off, sectors with the smallest negative impact are, in a relative sense, better-off. On balance, income will be redistributed in favour of visitors, horticulture and livestock interests and away from the timber-using industries, households, and automotive sales and services interests.

	Sectors
Relatively worse-off	Household
	Automotive sales and services
	Financial services
	Retail and wholesale trade
	Construction
Relatively better-off	Visitors
	Horticulture
	Livestock
	Other manufacturing
	Commercial fishing

Another less personal indicator of income distribution involves a measure of income inequality among classes. Conventionally, this means applying the concept of a Lorenz Curve to the problem of quantifying income equality. Figure 7 illustrates these curves for selected Montana counties and for the state as a whole in 1970. Data are normally available in the Census of Population to define a curve for a local area at some point in time. The analyst must estimate a new curve to determine the effect of a change in management activity on income equality. This may, again, be a subjective assessment. No systematic procedure is known that would facilitate this effort.

Figure 7 - Graph of cumulative income distribution

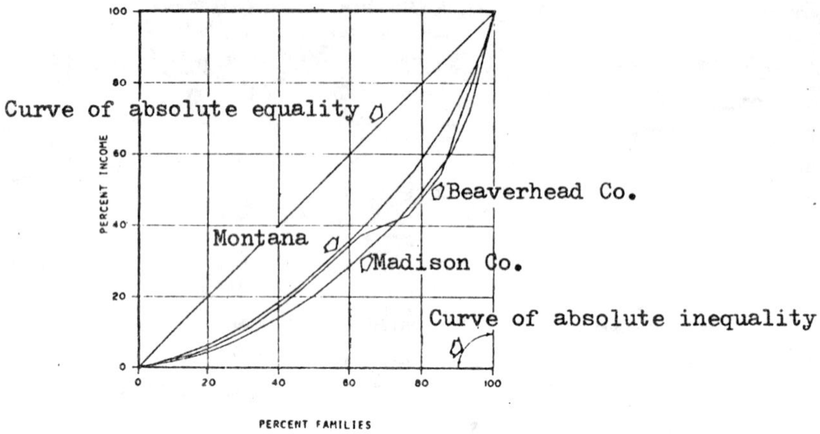

Source: U.S. Forest Service, 1974

However, it may be useful to indicate the general direction of such an analysis. The data below (USDC, 1973) show the change in income distribution resulting from 80 workers being displaced:

Income Class	Before Change			After Change		
	Families	%	Σ	Families	%	Σ
<$1 000	523	2.8	2.8	603	3.2	
1-1 999	786	4.1	6.9			
2-2 999	916	4.8	11.7			
3-3 999	902	4.7	16.4			
4-4 999	915	4.8	21.2			
5-5 999	1 191	6.3	27.5			
6-6 999	1 400	7.4	34.9	1 364	7.2	
7-7 999	1 788	9.4	44.3	1 744	9.2	
8-8 999	1 623	8.5	52.8			
9-9 999	1 499	7.9	60.7			
10-11 999	2 496	13.1	73.8			
12-14 999	2 359	12.4	86.2			
15-24 999	2 146	11.3	97.5			
25-49 999	435	2.3	99.8			
50 000+	38	0.2	100.0			
Total	19 017			19 017		

Forty-four workers were displaced from the timber-using industry where average incomes lie in the $7 000 to $7 999 class, while the remaining 36 workers were withdrawn at county average wage levels (USDC, 1971). Only the sectors changed are shown. A major assumption is that each employee represents one family unit. These data could be plotted on a graph such as Figure 7 shown earlier. It is clear that if these data were calculated and plotted, the new curve would show less income equality. While this change is admittedly slight (due to the specifics of this problem), the principle is more general and warrants consideration.

6. IMPACTS ON ECONOMIC EQUILIBRIUM

Introduction of a new forestry project or activity may have an impact on economic equilibrium or stability of the local area. Stability refers to the ability of the area to maintain its economic viability over time. It may seem difficult to separate indicators of equilibrium or stability from economic activity. An example may help clarify the distinction. The Western U. S. is dotted with ghost towns - the remains of once thriving mining communities. One could imagine decisions favouring mining years ago that ranked high on indicators of economic activity - income and employment of the time. But in terms of maintaining area equilibrium, these decisions may, in retrospect, be judged as somewhat lacking. Stability considerations add a long-range time element to impact analysis. Two broad indicators of future community stability seem worthy of note.

6.1 Economic Diversity

A fundamental axiom in the field of ecology is that the more diverse the ecosystem, the more stable it is. The analogy can be made to community equilibrium/stability and economic diversity. Ecosystem diversity insulates and ensures permanence of the system against natural catastrophy. So it is with economic systems. After the gold and silver were extracted, many communities were left without an economic base. The collapse of mining signalled the demise of the community, and ghost towns resulted.

The existence of "ghost towns" dramatizes the fact that economic processes are seriously jeopardized in an unstable environment. The most useful measure of economic diversity seems to be the distribution of economic activity among sectors. Distribution of employment would serve as an indicator. A management alternative would be judged to promote stability if it enhanced diversity as measured by a more equal distribution of employment among sectors. By this measure, to make a timber-oriented community more timber-dependent would not be desirable. Clearly, if carried to an extreme, this indicator may be incompatible with economic base theory. Diversity would tend to discourage further specialization; development of export or basic industries would not be emphasized. Therefore, area growth would be diminished.

Consider an example of one possible approach to a diversity calculation. Upon reflection, it will be discovered that diversity has a statistical counterpart - variance. A system of numbers that are all very similar will have relative low variance. Since similarity (a more equal distribution) is suggestive of more diversity, it follows that relatively lower variance levels will be associated with relatively more diversity, and hence stability. The analyst may conclude that the standard expression for statistical variance might be a useful expression of a diversity index.

$$\sigma^2 = \frac{\sum (X_j - \bar{X})^2}{N} = \text{Diversity Index}$$

Given the data in the listing below, the diversity index for the status quo is calculated to be about 139, Alternative 1 to be 1 667 and Alternative 2 as 50. On the basis of these calculations, Alternative 2 would be found to be the most diversified and hence most stable. Clearly, the above expression for variance emphasized deviation from a mean, more than would some other measures. The analyst should adopt some measure of diversity. It need not correspond to the variance expression shown above.

	Expected Employment		
Sector	Status Quo	Alt. 1	Alt. 2
A	25	25	35
B	25	25	35
C	50	75	50
	100	125	120

Another approach to economic diversity deals with industrial location. How will decisions today affect long-range expansion or contraction of industrial activity? The environment created for industrial location in the local area will certainly affect community equilibrium and stability.

The body of literature existing in the area of industrial location is wide-ranging (see Hoover, 1948; Mueller and Morgan, 1962). While much more restrictive, the literature concerning forest-based industries is still impressive. Most of these studies address the question of which factors (taxes, land prices, etc.) influence the decision of a firm to locate in a particular area (e.g. Hagenstein, 1964). Availability of wood, raw material, labour and transportation is often most important in timber industry location decisions. McMillan (1965) provides a reasonable synopsis:

> If an industry is resource oriented, it must place prime importance on raw materials. If the manufactured product embodies high labour costs or highly skilled labour, labour market conditions occupy a position of prime importance. No plant can justify its existence without a place to sell its output. Therefore, markets must rank high. But these are not determinants of a particular location. Instead, they are prerequisites to operation. Transportation may also fall into this category of prerequisites. (emphasis added)

While communities can do little about raw materials, markets, population and certain forms of transportation, public land managing agencies can certainly affect one - raw materials.

There are two important implications of forest management decisions on industrial location. First, these decisions need not affect location of all resource-oriented activities in the same way. Given a resource base and some capacity to produce outputs, a decision against one resource output may at the same time be a decision in favour of another - to increase timber availability may discourage industrial location of recreation industries. The consequence is unavoidable. As a corollary, since many factors other than timber raw material affect industrial location, timber management decisions may not be sufficient to create or maintain a desirable environment for industrial location. The second implication relates to the concept of "comparative advantage". Assume a situation involving only one resource-oriented industry, a timber-based industry. Assume further that a comparative advantage exists for this industry due to availability of wood raw materials. A decision to restrict this availability could have negative location consequences by eliminating the major (and possibly the only) reason firms locate in that particular area. The point is that major forest management decisions cannot avoid long-range industrial location implications.

6.2 Community Adjustments

It is important for the decision-maker to realize that a major decision on forest management projects may inextricably alter a community. The decision may lead to a fundamental change in the economic fabric of a local area. While communities may always be in a constant state of change, radical acceleration or departure from trends may have a strong impact on local equilibrium. The analyst should attempt to assess the implications of a forestry project in terms of adjustments in community lifestyle, social disorganization and local values.

Lifestyles - Lifestyle is the "way of living" chosen by individuals, groups and communities. More specifically, lifestyle is a composite of various elements resulting from the interaction of human beings with their physical and social environments. Included in these elements are use of time and attitude toward, and methods of interacting with other people and with the physical environment. Lifestyle is defined by the way these elements combine. Some consider lifestyle analysis to be synonymous with social impact assessments (Holden, 1975).

Lifestyles are also a reflection of the way people meet their physiological and psychological needs. People engage in activities they value; those who enjoy outdoor activity may be concerned about the effects of resource allocation decisions on national forest recreation. Those who make their living from mining, lumbering or grazing are likely to oppose decisions that remove land from these uses. Conflict of lifestyle, and hence land use, is inevitable. The potential conflict resulting from resource allocation decisions is largely one of local versus regional and national interests, as well as group versus group. It is important to understand that the needs and desires of local people are usually different from those of outsiders. While outsiders venture to the local area to meet certain needs, locals meet almost all needs in this area. As outsiders demand increase, lifestyles of the local population can become threatened.

In an overview of social impact assessment, Gale (1977) describes lifestyle measurement as follows:

> The term 'ways of life' or 'lifestyle' represents a way of characterizing a cluster of specific social variables. Use of this social impact category is a good way to avoid a variable-by-variable description of different groups within a community. Summaries such as 'the lifestyle of those expected to move into the area under the action alternative ...' reflect a gathering together of a number of variables.
>
> There is no 'standard set' of five or ten variables used consistently to describe a way of life or lifestyle. In identifying lifestyles, the focus is on those three or four variables or components which, in a particular situation, best characterize the relatively distinct way in which a certain group of people go about their daily activities.
>
> 'Distinct' is a key word in understanding lifestyles and in assessing impacts on them. For identifiable lifestyles to exist, it is not necessary for two groups to differ in almost every aspect of their daily lives. More typically, different lifestyles emerge as an increasing number of characteristics differ, although some common elements will remain.

Lifestyle impact analysis is still in its infancy. Wolf (ca 1974) terms the current state of the art as "explosive" and predicts "orders-of-magnitude" improvement in the near future. Present efforts seem to focus on descriptive analysis - often termed ethnographic studies (ISR, 1974). Predictive analyses are not commonplace and the analyst of adjustments in community lifestyle must often rely on informed judgements. Fortunately, significant recent work by Gale (1975) and Freeman (ca 1976) are adding to the set of tools and concepts needed to make these assessments.

Social Disorganization - The disorganization of a local area's social fabric together with lifestyle adjustments are primarily concerns of sociologists. Nevertheless, both strongly influence community longevity. It is difficult to imagine economic stability in the absence of social stability. Social disorganization refers to stress, to the point of severance, placed on the internal linkages that bind institutions together.

Social scientists group a wide variety of elements under the heading of social disorganization. The theme common to these elements is disruption of normal social processes. For example, consider the possible increase in the unemployment rate discussed earlier. Social disruptions and turmoil associated with massive unemployment are obvious. Further, a type of multiplier effect might exist, entailing additional stresses on community support facilities, family stability and overall morale. Major forest management decisions can lead to social disruption by changing the political power structure and reordering the economic structure of the local area.

Local Values - One of the most striking features of local economic impact studies is the cavalier way in which some analysts designate "good" impacts and "bad" impacts. More employment is "better" than less. Increases in the level of economic activity are "good. These "values" are implied in many analyses. The analyst should realize that impact indicators must be viewed as neutral. Implying goodness or badness is a value judgement. Analysts, probably inadvertently, have fallen into the trap of making value judgements about indicators.

Since the focus of most social impact analyses is the local area, it seems reasonable that the local people should make value judgements. Communities differ. The values shared by people in one area may not be shared by those in another. A "sense of community" may be at issue. The way people in a local area see themselves and their community should be considered. Take a small, rural, conservative, close-knit community in the Southwestern U.S. This community might react very differently to the prospect of 50 new jobs which might be filled by "outsiders" than would a larger metropolitan area. Simply stated, "goodness" or "badness" of indicator measurements should be evaluated, not assumed.

7. IMPACTS ON LOCAL GOVERNMENT

The final area of distributional consequence considered involves the relationship between the forestry project and governmental entities located in the local area. Many of the distributional consequences already discussed will eventually affect the local government. Those considered now are of more direct consequence. Of the many ways a change in forest management activities could affect these governments, fiscal impacts - both revenues and expenditures - are of prime importance. Since there is no uniformity as to either the nature of these fiscal impacts nor the analysis that would be appropriate, the following discussion will provide an overview of the types of issues with which the analyst must deal. Two general indicator areas will be considered.

7.1 Intergovernmental Payments

The relationship between different levels of government (federal, regional, local) will vary not only as a function of the nation in question, but also as a function of specific governmental units and agencies within a social economy. The principle of intergovernmental payments (transfers) has been a long established practice. Some of these payments are the result of specific policy actions designed to accomplish some social purpose. Some just happen. They are the natural result of government exercise of its proprietary power. The analyst should make a careful assessment of these consequences. This will necessarily involve a detailed analysis of the specific circumstance being evaluated. Three general types of intergovernmental payments may be found.

Revenue Sharing Payments - While the concept of one level of government (federal or state) sharing its revenues with a lower level (state or local) has been the subject of recently increased political interest in some countries, it has existed for a long time. For example, the practice of sharing revenues from the sale of public lands in the United States began in 1802 when the U.S. Congress provided that the state of Ohio would receive three percent of the net proceeds of the revenues from the sale of public lands in the state (EBS, 1970). Possibly the first form of revenue sharing in the United States that remains in effect today originated in 1908 with a law requiring that 25 percent of the net receipts generated by national forests be returned to countries in which the forests are located (U.S. Code, 1908). Other acts have been passed for lands administered by other agencies. The legislative history of these acts reflects that payments to state and local governments were intended as compensation by the federal government, since the lands in question were not available for purpose of local property taxation (PLLRC, 1970). To measure impacts on local government associated with a system of payments-in-lieu-of-taxes, the analyst must first evaluate existing policy - both statutory and administrative. Completing this, the analysis can proceed.

Consider an extremely simplified example of the types of calculations that ought to be made when a forestry project on a U.S. national forest will alter the level of money receipts. The increased receipts may be possibly due to a forestry project that increases receipts from grazing permits because of increased forage available on forest lands. In reality, actual payments to the counties would be determined by the provisions of the 1908 law previously discussed as modified by the 1976 Forest Management Act (U.S. Code, 1976a) and as supplemented by the In-Lieu Tax Act of 1976 (U.S. Code, 1976b). But let us ignore these details and work through the outline of a typical analysis. Assume that annual grazing receipts were to increase from $400 000 to $600 000. The other parameters of this example are shown in the following listing:

- Change in receipts - $200 000
- Payment basis - 25% of receipts
- Change in payments $50 000
- Distribution of national forest land - County A: 80%
- County B: 20%
- Current county revenues - County A: $ 5 000 000
- County B: $10 000 000
- Use of payments - Schools and roads

Given these data, payments to County A would decrease by $40 000 (80% x $50 000) and assuming no other change in revenue structure, the new level of revenues to the county would be 99 percent of the old (($4 960 000/5 000 000) x 100). Knowledge of the existing budget for schools and roads could then be used to assess the impact of receipt reduction on these budget items.

In practice, measurement of changes in payments-in-lieu-of-taxes is a very easily measured consequence of a forestry project. At least this is so in the United States. The reason for this is that the data needed to accomplish the analysis are readily available in public records. But the analyst should be aware of two problem areas. The first deals with identifying the change in receipts. The word "net" seems appropriate to highlight the fact that a change in one area of management may be associated with a change in some other area. For example, an increase in grazing receipts may be associated with a decrease in receipts due to timber harvest. The analyst should consider the relationship between these changes. They may or may not be offsetting. The second problem deals with estimating the revenue change associated with the forestry project. One common approach is to use proportions. If timber harvest is to decrease by 50 percent, revenues to the county will decrease accordingly. An alternative procedure is to vary the timber harvest level and then calculate revenue changes on the basis of average stumpage price received over the past several years. Depending on the particular location, either of these methods may be extremely risky. An analysis of U.S. Forest Service revenues in Montana over the 1960-73 period found a correlation coefficient of 0.29 between value and volume of timber cut, while a correlation of 0.97 was found between value and adjusted stumpage price (Schuster, 1976b). This suggests that the prudent analyst should seriously investigate the price element in timber revenue determination.

In-Kind Payments - The second type of intergovernmental payment concerns the contribution made by public agencies that tend to relieve local government of a financial burden. For example, development of a state campground may diminish or eliminate the need for a county to develop a park system. No money is transferred; consequently, the cost savings incurred by the local government are termed payments "in-kind" by the public agency. The key to determining the magnitude of these payments is identifying those activities assumed by the public agency that would normally be accomplished by the local government. There is no uniform agreement as to the list of these activities. The analyst must rely on the advice and consultation of the decision-maker and officials in local government.

Quantification of in-kind payments is not an exact process. The problem is one of data availability. Only a very few studies have been done to determine the levels of in-kind payments; study results are generally applicable to state and multistate areas only. Table 12 shows selected results of some of these studies. If evaluation of in-kind payments is to proceed, the analyst is forced to rely on aggregate levels of payments per acre. Data reflecting a stratification of payments relative to type of management activity are preferred over aggregates. In this way, payments associated with timber management could be applied to programme changes in that area. In-kind payments should be converted to the same base as the management programme. That is, if timber harvest is to be modified, the payments should be expressed in an acceptable unit - dollars per board foot. This procedure implies a strict proportionality that likely does not exist. Yet, the data base does not allow a more refined analysis.

Other Payments - A final area of intergovernmental payments involves secondary payments, payments associated with other governmental programmes. Local governments often receive funds from other government levels on some type of formula or matching basis. If a change in management activity affects the formula, receipts will be altered. For example, a local government may receive aid for education from the state on a student enrolment basis. If management activity leads to unemployment, out-migration of population and withdrawal of children from the local school system, a loss of state school funds is likely. Of course, the reverse could also result from an opposite change in management activity. And finally, when a management activity affects economic activity, it will likely also affect taxes collected by the local government. This change in revenue will change the availability of local matching funds and project funds (e.g. U.S. Bureau of Outdoor Recreation funds for community recreation projects). Measuring this impact indicator would require analysis of a specific local government and its programme involvement.

7.2 Cost to Local Government

A controversy exists in certain areas of the country relative to the financial support relationship between local government and public agencies, Shannon (1975) reports that "many county commissioners believe that county maintenance costs incurred as a direct result of management activities on national forests are far larger and are far more stable than the financial support received directly by county government from the national forests." There are probably many ways in which the mere presence of a major land management agency results in costs to local government. For example, because timber harvest activities require use of heavy logging trucks, county roads must be built and maintained to a higher standard than otherwise necessary. Unfortunately, identification of these cost items is largely at the speculative stage. Since few comprehensive analyses of this topic (EBS, 1970) have been done, generalized assessment of these costs is well beyond the state of the art.

Table 12. Estimated Forest Service Contributions In-Kind per Acre, by Type, FY 1962 and 1952

Region	1962				1952 Total
	Fire Control	Forest Highways	Road, Trail and Other	1962 Total	
		(cents)			
Northern	3.5	18.4	4.1	26.0	15.7
Rocky Mountain	1.9	17.5	1.5	20.9	11.9
Southwestern	1.5	15.9	4.0	21.4	21.3
Intermountain	2.5	12.1	3.5	18.1	15.2
California	85.4	24.4	28.4	138.2	54.6
Pacific Northwest	24.3	28.7	8.1	61.1	34.6
Eastern	6.4	17.3	5.9	29.6	19.9
Southern	10.3	16.8	6.8	33.9	23.1
North Central	11.2	14.5	18.6	44.3	23.8
All Regions	16.7	18.7	8.1	43.5	-

Source: U.S. Forest Service, 1965

Another aspect to local government costs should be considered: transaction costs. These are the added costs that must be borne by local government during the period during which adjustments in the local area are made - the transition period. For example, additional pressures may be placed on community service agencies as a result of additional employment. Pressures may be placed on local schools or law enforcements units because of accelerated social disorganization. Because understanding of these transaction costs, especially regarding forest management activity, is totally undeveloped, no known measures exist to evaluate this indicator.

8. CONCLUDING REMARKS

Evaluation of the distributional consequences of forestry projects is an important component in an overall assessment of the desirability of one project relative to another. But it is not the only component and there is no reason to believe it the most important component. Clearly, many other consequences are also important, including economic efficiency consequences and environmental quality consequences. The role and importance of distributional consequences in selecting between alternative forestry projects is a function of the decision-maker's goals and objectives. These not only provide the context for distributional analyses, but they also specify the content of the analysis. Since goals and objectives vary, this paper has adopted a rather neutral position and simply attempted to provide a general overview for analysis. It outlined some considerations, issues, problems and methodologies that often will be encountered in a distributional analysis.

Distributional analyses call for a high degree of ingenuity and adaptability on the part of the analyst. This is partly due to the fact that, at least conceptually, each analysis is different, tailor-made to the specific circumstance being evaluated. While the range of items to be evaluated in a distributional analysis is almost limitless, the impacts evaluated must be limited for very practical reasons of time and money. The analyst must translate the information needs of the decision-maker into operational analyses. Upon completion of the analysis, the results must be translated again into information for decision-making. This is not simply the problem of communication. It is also a problem of interpretation. Initial study results must be interpreted and evaluated with the context of the social and economic structure of the local economy in mind. The response to a given set of distributional consequences will vary from one economy to another. Local areas differ. Their internal linkages, structures, and value systems defy generalization (Dickerman and Butzer, 1975). This sharply reduces the analyst's ability to extrapolate knowledge of the distributional consequences of forestry projects from one area to another. Each analysis must start anew.

The other major factor calling for ingenuity and adaptability from the analyst is that the state of analytical capability is sadly deficient. Although several sophisticated techniques exist, they are only applicable to a relatively small set of consequences. And even for these, the needed data base might not exist. Analytical tools to assess some of the other consequences remain either undeveloped or underdeveloped. The analyst must adapt to these circumstances. Alternative data sources may need to be discovered. Major assumptions may need to be made. The project analyst may find it necessary to call upon the talents of political scientists, sociologists and other social scientists to enhance the quality of the distributional analysis. Ability to recognize deficiencies in data, deficiences in analytical capabilities, and deficiencies in personal expertise will probably increase with experience. Ability to effectively deal with these deficiencies will depend on the analyst's ingenuity and adaptability.

Over time, the quality of distributional analyses will likely improve. Better data will become available. Analytical techniques will be developed and refined. Analyst expertise will improve. The process takes time. Hopefully, this paper represents a positive part of that process.

REFERENCES CITED

Barkley, P.W., and T.H. Allison, Economic base studies in resource administration.
1968 Land Econ. 44:470-479

Bureau of Economic Analysis (BEA), Industry-specific gross output multipliers for BEA
1977 economic areas. U.S.D.C., Regional Economic Analysis Division. 135 pp.

Business-Economics Advisory and Research, Inc. (BEAR), The role of forest land in
1970 Washington economy. Wash. Forest Prot. Assoc., Seattle, Wash. 108 pp.

Bell, Enoch F., A technical critique of economic impact analysis. Unpub. pap. presented
1976 to Western Forest Economists Meeting, 3-5 May 1976. Wemme, Ore. 7 pp.

Bell, Enoch F., Estimating effect of timber harvesting levels on employment in western
1977 United States. U.S.D.A., Forest Serv. Res. Note INT-237, 11 pp. Intermountain Forest and Range Exp. Sta., Ogden, Utah.

Bentley, William R., Forest Service timber sales: a preliminary evaluation of policy
1968 alternatives. Land Econ. XLIV(2):205-218.

Bouderville, J.R., Problems of Regional Economic Planning. Edinburgh Press, Edinburgh, U.K.

Consulting Services Corporation (CSC), Study of impact of public lands on selected
ca 1969 regional economics. Nat. Tech. Inform. Serv. Rep. PB190164. 336 pp.

Convery, Frank J., Unit planning and local economic impacts of alternative forest
1973 management practices. Duke Univ., School Forest. Tech. Pap. 1. 27 pp.

Coppedge, Robert O., and Russell C. Youmans, Income multipliers in economic impact
1970 analysis. Oreg. State Univ. Coop. Ext. Serv. Spec. Rpt. 294. 3 pp.

Darr, David R., and Roger D. Fight, Douglas County, Oregon: potential economic impacts
1974 of a changing timber resource base. U.S.D.A. Forest Serv. Res. Pap. PNW-179.

Davis, Lawrence S., and William R. Bentley, The separation of facts and values - in
1967 resource policy analysis. J. Forest. 65:613-620.

Dickerman, Alan R., and Stanley Butzer, The potential of timber management to affect
1975 regional growth and stability. J. Forest. 75:268-269.

Drake, Ronald, Stanley Randall, and Michael Skinner, Evaluation of economic impacts of
1973 Forest Service programmes in northern New Mexico: development of analytical tools. Unpub. Rep., U.S.D.A. Econ. Res. Serv., Southwest Resource Groups. Berkeley, Calif. 154 pp.

EBS Management Consultants, Inc., Revenue sharing and payments in lieu of taxes on the
1970 public lands - Vol. I. (Prepared for Public Land Law Review Commission) Nat. Tech. Inform. Serv. Rep. PB195185. 201 pp.

Ferguson, Sheryl, and Clynn Phillips, 1975 Economic impact study of a national park proposal in the Sawtooth area of Idaho. Unpub. pap. Univ. Wyo. Wat. Resources Res. Inst. 9 pp.

Field, Richard C., and Frank J. Convery, 1976 Estimating local economic impacts in land-use planning. J. Forest. 74(3):155-156.

Fowler, Kenneth, 1974 Impacts of projected timber harvest in Humboldt County. Ph.D. Thesis. University of California, Berkeley. 110 pp. University Microfilms, Ann Arbor, Mich. (Diss. Abstr. 74-28,416).

Fox, Karl A., and T. Krishna Kumer, 1966 Delineating functional economic areas. In Research and Education for Regional and Area Development. Iowa State Univ. Press, Ames, Iowa. pp. 13-55.

Freeman, David M., ca 1976 Management strategy impact analysis - a procedure for displaying the distribution of social well-being. Unpub. Pap. U.S.D.A. Forest Serv., Region 2, Denver, Colo. 31 pp.

Gale, Richard P., 1975 The U.S. Forest Service and social impact assessment: a staff paper. Unpub. Rep. U.S.D.A. Forest Serv., Willamette Natl. For., Eugene, Ore. 141 pp.

Gale, Richard P., 1977 Social impact assessment: an overview. U.S. Forest Serv., Wash., D.C. 39 pp.

Galloway, L.E., 1967 Mobility of hired agricultural labour: 1957-1960. J. Farm Econ. 49(1):32-52.

Gamble, Hays B., 1967 The economic structure of Sullivan County, Pennsylvania. Penn. State Univ. Agric. Exp. Sta. Bull. 473. 60 pp.

Goldman, George E., and Anthony T. Nakazawa, 1974 A technical guide to the construction of county input-output models: the Santa Barbara case. Univ. Calif. Div. Agri. Sci. Spec. Pub. 8001. 73 pp.

Hagenstein, Perry R., 1964 The location decisions for wood using industries in the northern Appalachians. U.S.D.A. Forest Serv. Res. Pap. NE-16, 36 pp. Northeast Forest Exp. Sta., Upper Darby, Pa.

Hall, Ellen, 1977 Regional Industrial Multiplier system: a guide for river basin analysis. U.S.D.A. Forest Serv., State and Private Forestry, Doraville, Ga. 36 pp.

Herbst, John E., 1972 Timber harvest on public lands: its effects on a local economy. Unpub. Rep. U.S.D.A. Forest Serv., Region 3, Ogden, Utah. 24 pp.

Holden, John D., 1975 On the use of social sciences in land-use planning. Unpub. Rep. U.S.D.A. Forest Serv., Region 1, Missoula, Mont. 39 pp.

Hoover, Edgar M., 1948 The Location of Economic Activity. McGraw Hill, New York.

Hoover, Edgar M., 1971 An Introduction to Regional Economics. Alfred A. Knopf, Inc., New York.

Hughes, Jay M., 1970 Forestry in Itasca County's economy - an input-output analysis. Univ. Minn. Agric. Exp. Sta. Misc. Rep. 95, Forest Serv. 4. 10 pp.

Institute for Social Research (ISR), 1974 Appendix A: Methodology. In A Comparative Case Study of the Impact of Coal Development in the Way of Life of People in the Coal Areas of Eastern Montana and Northeastern Wyoming: Final Report. Univ. Montana, Missoula, Mt. pp. 150-185.

Isserman, Andrew M., 1977 The location quotient approach to estimating regional economic impacts. Amer. Inst. Planners J. 43:33-41.

Johnson, Maxine C., 1972 Wood products in Montana. Mont. Bus. Quart. 10(2):1-41.

Kaiser, Harold F., and George F. Dutrow, 1971 Structure and changes in the southern forest economy, 1958-1967. U.S.D.A. Forest Serv. Res. Pap. 50-71, 18 pp. Southern Forest Exp. Sta., New Orleans, La.

Leven, Charles L., 1970 A framework for the evaluation of secondary impacts of public investments. Amer. J. Agric. Econ. 52(5):723-729.

Marty, Robert, 1975 Comprehensive analysis of public forestry project and programme alternatives. J. Forest. 73(11):701-704.

Mathur, Vijay K., and Harvey S. Rosen, 1974 Regional employment multiplier. Land Econ. 50(2):93-96.

McKillop, William, 1974 Economic impacts of an intensified timber management programme. U.S.D.A. Forest Serv. Res. Pap. WO-23, 16 pp. Wash., D.C.

McMillan, T.E., Jr., 1965 Why manufacturers choose plant locations vs. determinants of plant locations. Land Econ. 41:239-246.

Miernyk, W.H., 1965 The Elements of Input-Output Analysis. Random House, New York.

Mueller, Eva, and James Morgan, 1962 Location decisions of manufacturers. Amer. Econ. Rev. 52:204-217.

Office of Management and Budget (OMB), 1972 Standard Industrial Classification Manual. U.S. Government Printing Office, Wash., D.C.

Okum, Arthur, 1975 Equality and Efficiency - The Big Tradeoff. The Brookings Inst., Wash., D.C.

Public Land Law Review Commission (PLLRC), 1970 One Third of the Nation's Land. U.S. Government Printing Office, Wash., D.C.

Polzin, Paul E., and Dennis L. Schweitzer, Economic importance of tourism in Montana.
1975 U.S.D.A. Forest Serv. Res. Pap. INT-171, 19 pp. Intermountain Forest and Range Exp. Sta., Ogden, Utah.

Rafsnider, Giles T., and Leonard Kunin, A 1967 input-output model of the Idaho economy.
1971a Unpub. Rep. State Planning and Community Affairs Agency, Boise, Idaho. 25 pp.

Rafsnider, Giles T., and Leonard Kunin, The economics of varying timber harvest levels
1971b on U.S. Forest Service lands in Idaho: an application of the 1967 input-output model of Idaho's economy. Unpub. Rep. State Planning and Community Affairs Agency, Boise, Idaho. 16 pp.

Reimer, Donnie Ray, Development of a Leontieff input-output model and its application in
1969 the analysis of the economic impact of a new forest industry on the economic growth and development of an underdeveloped micro-region in southern Indiana. Ph.D. Thesis. Purdue University. 167 pp. University Microfilms, Ann Arbor, Mich. (Diss. Abstr. 70-03, 965).

Schuster, Ervin G., Employment adaptability in the forest industry. Unpub. Pap. presented
1976a at Workshop on Analysis of Forest-Related Sectors in Regional and Local Economies, Society of American Foresters, 7-9 September 1976, Keystone Center, Colo. 22 pp.

Schuster, Ervin G., Local economic impact: a decision variable in forest resources
1976b management. Mont. Forest and Cons. Exp. Sta. Study Rep. 104 pp.

Schuster, Ervin G., William D. Koss, and E. Bruce Godfrey, Timber cut, employment and
1975 wages: multipliers for Idaho's timber using industry. Univ. Idaho Forest, Wildlife and Range Exp. Sta. Tech. Rep. 1. 15 pp.

Shannon, Richard, An examination and analysis of Forest Services payments to county
1975 government in the timbered counties of Montana. Unpub. Res. Prospectus. Univ. Montana, School of Forestry. 3 pp.

Stevens, Joseph, Distributional effects of employment related to alternative timber
1975 harvest flows (approximate title). Oreg. State Univ. Agric. Econ. Dept.

Stevens, Joe B., The Oregon wood products labour force: job rationing and worker adapta-
1976 tions in a declining industry. Unpub. Rep. Oreg. State Univ. Agric. and Res. Econ. Dept. 123 pp.

Tiebout, E.M., The Community Economic Base Study. Supplementary Paper 16, pp. 46-49.
1962 Committee for Economic Development, New York.

Ullman, Edward L., Michael F. Dacey, and Harold Brodsky, The economic base of American
1971 cities. Cent. Urb. and Reg. Res. Mono. 1. Seattle, Wash.

U.S. Code. Title 16, Par. 500.
1908

U.S. Code. 1937 Title 43, Par. 1181.

U.S. Code. 1939 Title 42, Par. 503A.

U.S. Code. 1976a Title 16, Par. 1601-1610.

U.S. Code. 1976b Title 32, Par. 1601-1607.

U.S. Department of Commerce, 1969 Input-output structure of the U.S. economy: 1963. Survey Current Bus. 49(11):16-47.

U.S. Department of Commerce, 1971 County business patterns 1970. CBP-70-39.

U.S. Department of Commerce, 1973 1970 Census of Population. Vol. 1, Pt. 39.

U.S. Forest Service, 1965 National forest contributions to state and local governments, 1962. In-service rep. Wash., D.C. 38 pp.

U.S. Forest Service, 1974 Socio-economic overview. Beaverhead National Forest R1-74-005, Darby, Mt., 96 pp.

U.S. Forest Service, 1975 Planning area guides. Unpub. Pap. Region 1, Missoula, Mt. 6 pp.

U.S. Forest Service, 1977 Land management planning notebook - appendix (draft). Region 6, Portland, Ore. 62 pp.

Waggener, Thomas R., 1970 Statement on proposed scenic management areas, U.S. Forest Service, North Cascades recreation plan. Unpub. Pap. U.S.F.S., Region 6, Portland, Ore. 26 pp.

Waggener, Thomas R., 1972 Measuring the impact of a reduction. In Timber Supply and the Environment. Proc. Current Issues Conf., Univ. Oregon, Eugene, Ore. pp. 19-25.

Waggener, Thomas R., 1976a Income redistribution considerations in economic impact analysis. Unput. Pap. presented at Workshop on Analysis of Forest-Related Sectors in Regional and Local Economics, Society of American Foresters, 7-9 September 1976, Keystone Center, Colo. 22 pp.

Waggener, Thomas R., 1976b Philosophical bases for local economic impact analyses. Unpub. Pap. presented to Western Forest Economists Meeting, 3-5 May 1976. Wemme, Ore. 12 pp.

Waggener, Thomas R., Community stability as a forest management objective. J. Forest.
1977 75(11):710-714.

Wall, Brian R., and Daniel D. Oswald, A technique and relationships for projection of
1975 employment in the Pacific Coast forest products industry. U.S.D.A., Forest Serv. Res. Pap. PNW-189, 49 pp. Pacific Northwest For. and Range Exp. Sta., Portland, Ore.

Wolf, C.P., Social impact assessment: the state of the art. Unpub. Rep. Army Corps
ca 1974 of Engineers, Institute for Water Resources. Fort Belvoir, Va., 43 pp.

Youmans, R.C., and H.H. Stoeverner, Generalizations of regional input-output models.
ca 1973 Unpub. Rep. U.S.D.A. Forest Serv. Int. Forest and Rang Exp. Sta., Ogden, Utah, 41 pp.

YoungDay, Douglas J., Who benefits from cheap wood? Unpub. Pap. presented at Western
1975 Forest Economists Meeting. Wemme, Ore.

Zinn, Gary W., A model for analyzing the contributions of forestry to a region. Ph.D.
1972 Thesis. Univ. of New York, College Environmental Sciences and Forestry. 247 pp. University Microfilms, Ann Arbor, Mich. (Diss. Abstr. 72-23, 567).

Zinn, Gary W., Local economic impacts of forest activities: some prospects for "uncon-
1976 ventional" analysis. Unpub. Pap. presented to Western Forest Economists Meeting, 3-5 May 1976. Wemme, Ore. 12 pp.

ECONOMIC ANALYSIS OF WATERSHED PROJECTS: Special Problems and Examples

by

Hans M. Gregersen and K. Brooks [1]

---

[1] H.M. Gregersen is Professor of Forestry and Agricultural and Applied Economics at the University of Minnesota, and K. Brooks is Associate Professor of Forest Hydrology at the same University. An earlier version of this paper (FO:MISC/78/18, May 1978) was presented at the Twelfth Session of the Working Party on the Management of Mountain Watersheds, Rome, 18-23 September 1978.

1. INTRODUCTION

Most forestry projects involve implications in terms of changes in water quality and/or quantity. Thus, economic analyses of most forestry projects should include explicit consideration of watershed related impacts and of potential activities to achieve acceptable watershed protection standards. In some cases the major objectives of a project may be water related and constitute the reason why the project is being considered and proposed. In other cases, water related concerns may merely enter the project analysis in the form of constraints on other project activities. In either case, the economic analyst should have something to say about water related impacts in his analysis.

The above is not to say that the same objectives and/or constraints apply in all situations. Thus, for example, project objectives will vary widely with climatic and landform conditions as well as other factors. Appendix I presents a schematic overview of objectives which tend to dominate in different climatic regions.

Economic analyses of watershed projects are no different in principle or concept than analyses of any other type of project. Thus the general concepts and guidelines presented in FAO's Economic Analysis of Forestry Projects [1] (EAFP) are valid for watershed related projects. However, some analytical issues and empirical problems are particularly important for such projects. Some of these issues and problems relate to economic factors, and they are the main subject of this paper. Others relate primarily to technical factors and their treatment is properly the task of hydrologists, engineers and other technical specialists. Thus, we do not discuss them further here, other than in terms of how the economist can interact with these other specialists in determining what physical input-output information is needed in order to carry out an economic analysis. The basic point is that the physical relationships must be quantified before an economic analysis can be carried out. Thus, the present discussion proceeds under the assumption that such information can be generated. Given the fact that the lack of such information is in practice the major bottleneck encountered in most watershed project appraisals, it may seem that this assumption is made to avoid a major problem. In fact, it is made to emphasise the point that the economist cannot solve the information and data problems associated with watershed projects. What he can do is to (a) suggest a systematic approach to identifying direct and indirect negative and positive impacts associated with a project and (b) point out what information and data are needed for him to be able to value these various impacts. The remainder of this discussion explores these two questions within the overall framework for project analysis set out in EAFP.

The specific points selected for further discussion are the following: [2]

1. Consideration of alternative means for achieving goals

2. Determination of project scope and context

---

[1] FAO Forestry Paper No. 17

[2] The fact that these are the only six points listed does not mean that they are the only ones of concern to the analyst of a watershed related project. For a more systematic discussion of the entire range of issues encountered, the reader is referred to EAFP.

3. Identifying costs for watershed projects
4. Identifying benefits for watershed projects
5. Treatment of benefits and costs in multiple purpose projects
6. Presenting cost and benefit information in an appropriate form.

In order to provide common empirical reference points during the discussion of each of these, two case study analyses of projects involving watershed considerations are summarized in Section 2. These two examples are then referred to in Section 3, which provides a discussion of the six points listed above. The reader who merely wants an overview of the issues can skip Section 2 (the cases) and go directly to Section 3.

Finally, in order to illustrate the types of empirical information that are required and examples of watershed related project analyses that are already available, a summary of some of the relevant documentation on this subject available from the United States is presented in Section 4. This section also illustrates a number of the general points discussed in Section 3.

2. EXAMPLES

The first of the examples is an economic analysis of alternative logging systems. The objective of the analysis is to find that system that maximizes net revenue subject to constraint on maximum allowable sediment discharge. It is an example of an economic analysis to provide information for an operational decision where water related concerns are entered as a constraint.

The second example illustrates in summary form an economic analysis of a major watershed project designed to reduce the rate of sedimentation in a reservoir, thereby extending the useful life of the reservoir and producing additional downstream benefits. The project also involves several other elements, including wood production in combination with watershed protection, pasture improvement and general improvement of upstream agriculture.

Example No. 1: Watershed Considerations as a Constraint in a Project

The growing worldwide concern for the environment makes this type of problem and this example relevant.

A 20 ha wood lot is to be harvested. The lot occupies land along a river with an average slope of 20-30%. A clearcut will not be allowed by regulatory agencies in order to prevent erosion and decrease resulting sediment flows. For this reason a selective cut will be made. However, it is anticipated that with standard logging techniques about 4 tons of sediment per hectare will enter the river the first year after the harvest. This amount of sediment is considered unacceptable by authorities and they will not issue the harvesting permit unless measures are taken to reduce sediment to no more than 2 tons per hectare. Thus, the forest manager must find an alternative that will reduce sedimentation of the river by at least 2 tons/ha/yr at the lowest cost possible, i.e. he is searching for the least cost alternative for logging the area that will meet the constraint.

Harvestable volume on the woodlot is 300 m^3/ha which can be sold for \$10/m^3.

If all 20 ha had been harvested using standard methods, it is estimated that the following costs and returns would have obtained:

Returns: 300 m^3/ha x \$10/m^3 x 20 ha equals \$60,000

Costs:	labour: 1 000 man hours x \$2.00/hr equals	\$2 000
	tractor: 250 hours x \$25/hour equals	6 250
	loading/transport: 120 hours x \$20/hr equals	2 400
	total cost:	\$10 650

Net revenue: \$60 000 minus \$10 650 equals \$49 350

However, as mentioned the standard method is not acceptable because of the high sediment discharge associated with it. Two alternatives are proposed that would meet the maximum discharge restriction.

The first feasible alternative consists of leaving a 25 m wide buffer strip (no cutting) along the river. The woodlot has a shoreline of 1 600 m, therefore, cutting would be reduced to a total of 16 ha instead of 20 ha. This means a loss of 4 ha of timber or 300 m3 x 4 ha x \$10/m3 which equals \$12,000 of revenue foregone. This is considered a cost for this alternative. It is assumed that other costs would be reduced by 20 percent since only 16 ha could be harvested. Thus, costs other than revenue foregone would decrease to \$8 520 (20 percent less than \$10 650). Total cost of this alternative would be \$20 520 (\$8 520 plus \$12 000).

The second alternative which meets the sediment discharge requirements consists of establishment of 40 m filter strip in which no machines are allowed. All commercial timber (i.e., 300 m^3 per ha) on this 6.4 ha filter strip can be cut but must be winched out at a higher cost. On the 6.4 ha of the filter strip costs are estimated to be \$8 094. For the remaining 13.6 ha costs will drop to an estimated \$7 242 to reflect reduction in area logged. Thus, total cost of this alternative will be \$15 336.

Assuming that these are the only two alternatives considered that meet the sediment discharge restriction, we would choose the lowest cost alternative or the filter strip approach. Revenue would be \$60 000 as before and cost would be \$15 336, for a net return of \$44 664, which compares with a net return of \$39 480 in the buffer strip alternative. The information generated in this analysis further indicates that the cost of the sediment discharge restriction would be \$49 350 minus \$44 664 or \$4 686.

Example No. 2: Economic Analysis of a Watershed Protection and Management Project 1/

Background on Project

Project Title:

Watershed protection for the Sierra Reservoir.

Project situation:

Some years ago a reservoir was built along the Sierra river to provide storage of water for downstream use during periods of low flow. Downstream uses include irrigation on some 9 500 ha and domestic water use by the local population. It has been found after five years of operation that the reservoir is silting in at a much faster rate than initially anticipated, thus reducing effective capacity and ability to meet water requirements downstream. Siltation is occurring at a rate of 4 million m3 per year. Present reservoir capacity is down to 100 million m3. At the present rate of siltation, it will only be four years before capacity is reduced to a point where it can no longer meet estimated water requirements of downstream users. (Domestic water use is increasing at a rate of about 6.19 percent per year, while irrigation use is fairly constant.)

Project goal:

To prevent the reduction (or loss) of water related downstream benefits, 2/ the project would extend the effective capacity and life of the reservoir by reducing the rate of siltation from 4 million m3/yr to 1 million m3/yr. 3/

Project points of view

(a) Downstream users of water have a direct interest in maintaining the capacity of the reservoir so that they can continue to receive water during the dry periods when river flow is inadequate to meet requirements;

(b) Upstream users of the land which would be affected by the various conservation measures proposed for the project are interested in how such measures would affect them. If effects are negative, some form of compensation may be included in the project plan;

(c) The nation at large is concerned with increased crop consumption, improved welfare of domestic water users, and losses or gains incurred by upstream land users.

The point of view adopted in the analysis is primarily that of the nation, although the other two viewpoints are also considered.

1/ Adapted from a project in the Andean foothills of a South American country.

2/ The benefits that would be lost without the project include crop values and health and satisfaction associated with domestic water use.

3/ Since there was apparently no problem of flood damage with or without the project, flood prevention was not included as a goal. It could be added in as a goal and treated in exactly the same way, if it was a problem.

Identification and Valuation of Project Costs

To accomplish the project goal, the following project components have been proposed in the technical design and analysis:

(1) Establish protection forest on the most critical areas where no other activity should take place because of slope or critical nature of soil protection.

(2) Establish protection/production forests on areas that need permanent protection but which are less critical so that some forest utilization can take place on a controlled basis.

(3) Build terraces on some of the most critical areas with very unstable soils.

(4) Manage and maintain pasture lands on a rotation, based on their carrying capacity and ability to regenerate. This will primarily involve control and policing activities together with technical assistance.

(5) Establish forest management on existing natural forest areas. This would include control on harvest and other activities, watershed protection inputs into access road establishment, inventory and other information gathering activities.

(6) Establish an overall watershed management and administration unit within the regional government to supervise and control implementation of an integrated watershed management programme for the whole watershed, including the above elements. Include extension services for local farmers.

In the project documentation, appropriate technology, input requirements and timing for each of the project components were analysed. Based on an initial survey of the total watershed of 17 500 ha, the scale of each of the project components was determined, as shown in Table 1. Average input requirements per ha were estimated and applied to the total areas to arrive at total labour, equipment, and other input requirements. These input requirements together with unit value estimates were then used by the economist in valuing the project costs, which are summarized on lines 4 through 8 of Table 4.

In developing economic values for inputs, only unskilled labour was shadow priced. Other inputs were valued in the economic analysis at their financial or market price values.

A project period of 26 years was considered appropriate, considering the relevant social discount rate of 12 percent. 1/

1/ For discussion of choice of project period (see EAFP).

Table 1

Areas associated with each project component

Component	Hectares
Protection plantings	760
Protection/production plantings	870
Terraces	320
Pasture use control	3 850
Natural forest management	3 160
Watershed planning & Adm.	(17 500) 1/

1/ Including the parts of the watershed not requiring direct action.

Identification of benefits

Reservoir demand (i.e. the demand on water from the reservoir which would not be available without it) 1/ is estimated at 86 million m3 in the first year (year 0) of the project as shown on the first line of columns 5 or 6 of Table 2. The capacity of the reservoir is 100 million m3 at present (start of project) and is decreasing by about 4 million m3 per year due to siltation. (See Col. 2 of Table 2. Thus, in about four years from the present the estimated capacity of the reservoir without the project would just be equal to demand. From then on, the reservoir would not meet the requirments for water from it.

With the project, it is estimated that the rate of siltation can be reduced to about 1 million m3 per year. Thus, the reservoir will be able to meet requirements for a longer period of time, although eventually, even with the project, demand for water will outstrip the capacity of the reservoir. (This will occur in year 10. Compare Cols. 3 and 6).

A first reaction might be to use the difference between the without and with project capacities as shown in Col. 4 as a measure of benefits. However, this would overstate benefits, since even without the project, the reservoir could satisfy demand for four more years. With or without the project, the benefits would be the same during those first four years and, thus, the benefits due to the project would be zero during that period (years 0-3). For the next six years (years 4-9) capacity with the project would still be above demand. Thus, with the project, the benefits due to the project for this period would be the difference between estimated demand and supply without the project, or the demand deficit which would start to be felt in year 4 if the project were not undertaken. (This is the difference between row items in Cols. 5 and 6). In year 10 demand would start to outstrip supply even with the project. Thus, from year 10 and on to the end of the project, the appropriate benefit figures would be the differences in capacity with and without the project (i.e. the difference between cols. 2 and 3). Using the above approach, the increased water use due to the project is identified and shown for each year in Col. 7 of Table 2.

The figures shown in Cols. 5 and 6 are gross figures which include evaporation from the reservoir. Since the evaporation would be approximately the same with and without the project, there is no need to adjust the figures shown in Col. 7. They represent net increases in effective water use.

In addition to the direct benefits associated with increased reservoir capacity, there will be some timber related benefits from the combined production/protection plantings. Based on experience elsewhere, these are expected to be as shown in Table 3. In years 6 through 10 there will be some minor thinning volumes available and in years 17 through 21 there will be final harvest volumes available.

In addition to the water and timber related benefits, the following indirect benefits were identified but not quantified in the study:

(a) Eventual increases in livestock production due to regulation of grazing on watershed lands. (At present, many of the pastures are marginal due to overgrazing). The project would restore these lands.

(b) Aesthetic values will increase as the land is rehabilitated.

1/ i.e., release of water in dry season to meet requirements during that period. It does not include the water used that would have been available without the reservoir, i.e., the requirements which would have been met from normal precipitation and river flow without it.

Table 2 - Watershed Project: Identification of Water Benefits

(millions of m3/yr)

(1)	(2) Reservoir Capacity	(3)	(4)	(5) Reservoir Use	(6)	(7)
Year	without project	with project	Difference with & without project	without project 1/	with project 2/	Difference in use with & without 3/
0	100	100	0	86.0	86.0	0
1	96	99	3	86.4	86.4	0
2	92	98	6	86.8	86.8	0
3	88	97	9	87.2	87.2	0
4	84	96	12	84	87.7	3.7
5	80	95	15	80	88.2	8.2
6	76	94	18	76	88.7	12.7
7	72	93	21	72	89.2	17.2
8	68	92	24	68	89.8	22.8
9	64	91	27	64	90.4	26.4
10	60	90	30	60	90	30.0
11	56	89	33	56	89	33
12	52	88	36	52	88	36
13	48	87	39	48	87	39
14	44	86	42	44	86	42
15	40	85	45	40	85	45
16	36	84	48	36	84	48
17	32	83	51	32	83	51
18	28	82	54	28	82	54
19	24	81	57	24	81	57
20	20	80	60	20	80	60
21	16	79	63	16	79	63
22	12	78	66	12	78	66
23	8	77	69	8	77	69
24	4	76	72	4	76	72
25	0	75	75	0	75	75

1/ Constrained by demand for water during first 4 years then constrained by capacity as demand outstrips supply

2/ Constrained by demand for first 10 years then constrained by capacity as demand outstrips capacity even with the project

3/ This is the measure due to the project, i.e. the difference in use with and without the project

Table 3 - Inputs and Outputs - Production Forest Component

	0	1	2	3	4	5	6	7	8	9	10	11–16	17	18	19	20	21
Ha planted	174	174	174	174	174								174	174	174	174	174
Thinning harvest																	
ha							174	174	174	174	174						
m^3/ha							20	20	20	20	20						
Total m^3							3480	3480	3480	3480	3480						
Final Harvest																	
ha													174	174	174	174	174
m^3/ha													525	525	525	525	525
Total m^3(1000)													91.3	91.3	91.3	91.3	91.3

(c) Access roads required for protection and other watershed management activities will permit faster and cheaper access by farmers to markets and increased mobility for extension personnel so they can reach more farmers.

(d) The project is expected to result in an increase in water quality in addition to quantity. A reduction in suspended loads carried over the reservoir dam will decrease the need for maintenance on individual irrigation installations.

Valuation of benefits

Based on studies of crop increases made possible by irrigation, it was estimated that irrigation water flowing out of the reservoir would return a net of P2 per m3 of water 1/. Since (1) the major portion of the water is used for irrigation, (2) there was no feasible way of placing a value on the water used for domestic purposes, and (3) there is no feasible way of allocating the increased water made possible by the project to irrigation and domestic use, it was decided to value the domestic water at the rate used for irrigation, namely P2 per m3. This was recognized to be a conservative estimate. Using this value per m3 and the water increase figures in Col. 7 of Table 2. the corresponding annual water related benefits from the project were determined as shown in row 1 of Table 4.

The wood production benefits were valued at P290 per m3 on the stump. This value was a parity price based on the value of imported wood. 2/ The parity price was adjusted down by 10 percent to reflect the lower quality of project wood. Total wood production benefits are shown on line 2 of Table 4.

Other benefits were not valued due to inadequate data or to the inappropriateness of attempting to quantify values, e.g., for the aesthetic benefits.

Comparing costs and benefits

As indicated on line 9 of Table 4, there is a net cost involved in the project for the first four years, after which the value flow turns positive and increases steadily over the life of the project. Using a rate of discount of 12 percent, we arrive at a Net Present Worth (NPW) for the project of some P292 million. 3/ The rate of return (EROR) of the project would be well in excess of 50 percent.

The high returns to this project can be explained quite easily. Since the reservoir was already in place and its cost represented "sunk costs", they were not included in the analysis of the project. Thus, the small amount of additional expenditure required for the watershed protection activities (the project) were compared with the returns which actually include the total incremental benefits from the reservoir. Obviously, if one were analysing a new reservoir project, the situation would be quite different, since the substantial expenditure for the reservoir would have to be added into the cost stream for the project, while the benefits would remain approximately the same.

1/ i.e., after subtracting from final crop value all costs back to the reservoir, e.g., farming costs, marketing, water distribution, etc., but excluding any sunk costs.

2/ The approach used to derive such parity values is described in EAFP.

3/ The rate of discount was given to the project planners by the national planning office and represents the rate used for evaluating all public projects in the country.

Table 4 -Value Flow Table

(millions of pesos)

		0	1	2	3	4	5	6	7	8	9	10	11	12	13	14	15	16	17	18	19	20	21	22	23	24	25
	BENEFITS																										
1	Irrigation & domestic use	0	0	0	0	7.4	16.4	25.4	34.4	45.6	52.8	60	66	72	78	84	90	96	102	108	114	120	126	132	138	144	150
2	Wood production							1.0	1.0	1.0	1.0	1.0							26	26	26	26	26				
3	Total	0	0	0	0	7.4	16.4	26.4	35.4	46.6	53.8	61	66	72	78	84	90	96	128	134	140	146	152	132	138	144	150
	COSTS																										
4	Planting protection forest	3.4	4.5	1.1																							
5	Planting production forest	1.3	1.6	1.6	1.6	1.6	0.3												1.3	1.6	1.6	1.6	1.6	0.3			
6	Terrace construction	0.8																									
7	Management costs 1/	1.0	1.0	1.0	1.0	1.0	1.0	1.0	1.0	1.0	1.0	1.0	1.0	1.0	1.0	1.0	1.0	1.0	1.0	1.0	1.0	1.0	1.0	1.0	1.0	1.0	1.0
8	Total	6.5	7.1	3.7	2.6	2.6	1.3	1.0	1.0	1.0	1.0	1.0	1.0	1.0	1.0	1.0	1.0	1.0	2.3	2.6	2.6	2.6	2.6	1.3	1.0	1.0	1.0
9	NET BENEFIT (COST)	(6.5	7.1	3.7	2.6)	4.8	15.1	25.4	34.4	45.6	52.8	60	65	71	77	83	89	95	125.4	131.4	137.4	143.4	149.4	130.7	137	143	149
10	Present Value at 12%	(6.5	6.3	2.9	1.8)	3.0	12.9	12.9	15.6	18.4	19.0	19.3	18.7	18.2	17.6	17	16.3	15.5	18.3	17.1	15.9	14.9	13.8	10.8	10.1	9.4	8.8
11	NPW at 12%	NPW at 12% P292 million																									

1/ including protection and extension services, maintenance and administration

3. DISCUSSION OF ISSUES

The two examples presented are representative of the types of economic analysis one encounters for watershed related projects. The following discussion outlines some of the major issues which arise concerning these examples and points to watch when applying the guidelines presented in EAFP to watershed related projects.

3.1 Considering Alternative Means for Achieving Project Goals

One of the basic points made in EAFP is that project planners should explore alternative means for achieving given project goals. If only one alternative is presented to the decision-maker, his only decision is whether to accept or reject it. On the other hand, if information is presented which permits him to look at a range of alternative means for achieving a goal, then he can more thoroughly consider and weigh the implications of different courses of action.

In Example 1 two alternatives to the standard logging approach were considered explicitly in the analysis. If other known alternatives had been available then they should also have been considered. In this case, the objective was to find the lowest cost alternative that met the maximum allowable sediment discharge restriction or constraint. Thus, one should note that costs and benefits for the standard logging approach were used only as a basis for comparison since it was, by definition, an unacceptable alternative due to the fact that it did not meet the constraint. Thus, actually only two alternatives were compared, the buffer strip one and the filter strip one. If others had been available (technically defined), they could very easily be included in the analysis.

The appraisal did not consider alternatives in the case of Example 2. However, there appear to be two which might have been considered. The first is the use of dredging at some future date to maintain reservoir capacity equal to demand. The second is the expansion of the reservoir to increase capacity so it can meet demand even when siltation occurs. In addition, the report on which this example is based did not discuss alternative technologies and scales for project components, nor did it go into the relative advantages of alternative timings of project activities to more efficiently achieve the goal of the project. Finally, although some of the project components were separable in terms of costs, the analysts did not have information on which to base a separation in terms of benefits. Thus, components were analysed separately and it was not possible to evaluate alternative combinations of project activities to find a more efficient overall solution for meeting the goals.

Based on the information available to the authors, it is difficult to state whether, in fact, additional alternatives were explored in the early stages of designing the project described in Example 2. However, the point to be emphasized here is that alternatives should be considered and that there should be an economics input at the early stage of project identification and formulation of alternatives. In other words, if possible, project planners should avoid discarding alternatives at an early stage on purely technical grounds. What may appear to be an inferior alternative to the technical expert may not be so from an economic point of view, given relative factor costs existing in the country in question. Initial - albeit rough - economic calculations can be extremely useful in terms of judging the initial set of alternatives and limiting them to those which will be studied in greater detail.

3.2 Determining Project Scope and Context

A major question facing project planners is what to include and what not to include within the scope of a given project.

From a practical point of view, it boils down to a question of where to cut off the endless chain of effects or impacts associated with a given project. The theoretical answer is: "Include all the impacts." The practical answer is: "Include all those impacts which you can identify and which appear to be large enough relative to the direct and immediate impacts to make a difference in the cost and benefit flows." The objective of a project evaluation is to generate the information needed to make a sound decision as to whether or not the project has benefits exceeding costs and, if so, whether the benefits exceed the costs by a large enough margin to make it worthwhile to commit scarce resources to the project rather than to some alternative use. If the direct benefits associated with a project are large enough relative to costs to make the project worth undertaking from an economic point of view, then spending a large amount of effort and funds on further analysis of all the various indirect impacts will not be worthwhile. However, if the project is marginally unacceptable, then there is a much stronger case for detailed analysis of indirect impacts. No general guidelines can be put forth here on how to determine the appropriate cutoff for considering indirect impacts. That will depend on each project situation, the knowledge of the project planners and staff specialists, the cost and time involved in generating information on indirect impacts, and the objectives of the institution sponsoring the analysis.

In the case of Example 1, the scope was very narrow, mainly due to the fact that the project involved a very small area and probably had insignificant indirect impacts. The example illustrates well the type of brief, uncomplicated analysis associated with operational decisions. Once this particular situation had been analysed and the best logging method chosen (the lowest cost method that met the constraint) it is likely that that method was accepted and used for other similar logging situations without further analysis, i.e., this simple analysis served as the basis for developing an operational guideline for logging that says: "In situations of river side logging, a filter strip system is the cheapest alternative logging system which meets the specified maximum allowable sediment discharge constraint."

In the case of Example 2, the project scope included the major impact elements, with the exception that there was no consideration given to how the project would affect the farmers upstream on the watershed lands who would have to change their operations due to conversion of land to forest or due to curtailment of grazing on critical watershed lands. Similarly, there was no quantitative analysis of the positive impacts on farm economies associated with the improved road network and the increased mobility and availability of extension services. Ideally, these should have been included in the analysis, and one would expect - even without having information on the project background and area - that it would have been possible to provide some more explicit treatment of these impacts.

The question of project scope is closely related to other aspects of project definition: (1) project points of view, and (2) cost and benefit identification.

Concerning project points of view, Example 1 can be identified with two: the logging operator (or company involved with logging the area), and the public point of view concerning sediment discharge. In this case, the public point of view has been expressed in terms of the maximum allowable discharge regulation and thus does not need to receive further consideration in the analysis.1/ The logging operator or company point of view (assuming that this is a private entity involved) is really the point of view from which the analysis is carried out, i.e., the question is: "What is the minimum cost we have to incur to achieve the constraint?" If the public sector is doing the logging, the question remains the same from an economic efficiency point of view.

Example 2 is somewhat more complex in terms of points of view. As stated in the text, there are three points of view identified, namely the downstream water users, the upstream land users and the national point of view which incorporates the other two points of view within an overall objective function. 2/ The downstream users' point of view defines the scope of the project at that end: the project should be defined broadly enough to include the necessary downstream costs to achieve the benefits accruing to the downstream users.

1/ Unless, of course, the analyst is also asked to look explicitly at the costs and benefits associated with different levels of sedimentation. This, however, is a separate question.

2/ As mentioned in EAFP this objective function relates to project impact of aggregate consumption.

On the other hand, the upstream land users' point of view defines the scope of the project at that end: the project should be defined broadly enough to include those costs and benefits for that group that occur because of the project. As mentioned earlier, there did not appear to be adequate consideration given to this point of view and the associated costs and benefits.

Consideration of points of view helps the analyst in identifying the appropriate scope and in identifying relevant costs and benefits for use in the economic, financial and social analysis of the project. The following two sections discuss cost and benefit identification in terms of economic efficiency analysis.

3.3 Identifying Costs

One can specify three main categories of costs involved in watershed projects. These are:

Structures and work costs: These include costs of dams, gully plugs, construction of contour furrows or terraces, channel construction or improvement, road relocation, retainer walls, etc., and maintenance of these structures and facilities.

Vegetation manipulation costs: These mainly include costs of removal of vegetation and planting and management costs associated with the establishment of new vegetation.

Value of outputs foregone: Even eroded or deteriorated lands may be producing values through grazing, subsistence farming, etc. These activities may have to be curtailed for a period of time in order to restore land to some higher level of productivity. The value of such production foregone should be included as a project cost. In the case of a protection project, timber harvested per unit area may be reduced due to the introduction of buffer strips along rivers, streams, roads, etc. Selective harvest may have to be imposed on steep hillsides which may in turn reduce the present value of harvests. This reduction is a cost.

The first two categories of costs are quite obvious, and both examples in Section 2 treated these in an adequate fashion. The third category - value of outputs foregone - is also relevant to both cases. In Example 1, it can be noted that the analyst treated the value of timber foregone through creation of a buffer strip as a cost. He could also have merely reduced the total benefit figure by this amount, thus treating this value foregone in terms of benefits. Either way would have produced the same result, since the objective was to arrive at the alternative with the highest net return.

In the second example, there were values of outputs foregone from changes in land use that should have been considered but were not, as explained in the previous section. This supports the point made earlier that project scope points of view and cost and benefit identification are closely interrelated. Since the upstream land users' point of view was not adequately defined, the analyst also missed identifying explicitly changes in value of output associated with upstream land use due to restriction of grazing on some lands and shift in land use from agriculture to forestry on other lands.

In identifying costs (or project inputs) it is essential that the "with and without" test be applied. Basically, this means that the analyst asks and answers the following question: "What would the situation likely be without the project over the period of years considered for the project and what would the situation likely be with the project?" Only the difference should be attributed to the project. This approach applies to both costs and benefits (inputs and outputs). The particular point to emphasize here is that the "with" and "without" project comparison is not the same as the "before" and "after" project comparison for most types of watershed projects. It is likely that the situation as it exists before the project is introduced, would change over the life of the project; and it is necessary to estimate how the situation would develop over time without the project, since only the difference between costs and benefits without and with the project can be attributed to the project.

For example, in the case of costs, one might be tempted to include as part of a larger watershed improvement project the maintenance costs for a road system that is being affected by erosion processes as a cost. But, assume that the road is essential to the communications of the region in question and, regardless of whether or not the project is undertaken, the road would have to be maintained and kept open for the region. Thus, the road maintenance would be undertaken with or without the broader watershed improvement project. In this case, the costs of road maintenance should not be included as a project cost. If, with the project, the road maintenance costs can be reduced, then this reduction in costs should be included as a benefit due to the project.

Similarly, assume that the present level of use of a given land area will have to be curtailed as part of an overall watershed protection project. As mentioned above, the value foregone due to this reduction in use should be treated as a cost. But suppose that use of the land would have increased over time without the project. Then the analyst will have to estimate what that increase in use would have been in order to arrive at correct estimates of the costs, due to impact of the project on use of the given area of land.

3.4 Identifying Benefits

Similar arguments stated above apply to benefits. For example, over time, without the proposed watershed project, soil conditions might deteriorate, erosion might increase, etc. The analyst has to make sure that these changes are taken into account. Figure 1 illustrates a typical situation. As noted, at time 0, production is at level X. Without the project, conditions would deteriorate until in year n production would have decreased to Y. With the project, it is estimated that production will increase to Z. The point to note here is that both Z minus X and X minus Y are legitimate benefits to be attributed to the project. Thus, the analyst will not only need to estimate the increase in production which will be possible (i.e. Z-X), but he will also have to make an estimate of the losses which will be avoided (i.e. X-Y). Example 2 illustrates this point.

Application of the "with and without" test also brings out another point related to benefit identification and valuation (which is also illustrated by Example 2). The point is that merely because a project changes some physical dimension in a positive way, this does not necessarily mean that there is a benefit involved. In Example 2, the project starts immediately to reduce the level of siltation in the Sierra reservoir and thereby increases the effective capacity of the reservoir. However, even without the project the level of capacity of the reservoir is in excess of demand and will continue to be so for the next 4 years (see Table 2). Applying the "with and without" test, the analyst can see that consumption of water (the relevant benefit parameter) will remain the same with or without the project for this period (see Cols. 5 and 6, Table 2). Thus, the benefits (losses avoided) due to the project will be zero during the first four years, or until the capacity of the reservoir without the project would have fallen below requirements for water. This point applies more broadly to many different types of watershed projects.

The above point relates to the fact that in an economic efficiency analysis, benefits have to be measured in terms of human consumption. Thus, for example, the hydrologist may provide an estimate of tons or cubic metres of soil loss that can be avoided by undertaking a given project. But this information is not enough for an economic analysis. In order to value the benefits from the project, such losses avoided have to be translated into a schedule of crop or other consumption losses avoided. Thus, agricultural experts have to come up with a relationship between soil loss and crop production or soil loss and production of some other consumption item. This consumption loss can then be valued and used as the benefit in the economic efficiency analysis.

Most hydrology projects are undertaken by the public sector, the main reason being that "water" is seldom sold in the market place. While farmers using irrigation are often required to pay some amount for water they use, and households and industries often pay a nominal fee for domestic and industrial water, in general water is not priced in the market in the traditional sense. 1/ Thus, measures of willingness to pay for water, which are the appropriate measure of the value of benefits in an economic analysis, are seldom available (see EAFP, Chapters 2 and 5). Similarly, many of the inputs required for hydrology projects are often not priced in the market. Thus, it becomes necessary to estimate shadow prices for such non-market priced benefits and costs. We mention this issue here, since it is important in the economic analysis of watershed projects. However, the process of shadow pricing is rather complicated and is discussed elsewhere. The reader is referred to detailed treatment of the subject in EAFP and in a recent work done for the Organization for Economic Cooperation and Development. 2/

1/ I.e., water prices are administratively set and not determined on the basis of the interaction of supply and demand.

2/ See OECD (1979)

Figure 1

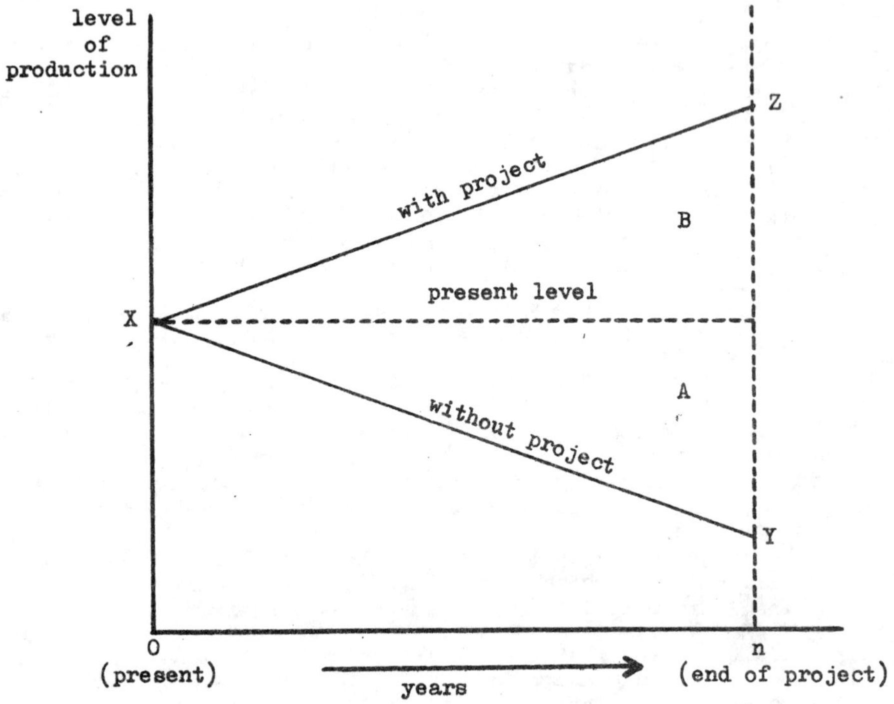

A - losses avoided

B - production increases over present level

3.5 Treatment of Benefits and Costs in Multiple Purpose Projects

A point worth mentioning here is the need to use care and caution in identifying costs and benefits associated with multiple purpose projects which include a watershed management element. For example, in some cases, trees planted on denuded lands as part of a watershed protection or restoration project will also be managed for controlled harvest for fuel or other products. In such cases, both types of benefits will have to be included in the analysis. 1/ Proper allocation of tree planting costs to the watershed benefits and the wood output benefits is difficult. If timber production is the main objective of the project with watershed protection or restoration as a secondary purpose, then one practical approach would be to allocate the basic costs to the timber objective. Any additional costs of vegetation management to achieve the constraint or watershed objective would be allocated to the watershed component of the project. Similarly, in the case of logging road redesign to meet certain watershed constraints or objectives, the equivalent of the minimum road cost to get the timber out would be attributed to the timber element, while the additional costs associated with higher standards to meet the watershed objectives would be allocated to the watershed element.

In the case of a primary purpose watershed project, the cost of tree planting or other activities would be associated with the primary purpose and benefits, while timber benefits would be treated as secondary benefits. As mentioned earlier, it is important in such cases to remember to subtract any secondary costs associated with the timber production up to the point of valuation of the timber (e.g., stumpage level, delivered log level, etc.).

3.6 Timing of Costs and Benefits

Most watershed projects tend to be longer term projects in the sense that the inputs occur over a considerable period of time and the benefits accrue over an even longer period of time. Further, benefits and costs are constantly changing over time.

A main problem is to develop a sound estimate of the timing of the benefits. Restoration projects generally take time to implement. Full productivity is restored slowly in most cases. For example, if trees are planted on a deteriorated watershed, the full protective effect on erosion control will take some time to achieve.

In order to keep track of the project assumptions regarding the build-up to project benefits and costs over time, it is essential to use appropriate physical flow tables and, ultimately, properly designed value flow tables. (Such tables are shown as Tables 3 and 4 in Example 2 in this paper.)

3.7 Treatment of Uncertainty

Watershed related projects are particularly subject to great uncertainty in terms of the values of costs and benefits used. Thus, it is important that project appraisals include explicit treatment of uncertainty. Neither of the two examples presented earlier did so, and that is perhaps a typical situation found in most economic appraisals.

There are some simple techniques, such as sensitivity analysis and break-even analysis, which can be applied rather easily and cheaply in most cases. Basically, sensitivity analysis involves varying assumptions concerning the values of key parameters and then testing the sensitivity of the chosen measures of project worth to such changes. A breakeven analysis is aimed at identifying values of key parameters which would switch the profitability of a project from acceptable to unacceptable levels.

1/ Of course, any associated costs involved in harvest will have to be subtracted, if roadside value for the harvest is used instead of stumpage value.

4. REVIEW OF UNITED STATES LITERATURE

The previous section provided a general view of some of the main problems to watch for in carrying out an economic analysis of a major purpose watershed project or a project which includes water related considerations as constraints. In the present section, a brief review is provided of some of the most relevant work done in the United States dealing with watershed project economics.

The need for watershed management practices developed in the United States largely because of erosion, sedimentation, and flooding which resulted from (a) overgrazing of western rangelands, (b) uncontrolled wild fires and (c) careless logging operations (Bailey and Croft, 1937; Trimble and Weitzman, 1953; Haupt, 1959; Rice et al., 1963; Packer and Christensen, 1964; and Packer, 1967). Thus early watershed management practices were synonymous with protection and restoration (Packer and Laycock, 1969). The protective influence of vegetation on the processes of erosion, flooding and sedimentation has been recognized for some time. The effects of vegetation and land management activities on water yield, on the other hand, has been a subject of considerable misunderstanding. Early thinking on this matter suggested that because dense forested watersheds were headwater areas of most flowing streams and rivers, such forest cover was essential to the production of water (Satterlund, 1972). Experimental evidence has suggested just the opposite. Forest vegetation consumes large quantities of water by means of transpiration. This results in a loss of water to soil moisture and streamflow. The inference from such knowledge is that forest harvesting or vegetation removal in general will tend to increase water yield. This stimulated watershed research throughout the United States (Hoover, 1944; Dortignac, 1965; Hibbert, 1965). The possibilities of watershed management directed towards increasing water yields had important implications to land management in the water-poor regions of the United States.

Regardless of whether goals of watershed management are to protect or to increase the water resource, or both, the first step in economic analysis is to determine distinctive production functions for various management practices (Lloyd, 1969). Several examples of physical relationships or production functions are discussed below in terms of the following broad management goals: (1) watershed protection and (2) watershed restoration and water yield improvement.

4.1 Watershed Protection

Vegetation protects soil from the energy of raindrop impact, minimizes overland flow, and along with the soil-binding benefits of root systems on steep slopes, reduces erosion, sedimentation and nutrient losses from watersheds (Satterlund, 1972; UNESCO).

1972). Conversely, substantial removal of vegetation by fires, shifting cultivation (agricultural expansion), timber harvesting, grazing, road construction, or urbanization, can increase soil disturbance resulting in soil and nutrient losses from the watershed, sedimentation of downstream receiving waters, and more frequent occurrence of floods. Packer and Laycock (1969) summarized impacts of a variety of land uses and indicated that the density of plant and litter cover were the most important factors affecting soil erosion and overland flow on range lands. In forestry, poor construction of logging roads and improper skidding practices can be major causes of erosion and sedimentation.

Because of the undesirable consequences of livestock overgrazing, overpopulations of wildlife, poor road construction, improper skidding practices, extensive wildfires, excessive recreation use and related activities, land management agencies have developed guidelines to minimize adverse impacts. In some cases, such guidelines are imposed by law. Limited cause-and-effect relationships and hydrologic information are usually available to establish these guidelines for a specific area. Typically such guidelines are based on experiments from a few intensively studied watersheds. Such guidelines usually result in regulations concerning the maximum slope and location for logging roads, rules for building roads in proximity to water courses, and requirements for the maintenance of buffer strips of vegetation along water courses to reduce sedimentation of streams (Lantz, 1971; U.S. Department of Interior, 1970). In some cases (e.g., Korea) land use laws establish maximum slopes on which land clearing can take place.

4.2 Erosion, Sedimentation and Flooding

Erosion and sediment control methods used in eastern and western areas of the United States have been evaluated by Thronson (1973). Cost data were applied to theoretically predicted soil losses. Over 25 control methods were examined. The principal cost elements consisted of labour, equipment and materials. Costs per cubic yard of soil retained by conservation methods were compared with costs of several methods of sediment removal in stream channels and reservoirs. Annual cost figures were based on control effectiveness and economic life of each project. Sediment removal was in general found to be more costly than erosion control. Associated costs such as fisheries habitat damage and loss of site productivity would be important but were not quantified here. In order to identify both direct and indirect costs associated with watershed maintenance or restoration projects, specific examples or case studies will be discussed.

The annual fire-flood phenomena in southern California illustrates the importance of maintaining deep-rooted vegetation, in this instance chaparral, on steep mountainous slopes. Frequent fires in the chaparral watersheds above the densely populated areas result in severe flooding and erosion (Rice *et al.*, 1963; Corbett and Rice, 1966; Packer and Laycock, 1969). The adverse economic consequences of such fires have been used to justify extensive contour-furrowing and contour-trenching and seeding to minimize runoff and erosion and to speed up the reestablishment of a vegetative cover. Such mechanical or structural solutions, although costly, are often the only effective means to rehabilitate severely eroded, steep-sloped watersheds. The costs associated with fire prevention and other management should be measured against losses of watershed productivity, costs of sedimentation and losses of life and property in addition to reclamation costs.

Overgrazing of high elevation watersheds in northern Utah caused substantial mudslides and debris-laden floods that resulted in substantial losses of property and life (Bailey and Croft, 1937; Bailey and Copelund, 1961). These watersheds were subsequently contour trenched and revegetated. Considerable success was achieved in containing erosion and to a certain extent, in reducing flash floods. In this example, as with the fire-flood problem in southern California, the benefits derived from proper watershed management cannot be entirely quantified in economic terms. For example, values of human lives saved and wildlife habitat and esthetics damage avoided, cannot readily be quantified.

In some instances natural levels of erosion and sedimentation are excessive and cannot be controlled with watershed maintenance or restoration projects. Such was the case in southeastern Utah where considerable contour furrowing, gully plug emplacement and range reseeding had little impact on erosion and sedimentation (Workman and Keith, 1975). The area had been overgrazed in the past, but the extremely high levels of natural or geologic erosion minimized the impacts of rehabilitation efforts. The purpose of this project was to establish a more dense vegetative cover to reduce sedimentation of the lower Colorado River and thereby extend the life of Lake Powell and Glen Canyon Dam. Benefits in terms of a reduction in municipal water treatment costs downstream of the project were also evaluated. The project costs averaged US$5.45 to US$13.31 per acre and had project lives of from 7 to 10 years which resulted in only an 11 percent reduction in sedimentation. Workman and Keith (1975) looked at minimum possible project costs and compared them with maximum possible benefits to provide the project with "every conceivable benefit of the doubt." In this analysis the benefits consisted of (a) avoidance of water treatment costs of downstream communities with an estimated maximum annual value of US$49 030, and (b) the extended flow of goods and services associated with the extension of the lives of Lake Powell and Glen Canyon Dam, which included irrigation water, electricity, flood control, and recreation. The total maximum annual value of benefits was US$48 749 030. A benefit cost ratio was then calculated using a 7 percent discount rate to aggregate future benefits in present value terms. The present values of benefits in the absence of erosion control for 200 years was then subtracted from the present value with erosion control. This maximum present value of benefits was then divided by the cost of treating the entire 1 280 000 acres of frail watersheds. A benefit-cost ratio of 0.12 resulted. Thus, restoration treatments on these frail desert soils in the Upper Colorado River Basin would only return US$0.12 for every $1 spent on treatment.

Green (1971) investigated economic impacts of grazing on erosion and sedimentation in terms of site deterioration as well as effects on the life of a downstream reservoir in north central Utah. The Joes Valley Reservoir, a US$7.5 million project with 62 500 acre-ft. storage capacity was designed for 8 500 acre-ft. of dead storage for sedimentation. With an expected life of 100 years, the project was designed to handle sedimentation rates of about 85 acre-ft. per year on the average. Natural levels of sediment production (approximately 22 acre-ft./year) were determined along with livestock carrying capacity-sediment production relationships. Reduced grazing levels of 500 animal unit months (AUM's) and associated range management practices without rehabilitation measures such as furrowing resulted in reductions of 5 acre-ft. of sediment per year. The removal of 5 acre-ft. of sediment from the reservoir would cost US$6 630 to US$19 890 which more than compensated for losses of receipts from grazing of 500 AUM's on federal forest lands. The costs of contour furrowing levels needed to reduce sedimentation for an 80-year period were also developed and contrasted to costs of grazing level reduction.

The costs of site deterioration caused by several different logging systems were compared recently by Klock (1976) in the Pacific Northwest. Unit cost figures for different logging systems were considered essential to quantify effects on reduced site productivity as well as effects on sedimentation, water quality in terms of domestic, irrigation and esthetic values, fisheries habitat, and need for control structures and associated seeding of distrubed sites. Costs, although indirect, "... should be added to the direct operational costs when systems of log removal for a particular forest site are evaluated" (Klock, 1976)1/

Onsite damages of forest lands were estimated in terms of the fertilizer or nutrient replacement needed to maintain productivity. The following approximation for productivity damage assessment (PDA) was suggested:

$$\text{PDA} = \frac{\text{Area x Erosion depth x (\% total N) x A x B}}{\text{Timber Volume Assessed}}$$

where: % total N = percent total nutrient, such as nitrogen in the soil

A = fertilizer conversion rate in pounds of N per pound of fertilizer

B = cost in dollars of fertilizer per pound per acre.

Such data as inches of soil loss (erosion) per activity, biomass productivity estimates, and soil nutrient make-up, essential to such an analysis would probably be limited in most field situations. One could hardly argue with this approach conceptually, however, in terms of evaluating the effects of erosion on forage or timber productivity because of the long time period required for natural processes to replenish essential plant nutrients.

Downstream damages considered by Klock consisted of (a) detrimental effects on salmon spawning stream beds and (b) costs of sediment removal from channels by dredging. Excessive sedimentation not only affects fish directly but also covers spawning beds which may reduce salmon and steelhead reproduction (Brown, 1974). The value of salmon spawning streams ranged from US$10 000 per acre in Lost Creek, Oregon to as much as US$3.0 million per acre on the Fraser River in British Columbia (Klock, 1976). Costs of sediment removal on the other hand ranged from US$0.25 to US$18.30 per cubic yard depending on method and placement requirements for dredge spoils (see Table 5). A method of assessing sediment costs per thousand board feed of merchantable timber was suggested as follows:

$$\text{Sediment cost} = \frac{\text{Area x Sediment Depth x Removal Cost}}{\text{Timber Volume Assessed (MBF)}}$$

When considering only yarding effects, Klock estimated that one inch of soil loss created a sedimentation assessment of US$62.53/1000 bd. ft. of saw logs in one case. This estimate of sedimentation did not even consider road construction which could result in even higher levels. A summary of Klock's results is presented in Table 5.

A closely related and often mentioned benefit of watershed management in addition to those previously described, is flood control. The beneficial influence of vegetative cover, particularly a dense forest cover on overland flow and flooding has been discussed in some detail by Lull and Reinhart (1972) for the eastern United States. In order to quantify such benefits for an economic analysis, a flood damage-frequency analysis before and after removal of vegetative cover by some land use practice should be considered, as accomplished in flood control project evaluations. However, unless large land areas are drastically affected, say by wildfires or extensive clearcuts, the discharge-frequency relationship would show little effect. Even if 30 percent or more of a watershed is

1/ Swanson and McCallum (1969) in a similar study evaluated soil losses from agricultural watersheds in terms of agricultural output-yield reductions.

denuded, the effect on flooding frequency would be expected to be more pronounced for annual maximum peak flows and have little influence on the large destructive floods of 100-year recurrence interval or greater. Hollis (1975) indicated that paving 20 percent of a watershed increased the probability of only small floods of a return period of 10 years or less, but that paving 30 percent could result in a substantial increase in floods with a 100-year return period. Therefore, the percentage of the watershed disturbed is a critical factor in estimating effects on flooding potential.

Harr, et al. (1975) found that road construction and poorly designed drainage systems significantly increased peak streamflow when 12 percent or more of watersheds were affected in the Oregon Coast Range. With proper culvert design and bridges in head-water areas, peak flows were not affected. Road construction guidelines as described by Trimble (1959) and USDI (1970) are typically established to reduce erosion and sedimentation which in turn minimizes flooding potential. The costs of higher road standards can be compared to erosion reduction benefits.

A factor that has been observed, yet is difficult to quantify, is the channel constriction process associated with accelerated erosion and sedimentation. Not only does sediment tend to accumulate and diminish the capacity of a channel to transmit flow, but there is a tendency for sediment and debris to build up and form small "detention" dams within a stream system. Such a debris-dam system is perpetuated by several "average" flow years. The consequence of both factors is that when a less frequent but more intense storm and resultant stream discharge occurs, either (1) flood waters escape the sediment filled channel much more quickly than before or (2) the buildup and wash out phenomena of a cascade of sediment and debris dams may accelerate the velocity, increase the total discharge over a shorter time interval, and increase the debris-carrying capacity of the stream. The consequences are greater damages downstream. Therefore, the erosion and sedimentation factors may be important in flood damage from wildland watersheds just as the "quick response" direct runoff which occurs from a denuded watershed.

4.3 Water Quality Considerations

In order to ascertain the benefits of watershed managment practices for the purpose of maintaining a high quality water yield, the value of such high quality water has to be estimated. One approach may be to consider the environmental impacts of water pollution from land management practices. Unger et al. (1973) considered environmental impacts of water pollution in terms of health, esthetics, and production. The economic consequences of water pollution under the health category include medical service demands, loss of man hours because of illness, human life lost, or the costs of making water supplies safe for human consumption. Impacts on esthetics may be reflected in terms of private property devaluation, alterations of recreation opportunities or social values of "quality of life". Other economic consequences include changes in industrial water treatment costs, changes in types of industries and employment, regional and area dislocations, input costs, income redistribution and final product prices. Jordening and Allwood (1973) also considered both on-site and off-site costs of water pollution. Costs of water pollution were considered in terms of opportunity costs, damage costs and reduced efficiency of productivity or increased production costs. The costs of water pollution were then schematically compared to the costs of water pollution control as illustrated in Figure 2.

For industrial and municipal water quality evaluations the sources of pollutants, technology of water treatment, and costs are easier to evaluate than the "non-point" sources of pollutants from activities on wildland watersheds. Although most previous research has focused on management impacts on erosion and sedimentation, several studies have indicated significant impacts of land use activities on stream temperatures and nutrient levels of receiving waters. (These effects are included in the water "quality" category). Such studies have resulted in the establishment of management guidelines.

"Buffer strips" or streamside vegetation of a specified width have become requirements in timber harvesting operations to minimize impacts on stream temperature as well as sedimentation. Brown and Krygier (1970) reported that removal of streamside vegetation can raise water temperatures above acceptable levels for the survival of salmon and steelhead trout. Maintaining buffer strips along stream channels involves a pollution abatement cost equal to the value of timber withdrawn from harvest and any increased costs of harvest. Dykstra and Froehlich (1976) evaluated stream protection costs in Oregon and found that maintaining buffer strips 55 and 150 feet wide along stream channels resulted in timber volume foregone of 0 to 6 and 6 to 17 percent, respectively. Buffer strip requirements for protecting water quality may result in a significantly larger land base requirement to produce a given volume of timber. Such costs need to be compared with the benefits before deciding on regulations.

Dykstra and Froehlich (1976) also contrasted costs of removing debris from channels following timber harvesting operations with buffer strip costs. The removal of debris from channels following harvesting was reported to cost from $100 to $500 per 100 feet of channel, depending on the stream gradient, width and quantity of debris. Such costs represent "costs of pollution" in Figure 2.

Figure 2

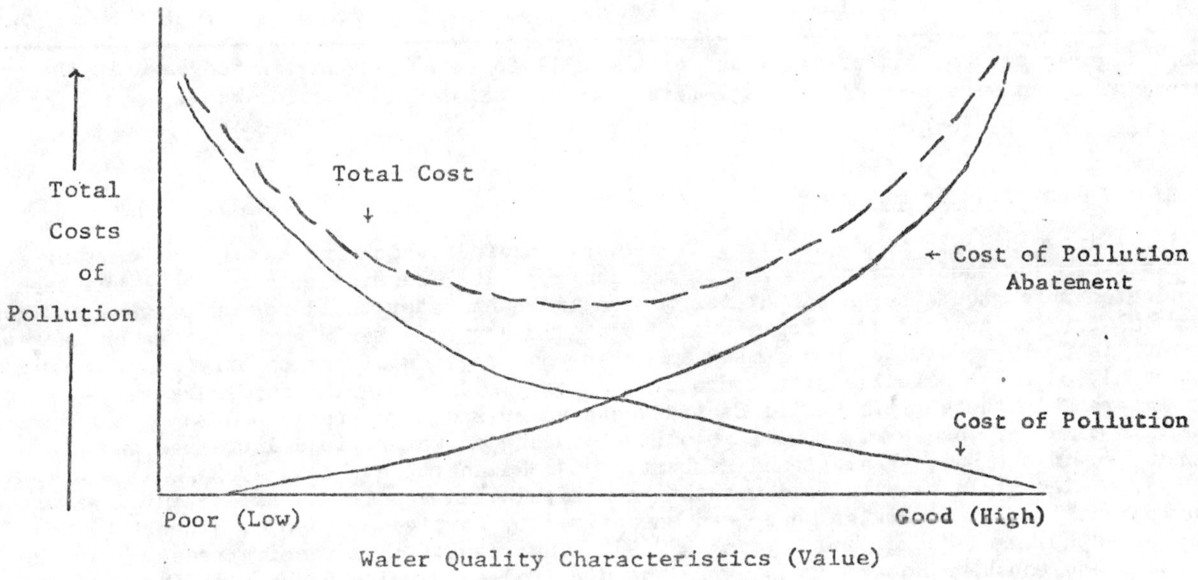

Total Cost Determination for Pollution Control
(From Jordening and Allwood, 1973)

Table 5

Some cost estimates of tree removal per 1 000 board feet of merchantable lumber for several yarding systems used in the fire-affected study area

System	Erosion	Direct cost 1/	Indirect assessment Productivity	Sedimentation	Total "cost"
	Inches		-Dollars-		
Tractor, slopes 0-30%	0.20	34.85	3.30	12.50	50.65
Tractor, slopes 30-50%	.80	34.85	10.60	50.00	95.45
Tractor (over snow), slopes 0-40%	.08	38.60	1.50	5.00	45.10
Cable skidding	1.50	35.00	17.00	93.80	145.80
Skyline (Wyssen)	.04	52.73	.80	2.52	56.05
Helicopter	.04	74.98	.80	2.52	78.30

Cost figures are for timber delivered at the mill for each yarding method used in the study area and were provided by Pack-River Lumber Company, Peshastin, Wash.

Source: Klock, 1976

4.4 Water Yield Improvement

Since the 1928 study at Wagon Wheel Gap, Colorado which indicated that cutting forests can increase streamflow (Bates and Henry, 1928), considerable research has been conducted throughout the United States to quantify the water yield response to vegetation removal for various climatic regimes and vegetation types. Early emphasis was on increasing water yields by vegetation management in water-poor regions of the country such as Arizona (Barr, 1956). The potential for increasing water yield by vegetation management was also of interest in the eastern United States where water shortage problems often occur during certain times of the year. Several studies focusing on water yield increases have been conducted in the East (Hewlett and Hibbert, 1961; Reinhart, et al., 1963; Lull and Reinhart 1967). Water yield results from vegetation manipulation studies on experimental watersheds throughout the United States have been summarized by Dortignac, (1965), Hibbert (1965), Hoover (1969), and Packer and Laycock (1969). The results of intensive research in the central and southern Rocky Mountains by the U.S. Forest Service concerning the potential of water yield increases in all major vegetation types have been documented in a series of "status-of-our-knowledge" reports (Hibbert et al., 1974; Ffolliott, 1974; Ffolliott and Thorud, 1974; Rich and Thompson, 1974; Baker, 1975; Ffolliott and Thorud, 1975; Leaf, 1975a and 1975b; Orr, 1975; Sturges, 1975; and Springfield, 1976). A good information base has thus been established in the Rocky Mountain region which should provide insight into physical relationships and responses to management schemes to allow for economic evaluations of various product-mix alternatives.

Most of the above mentioned studies indicated that reductions in forest vegetation or conversion from deep-rooted species to grass vegetation resulted in increased water yields. Reforestation decreased water yields. Knowledge of water yield responses for all vegetation types and climatic regions has not been determined; however, rough estimates of water yield response can be estimated from previous studies. As Satterlund (1972) indicates a question which needs to be answered before management practices become implemented is "what balance between water and other goods and services is desirable?"

In order to answer such a question, estimates of multiple-use production relationship, including supplementary, complementary and competitive relationships need to be identified and costs and benefits determined for each alternative.

When wildland watersheds are to be managed with increased water yield as a goal, it is essential that water yield responses be evaluated in a multiple use framework. Such an analysis is needed for economic studies because costs and benefits of management schemes cannot be determined if the physical responses (production functions) of such management are not known. As pointed out by Clawson (1974), compatible and incompatible uses of land must be identified and tradeoffs quantified. For example, when the output of one product is increased what effect does it have on other outputs or services. Some of the more recent investigations have attempted to quantify such relationships.

O'Connell and Brown (1972) developed production functions for ponderosa pine watersheds in north-central Arizona by evaluating effects of different levels of strip-cutting on water yield (acre-ft.) wood (bd. ft.), herbage (Lb), and sediment production in tons. The results are summarized in Table 6.

Ffolliott and Thorud (1975) provided an extensive and detailed summary of water yield improvements for the following vegetation zones in Arizona: alpine, mixed conifer, aspen, ponderosa pine, pinyon-juniper, chaparral, grassland, desert shrub, and riparian association. The mixed conifer, ponderosa pine and chaparral vegetation zones were considered to provide the greatest opportunities for water yield increases. Twelve alternatives for state-wide water yield improvement were contrasted. Two of these alternatives are shown in Table 7 for illustration: For each vegetation zone, water yield improvement opportunities as a result of various levels of clearing were examined. In addition, timing and peakedness of streamflow, erosion and sedimentation, water quality indicators, timber, herbage (forage), fish and wildlife, and recreation opportunities were also considered where appropriate. Thus the basic production function elements needed for a product mix evaluation and economic analysis were presented for each vegetation zone.

Even in vegetation zones where water yield (quantity) improvement opportunities do not exist, the management decision should consider impacts on water quality, regimen and soil erosion. Table 8 indicates in general how various land-use or management activities may affect water yield characteristics. Obviously many assumptions are implied, but it may be useful in an initial assessment of components of a multiple product mix for any given area. For economic analysis, however, such relationships need to be quantified both in terms of physical response and in economic value.

Satterlund (1972) discussed earlier work by Worley and Miller (1969) in which only two products, timber and water, were evaluated with the goal of increasing water yield without decreasing timber yield. The product-product relationship for various management alternatives is shown in Table 9. Two outputs, timber and water were contrasted. The basis of selecting the management alternative was that first the alternative must be better in at least some aspect with no reduction in the other. Alternatives 6 and 7 are thus best. The decision to implement 6 versus 7 depends on whether the value of water is worth US$5.50 per acre-ft. ((US$3.50-4.50) + (0.40-0.22 acre-ft.)).

Table 6

Estimates of Average Annual Production of Different Levels of Strip Cutting of Ponderosa Pine in Arizona

Production	Percent Strip Cut		
(Avg. Annual Inc. Per Acre)	33	60	100 (clearcut)
Water (ac. ft.)	0.08	0.12	0.16
Wood (bd. ft.)	160	116	36
Herbage (lb.)	150	225	825
Sediment (tons)	0.07	0.04	0.02

Production is average annual increase per acre over the existing management level for a 90-year period.

Source: O'Connell and Brown, 1972

Table 7

Alternative Water Yield Improvement Schemes by Vegetation Manipulation

1.	Mixed conifers - convert 1/3	=	8 273 ac. ft./year
	Ponderosa pine - clear 1/3	=	342 999 ac. ft./year
	Chaparral - convert 40%	=	251 289 ac. ft./year
	Total	=	602 561 ac. ft./year
2.	Mixed conifers - convert 2/3	=	41 370 ac. ft./year
	Ponderosa pine - clear 2/3	=	685 997 ac. ft./year
	Chaparral - convert 60%	=	502 577 ac. ft./year
	Total	=	1 229 949 ac. ft./year

Source: Ffolliott and Thorud, 1975

Table 8

Relative effects of watershed activities on water yield characteristics

Watershed Management and Related Land Use Activities	Water Yield and Related Characteristics								
			Water Quality Indicators						
	Quantity of Water Yield	Soil Loss Erosion	Temp.	Dissolved Oxygen	Nutr.	Sediment	Disease Org. & Colifirm	Flooding Opp.	Timing of Streamflow
1. Grazing	+	+	0	–	+	+	+	+	+ (?)
2. Timber cutting:									
Thinning	+	+	?	?	?	?	0	+	+ (?)
Clearing	+	+	+	?	?	?	0	+	+
3. Road construction	+	+	?	?	+	+	0	+(?)	+
4. Herbicide applications	+	?	+(?)	?	+	?	0	?	+
5. Fire	+	+	+	–	?	+	0	+	+
6. Recreation use	?(+)	+	?	–	+	+	+	+(?)	+
7. Conversion from trees or brush to grass	+	+(?)	+(?)	?	?	?	0	+	+
8. Conversion from grass to brush or trees	–	–	–	?	?	?	0	–	–
9. Urbanization	+	+	+	–	+	+	+	+	+

Expected responses: + = increase, – = a decrease, ? = questionable effect; and "0" indicates no effect expected.

Table 9

Relationships between water yield and timber yield for selected management alternatives.

	Management Alternative	Annual Yield Per Acre	
		Timber ($)	Water (Acre-ft.)
1.	Harvest only, clearcut	3.5	0.20
2.	Harvest only, shelterwood	3.7	0.18
3.	Commercial thinning, clearcut	4.0	0.22
4.	Commercial thinning, shelterwood	4.5	0.20
5.	Pre-commercial thinning, clearcut	4.0	0.24
6.	Pre-commercial thinning, shelterwood	4.5	0.22
7.	Convert moist sites to grass and pre-commercial thinning, clearcut remaining timber...	3.5	0.40
8.	Ibid, shelterwood	3.8	0.32

Source: Worley and Miller, 1969, as presented by Satterlund, 1972

Realistic evaluations of management alternatives for increasing water yield on wildland watersheds should involve a multiple-product analysis far more complex than the preceeding examples. Brown (1976) illustrated a procedure for analyzing multiple-product alternatives (Figure 3). As discussed by O'Connell (1971), to properly model the economics of management of multiple resources: (1) value and cost data should be determined by sound methodology, (2) local, regional and national impacts should be evaluated on the basis of income, employment, social and political criteria, (3) negative and positive effects on market and non-market outputs should be included, (4) marginal and incremental analysis should be used throughout, (5) a series of benefit-cost ratios should be determined for each alternative, and (6) "an economic optimum mix is obtained when the marginal rate of substitution is equal to the inverse ratio of the prices." A detailed economic analysis of the 8.4 million acre Salt-Verde watershed in central Arizona was conducted with the above factors in mind (Brown, O'Connell and Hibbert, 1974; O'Connell, 1974).

<u>Figure 3.</u> <u>Alternative Analysis Procedure</u>

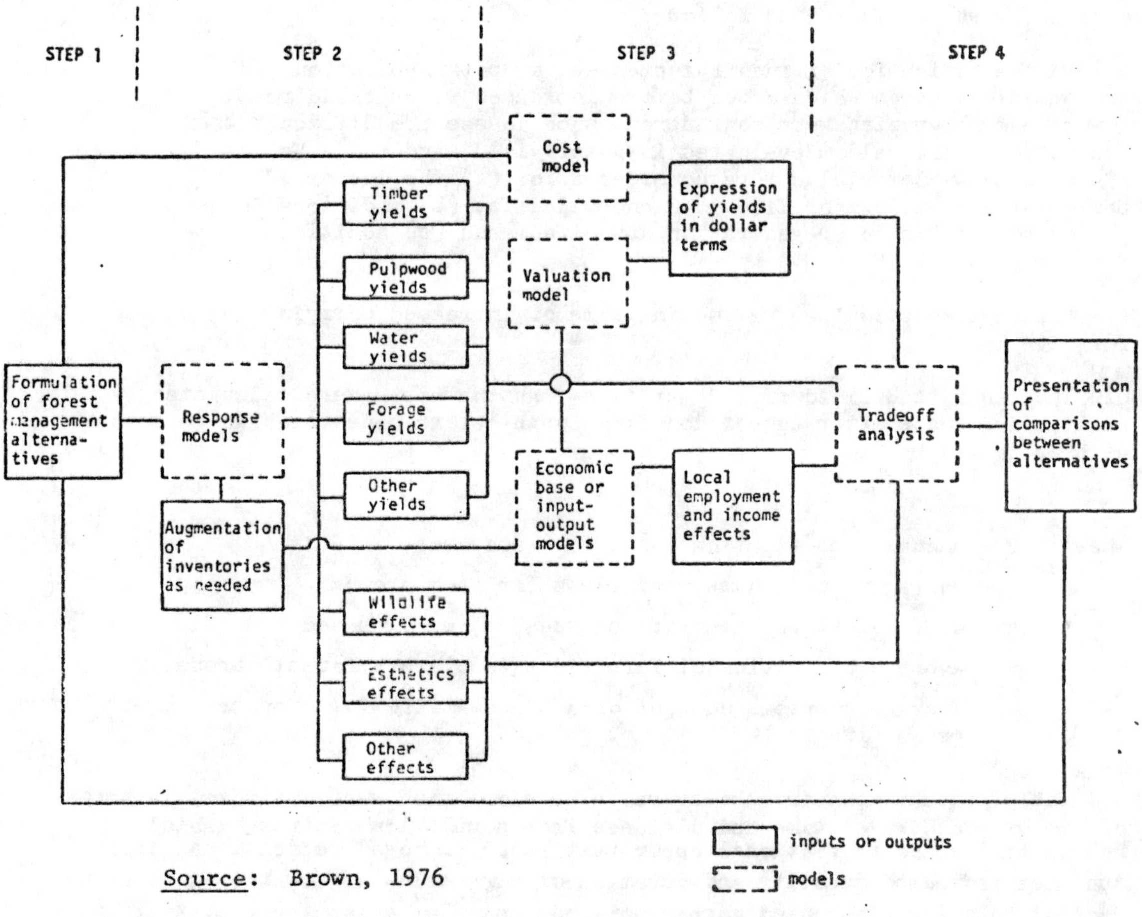

Source: Brown, 1976

The purpose of the Salt-Verde study was to determine the economic feasibility of converting chaparral vegetation to grass in order to increase water yield and forage production for livestock, and to reduce the costs of fire-fighting (Brown, O'Connell and Hibbert, 1974). Costs of several methods of conversion were determined. Primary costs and benefits estimated over a 50-year period were expressed as present value (PV):

$$PV = \sum_{t=1}^{50} Y_t/(1+i)^t$$

where: Y_t = cost or benefit in year t

i = discount rate (6 7/8 percent as recommended by U.S. Water Resources Council, 1973)

t = time in years

Present management was then contrasted to the various alternative impacts on water runoff, forage production, fires, recreation, soil losses, wildlife and esthetics. Each of these products was evaluated as follows:

a. In the estimation of runoff increases, evapotranspiration and transmission losses which occur before increased water could reach downstream reservoirs were considered (such losses are difficult to quantify and are seldom evaluated in water yield studies). Values of increased water yields were considered for (a) the increased period of streamflow for livestock and wildlife, (b) additional water for hydroelectric power generation downstream and (c) additional water for irrigation.

b. Forage production was valued in terms of increased carrying capacities for livestock.

c. The benefits derived from lower fire-fighting costs were calculated by subtracting costs after conversion from costs before conversion where costs were estimated by:

$$C_i = N_{ik} P_k c_k$$

where: C_i = annual fire-fighting costs for each area;

N_i = average annual number of fires for each area;

k = U.S. Forest Service fire classes, A, B, C, D and E

P_k = average proportion of fires in size class k for all areas

c_k = average suppression cost of a class k fire for chaparral or grass fires.

d. The effects of chaparral conversion on recreation were estimated in terms of impacts on "travel time and distance from population centres, vehicle accessibility and recreational opportunities". Improved recreational opportunities increased quantity and duration of streamflow at downstream riparian recreation sites, increased access to ponderosa pine areas for foot traffic, and improved hiking and rock hunting conditions were estimated for conversion practices. The values of such benefits, however, were not presented.

e. Wildlife habitat was considered to be improved if conversion was limited to extremely dense stands, if no more than 50 percent of a stand was converted and if converted portions were in small openings spread throughout the stand. Again dollar values were not attached to such benefits.

f. The effects of different conversion methods and patterns were evaluated on esthetics but were not quantified.

Brown, O'Connell and Hibbert (1974) looked at conversion alternatives for 139 areas from both the benefit-cost ration (B/C) approach and the benefit-cost difference (B-C) which gives a measure of net benefits. Because of the recognized uncertainties in estimated changes in water yield, forage production, and incidence of wildfires, benefit cost analyses were performed using "...the lowest reasonable estimates of the above benefits." The results of their analysis are summarized in Table 10. Problems with such benefit-cost or net benefit analysis are encountered when dollar values cannot be attached to some of the benefits or costs.

Table 10

Average annual impacts (before conversion minus after conversion) for chaparral areas with B/C greater than one.

Impacts	Alternative I (best estimate)	Alternative I (low estimate)	Alternative II
Value Determined impacts/acre			
Water (acre-ft. off-site)	0.21	0.23	0.26
Forage (AUM)	0.22	0.06	0.24
Fire ($)	0.34	0.09	0.40
Economic effects (annuity)1/			
Gross Return ($)	4.49	3.34	5.18
Cost ($)	1.98	1.95	3.31
Net Return ($)	2.51	1.39	1.87
Non-Value Determined impacts (long term average impacts)			
Recreation	some areas +, other no effect		
Soil	0 or + with proper mgt., negative with improper mgt.		
Wildlife habitat	+ with proper mgt., negative with improper mgt.		
Esthetics	+ or negative		

1/ Annuity = PV $i\ (1-i)^t/(1+i)^t-1$

Source: Brown, et. al., 1974

5. CONCLUSIONS AND RECOMMENDATIONS

The present paper presents an overview of some special problems associated with economic analyses of watershed projects. No attempt is made to provide systematic, detailed guidelines for project analysis, since these are covered in FAO's Economic Analysis of Forestry Projects (EAFP). The paper presents some examples and case studies of economic analyses of watershed projects and provides insights into how the analyst can consider watershed elements when they are imposed as constraints on projects that have other goals (e.g., wood production). The paper also surveys experience from the United States related to watershed project cost and benefit identification, valuation and comparisons.

A question remains: What lessons and conclusions can be drawn in terms of how the economist can work more effectively with hydrologists, foresters, agromists and other technical specialists in attempting to provide improved analyses of watershed projects? Based on the discussion in this paper and a review of a number of watershed project appraisals it would appear that following points are relevant in answering this question:

(1) In general, it would appear that the weakest link - or the major problem - in carrying out an appraisal of a watershed project relates to the identification and quantification of the physical input-output relationships and the costs and benefits involved. Once costs and benefits have been appropriately identified and quantified in physical terms, there do not appear to be any special problems involved in valuing them and comparing them in terms of the measures of project worth commonly used. With regard to this point, it would appear - as is indicated in a review of U.S. experience - that there are a lot more data available on input-output relationships than is generally thought and used in projects. The problem is that very little has been done to bring this information together in a practiced form that can be used by the general project planner. Thus there is a need to spend a lot more time and effort in developing comparative studies and translating highly technical information into practical guidelines that can be used by project planners.

We fully recognize that the technical specialist and researcher may argue that each case is a different one and that it is impossible to transfer the experience from one situation to another situation. While we agree that there is seldom a situation where experience from one project fits perfectly the conditions for another project, we also suggest that most analysts are dealing with averages and orders of magnitude in their attempts to analyze new projects, particularly in developing countries.

(2) Economists and the other technical specialists have to interact at all stages in the project planning process, for the economist cannot carry out an economic analysis unless he has the basic physical input-output information on which to base his analysis. The economist has to make known at an early stage his information needs. If he does not, then he can rightly be criticized. However if he has made his needs known and the appropriate information is not made available, then the primary responsibility for generating the needed information lies squarely on the shoulders of the hydrologist and other technical specialists. This is not within the economist's area of competence. His main responsibility starts when the appropriate information has been generated. We stress the word "appropriate" since in a number of cases it has been observed that a great deal of information has been accumulated for a project, but it is not the right information for the purposes of quantifying and valuing costs and benefits. Thus, for example, it is not enough to have information on average per ha. soil losses under various conditions. The agronomist and soil experts must make a specific link between soil loss and crop loss, for benefits in this case have to be specified in terms of consumption losses avoided. We do not "consume" soil,

we consume the products grown on it. In order to value such product losses avoided through implementation of a watershed project, we will need to link soil loss to crop production changes. The same agrument holds for other types of relationships. Previous examples in this paper illustrated this point for several types of watershed projects.

With the above in mind, we strongly recommend that if an economic analysis is to be carried out for a watershed project then the economist should be included in the planning process at an early stage so he can make his information needs known. It may well be that the information he needs cannot readily be generated with available time and funds. In such cases, it will not be possible to carry out an economic analysis that considers both costs and benefits. Rather, the economist will have to stick to a cost-effectiveness analysis or some other types of partial analysis. Or, at the extreme, he will have to state that an economic analysis is not possible, given the present state of knowledge and data availability. However, at this point we should stress again that, in many cases, more information is available than is generally thought and used. It would be well worth-while to spend some time and effort on bringing together such information in a readily understood and accessible fashion.

In the present paper we have attempted to review some of the technical and economic elements involved in watershed project planning. There are a number of detailed references which provide insights into specific aspects of watershed project planning, mainly from the technical point of view. These are cited in the paper and included in the list of references. A more thorough integration of the available technical information with the economist's approach to project analysis is needed.

REFERENCES AND LITERATURE CITED

Bailey, R.W. and O.L. Copeland, Jr. Vegetation and engineering structures in flood
1961 and erosion control. Paper presented before 15th Congress of Int. Union of Forest Res. Org., Vienna, Austria.

Bailey, R.W. and A.R. Croft, Contour trenches control floods and erosion on range lands.
1937 Emergency Cons. Work Forestry Pub. 4 22 pp.

Baker, M.B., Modeling management of ponderosa pine forest resources. In Symp. on Water-
1975 shed Management, Logan, Utah. Proc. p. 478-493 ASCE.

Banker, M.B., Jr., and H.E. Brown, Multiple use evaluations on ponderosa pine forest
1974 land. 18th Annual Arizona Watershed Symp. Proc. Arizona Water Commission pp. 18-25.

Barr, George W. (ed.), Recovering rainfall. Report of the Arizona Watershed Programme.
1956 University of Arizona, Department of Agricultural Economics, Tucson. Part I, 33 p. Part II, 218 p.

Bates, C.G. and A.J. Henry, Forest and streamflow experiments at Wagon Wheel Gap,
1928 Colorado. U.S. Weather Bureau Monthly Weather Rev. Suppl. No. 30. 79 pp.

Beardsley, W., Economic values of recreation benefits determined by three methods. USDA
1970 Forest Service Research Note RM-176 Rocky Mtn. Forest and Range Exp. Sta. pp. 4.

Boughton, W.C., Effects of land management on quantity and quality of available water -
1970 a review. Australian Water Resources Council, Research Report 120. 330 p. SWRA W71-0254

Brown, G.W., Forestry and Water Quality. Oregon State University Book Stores Inc.,
1974 Corvallis, Oregon.

Brown, G.W. and J.T. Krygier, Effects of clear-cutting on stream temperature. Water
1970 Resources Research, Vol. 6 (4): 1133-1139.

Brown, H.E., Evaluating watershed management alternatives. American Society of Civil
1971 Engineers, Irrig. and Drainage Division 97 (IR1): 93-108.

Brown, H.E. et al. Opportunities for increasing water yields and other multiple use
1974 values on ponderosa pine forest lands. USDA Forest Ser. Rocky Mtn. Sta. RM-129. 36 p.

Brown, T.D., Alternatives analysis for multiple use management: A case study. U.S.
1976 Forest Service Rocky Mtn. Forest and Range Exp. Sta. RM-176, Ft. Collins, Colo. 16 p.

Brown, T.D., P.F. O'Connell and A.R. Hibbert, Chaparral conversion potential in Arizona
1974 II: An economic analysis. U.S. Forest Service Rocky Mtn. Forest and Range Exp. Sta. RM-127, Ft. Collins, Colo. 28 p.

Clary, W.P. Pinyon-juniper control - does it pay? Arizona Watershed Symp. Proc. 18: 26-29
1974

Clary, W.P., 1975 Multiple use effects of manipulating pinyon-juniper. In: Symposium on Watershed Management, Logan, Itah, Proceedings. p. 469-477.

Clawson, M., 1974 Economic trade-offs in multiple-use management of forest lands. American Journ. Agricult. Economics. 56 (5): 919-926.

Corbett, E.S. and R.M. Rice, 1966 Soil slippage increased by brush conversion. USDA Forest Service Research Note PWS - 128, 8 pp.

Cotner, M.L. and D.A. Jameson, 1959 Costs of juniper control: bulldozing vs. burning individual trees. Rocky Mtn. Forest and Range Experiment Station, Station Paper 43. 14 p.

Daves, Thomas E., 1974 Economics of small watershed planning. Tech. Bulletin 295. Agricultural Experiment Station, University of Minnesota.

Dixon, J.W., 1964 Water resources Part I. Planning and Development. In: Handbook of Applied Hydrology. Ed. U.T. Chow. McGraw Hill Book Co. pp. 26-1-26-29.

Dortignac, E.J., 1965 Forest water yield management opportunities. Proc. International Symp. on Forest Hydrology. p. 579-591.

Dykstra, D.P. and H.A. Froehlich, 1976 Stream protection: what does it cost? In: Loggers Handbook. Vol. XXXVI. Shultz/ Wack/Weir, Portland, Oregon. pp. 25-28.

Ffolliott, P.F., 1974 Potentials for water yield improvement by vegetation management in Arizona. In: 18th Annual Arizona Watershed Symp. Proc. Arizona Water Commission, Phoenix, Ariz. pp. 47-51.

Ffolliott, P.F. and D.B. Thorud, 1974 Vegetation management for increased water yield in Arizona. Arizona Agricultural Exp. Sta., Tech. Bulletin 215, 38 p.

Ffolliott, P.F. and D.B. Thorud, 1975 Water yield improvement by vegetation management - Focus on Arizona. School of Renewable Natural Resources, University of Arizona NTLS PB-246-055, 5285 Port Royal Road, Springfield, Virginia 22161.

Fogel, M.M., 1971 Evaluating the effects of water yield management. International Seminar for Hydrology Professors. 3rd Purdue Univ.

Gary, H.L., 1975 Watershed management problems and opportunities for the Colorado front range ponderosa pine zone: the status of our knowledge. USDA Forest Research Paper RM-139. 32 pp.

Gittinger, J.P., 1972 Economic Analysis of Agricultural Projects. The Johns Hopkins University Press, 221 pp.

Green, Alan W., 1971 Some economic considerations of watershed stabilization on National Forests, USDA Forest Serv. Res. Pap. INT-92, 10 p.

Hansen, Edward A., 1975 Does canoeing increase streambank erosion? U.S. Forest Service Res. Note NC-186, North Central For. Exp. Sta.

Harr, R.D., 1976 Forest practices and stream flow in Western Oregon. USDA Forest Service General Technical Report PNW-49, Pacific Northwest Forest and Range Experiment Station, Portland, Oregon, 18 pp.

Harr, R.D., W.C. Harper and J.T. Krygier, 1975 Changes in storm hydrographs after road building and clear-cutting in the Oregon coast range. Vol. 11 (3): 436-444.

Haupt, H.F., 1959 Road and slope characteristics affecting sediment movement from logging roads. Journ. Forestry 57: 329-332.

Heede, B.H. and J.G. Mufich, 1974 Field and computer procedures for gully control by check dams. Journ. of Environmental Mgt. 2 (1): 1-49.

Hewlett, J.D. and A.R. Hibbert, 1961 Increases in water yield after several types of forest cutting. Int. Assoc. Sci. Hydrol. 6: 5-17.

Hibbert, A.R., 1965 Forest treatment effects on water yields. Proc. International Symposium in Forest Hydrology. p. 527-544.

Hibbert, A.R., E.A. Davis and D.G. Scholl, 1974 Chaparral conversion potential in Arizona. I: Water yield responses and effects on other resources. USDA For. Serv. Research Paper RM-126, Ft. Collins, Colorado. 36 p.

Hollis, G.E., 1975 The effect of urbanization on floods of different recurrence interval. Water Resources Research 11 (3): 431-434.

Hoover, M.D., 1944 Effects of removal of forest vegetation upon water yields. Trans. Amer. Geophys. Union. Vol. 6: 969-975.

Hoover, M.D., 1969 Vegetation management for water yield. Symp. on Water Balance in No. America, Banff, Alberta, Water Balance Modification - Increases the Yield, Proceeding Series 7. p. 191-195.

Jensen, N.E., 1972 Pinyon-juniper woodland management for multiple use benefits. Journ. Range Mgt. 25: 231-235.

Jordening, D.L. and J.K. Allwood, 1973 Research needs and priorities: Water pollution control benefits and costs. U.S. Environmental Protection Agency, EPA-600/5-73-008b, 198 pp.

Kelso, M.M. and L. Mack, 1964 The value of additional surface water to agriculture in the Salt River project. 8th Ann. Arizona Watershed Symposium Proc. Tempe, Arizona. pp. 40-47.

Klock, G.D., 1976 Estimating two indirect logging costs caused by accelerated erosion. USDA Forest Service Central Technical Report PNW-44, Pacific Northwest Forest and Range Experiment Sta., Portland, Ore.

Lantz, R.L., 1971 Guidelines for stream protection in logging operations. A report of the Research Division, Oregon State Game Commission, 29 pp.

Leaf, C.F., 1975a Watershed management in the Rocky Mountains Subalpine Zone. USDA Forest Service Research Paper RM-157, Rocky Mtn. Forest and Range Exp. Sta. 31 pp.

Leaf, C.F., 1975b Watershed management in the central and southern Rocky Mountains: A summary of the status of our knowledge by vegetation types. USDA Forest Service Research Paper RM-142. 28 pp.

Leopold, Luna B., Hydrologic research on instrumented watersheds. Paper presented at
1970 International Association of Scientific Hydrology - UNESCO Symposium on the Results of Research on Representative and Experimental Basins, Wellington, New Zealand, Dec. 10.

Lloyd, Duane R., Economics of multiple-use. Conference on multiple-use of southern
1969 forests. Pine Mountain, Georgia. Proceedings p. 45-54.

Lull, H.W. and K.G. Reinhart, Increasing water yield in the Northwest by management of
1967 forested watersheds. USDA Forest Service Research Paper NE-66. 45 pp.

Lull, H.W. and K.G. Reinhart, Forests and Floods in the Eastern United States. USDA
1972 Forest Service Research Paper NE-226. 94 pp.

Lull, H.W. and W.E. Sopper, Hydrologic effects from urbanization of forested watersheds
1969 in the Northeast. USDA Forest Service Research Paper NE-146. 31 pp.

Meeham, W.R., W.A. Farr, D.M. Bishop and J.H. Patric, Some effects of clearcutting on
1969 salmon habitat of two southeast Alaska streams. Institute of Northern Forestry, USDA Forest Service Research Paper PNW-82, Pacific Northwest Forest and Range Exp. Sta., Juneau, Alaska, 45 pp.

Miller, W. and Byers, D., Economic efficiency vs. environmental quality in small water-
1972 shed development. Amer. Water Resource Assoc. National Symp. on Watersheds in Transition CSU Proc. p. 63-67.

Muhammad, I., Watershed management and soil erosion problems in West Pakistan. Pakistan
1968 Journal of Science 20 (5/6): 234-256.

North, R.M. et. al., Survey of economic-ecological impacts of small watershed development.
1974 Georgia Intitute of Technology, Atlanta, Environmental Resources Center. 138 p. NTIS As PB-235 407-4WN.

O'Connell, P.F., Economic modeling in natural resource planning. Arizona Watershed
1971 Symp. Proc. 14: 31-38.

O'Connell, P.F., Detailed studies in the Salt-Verde Watershed. In: 18th Annual Arizona
1974 Watershed Symposium, Proceedings Arizona Water Commission, Phoenix, Arizona, pp. 52-57.

O'Connell, P.F. and H.E. Brown, Use of production functions to evaluate multiple use treat-
1972 ments on forested watersheds. Water Res.,Res. 8 (5): 1188-1197.

O'Hayre, A.P., A hydronomic analysis of resource management alternatives: A case study
1975 of Itasca County. Ph. D. Dissertation, University of Minnesota.

O'Hayre, A.P., J.C. Clausen and A.C. Mace, Jr., A hydronomic analysis of forest manage-
1975 ment alternatives for environmental quality: A case study of Itasca County WRRC Bul. 83, Water Resources Research Center, University of Minnesota, Minneapolis, Minn. 124 pp.

OECD (Organization for Economic Cooperation and Development), April, 1979. Guide (Parts
1979 I and II) Cooperative Action Programme: Joint Activity on Multipurpose Hydraulic Projects. Paris.

Orr, H.K., Watershed management in the Black Hills: The status of our knowledge. USDA Forest Serv. Research Paper RM-141. 12 pp. Ft. Collins, Colorado.

Packer, P.E., Criteria for designing and locating logging roads to control sediment. Forest Science 13:2-18.

Packer, P.E. and C.F. Christensen, Guides for controlling sediment from secondary logging
1964 roads. Intermountain Forest and Range Exp. Sta. and Northern Region, USDA Forest Service, 42 pp.

Packer, P.E. and W.A. Laycock, Watershed management concepts in the United States:
1969 Concepts and principles. Lincoln Papers in Water Resources No. 8: 1-22, Lincoln College, Canterbury, New Zealand

Pearson, H.A., Range and Wildlife Opportunities. Proc. Symp. on Management of Young
1974 Pines, p. 19-27.

Peskin, H.M. and E.P. Seskin, Cost Benefit Analysis and Water Pollution Policy. Urban
1975 Institute, Washington, D.C. 370 pp.

Reinhart, K.G., A.R. Eschner and G.R. Trimble, Jr., Effect on steamflow of four forest
1963 practices in the mountains of West Virginia. USDA Forest Service Research Paper NE-1. 79 pp.

Rice, R.M., R.P. Crouse, and E.S. Corbett, Emergency measures to control erosion after
1963 a fire on the San Dimas Experimental Forest. Proc. Federal Interagency Sedimentation Conference, USDA Misc. Publ. 970: 123-129.

Rich, L.R. and J.R. Thompson, Watershed management in Arizona's mixed conifer forests:
1974 The status of our knowledge. USDA For. Serv. Research Paper RM-130, Ft. Collins, 15 p.

Rosenberry, P.E., A.B. Daugherty, and G.A. Parelis, Technological change and the economics
1968 of conservation. Journ. Soil and Water Conservation 23 (4): 123-126.

Sassman, R.W., J.W. Barrett, and J.G. Smith, Economics of thinning stagnated pine sapling
1972 stands in the pine-grass areas of central Washington. U.S. Forest Service Pacific NW For. and Range Exp. Sta. Res. Paper PNW-144. 17 pp.

Satterlund, D., Wildland Watershed Management. The Ronald Press Company, New York. 370 pp.
1972

Schramm, G. et al., Decision making under uncertainty: Economic evaluation of stream-
1974 flow forecasts. Univ. of Michigan, Ann Arbor, School of Natural Resources. 443 p. NTIS PB-236 417-2WM.

Springfield, H.W., Characteristics and management of southwestern pinyon-juniper ranges:
1976 The status of our knowledge. USDA For. Serv. Res. Pap. RM-160, 32 p. Rocky Mtn. For. and Range Exp. Sta., Ft. Collins, Colo.

Sturges, D.L., Hydrologic relations on undisturbed and converted big sagebrush lands:
1975 The status of our knowledge. USDA Forest Service Research Paper RM-140.

Swanson, E.R. and D.E. McCallum, Income effects of rainfall erosion control. Journ.
1969 Soil and Water Conservation.

Thronson, R.E., Comparative costs of erosion and sediment control, construction activities.
1973 U.S. Environmental Protection Agency, Office of Water Program Operations, Water Quality and Non-Point Source Division, 205 p.

Turner, J.M. and F.R. Larson, Cost analysis of experiments on ponderosa pine watersheds.
1974 USDA Forest Service Res. Paper RM-116, 12 pp.

Trimble, G.P., Jr., Logging roads in northeastern municipal watersheds. Journ. Amer.
1959 Water Works Assoc. 51: 407-410.

Trimble, G.R. and S. Weitzman, Soil erosion on logging roads. Proc. Soil Science
1953 Society of America 17: 152-154.

UNESCO, Influence of man on the hydrologic cycle: guidelines to policies for the safe
1972 development of land and water resources. In: Status and Trends of Research in Hydrology. 1965-74. pp. 31-70.

Unger, S.G., M.J. Emerson, and D.L. Jordening, State-of-act review: Water pollution
1973 control benefits and costs. U.S. Environmental Protection Agency, EPA-600/5-73-008a, 121 pp.

U.S. Bureau of Reclamation, Region 5., Report on clearing and controlling phreatophytes,
1963 Cost Data Division of Irrigation, Amarillo, Texas. 14 p.

U.S. Department of Agriculture, Soil Conservation Service, Economics Guide for Watershed
1964 Protection and Flood Protection. Washington, D.C.

U.S. Department of Interior, Industrial Waste Guide, Logging Practices. Federal Water
1970 Pollution Control Administration, Northwest Region, Portland, Oregon. 40 pp.

Welch, H.W. and G.D. Lewis, Assessing environmental impacts of multiple use land manage-
1976 ment. Journ. of Environmental Management 4: 197-209.

Workman, J.P. and J.E. Keith, Economics of soil treatments in the Upper Colorado. In:
1975 Symposium on Watershed Management, Logan, Utah, Proceedings. p. 591-596. ASCE.

Workman, J.P. and C.R. Kienast, Pinyon-juniper manipulation - some socio-economic
1975 considerations. In: The Pinyon-Juniper Ecosystem: A Symposium. p. 163-177. USU College of Natural Resources, Logan Agr. Exp. Sta. 194 p.

Worley, David P., Economic evaluation of watershed management alternatives - The Beaver
1966 Creek Watershed - Arizona. New Mexico Water Conference 11th, University Park, Proceedings.

Worley, D.P. and Miller, A procedure for upstream watershed economic evaluation. 8th
1964 Ann. Arizona Watershed Symposium, Tempe, Arizona, pp. 36-39.

Worrell, A.C., Economics of American Forestry. John Wiley and Sons. New York.
1959 441 pp., illus.

APPENDIX 1

WATERSHED MANAGEMENT OBJECTIVES RELEVANT FOR DIFFERENT CLIMATIC REGIONS

The following table presents a general view of the variety of watershed management objectives which may be relevant for different climatic regions. Obviously, there are exceptions to such a generalized ranking. But it does indicate some general considerations. Specific site characteristics and other factors such as proximity to population centers and level of economic development may in some situations reorder the watershed management priorities listed. For example in a "mid-latitude mixed forest with abundant rainfall and water supplies, the management of municipal watersheds for maintaining or perhaps increasing the quantity of water yield to satisfy the demands of an increasing population may be a major objective. Conversely, the goal of increasing the quantity of water in many desert ecosystems may be unrealistic from a watershed management viewpoint because of the lack of opportunities to do so.

SUGGESTED IMPORTANCE OF WATERSHED MANAGEMENT OBJECTIVES FOR DIFFERENT CLIMATE - VEGETATION REGIMES OF THE WORLD.

Climate 1/	Typical Annual Precipitation (cm)	Comments	WATERSHED MANAGEMENT OBJECTIVES 2/ Increase Quantity	Reduce Flooding	Drainage of Wetlands	Minimize Erosion & Sedimentation	Improve Water Quality	Minimize Wind Erosion
Boreal Forest	25 to 75	6 months below freezing	3	2	1	3	2	4
Mid-Latitude Mixed Forest	50 to 200	hot summers cold winters	2	2	3	1	2	4
Mid-Latitude Coastal Evergreen Forest	40 to 500	mild, humid high winter precip.	3	2	4	1	2	4
Mediterranean Scrub Woodland	40 to 90	warm, dry summers wet winters	1	2	4	2	3	3
Mid-Latitude Steppe	25 to 85	wet, hot summers cold winters	3	2	4	1	2	2
Desert	0 to 25	evapotranspiration exceeds precip.	2 3/	2	4	2	3	2
Tropical Rain and Semi-deciduous Forests	165 to 760	rainfall evenly distributed, warm, wet	4	2	3	1	3	4
Tropical Savanna and Thorn Scrub Woodland	75 to 130	pronounced wet and dry periods	2	2	4	1	2	3
Mountain								
- Precipitation limiting in adjacent areas	variable	generally an increase in precip. with elevation	1	2	4	2	2	4
- Precipitation abundant	variable		3	2	4	1	2	4

1/ As classified by Rumney (1968); Polar ice cap and tundra climates were not considered in this analysis.

2/ Objectives: 1=Primary; 2=Secondary; 3=Usually not of major importance; 4=Little or no importance. It should be noted that such objectives are highly dependent upon individual watershed characteristics including topography, slopes and soils and other factors such as land use, population and level of economic development.

3/ Although the goal of increasing water supplies in desert areas is of top priority, watershed management opportunities for increasing water yield and extremely limited and, therefore, not considered a primary objective of management.

DISCOUNTING PROCEDURES FOR PROJECT APPRAISALS:

Common Problems and Appropriate Solutions

by

Thomas W. Houghtaling and Hans M. Gregersen [1]

---

[1] T.W. Houghtaling is a forest economist with a public utilities company in Minnesota, and at the time this paper was prepared was Research Associate at the University of Minnesota. H.M. Gregersen is Professor of Forestry and Agricultural and Applied Economics at that University.

Definitions

ROR - Internal Rate of Return (general)

ERR - Economic Internal Rate of Return

FRR - Financial Internal Rate of Return

NPW - Net Present Worth

PV - Present Value

1. INTRODUCTION

Forestry projects nearly always involve a number of years between the time when the first project investment is made and when final outputs (benefits) from the project occur. Usual practice in appraising such projects is to calculate some measure(s) of project worth that take time into account. The most common measures calculated include the net present worth (NPW) and the internal rate of return (ROR). Both these measures are discussed in Chapter 9 of Economic Analysis of Forestry Projects 1/, in terms of both financial and economic analyses.

The purpose of the present paper is to explore in more technical detail solutions to some common problems encountered in using discounting and compounding formulas and procedures.

Specifically, the following three questions are dealt with in some detail:

(i) How should years or time intervals be defined and designated in project analyses? What mistakes commonly occur when years (time intervals) are improperly defined, and how can the common errors be avoided?

(ii) How should inflation be treated and how can we best treat situations where both growth rates and discount rates have to be treated simultaneously (e.g., real prices for timber may be increasing at 3 percent per year and we are using a 10 percent discount rate to calculate present values for a particular investment analysis)?

(iii) Finally, we touch on the question of the importance of uncertainty concerning future values (costs, prices, physical magnitudes) depending on a) the period to which the uncertainty refers and b) the discount rate used for present value calculations, or the estimated internal rate of return of the project.

2. PROPER DESIGNATION AND TREATMENT OF TIME INTERVALS

Conventional investment analysis is carried out on an annual basis, with rates of return expressed in percent per annum on a compound basis. Costs and benefits which occur at any time during a given year are presumed to occur at a single point in time, conventionally, at the beginning of the year. A year can be defined to begin on any date, i.e., January 1, April 1, June 15, etc. Often in forestry a "year" begins on the day of planting or some other initial investment activity. It is only by convention that a period of one year generally is chosen as the time interval for investment analysis. It would be equally valid (and perhaps slightly more accurate) to choose a shorter period, such as one quarter or one month. However, it is unlikely that the added complexity would be justified, except perhaps for short-term projects at extremely high discount rates. Cash flow tables normally also are constructed on an annual basis, as is illustrated by Table 1, for a one hectare fuelwood plantation in Korea. 2/

1/ FAO Forestry Paper No.17, FAO, Rome, 1979, hereafter referred to as EAFP.

2/ See Case Study No. 2, in Economic Analysis of Forestry Projects: Case Studies. FAO Forestry Paper No. 17, Supp. 1, FAO, Rome, 1979.

Table 1 - Korea Fuelwood Cash Flow Analysis

		YEAR																				
		0	1	2	3	4	5	6	7	8	9	10	11	12	13	14	15	16	17	18	19	20
[1]	BENEFITS (fuelwood)	-	-	-	6.5	13.0	26.0	52.0	65.0	65.0	65.0	65.0	65.0	65.0	65.0	65.0	65.0	65.0	65.0	65.0	65.0	1625.0
[2]	COSTS																					
[3]	Establishment	130.0	-	-	-	-	-	-	-	-	-	-	-	-	-	-	-	-	-	-	-	-
[4]	Supervision	2.4	2.4	2.4	2.4	2.4	2.4	2.4	2.4	2.4	2.4	2.4	2.4	2.4	2.4	2.4	2.4	2.4	2.4	2.4	2.4	-
[5]	Misc. (tools, etc.)	7.5	-	-	-	-	-	-	-	-	-	-	-	-	-	-	-	-	-	-	-	-
[6]	Harvesting	-	-	-	1.9	3.7	7.5	15.0	18.7	18.7	18.7	18.7	18.7	18.7	18.7	18.7	18.7	18.7	18.7	18.7	18.7	937.5
[7]	Total Cost	139.9	2.4	2.4	4.3	6.1	9.9	17.4	21.1	21.1	21.1	21.1	21.1	21.1	21.1	21.1	21.1	21.1	21.1	21.1	21.1	937.5
[8]	NET BENEFIT(cost)	(139.9)	(2.4)	(2.4)	2.2	6.9	16.1	34.6	43.9	43.9	43.9	43.9	43.9	43.9	43.9	43.9	43.9	43.9	43.9	43.9	43.9	687.5

2.1 Designating Years in Project Analysis

There is no single convention regarding the designation of year numbers in investment analysis. Consequently, some confusion arises in practice, which often leads to errors in analysis. In the case of forest plantation projects, the age of the stand of trees (the number of years since planting) has often been used as the basis for numbering years when planting occurs in the initial investment year. Thus, the initial investment year would be labeled "zero". In such cases the cash flow table year designation and the age of the plantation would coincide. Often site preparation or other costs attributable to a new plantation occur one or more years before planting. Thus, using year designations corresponding to the age of the plantation would involve the use of negative years (for investment costs occuring one or more years before planting). While such designation would not affect the results of the analysis, it could lead to unnecessary confusion among persons reviewing the work. The convention recommended here is to designate years according to the age of the initial investment outlay for the project, with year "zero" as the "first" year of the investment project. In general terms, "year n" is the (n+1)th year" in any analysis.

In many cases the project analyst does not "write up" the appraisal report. Instead, another person "back in the office" is assigned the task. Therefore, confusion over the designation of years is often not exclusively confined to those who read the appraisal report, but rather it surfaces within the appraisal report due to inconsistency in the designation of years by two or more individuals who work on the appraisal.

In the Korea fuelwood project (Table 1) the beginning of year zero corresponds to the date of planting, since no investments took place prior to the year of planting. Therefore, the age of trees removed at final harvest is 20.

The calculation of the net present worth (NPW) of a project is made easier by designating the first year of the investment (project) as year zero. Since NPW is normally expressed in terms of the first year (the "present" in terms of the project decision), the year number automatically corresponds to the number of years for which the net benefits (costs) need to be discounted. Each net benefit (cost) in Table 1 can be multiplied by the discount factor, $\frac{1}{(1+i)^n}$, where (i) equals the annual interest (discount) rate and (n) equals the year number shown in the column heading) to obtain the present value for each net benefit (cost) entry. The NPW of the Korea project (the sum of each year's net present value) at a 10 percent discount rate equals W170.0 thousand. 1/

2.2 Treating Equal Annual or Periodic Costs and Benefits

When a net benefit (cost) occurs annually or on a regular periodic basis for a sequence of years, the use of formulas to calculate the present value of such benefits or costs greatly simplifies the determination of the NPW, since the number of calculations is reduced. Table 2 summarizes the formulas needed to calculate the present and future values of annual and periodic "payments". 2/ The formulas in Table 2 are labeled based on the

---

1/ See EAFP, Chapter 9, for further discussion of NPW calculations.

2/ By convention these formulas are titled "payment" formulas. The word "payment" describes a sequence of two or more equal costs or benefits which occur annually or periodically.

assumption that interest rates are expressed and compounded on an annual basis. An amount of time other than one year could be used. The periodic formulas assume that two or more years (or other specified equal lengths of time) separate payments. The derivations of the formulas in Table 2 are presented in Appendix 1. The application of the present value formulas is illustrated below, using the Korea example.

2.3 "End of Year" vs. "Beginning of Year" Designations

"End of year" designations imply that annual payments start one year from the present and that periodic payments start one period from the present. This is equivalent to saying that annual (periodic) payments start at the beginning of the second year (period).

"Beginning of year" designations imply that annual (periodic) payments begin immediately and at the beginning of each successive year (period).

In both cases, in terms of discounting or compounding, the beginning of one year (period) is considered equivalent to the end of the previous year (period).

The only significance of destinguishing between "end" and "beginning" of year payments relates to the first year (period) of an annual(periodic) series of payments. More specifically, it is only of concern when the analyst uses one of the formulas for present value of an annual (periodic) series of payments. However, use of these formulas is common in forestry projects (e.g., calculating soil expectation values, present value of annual taxes, protection costs, etc.) so the difference is worth noting and is explained in more detail below.[y]

Quite often one encounters investment analyses where the first year is labeled "year 1" and the values (net benefits or costs) for that year are discounted back one year to arrive at NPW. The implicit assumption in such cases is that the analyst is using an "end of year" approach, i.e., assuming that any payment in a given year occurs at the end of that year. As explained above, we prefer to use the beginning of year assumption in assigning costs and benefits. Thus, we label the first year as "0" and do not discount the net benefit or cost value occurring during that year. Either approach can be used and will provide correct results, so long as the assumption is appropriately stated. It is a simple matter to adjust from one to the other.

---

[y] A common mistake in forestry is to estimate an annual land rental equivalent (A) of a given land value (V) by applying the formula $A = V(i)$ where i equals the discount rate. In this case the assumption underlying the formula is that rental payments start one year from the present (i.e., end of first year or beginning of second year). In reality, rental payments would start at the beginning of the first year, and the appropriate formula would be $A = V(i)/(1+i)$. Other common mistakes have been identified. Appropriate treatment of annual and periodic series are discussed in the following paragraphs.

Table 2 - Annual and Periodic Payment Formulas

	Payments Begin One Year (or Period) From Present		Payments Begin Immediately	
	Finite Number of Payments (1)	Infinite Number of Payments (2)	Finite Number of Payments (3)	Infinite Number of Payments (4)
1. DISCOUNTED ANNUAL PAYMENT FACTOR	$\left[\frac{(1+i)^n - 1}{i(1+i)^n}\right]$	$\left[\frac{1}{i}\right]$	$\left[1 + \frac{(1+i)^n - 1}{i(1+i)^n}\right]$ or $\left[\frac{(1+i)^{n+1} - 1}{i(1+i)^n}\right]$	$\left[1 + \frac{1}{i}\right]$ or $\left[\frac{1+i}{i}\right]$
2. COMPOUNDED ANNUAL PAYMENT FACTOR	$\left[\frac{(1+i)^n - 1}{i}\right]$	n.a.	$\left[(1+i)^n + \frac{(1+i)^n - 1}{i}\right]$ or $\left[\frac{(1+i)^{n+1} - 1}{i}\right]$	n.a.
3. ANNUAL CAPITAL RECOVERY FACTOR	$\left[\frac{i(1+i)^n}{(1+i)^n - 1}\right]$	n.a.	n.a.	n.a.
4. ANNUAL SINKING FUND FACTOR	$\left[\frac{i}{(1+i)^n - 1}\right]$	n.a.	n.a.	n.a.
5. DISCOUNTED PERIODIC PAYMENT FACTOR	$\left[\frac{(1+i)^{nt} - 1}{(1+i)^{nt}\left[(1+i)^t - 1\right]}\right]$	$\left[\frac{1}{(1+i)^t - 1}\right]$	$\left[\frac{(1+i)^{(n+1)t} - 1}{(1+i)^{nt}\left[(1+i)^t - 1\right]}\right]$	$\left[\frac{(1+i)^t}{(1+i)^t - 1}\right]$
6. COMPOUNDED PERIODIC PAYMENT FACTOR	$\left[\frac{(1+i)^{nt} - 1}{(1+i)^t - 1}\right]$	n.a.	$\left[\frac{(1+i)^{(n+1)t} - 1}{(1+i)^t - 1}\right]$	n.a.

i = rate of interest (discount) in decimal form
n = number of years or periods until last payment
t = number of years between periodic payments

2.4 Using "End of Year" Annual Payment Formulas

The NPW for the Korea project cash flow (Table 1) as derived above was obtained by discounting each year's net benefit (cost). A much easier method is to apply the finite discounted annual payment factor for a payment beginning in one year (row 1, column 1 of Table 2) to the year 1-2 net costs and the year 7-19 net benefits. These discounted values are expressed in terms of the year prior to the start of the net benefits (costs), i.e., years 0 and 6, respectively. The discounted value of net benefits for years 7-19 must be further discounted for 6 years to arrive at present value. The other 6 individual net benefits (costs) are discounted individually as before. Calculations are shown in Table 3. Step by step explanations of the procedures used in Table 3 are presented in Appendix 5.

2.5 Using "Beginning of Year" Annual Payment Formulas

Another equally valid way to calculate the above NPW would be to use the finite discounted annual payment factor for a "beginning of year" payment (row 1, column 3 of Table 3) for the same two sequences of annual payments. This method is shown in Table 4. Only columns 3 and 4 have changed from Table 3. The discounted values shown in column 4 of Table 4 are expressed in terms of the year in which the payments begin, not in terms of one year before the payments begin, as was the case in Table 3. Each method (Table 3 or Table 4) gives the same NPW, and it does not matter which method is used. However, it is obviously necessary to distinguish between the two methods. One of the common errors in financial or economic analysis of forestry projects is the failure to make such a distinction, which often leads to the use of the wrong number of years for discounting.

The distinction between the two designations can be put in another way. When using "end of year" designations, the present value of a series of annual payments starting in year n+1 will be expressed in terms of present value in year n. When using "beginning of year" designations, the present value of a series of annual payments starting in year n+1 will be expressed in terms of present value in year n+1.

2.6 Using "Periodic" Payment Formulas

To illustrate the use of periodic payment formulas, another example can be used. The cash flow is presented as Table 5. Two sets of periodic net payments occur. Net benefits of 90.2 occur seven times from year 6 through year 18 at regular intervals of two years, and net costs of 2.4 occur nine times from year 3 through year 19, also at intervals of two years.1/ As in the case of annual payments, periodic payments can be defined as beginning in one period after the base year of discount, or as beginning immediately. The method of calculating NPW assuming payments beginning "next" period ("end of period" payments) is illustrated in Table 6 using the formula from row 5, column 1 of Table 2. 2/ The other equally correct method is to use the discounted periodic payment factor for a payment beginning immediately ("beginning of period" payments). This method of calculating NPW, using the formula from row 5, column 3 of Table 2, is shown in Table 7.

1/ To be consistent with the annual cost treatment in the Korea cash flow above, net costs for years 1-2 are treated as annual costs, and net costs for alternate years between 3-19 are treated as periodic costs.

2/ The procedures used in Table 6 are similar to the procedures used in Table 3, which are more fully described in Appendix 4.

Tables 3 and 4 - Two Methods of Calculating Net Present Worth (NPW) for Korea Fuelwood Project

Table 3

Discounted Value of Annual Payment Expressed in Terms of the Year Prior to the First Payment

(i=10%)

	Year	Net Benefit or (cost)	Discounted Value of Annual Payment: Year	Discounted Value of Annual Payment: Value	Present Worth (Year 0)
	[1]	[2]	[3]	[4]	[5]
[1]	0	(139.9)			(139.900)
[2]	1-2	(2.4)	0	(4.165)	(4.165)
[3]	3	2.2			1.653
[4]	4	6.9			4.713
[5]	5	16.1			9.997
[6]	6	34.6			19.531
[7]	7-19	43.9	6	311.837 1/	76.024
[8]	20	687.5			102.192

Net Present Worth (NPW) = 170.0

1/ This value of 311.837 can be added to the year 6 net benefit of 34.6. The result (346.437) can then be discounted 6 years to get 195.555. This method will save one calculation.

Table 4

Discounted Value of Annual Payment Expressed in Terms of the Year in Which Payment Begins

(i=10%)

	Year	Net Benefit or (cost)	Discounted Value of Annual Payment: Year	Discounted Value of Annual Payment: Value	Present Worth (Year 0)
	[1]	[2]	[3]	[4]	[5]
[1]	0	(139.9)			(139.900)
[2]	1-2	(2.4)	1	4.582	(4.165)
[3]	3	2.2			1.653
[4]	4	6.9			4.713
[5]	5	16.1			9.997
[6]	6	34.6			19.531
[7]	7-19	43.9	7	343.021	176.024
[8]	20	687.5			102.192

Net Present Worth (NPW) = 170.0

Table 5 - Mythopia Cash Flow Analysis

		YEAR																				
		0	1	2	3	4	5	6	7	8	9	10	11	12	13	14	15	16	17	18	19	20
[1]	BENEFITS (fuelwood)	-	-	-	-	32.5	-	130.0	-	130.0	-	130.0	-	130.0	-	130.0	-	130.0	-	130.0	-	1625.0
[2]	COSTS																					
[3]	Establishment	130.0	-	-	-	-	-	-	-	-	-	-	-	-	-	-	-	-	-	-	-	-
[4]	Supervision	2.4	2.4	2.4	2.4	2.4	2.4	2.4	2.4	2.4	2.4	2.4	2.4	2.4	2.4	2.4	2.4	2.4	2.4	2.4	2.4	-
[5]	Misc. (tools, etc.)	7.5	-	-	-	-	-	-	-	-	-	-	-	-	-	-	-	-	-	-	-	-
[6]	Harvesting	-	-	-	-	9.4	-	37.4	-	37.4	-	37.4	-	37.4	-	37.4	-	37.4	-	37.4	-	937.5
[7]	Total Cost	139.9	2.4	2.4	2.4	11.8	2.4	39.8	2.4	39.8	2.4	39.8	2.4	39.8	2.4	39.8	2.4	39.8	2.4	39.8	2.4	937.5
[8]	NET BENEFIT (cost)	(139.9)	(2.4)	(2.4)	(2.4)	20.7	(2.4)	90.2	(2.4)	90.2	(2.4)	90.2	(2.4)	90.2	(2.4)	90.2	(2.4)	90.2	(2.4)	90.2	(2.4)	687.5

Tables 6 and 7 - Two Methods of Calculating Net Present Worth (NPW) for Mythopia Project

Table 6

Discounted Value of Annual/Periodic Payment Expressed in Terms of Year Prior to First Pay.

(i=10%)

	Year[1/]	Net Benefit or (cost)	Discounted Value of Annual/Periodic Pay. Year	Discounted Value of Annual/Periodic Pay. Value	Present Worth (Year 0)
	[1]	[2]	[3]	[4]	[5]
[1]	0	(139.9)			(139.900)
[2]	1-2	(2.4)	0	(4.165)	(4.165)
[3]	3-19 (p)	(2.4)	1	(9.373)	(8.521)
[4]	4	20.7			14.138
[5]	6-18 (p)	90.2	4	316.417	216.117
[6]	20	687.5			102.192

Net Present Worth (NPW) = 180.761[2/]

Table 7

Discounted Value of Annual/Periodic Payment Expressed in Terms of Year in Which Pay. Begins

(i=10%)

	Year[1/]	Net Benefit or (cost)	Discounted Value of Annual/Periodic Pay. Year	Discounted Value of Annual/Periodic Pay. Value	Present Worth (Year 0)
	[1]	[2]	[3]	[4]	[5]
[1]	0	(139.9)			(139.900)
[2]	1-2	(2.4)	1	(4.582)	(4.165)
[3]	3-19 (p)	(2.4)	3	(11.341)	(8.521)
[4]	4	20.7			14.138
[5]	6-18 (p)	90.2		382.864	216.117
[6]	20	687.5			102.192

Net Present Worth (NPW) = 180.761

1/ The letter "p" following a group of years refers to a periodic payment occurring every two years.

2/ The NPV under this harvest assumption is greater than under the original assumption due to earlier harvesting.

2.7 Expressing Benefits and Costs in Annual Terms

For some projects it is useful to express a single payment as an equivalent sequence of annual payments at some specified discount rate, e.g., an investment cost which is annualized over the life of the investment. For example, in many forest industry projects output is relatively constant for a sequence of years. To determine whether the project is justified, annual benefits can be compared with annual costs (including annualized fixed costs). Also, for such industrial projects it is useful to know in any given year what the average cost of production is (including interest on and depreciation of fixed assets).

Multiple payments can also be expressed in annual terms. For example, if harvest benefits occurred in years 15, 25, and 30, they could be expressed in annual terms over the entire rotation. To do so, one must first calculate the present value of these harvest benefits, and then "annualize" that single value, applying the same discount rate used to calculate the present value. Another example of annualizing a group of costs and benefits is illustrated by the calculation of soil rent. First, the soil expectation value (the NPW of a project over an infinite period, excluding land cost) is calculated, and then soil rent is defined as the annual equivalent of the soil expectation value.

Very often the annual capital recovery factor is used to determine the amount of equal installments (beginning in one year) necessary to repay a loan granted "today". (See "Treatment of loans", below). In the case of a loan it makes no sense to begin repaying the loan immediately since it would serve only to reduce the amount of the loan granted. The convention of using this annual capital recovery factor (which assumes payments to begin in one year) applies also to cases where single payments of equity capital are annualized (see Appendix 1, page 3, footnote 1).

The conventional method of annualizing a payment is to use the annual capital recovery factor (row 3, column 1 of Table 2). By multiplying the single payment by this factor, the "equivalent" sequence of annual "end of year" payments is determined. For example, in the Korea cash flow table (Table 1) it is evident that the "establishment" and "miscellaneous" costs of W137.5 all occur in year zero. At a 10 percent discount rate this single cost is equivalent to an annual cost (years 1-20) of W16.2, using the annual capital recovery factor.

2.8 Treatment of Loans

Often a financial investment analysis will include a loan. In a financial cash flow table, the loan amounts received are treated as receipts and the loan amounts plus interest repaid are treated as disbursements. The loan is treated exactly like any other receipt or disbursement. In many cases, the exact loan schedule is not detailed in the information given the analyst. Instead, the loan disbursement schedule is given together with the loan rate of interest, then the repayment schedule is defined only by the number of years of no repayments (grace period), if any, the number of years of repayment of interest (sometimes in precalculated amounts), and the number of equal installments of repayment of capital and interest in the final years of the loan. The analyst must then make a detailed schedule from the information given.[1] When making a loan schedule, the analyst needs to define carefully the year numbers. Otherwise, it is likely that the loan schedule will contain errors. (This is a common source of error in the projects reviewed.)

---

[1] Often the "grace period" of a loan refers to the number of years during which no principal is repaid (interest may or may not be repaid during this "grace" period).

A case study of Philippine smallholder forestry provides an example of a loan disbursement and repayment schedule, including a grace period on repayment. Table 8 reproduces the cash flow schedule from this project for an individual smallholder who receives a loan of 1 500 pesos for each of the first four years of the project.1/ Line 1 in Table 8 shows the loan as a receipt by the smallholder. He has a three year grace period beyond year 3 before he starts repaying interest and principal. Then he pays off the interest on the first 7 years during the 8th and 9th years of the project and finally repays the remaining balance (principal plus current interest) in 6 equal installments (calculated using the capital recovery factor, described above) during the 10th to the 15th years.

Table 9 shows the calculations necessary to determine the schedule of repayments, considering the grace period and using a 12 percent interest rate. When there is no grace period, one can merely use the capital recovery formula shown in Table 9 to determine the appropriate equal annual repayments.

A useful method to check on the accuracy of the loan schedule is to calculate the loan's "internal rate of return" (ROR). 2/ The ROR for the loan must equal the loan rate of interest if the schedule was done correctly. Alternatively, the NPW of the loan can be calculated using the loan rate of interest as the discount rate. The NPW must then equal zero. Such a check will detect most computational errors and most errors in timing. However, if timing of the entire schedule is incorrect, (but the timing and calculating within the schedule are correct) this check will not detect the error. Therefore, it is necessary to make sure that the actual loan starting date, ending date and the numbers of disbursements and repayments are correct.

3. DEVELOPMENT OF COMPOSITE DISCOUNT RATES

Most investment analyses are conducted using real values rather than current (nominal) values, which include general price inflation. Discount rates used and internal rates of return calculated are real rates. Rates of return on alternative investments, such as bank accounts and bonds, are commonly given in current terms (they include effects of general price inflation) and therefore, are not directly comparable with real rates of return calculated for projects or alternative investments. 3/ The rate of discount (i) used in the compound and discount formulas (Table 2) can be defined as either a current rate (v) or a real rate (r). The formulas in these tables are used in the same manner regardless of the assumption made. However, it is important to ensure that if a current rate is chosen all values entered into the cash flow are current values, and if a real rate is chosen, all cash flow values are real values.

1/ Case Study No. 1 in FAO Forestry Paper No. 17, Supp. 1

2/ The loan's ROR is calculated by equating the present value of loan disbursements with the present value of loan receipts. At one discount rate (the loan's ROR) such an equality will hold.

3/ Increasing general price inflation reduces the real rate of return earned/paid on "fixed yield" investments (i.e., loans, bank accounts, bonds, etc.). The real rate of return on such investments varies inversely to the rate of inflation. See H. Gregersen, "Effect of Inflation on Evaluation of Forestry Investments." Journal of Forestry, Vol. 73, No.9, pp. 570-572, 1975.

TABLE 8 Financial Cash Flow, 10 ha. Plantation

Philippine Smallholder Plantation Project
(value in constant pesos)

	YEARS 0	1	2	3	4	5	6	7	8	9	10	11	12	13	14	15	16	17	18	19
Cash Receipts																				
1. Bank loans	1500	1500	1500	1500	-	-	-	-	-	-	-	-	-	-	-	-	-	-	-	-
2. Sales	-	-	-	-	-	-	-	5523	6174	6174	6810	6810	7434	7434	8046	6174	6174	6174	6174	6174
Total	1500	1500	1500	1500	-	-	-	5523	6174	6174	6810	6810	7434	7434	8046	6174	6174	6174	6174	6174
Cash Disbursements																				
3. Land preparation	590	590	590	590	-	-	-	-	-	-	-	-	-	-	-	-	-	-	-	-
4. Purchase seedlings	130	130	130	130	-	-	-	-	-	-	-	-	-	-	-	-	-	-	-	-
5. Lining/Dig/Plant	300	300	300	300	-	-	-	-	-	-	-	-	-	-	-	-	-	-	-	-
6. Replanting	130	130	130	130	-	-	-	-	-	-	-	-	-	-	-	-	-	-	-	-
7. Fertilizing	300	300	300	300	-	-	-	-	-	-	-	-	-	-	-	-	-	-	-	-
8. Weeding	550	550	550	550	-	-	-	275	275	275	275	275	275	275	275	275	275	275	275	275
9. Singling	-	-	-	-	-	-	-	-	100	100	100	100	100	100	100	100	100	100	100	100
10. Crops/Livestock	-	-	-	-	-	-	-	-	-	-	-	-	-	-	-	-	-	-	-	-
Total	2000	2000	2000	2000	-	-	-	275	375	375	375	375	375	375	375	375	375	375	375	375
Loan Payments																				
11. Accum. interest	-	-	-	-	-	-	-	3200	3050	-	-	-	-	-	-	-	-	-	-	-
12. Princ. & interest	-	-	-	-	-	-	-	-	-	1459	1459	1459	1459	1459	1459	-	-	-	-	-
Total	-	-	-	-	-	-	-	3200	3050	1459	1459	1459	1459	1459	1459	-	-	-	-	-
Cash Balance After Loan Payments	(500)	(500)	(500)	(500)	-	-	-	2048	2749	4340	4976	4976	5600	5600	6212	5799	5799	5799	5799	5799

Table 9

Derivation of Farmer Loan Repayments

Year	Loan Amount 1/	Interest Due At End of Year 7 (Beginning of year 8)
0	P 1,500	P 1816
1	P 1,500	P 1461
2	P 1,500	P 1143
3	P 1,500	P 860
	Total	P 5280

Interest due beginning year 7	P 5,280
Amount paid in year 7	P 3,200 2/
Remaining interest due	P 2,080
Principal outstanding (4x$1 500)	P 6,000
Total loan outstanding	P 8,080
One year's interest 12%	P 970
Total interest due 3/ in year 8	P 3,050

The annual repayments of the loan for years 9 - 14 (line 11, Table 5) are derived as follows:

$$A = P\ 6{,}000 \left[\frac{.12(1.12)^6}{(1.12)^6 - 1} \right]$$

$$= P\ 6{,}000\ (0.2432) = P\ 1459$$

The factor in parentheses is called the "capital recovery multiplier", and is given as follows:

$$\left[\frac{i(1+i)^n}{(1+i)^n - 1} \right]$$

The value of this multiplier can be found in most tables of compound and discount factors.

1/ Assuming 2 ha planted per year at a total annual cost equal to P 2000 (Table 2) 75 percent of this amount, or P 1500, is covered by the Bank loan.

2/ Arbitrarily set at this level in order to keep interest payments in years 7 and 8 approximately equal.

3/ P 2080 plus P 970.

3.1 Using Current (Nominal) Values (i.e., Including Inflation)

When a project is being implemented, what is actually paid for inputs and received for outputs is in current price terms. Therefore, general price inflation must somehow be considered in the appraisal in order to determine the actual cash flow. Thus, whether or not the project analysis is carried out on a current or real value basis, current values must be taken into account for budgeting pruposes. When the analysis is expressed in current value terms, each entry in the cash flow table can be used directly for budgeting pruposes. However, it is a simple matter to adjust a real value cash flow to a current value cash flow and vice versa.

To convert real values and real discount rates to current terms, it is first necessary to estimate a general rate of price inflation for the project years. Since individual projects are usually too small to significantly affect the level of general price inflation, the expected inflation rate can be estimated independently of the project. 1/ The current values of inputs and outputs for any given year (y) can then be determined by multiplying the real values by $(1+f)^y$, with f equal to the estimated average annual inflation rate. 2/

If one wishes to combine the real value discount rate (r) with the inflation rate (f), then one can calculate a current value discount rate (v) as being equal to: $((1+r)(1+f)-1)$ This rate, v, would be used if prices were expressed in current value terms.

3.2 Using Real Values (i.e., Net of Inflation)

It is generally much easier to define (i) (the discount rate to be used) as a real rate (r) and also express each cost and benefit in real terms. The numbers entered into the cash flow table will be more easily determined, since no inflation factors need to be estimated. This eliminates a major potential source of error in cash flow tables; it makes expected real price increases more clearly visible, and the use of annual and periodic payment formulas will be greatly simplified. Also, real values in the cash flow table will probably be more meaningful to the decision-maker. Because of these reasons, it is recommended here that analyses be carried out in real value terms. Once the "real" cash flow is determined, the appropriate contingencies for inflation can be determined and allocated for budgetary pruposes.

1/ Over the long term, however, rates of inflation are difficult to estimate and are subject to a high degree of error. The analyst must recognize the limitation on the estimated rate(s). Also, the rates of inflation often vary between the domestic economy and the offshore (exporting) countries, which means that two or more rates of inflation must be estimated.

2/ The value (f) is an average rate of inflation. If the rate of inflation is expected to vary from year to year, it would be more accurate to discount each yearly value by dividing by:

$$(1+r)^y((1+f_0) \times (1+f_1) \times (1+f_2) \times \ldots \times (1+f_{y-1}))$$

where (f_0) equals the rate of inflation during year 0, (f_1) equals the rate of inflation during year 1, etc., and (y) equals the given year.

Whether current or real values are used, the analyst needs to establish a frame of reference for comparisons. A real value frame of reference is preferable, since it is easier to comprehend.1/

In summary, the analyst can use either real or current values in his analysis. The results of the analysis will be the same, so long as he consistently uses the same basis throughout the analysis. For the reasons given above, we recommend working with real values, The results can be converted easily to current value terms.

3.3 Introducing "Real" Value Rates of Growth or Decline

Many prices of inputs and outputs will not remain at their year zero levels, even in real terms. If real price increases are expected, e.g., for stumpage prices, then each real price observed in year zero must be multiplied by $(1+g_p)^n$, where (g_p) equals the estimated average annual growth rate in the price and (n) equals the number of years since year zero. If real prices are expected to decline, then (g_p) would be negative. Usually expected price growth rates are expressed in average annual terms, even though rates actually fluctuate from year to year. It is very seldom that such fluctuations can be predicted in advance. Therefore, only expected annual averages of (g_p) are used.

Sometimes constant relative quantity changes are expected to occur. The real value of a cost or benefit in year (n) would be $(a(1+g_p)^n(1+g_q)^n)$, where (a) equals the present price times quantity of the input or output, (g_p) and (n) are defined as above, and (g_q) equals the average annual quantity growth rate. In such cases a composite growth rate (g) can be defined which equals $((1+g_p)(1+g_q)-1)$. This composite (g) is simpler to use. 2/ All further references to the real growth rate, (g), unless otherwise stated, refer to this type of composite rate. If (g) has only one component, say (g_p), the other component (g's) can be defined as being equal to zero. In that sense, (g) is always a composite of component (g's).

The compounded and discounted annual and periodic payment formulas shown in Table 2 can be restated under the assumption that composite growth rates exist. Table 10 lists these revised formulas. The derivation of the formulas is presented in Appendix 2. The first four rows of Table 10 are relevant when the discount rate (i) is greater than the growth rate (g). In that event, a composite discount rate (d) equals $\frac{(1+i)}{(1+g)} -1$. In other words, the overall effect of compounding at a rate (g) and discounting at a rate (i) (where $i>g$) is a discount effect equal to a rate (d). The bottom four rows of Table 10 restate the factors listed in the top four rows for the case where the discount rate (i) is less than the growth rate (g). In that event, a composite compound rate (c) equals $\frac{(1+g)}{(1+i)} -1$.

1/ It is sometimes argued that current rates of interest, such as rates on savings accounts or government bonds, should be used as frames of reference (i.e., as alternative rates of return). This may be true in the short term, when the inflation rate is likely to remain unchanged. However, for long term projects such as in forestry, this will likely lead to confusion and may result in poor decisions.

2/ A composite (g) can be composed of any number of separate component (g's).

Table 10 - Annual and Periodic Payment Formulas, Assuming the Payments Increase at an Annual Rate (g)

	Payments Begin One Year (or Period) From Present		Payments Begin Immediately	
	Finite Number of Payments (1)	Infinite Number of Payments (2)	Finite Number of Payments (3)	Infinite Number of Payments (4)
1. DISCOUNTED ANNUAL PAYMENT FACTOR (i>g)	$\left[\frac{(1+d)^n - 1}{d(1+d)^n (1+g)}\right]$	$\left[\frac{1}{d(1+g)}\right]$	$\left[\frac{(1+d)^{n+1} - 1}{d(1+d)^n}\right]$	$\left[\frac{1+d}{d}\right]$
2. COMPOUNDED ANNUAL PAYMENT FACTOR (i>g)	$\left[\frac{(1+g)^{n-1}\left[(1+d)^n - 1\right]}{d}\right]$	n.a.	$\left[\frac{(1+g)^n\left[(1+d)^{n+1} - 1\right]}{d}\right]$	n.a.
3. DISCOUNTED PERIODIC PAYMENT FACTOR (i>g)	$\left[\frac{(1+d)^{nt} - 1}{(1+g)^t (1+d)^{nt}\left[(1+d)^t - 1\right]}\right]$	$\left[\frac{1}{(1+g)^t\left[(1+d)^t - 1\right]}\right]$	$\left[\frac{(1+d)^{(n+1)t} - 1}{(1+d)^{nt}\left[(1+d)^t - 1\right]}\right]$	$\left[\frac{(1+d)^t}{(1+d)^t - 1}\right]$
4. COMPOUNDED PERIODIC PAYMENT FACTOR (i>g)	$\left[\frac{(1+g)^{(n-1)t}\left[(1+d)^{nt} - 1\right]}{(1+d)^t - 1}\right]$	n.a.	$\left[\frac{(1+g)^{nt}\left[(1+d)^{(n+1)t} - 1\right]}{(1+d)^t - 1}\right]$	n.a.
5. DISCOUNTED ANNUAL PAYMENT FACTOR (i<g)	$\left[\frac{(1+c)^n - 1}{c(1+i)}\right]$	n.a.	$\left[\frac{(1+c)^{n+1} - 1}{c}\right]$	n.a.
6. COMPOUNDED ANNUAL PAYMENT FACTOR (i<g)	$\left[\frac{(1+i)^{n-1}\left[(1+c)^n - 1\right]}{c}\right]$	n.a.	$\left[\frac{(1+i)^n\left[(1+c)^{n+1} - 1\right]}{c}\right]$	n.a.
7. DISCOUNTED PERIODIC PAYMENT FACTOR (i<g)	$\left[\frac{(1+c)^{nt} - 1}{(1+i)^t\left[(1+c)^t - 1\right]}\right]$	n.a.	$\left[\frac{(1+c)^{(n+1)t} - 1}{(1+c)^t - 1}\right]$	n.a.
8. COMPOUNDED PERIODIC PAYMENT FACTOR (i<g)	$\left[\frac{(1+i)^{(n-1)t}\left[(1+c)^{nt} - 1\right]}{(1+c)^t - 1}\right]$	n.a.	$\left[\frac{(1+i)^{nt}\left[(1+c)^{(n+1)t} - 1\right]}{(1+c)^t - 1}\right]$	n.a.

c = composite compound rate
d = composite discount rate
g = annual growth rate of payment

i = rate of interest (discount) in decimal form
n = number of years or periods until last payment
t = number of years between periodic payments

In other words, the overall effect of compounding at a rate (g) and discounting at a rate (i) (where i<g), is a compound effect equal to a rate (c). Depending upon the relationship between (i) and (g), either a (c) or a (d) is calculated and used in the appropriate formula shown in Table 10. Actually, it is not mathematically necessary to define a separate (c) for the case where i<g since a negative (d) could be used in place of (c). However, it is more meaningful and useful to deal with positive rates defining a net effect of (i) and (g) as either a compound effect or a discount effect.

3.4 Using Composite Discount Formulas for Annual Payments

The use of the discount formulas found in Table 10 is demonstrated using the two alternate cash flows presented earlier, the Korea fuelwood case (Table 1) and the Mythopia example (Table 5). Table 11 is a modification of the Korea fuelwood analysis. Two of the original assumptions were changed. First, the real price of fuelwood was assumed to increase by 2 percent annually beginning in the first year. Second, the real wage of silvicultural labour was assumed to rise by 4 percent annually, also beginning in the first year.

The entries in the net benefit (cost) row of Table 11 have no consistent annual or periodic pattern, unlike the net benefit (cost) row of Table 1. It appears, then, that the NPW would be best determined by discounting each yearly entry back to year zero. However, an easier method exists. Table 12 illustrates the use of the annual payment factor, when the payment begins "next year", under the assumption of growth rates. The procedures used here are similar to the procedures used in Table 3, which are explained in Appendix 5. Each net benefit (cost) from Table 11 for years 0-20, except for years 7-19, is entered in column 3 as a net benefit (cost). In Table 12 the net benefits for years 7-19 have been split up into their three individual benefit and cost components, shown in rows 7-9. Column 4 shows the base year (year 7) value of each component. Fuelwood benefits are compounded at 2 percent annually, supervision cost is not compounded, and harvesting cost is compounded at 4 percent annually, as indicated by column 5. The composite discount rate (d) for each component is shown in column 6. Each single payment (rows 1, 3, 4, 5, 6 and 10) is discounted to year zero at a 10 percent discount rate (i= 10%) and entered into column 9. Each sequence of annual payments (rows 2, 7, 8 and 9) is discounted back to the year before the payment begins (the year is shown in column 7 and the discounted value in that year is shown in column 8). For the two annual payments without growth rates, the previously explained formula from Table 2 (row 1, column 1) was used (with i= 10%). For the other two annual payments (the ones which include growth rates), the formula from row 1, column 1 of Table 10 was used (where (d) equals the value in column 6). These discounted values are further discounted to year zero (using i = 10%), and the resulting present value is indicated in column 9 of Table 12. The sum of all of these present values equals the NPW, which is W148.4.

The NPW of the Korea fuelwood cash flow (Table 11) was calculated in Table 12 using the "growth" discounted annual payment formula which assumes that payments begin in one year (row 1, column 1 of Table 10). An equally valid alternative is to use the "growth" discounted annual payment formula which assumes that payments begin immediately (row 1, column 3 of Table 10). This was done in Table 13 for the same cash flow (from Table 11). The only difference between Table 13 and Table 12 is that the discounted values of the four annual payments (rows 2, 7, 8 and 9) are expressed in terms of the base year instead of one year earlier. Therefore, only columns 7 and 8 differ between the two tables. These discounted values are further discounted to year zero (using i = 10%), and the resulting present values are entered into column 9. Each entry in column 9 of Table 13 is, of course, identical to the corresponding entry in column 9 of Table 12.

The above may seem very complex and cumbersome. However, in practice, the modifications suggested reduce the complexity of the calculations needed to arrive at useable conceptually correct results, particularly when one is dealing with a fairly long project period.

3.5 Using Composite Discount Formulas for Periodic Payments

The use of discounted periodic payment formulas which assume growth rates can be demonstrated by using the cash flow exhibited in Table 14, which is a modification of the Mythopia cash flow (Table 5), used earlier to illustrate the use of periodic payment formulas. The same two growth rates assumed in Table 11 are also assumed in this case: the real price of fuelwood increases by 2 percent annually, beginning in the first year, and the real wage of silvicultural labour rises by 4 percent annually, also beginning in the first year.

The NPW of the Table 14 cash flow is calculated by separate methods in Table 15 and Table 16, which are constructed in the same format as Tables 12 and 13 discussed above. Each net benefit (cost) from Table 14 is either entered as a net benefit (cost) in column 3 of each table, or is broken down into component benefits and costs which are individually entered in column 4.[1] In Table 15 the discounted values of the annual and periodic payments are expressed in terms of the year or period prior to the start of the payments (the years shown in column 7). The discounted values of the periodic payments of Table 15 were calculated using the formula from row 3, column 1 of Table 10. In Table 16 the discounted values of the annual and periodic payments are expressed in terms of the year in which the payments begin (the years shown in column 7 of Table 16). The discounted values of the periodic payments of Table 16 were calculated using the formula from row 3, column 3 of Table 10. The entries in column 9 of each table are the present values of the numbers in columns 3 or 8. These present values are, of course, the same in both tables since each table only illustrates a different method to calculate the same NPW.

In cases where one or more growth rates exceed the discount rate, the discounted annual and periodic payment factors from rows 5 and 7 of Table 10 would be applied in the same manner as described above. In such cases a composite compound rate (c) would be calculated instead of a composite discount rate (d). The same procedures and format used in Tables 12, 13, 15 and 16 could still be used.

---

[1] In the cases where growth rates apply to these individual benefits and costs, the base year amount is entered into column 4.

Table 11 - Korea Fuelwood Cash Flow Analysis, Assuming Benefits Increase by Two Percent Annually and Harvesting Costs Increase by Four Percent Annually

	YEAR 0	1	2	3	4	5	6	7	8	9	10	11	12	13	14	15	16	17	18	19	20
[1] BENEFITS (fuelwood)	-	-	-	6.9	14.1	28.7	58.6	74.7	76.2	77.7	79.2	80.8	82.4	84.1	85.8	87.5	89.2	91.0	92.8	94.7	2414.7
[2] COSTS																					
[3] Establishment	130.0	-	-	-	-	-	-	-	-	-	-	-	-	-	-	-	-	-	-	-	-
[4] Supervision	2.4	2.4	2.4	2.4	2.4	2.4	2.4	2.4	2.4	2.4	2.4	2.4	2.4	2.4	2.4	2.4	2.4	2.4	2.4	2.4	-
[5] Misc. (tools, etc.)	7.5	-	-	-	-	-	-	-	-	-	-	-	-	-	-	-	-	-	-	-	-
[6] Harvesting	-	-	-	2.1	4.3	9.1	19.0	24.6	25.6	26.6	27.7	28.8	29.9	31.1	32.4	33.7	35.0	36.4	37.9	39.4	2054.2
[7] Total Cost	139.9	2.4	2.4	4.5	6.7	11.5	21.4	27.0	28.0	29.0	30.1	31.2	32.3	33.5	34.8	36.1	37.4	38.8	40.3	41.8	2054.2
[8] NET BENEFIT (cost)	(139.9)	(2.4)	(2.4)	2.4	7.4	17.2	37.2	47.7	48.2	48.7	49.1	49.6	50.1	50.6	51.0	51.4	51.8	52.2	52.5	55.6	362.5

Table 12 - Calculation of Net Present Worth (NPW) for Korea Fuelwood Project 1/ When the Disc. Value of Annual Payment is Expressed in Terms of the Year Prior to First Payment (i = 10%)

Item	Year	Net Benefit or (cost)	Individual Benefit or (Cost): Base Year Amount	Annual Growth Rate (%)	Composite Discount Rate (%)	Discounted Value of Annual Payment: Year	Value	Present Worth (Year 0)
[1]	[2]	[3]	[4]	[5]	[6]	[7]	[8]	[9]
[1] Net Cost	0	(139.9)						(139.900)
[2] Net Cost	1-2	(2.4)				0	(4.165)	(4.165)
[3] Net Benefit	3	2.4						1.803
[4] Net Benefit	4	7.4						5.054
[5] Net Benefit	5	17.2						10.680
[6] Net Benefit	6	37.2						20.998
[7] Fuelwood Benefit	7-19		74.7	2	7.843	6	583.863	329.575
[8] Supervision Cost	7-19		(2.4)	0	n.a.	6	(17.048)	(9.623)
[9] Harvesting Cost	7-19		(24.6)	4	5.769	6	(212.252)	(119.811)
[10] Net Benefit	20	362.5						53.833

Net Present Worth (NPW) = 148.4

Table 13 - Calculation of Net Present Worth (NPW) for Korea Fuelwood Project 1/ When the Disc. Value of Annual Payment is Expressed in Terms of the Year in Which Payment Begins (i=10%)

Item	Year	Net Benefit or (cost)	Individual Benefit or (Cost): Base Year Amount	Annual Growth Rate (%)	Composite Discount Rate (%)	Discounted Value of Annual Payment: Year	Value	Present Worth (Year 0)
[1]	[2]	[3]	[4]	[5]	[6]	[7]	[8]	[9]
[1] Net Cost	0	(139.9)						(139.000)
[2] Net Cost	1-2	(2.4)				1	(4.582)	(4.165)
[3] Net Benefit	3	2.4						1.803
[4] Net Benefit	4	7.4						5.054
[5] Net Benefit	5	17.2						10.680
[6] Net Benefit	6	37.2						20.998
[7] Fuelwood Benefit	7-19		74.7	2	7.843	7	642.249	329.575
[8] Supervision Cost	7-19		(2.4)	0	n.a.	7	(18.753)	(9.623)
[9] Harvesting Cost	7-19		(24.6)	4	5.769	7	(233.477)	(119.811)
[10] Net Benefit	20	362.5						53.833

Net Present Worth (NPW) = 148.4

1/ Assuming benefits increase by 2% annually and harvesting costs increase by 4% annually

Table 14 - Mythopia Cash Flow Analysis, Assuming Benefits Increase by Two Percent Annually and Harvesting Costs by Four Percent Annually

	-YEAR-																				
	0	1	2	3	4	5	6	7	8	9	10	11	12	13	14	15	16	17	18	19	20
[1] BENEFITS (fuelwood)	-	-	-	-	35.2	-	146.4	-	152.3	-	158.5	-	164.9	-	171.5	-	178.5	-	185.7	-	2414.7
[2] COSTS																					
[3] Establishment	130.0	-	-	-	-	-	-	-	-	-	-	-	-	-	-	-	-	-	-	-	-
[4] Supervision	2.4	2.4	2.4	2.4	2.4	2.4	2.4	2.4	2.4	2.4	2.4	2.4	2.4	2.4	2.4	2.4	2.4	2.4	2.4	2.4	-
[5] Misc. (tools, etc.)	7.5	-	-	-	-	-	-	-	-	-	-	-	-	-	-	-	-	-	-	-	-
[6] Harvesting	-	-	-	-	11.0	-	47.3	-	51.2	-	55.4	-	59.9	-	64.7	-	70.0	-	75.8	-	2054.2
[7] Total Cost	139.9	2.4	2.4	2.4	13.4	2.4	49.7	2.4	53.6	2.4	57.8	2.4	62.3	2.4	67.1	2.4	72.4	2.4	78.2	2.4	2054.2
[8] NET BENEFIT (cost)	(139.9)	(2.4)	(2.4)	(2.4)	21.8	(2.4)	96.7	(2.4)	98.7	(2.4)	100.7	(2.4)	102.6	(2.4)	104.4	(2.4)	106.1	(2.4)	107.5	(2.4)	362.5

Table 15 - Calculation of Net Present Worth (NPW) for Mythopia Project[2/] When Discounted Value of Annual/Periodic Payment is Expressed in Terms of Year Prior to First Payment (i=10%)

Item	Year[1/]	Net Benefit or (Cost)	Individual Benefit or (Cost): Base Year Amount	Annual Growth Rate (%)	Composite Discount Rate (%)	Discounted Value of Annual/Periodic Payment: Year	Value	Present Worth (Year 0)
[1]	[2]	[3]	[4]	[5]	[6]	[7]	[8]	[9]
[1] Net Cost	0	(139.9)						(139.900)
[2] Supervision Cost	1-19		(2.4)	0	n.a.	0	(20.076)	(20.076)
[3] Fuelwood Benefit	4		35.2					24.042
[4] Harvesting Cost	4		(11.0)					(7.513)
[5] Fuelwood Benefit	6-18 (p)		146.4	2	7.843	4	563.277	384.726
[6] Harvesting Cost	6-18 (p)		(47.3)	4	5.769	4	(200.397)	(136.874)
[7] Net Benefit	20	362.5						53.833

Net Present Worth (NPW) = 158.2

Table 16 - Calculation of Net Present Worth (NPW) for Mythopia Project[2/] When the Discounted Value of Annual/Periodic Payment is Expressed in Terms of Year in Which Pay. Begins (i=10%)

Item	Year[1/]	Net Benefit or (Cost)	Individual Benefit or (Cost): Base Year Amount	Annual Growth Rate (%)	Composite Discount Rate (%)	Discounted Value of Annual/Periodic Payment: Year	Value	Present Worth (Year 0)
[1]	[2]	[3]	[4]	[5]	[6]	[7]	[8]	[9]
[1] Net Cost	0	(139.9)						(139.900)
[2] Supervision Cost	1-19		(2.4)	0	n.a.	1	(22.083)	(20.076)
[3] Fuelwood Benefit	4		35.2					24.042
[4] Harvesting Cost	4		(11.0)					(7.513)
[5] Fuelwood Benefit	6-18 (p)		146.4	2	7.843	6	681.565	384.726
[6] Harvesting Cost	6-18 (p)		(47.3)	4	5.769	6	(242.481)	(136.874)
[7] Net Benefit	20	362.5						53.833

Net Present Worth (NPW) = 158.2

1/ "(p)" refers to periodic payment eccurring every two years

2/ Assuming benefits increase by 2% annually and harvesting costs increase by 4% annually

4. UNCERTAINTY AND TIME CONSIDERATIONS [1]

A usual point made about uncertainty (and risk) is that the further into the future a given event occurs the more uncertain is that event. For example, in projecting future stumpage prices, we can be fairly confident that an estimate of next year's price will be reasonably accurate unless totally unforeseen major events occur to disrupt trends. Such events are also less likely in a shorter period. We cannot be nearly as confident about stumpage price estimates 30 years from now. The same holds for final or intermediate product prices and for costs.

However, at the same time, errors in far future estimates are less important in NPW or ROR calculations than errors in near future estimates. The degree of importance depends on a) the discount rate used in NPW calculations, b) the project's implied ROR, and c) the period of years between the present and the year to which the estimate applies.

For example, assume an estimated stumpage price for pine of $30/m^3 some 40 years from now. A 20 percent error would put the range between $24 and $36 per m^3, or $6 on either side of the best estimate. However, in present value terms, discounting at 10 percent, the range would appear as follows:

Low	Best	High
$0.53/m^3	$0.66/m^3	$0.80/m^3

In other words, looked at in present value terms, the spread between high and low estimates is only $0.80 - $0.53 or $0.27/m^3, as compared to $12.00/m^3 in future or current value terms. (The difference, of course, is still 20 percent.) What looks like a large difference in absolute terms 40 years from now is a relatively small absolute difference in PV terms.

But now, suppose we had a faster growing plantation and we estimated a stumpage price of $20/m^3 some 15 years from now. The present value, at 10 percent, would be $4.79/m^3 and the range with a 20 percent error would be from $3.83 - $5.75 per m^3 or a spread of $1.92/m^3 in PV terms between high and low.

In this case, holding the discount rate constant, we can see that a given absolute (or percentage) error in an estimate is more important, in PV terms, the closer it occurs to the present. [2] Conversely, using a higher discount rate reduces the importance of an error in estimate of a value that occurs at a given time in the future.

In sum, on the one hand, uncertainty concerning estimates of future values tends to increase the further into the future we go. On the other hand, the further into the future a given value or event occurs, the less we have to be concerned about reducing the uncertainty surrounding the estimate of that value or event, because its impact on NPW or ROR will be less.

1/ See EAFP, Chapter 10, which deals with treatment of risk and uncertainty in project planning. Here we merely want to relate the subject to time considerations.

2/ Obviously, if the error is estimated for year zero prices, the PV of the error will be the same as its current value.

Appendix 1

Treatment of Annual and Periodic Payments

Geometric series

Compounded and discounted annual and periodic payment factors, capital recovery factors, and sinking fund factors are all derived from the expression for the sum of a geometric series. For any non-zero real numbers; (a) and (k) and for non-negative integers, (y) and (n), the following serires is denoted:

$$\sum_{y=0}^{n} aK^y = a + aK + aK^2 + \cdots\cdots + aK^y + \cdots\cdots + aK^n$$

Then:

$$K\sum_{y=0}^{n} aK^y = aK + aK^2 + aK^3 + \cdots + aK^y + aK^{y+1} + \cdots + aK^n + aK^{n+1}$$

Therefore:

$$\left[\sum_{y=0}^{n} aK^y\right] - K\left[\sum_{y=0}^{n} aK^y\right] = a - aK^{n+1}$$

$$\left[\sum_{y=o}^{n} aK^y\right](1-K) = a(1-K^{n+1})$$

(1) $$\sum_{y=0}^{n} aK^y = a\left[\frac{1-K^{n+1}}{1-K}\right]$$

Using the same notations above but excluding the first term of the series, the following series is denoted:

$$\sum_{y=1}^{n} aK^y = aK + aK^2 + \cdots + aK^y + \cdots + aK^n$$

$$= a\left[\frac{1-K^{n+1}}{1-K} - 1\right]$$

$$= a\left[\frac{1-K^{n+1} - 1+K}{1-K}\right]$$

(2) $$\sum_{y=1}^{n} aK^y = a\left[\frac{K(1-K^n)}{1-K}\right]$$

Determination of present value factors and future value factors for annual payments beginning "next year"

The discounted annual payment factor can be derived from either equation 1 or equation 2, depending upon when the annual payment begins. Most texts refer to the second case, i.e., payments beginning "next year" and continuing for a total of n years. Therefore, this is the case derived below (from equation 2). Once that factor has been determined, it is much easier to derive the first case (when the payment begins "this year" and continues for n more years) by using the case 2 factor rather than equation 1.

Let: a = annual payment beginning next year and continuing for a total of n years

n = number of annual payments (also equals the number of years until the last payment)

i = rate of interest (discount)

y = any given year

Then, the series, $\frac{a}{1+i} + \frac{a}{(1+i)^2} + \cdots + \frac{a}{(1+i)^y} + \frac{a}{(1+i)^n}$, equals the present value of an annual payment (a) received <u>next year</u> and continuing for n years. Substituting $\frac{1}{1+i}$ for K in equation 2, the following equation results:

$$\sum_{y=1}^{n} \frac{a}{(1+i)^y} = a\left[\frac{\frac{1}{(1+i)}\left[1 - \frac{1}{(1+i)^n}\right]}{1 - \frac{1}{(1+i)}}\right]$$

Multiplying the numerator and denominator by $\frac{(1+i)^{n+1}}{(1+i)^{n+1}}$:

$$= a\left[\frac{(1+i)^n\left[1 - \frac{1}{(1+i)^n}\right]}{(1+i)^{n+1} - (1+i)^n}\right]$$

$$= a\left[\frac{(1+i)^n - 1}{(1+i)^n\,(1+i-1)}\right]$$

(3) $$\sum_{y=1}^{n} \frac{a}{(1+i)^y} = a\left[\frac{(1+i)^n - 1}{i(1+i)^n}\right]$$

DISCOUNTED ANNUAL PAYMENT FACTOR (payment starting "next year")

The future value (in year n) of an annual payment (a) received <u>next year</u> and continuing for a total of n years is derived from equation 3 as follows:

$$\sum_{y=0}^{n-1} a(1+i)^y = a + a(1+i) + a(1+i)^2 + \cdots + a(1+i)^y + \cdots + a(1+i)^{n-2} + a(1+i)^{n-1}$$

$$= a\left[\frac{(1+i)^n - 1}{i(1+i)^n}\right](1+i)^n$$

(4) $$\sum_{y=0}^{n-1} a(1+i)^y = a\left[\frac{(1+i)^n - 1}{i}\right]$$

COMPOUND ANNUAL PAYMENT FACTOR (payment starting "next year")

<u>Determination of present value factors and future value factors for annual payments beginning "this year"</u>

Using equation 3, the case 1 factor (the present value of an annual payment beginning <u>this year</u> and continuing for n <u>more</u> years) is derived as follows:

Let: a = annual payment beginning <u>this year</u> and continuing for n <u>more</u> years

n = number of years until last payment (there are n+1 payments)

i = rate of interest (discount)

y = any given year

Then: $$\sum_{y=0}^{n} \frac{a}{(1+i)^y} = a + \frac{a}{(1+i)} + \frac{a}{(1+i)^2} + \cdots + \frac{a}{(1+i)^y} + \cdots + \frac{a}{(1+i)^n}$$

(5a) $$\sum_{y=0}^{n} \frac{a}{(1+i)^y} = a\left[1 + \frac{(1+i)^n - 1}{i(1+i)^n}\right]$$

DISCOUNTED ANNUAL PAYMENT FACTOR (Form a) (payment starting "this year")

$$= a\left[\frac{i(1+i)^n + (1+i)^n - 1}{i(1+i)^n}\right]$$

$$= a\left[\frac{(1+i)^n\,(1+i) - 1}{i(1+i)^n}\right]$$

(5b) $$\sum_{y=0}^{n} \frac{a}{(1+i)^y} = a\left[\frac{(1+i)^{n+1} - 1}{i(1+i)^n}\right]$$

DISCOUNTED ANNUAL PAYMENT FACTOR (Form b) (payment starting "this year")

Equation 5a is generally easier to remember (especially in relation to equation 3). And when using an electronic calculator which automatically gives on the "next year" (equation 3) form, equation 5a is easier to use. However, equation 5b is easier to use with an electronic calculator which doesn't automatically do compounding and discounting.

As before the compounded factor can be easily derived from the discounted factor. The future value (in year n) of an annual payment (a) received this year and continuing for n more years is derived from equations 5a and 5b as follows:

$$\sum_{y=0}^{n} a(1+i)^y = a + a(1+i) + a(1+i)^2 + \cdots + a(1+i)^y + \cdots + a(1+i)^{n-1} + a(1+i)^n$$

Using equation 5a:

$$= a\left[1 + \frac{(1+i)^n - 1}{i(1+i)^n}\right](1+i)^n$$

(6a) $$\sum_{y=0}^{n} a(1+i)^y = a\left[(1+i)^n + \frac{(1+i)^n - 1}{i}\right]$$ COMPOUNDED ANNUAL PAYMENT FACTOR (Form a) (payment starting "this year")

Using equation 5b:

$$= a\left[\frac{(1+i)^{n+1} - 1}{i(1+i)^n}\right](1+i)^n$$

(6b) $$\sum_{y=0}^{n} a(1+i)^y = a\left[\frac{(1+i)^{n+1} - 1}{i}\right]$$ COMPOUNDED ANNUAL PAYMENT FACTOR (Form b) (payment starting "this year")

As was the case with equation 5a, it may again be easier to remember equation 6a since it is composed of the sum of the future value of an annual payment beginning next year (and continuing for a total of n years) and the future value of a single payment made this year. However, equation 6b is as simple or simpler to use with electronic calculators.

Annual capital recovery factors and sinking fund factors

Two other useful factors can be derived from preceding equations. The capital recovery factor (equation 7) is the inverse of the discounted annual payment factor (when payment starts "next year"), equation 3. The sinking fund factor (equation 8) is the inverse of the compounded annual payment factor (when payment starts "next year"), equation 4.[1/]

(7) $$a = \sum_{y=1}^{n} \frac{a}{(1+i)^y}\left[\frac{i(1+i)^n}{(1+i)^n - 1}\right]$$ ANNUAL CAPITAL RECOVERY FACTOR

(8) $$a = \sum_{y=0}^{n-1} a(1+i)^y\left[\frac{i}{(1+i)^n - 1}\right]$$ ANNUAL SINKING FUND FACTOR

Determination of present value factors and future value factors for periodic payments beginning "next period"

Compound and discounted periodic payment factors can also be derived from equation 2. The present value of a periodic (every t years) payment (a) is derived below.

Let: a = periodic payment beginning in t years and continuing every t years for a total of n periods

t = time interval between periods

n = number of periodic payments (nt equals the number of years until the last payment)

i = rate of annual interest (discount)

y = any given period (yt equals any given year in which a payment occurs)

Then the series, $\frac{a}{(1+i)^t} + \frac{a}{(1+i)^{2t}} + \frac{a}{(1+i)^{3t}} + \cdots + \frac{a}{(1+i)^{yt}} + \cdots + \frac{a}{(1+i)^{nt}}$, equals the present value of a periodic payment. Substituting a composite periodic discount rate (p) equal to the rate of discount per period rather than per year, the same steps can be followed as were followed in the derivation of equation 3.

1/ Capital recovery and sinking fund factors are generally not meaningful in practical applications when payments begin "this year." However, they are theoretically as meaningful as equations 7 and 8 (derived below), and may have some practical use as well. These factors would be the reciprocals of equations 5b and 6b, respectively.

First p is defined: $p = (1+i)^t - 1$; $(1+p) = (1+i)^t$

Then, substituting $(1+p) = (1+i)^t$, the series,

$$\frac{a}{(1+p)} + \frac{a}{(1+p)^2} + \frac{a}{(1+p)^3} + \cdots + \frac{a}{(1+p)^y} + \cdots + \frac{a}{(1+p)^n},$$

is obtained, which is in the same form as the discount <u>annual</u> payment factor series. Therefore, equation 3 can be derived as shown earlier, but this time using p instead of i.

$$\sum_{y=1}^{n} \frac{a}{(1+p)^y} = a\left[\frac{(1+p)^n - 1}{p(1+p)^n}\right]$$

Substituting, $p = (1+i)^t - 1$, and $(1+i)^t = (1+p)$

(9) $$\sum_{y=1}^{n} \frac{a}{(1+i)^{yt}} = a\left[\frac{(1+i)^{nt} - 1}{(1+i)^{nt}\left[(1+i)^t - 1\right]}\right]$$

DISCOUNTED PERIODIC PAYMENT FACTOR (payment starting in t years)

The future value (in year nt) of a periodic payment (a) received in the next period (in t years) and continuing every t years for a total of n periods is derived from equation 9 as follows:

$$\sum_{y=0}^{n-1} a(1+i)^{yt} = a + a(1+i)^t + a(1+i)^{2t} + \cdots + a(1+i)^{yt} + \cdots + a(1+i)^{(n-2)t} + a(1+i)^{(n-1)t}$$

$$= a\left[\frac{(1+i)^{nt} - 1}{(1+i)^{nt}\left[(1+i)^t - 1\right]}\right](1+i)^{nt}.$$

(10) $$\sum_{y=0}^{n-1} a(1+i)^{yt} = a\left[\frac{(1+i)^{nt} - 1}{(1+i)^t - 1}\right]$$

COMPOUNDED PERIODIC PAYMENT FACTOR (payment starting in t years)

<u>Determination of present value factors and future value factors for periodic payments beginning "this year"</u>

Using equation 9, the present value of a periodic payment beginning <u>this year</u> and continuing every t years for n <u>more</u> periods is derived as follows:

Let: a = periodic payment beginning <u>this year</u> and continuing every t years for n <u>more</u> periods

t = time interval between periods

n = number of periodic payments after first payment (there are n+1 payments and nt equals the number of years until the last payment)

i = rate of annual interest (discount)

y = any given period (yt equals any given year in which a payment occurs)

Then: $$\sum_{y=0}^{n} \frac{a}{(1+i)^{yt}} = a + \frac{a}{(1+i)^t} + \frac{a}{(1+i)^{2t}} + \frac{a}{(1+i)^{3t}} + \cdots + \frac{a}{(1+i)^{yt}} + \cdots \frac{a}{(1+i)^{nt}}$$

$$\sum_{y=0}^{n} \frac{a}{(1+i)^{yt}} = a\left[1 + \frac{(1+i)^{nt} - 1}{(1+i)^{nt}\left[(1+i)^t - 1\right]}\right]$$

$$= a\left[\frac{(1+i)^{nt}\left[(1+i)^t - 1\right] + (1+i)^{nt} - 1}{(1+i)^{nt}\left[(1+i)^t - 1\right]}\right]$$

$$= a\left[\frac{(1+i)^{nt}\left[(1+i)^t - 1 + 1\right] - 1}{(1+i)^{nt}\left[(1+i)^t - 1\right]}\right]$$

(11) $$\sum_{y=0}^{n} \frac{a}{(1+i)^{yt}} = a\left[\frac{(1+i)^{(nt+t)} - 1}{(1+i)^{nt}\left[(1+i)^{t} - 1\right]}\right]$$ DISCOUNTED PERIODIC PAYMENT FACTOR (payment starting "this year")

Again, as before the compounded factor can be easily derived from the discounted factor. The future value (in year nt) of a periodic payment (a) received this year and continuing every t years for n more periods is derived from equation 11 as follows:

$$\sum_{y=0}^{n} a(1+i)^{yt} = a + a(1+i)^{t} + a(1+i)^{2t} + \ldots + a(1+i)^{yt} + \ldots + a(1+i)^{(n-1)t} + \ldots + a(1+i)^{nt}$$

$$= a\left[\frac{(1+i)^{(nt+t)} - 1}{(1+i)^{nt}\left[(1+i)^{t} - 1\right]}\right](1+i)^{nt}$$

(12) $$\sum_{y=0}^{n} a(1+i)^{yt} = a\left[\frac{(1+i)^{(nt+t)} - 1}{(1+i)^{t} - 1}\right]$$ COMPOUNDED PERIODIC PAYMENT FACTOR (payment starting "this year")

Periodic capital recovery factors and sinking fund factors

Theoretically, periodic capital recovery factors and periodic sinking fund factors are possible. They would be derived analogously to equations 7 and 8 from equations 9 and 10, respectively. However, in practice they are not in general use.

Determination of "infinite" factors

When annual or periodic payments continue for an infinite period (or for a sufficiently long finite period such that the "next" payment has no measurable present value) the equations for present values of such payments become much simplified. Four equations described in the preceding paragraphs can be simplified, equations 3, 5b, 9 and 11.

(3) $$\sum_{y=1}^{n} \frac{a}{(1+i)^{y}} = a\left[\frac{(1+i)^{n} - 1}{i(1+i)^{n}}\right] = a\left[\frac{1}{i} - \frac{1}{(1+i)^{n}}\right]$$ FINITE DISCOUNTED ANNUAL PAYMENT FACTOR (payment starting "next year")

When $n \to \infty$, the second term $\left(\frac{1}{i(1+i)^{n}}\right)$ approaches zero, leaving $\left[\frac{1}{i}\right]$.

(13) $$\sum_{y=1}^{n=\infty} \frac{a}{(1+i)^{y}} = a\left[\frac{1}{i}\right]$$ INFINITE DISCOUNTED ANNUAL PAYMENT FACTOR (payment starting "next year")

(5b) $$\sum_{y=0}^{n} \frac{a}{(1+i)^{y}} = a\left[\frac{(1+i)^{n+1} - 1}{i(1+i)^{n}}\right] = a\left[\frac{1+i}{i} - \frac{1}{i(1+i)^{n}}\right]$$ FINITE DISCOUNTED ANNUAL PAYMENT FACTOR (payment starting "this year")

When $n \to \infty$, the second term $\left[\frac{1}{i(1+i)^{n}}\right]$ approaches zero, leaving $\left[\frac{1+i}{i}\right]$

(14) $$\sum_{y=0}^{n=\infty} \frac{a}{(1+i)^{y}} = a\left[\frac{1+i}{i}\right] \text{ or } a\left[1 + \frac{1}{i}\right]$$ INFINITE DISCOUNTED ANNUAL PAYMENT FACTOR (payment starting "this year")

(9) $$\sum_{y=1}^{n} \frac{a}{(1+i)^{yt}} = a\left[\frac{(1+i)^{nt} - 1}{(1+i)^{nt}\left[(1+i)^{t} - 1\right]}\right] = a\left[\frac{1}{(1+i)^{t} - 1} - \frac{1}{(1+i)^{nt}\left[(1+i)^{t} - 1\right]}\right]$$ FINITE DISCOUNTED PERIODIC PAYMENT FACTOR (payment starting in t years)

When $n \to \infty$, the second term $\left[\frac{1}{(1+i)^{nt}\left[(1+i)^{t} - 1\right]}\right]$ approaches zero, leaving $\left[\frac{1}{(1+i)^{t} - 1}\right]$.

(15) $$\sum_{y=1}^{n=\infty} \frac{a}{(1+i)^{yt}} = a\left[\frac{1}{(1+i)^{t} - 1}\right]$$ INFINITE DISCOUNTED PERIODIC PAYMENT FACTOR (payment starting in t years)

(11) $$\sum_{y=0}^{n} \frac{a}{(1+i)^{yt}} = a\left[\frac{(1+i)^{(nt+t)} - 1}{(1+i)^{nt}\left[(1+i)^{t} - 1\right]}\right] = a\left[\frac{(1+i)^{t}}{(1+i)^{t} - 1} - \frac{1}{(1+i)^{nt}\left[(1+i)^{t} - 1\right]}\right]$$ FINITE DISCOUNTED PERIODIC PAYMENT FACTOR (payment starting "this year")

When $n \to \infty$, the second term $\left[\frac{1}{(1+i)^{nt}\left[(1+i)^{t} - 1\right]}\right]$ approaches zero, leaving $\left[\frac{(1+i)^{t}}{(1+i)^{t} - 1}\right]$.

(16) $$\sum_{y=0}^{n=\infty} \frac{a}{(1+i)^{yt}} = a\left[\frac{(1+i)^{t}}{(1+i)^{t} - 1}\right]$$ INFINITE DISCOUNTED PERIODIC PAYMENT FACTOR (payment starting "this year")

Appendix 2

Treatment of Annual and Periodic Payments When Payments Grow Annually

Determination of present value factors and future value factors for annual payments beginning "next year"

The usual case for a series representing the present value of annual payments beginning "next year" and continuing for a total of n years is: [1/]

$$\sum_{y=1}^{n} \frac{a}{(1+i)^y} = \frac{a}{(1+i)} + \frac{a}{(1+i)^2} + \ldots\ldots + \frac{a}{(1+i)^y} + \ldots\ldots + \frac{a}{(1+i)^n}$$

where,
- a = annual payment beginning next year and continuing for a total of n years.
- n = number of annual payments (also equals the number of years until the last payment)
- i = rate of interest (discount)
- y = any given year

However, this only represents the special case (perhaps the most common case) when (a) remains constant. Often an annual payment (beginning "next year") would increase at an annual rate (g), such that the following sequence would result.[2/]

$$a,\ a(1+g),\ a(1+g)^2,\ \ldots,\ a(1+g)^{y-1},\ \ldots,\ a(1+g)^{n-1}$$

The present value of all (n) payments is represented by the following series:

$$\frac{a}{(1+i)} + \frac{a(1+g)}{(1+i)^2} + \frac{a(1+g)^2}{(1+i)^3} + \ldots + \frac{a(1+g)^{y-1}}{(1+i)^y} + \ldots + \frac{a(1+g)^{n-1}}{(1+i)^n}$$

Factoring out $\left[\frac{a}{1+g}\right]$, the series becomes:

$$\left[\frac{a}{1+g}\right]\left[\frac{1+g}{1+i}\right] + \left[\frac{a}{1+g}\right]\left[\frac{1+g}{1+i}\right]^2 + \left[\frac{a}{1+g}\right]\left[\frac{1+g}{1+i}\right]^3 + \ldots + \left[\frac{a}{1+g}\right]\left[\frac{1+g}{1+i}\right]^y + \ldots + \left[\frac{a}{1+g}\right]\left[\frac{1+g}{1+i}\right]^n$$

Three possible relationships exist between (i) and (g).

1. If $i > g$, then a composite discount rate (d) can be defined.
2. If $i = g$, then a direct summing of terms is possible, and the present value of the payments becomes; $n\left[\frac{a}{1+g}\right]$ while the future value of the payments becomes; $n\left[\frac{a}{1+g}\right](1+i)^n$.
3. If $i < g$, then a composite compound rate (c) can be defined.

The composite discount rate (d) equals: $\left[\frac{(1+i)}{(1+g)} - 1\right]$.
Substituting $\left[\frac{1}{(1+d)}\right]$ into the above series for $\left[\frac{1+g}{1+i}\right]$, the following series results:

$$\left[\frac{a}{1+g}\right]\left[\frac{1}{(1+d)}\right] + \left[\frac{a}{1+g}\right]\left[\frac{1}{(1+d)^2}\right] + \left[\frac{a}{1+g}\right]\left[\frac{1}{(1+d)^3}\right] + \ldots + \left[\frac{a}{1+g}\right]\left[\frac{1}{(1+d)^y}\right] + \ldots + \left[\frac{a}{1+g}\right]\left[\frac{1}{(1+d)^n}\right]$$

This series is in the same form as the series used to develop equation 3 in Appendix 1 . Instead of annual payment (a) and discount rate (i), there is annual payment $\left[\frac{a}{1+g}\right]$ and composite discount rate (d). Therefore, the following equation (analogous to equation 3 in Appendix 1) for the present value of a string of annual payments can be derived:

$$\sum_{y=1}^{n} \frac{\left[\frac{a}{1+g}\right]}{(1+d)^y} = \frac{a}{1+g}\left[\frac{(1+d)^n - 1}{d(1+d)^n}\right]$$

1/ See Appendix 1 , page 2 .

2/ The rate (g) could also represent an annual decrease in (a). In that case (g) would be negative. If two or more growth rates affected (a), (g) would be a composite of all growth rates.

(A) $$\sum_{y=1}^{n} \frac{\left[\frac{a}{1+g}\right]}{(1+d)^y} = a\left[\frac{(1+d)^n - 1}{d(1+d)^n(1+g)}\right]$$

$i > g$
DISCOUNTED ANNUAL PAYMENT FACTOR
(payment begins "next year")

The future value equation is derived from equation A as follows:

$$\sum_{y=0}^{n-1} a(1+g)^{n-y-1}(1+i)^y = a(1+g)^{n-1} + a(1+g)^{n-2}(1+i) + a(1+g)^{n-3}(1+i)^2 + \ldots + a(1+g)^{n-y-1}(1+i)^y + \ldots + a(1+g)(1+i)^{n-2} + a(1+i)^{n-1}$$

$$\sum_{y=0}^{n-1}\left[\frac{a(1+g)^{n-1}(1+i)^y}{(1+g)^y}\right] = \sum_{y=0}^{n-1} a(1+g)^{n-1}(1+d)^y = a\left[\frac{(1+d)^n - 1}{d(1+d)^n(1+g)}\right](1+i)^n$$

$$= a\left[\frac{(1+i)^n\left[(1+d)^n - 1\right]}{d\left[\frac{(1-i)^n}{(1+g)^n}\right](1+g)}\right]$$

(B) $$\sum_{y=0}^{n-1} a(1+g)^{n-1}(1+d)^y = a\left[\frac{(1+g)^{n-1}\left[(1+d)^n - 1\right]}{d}\right]$$

$i > g$
COMPOUNDED ANNUAL PAYMENT FACTOR
(payment begins "next year")

The composite compound rate (c) equals $\left[\frac{(1+g)}{(1+i)} - 1\right]$.[1]

Substituting (1+c) into the original series for $\left[\frac{1+g}{1+i}\right]$, the following series results:

$$\left[\frac{a}{1+g}\right](1+c) + \left[\frac{a}{1+g}\right](1+c)^2 + \left[\frac{a}{1+g}\right](1+c)^3 + \ldots + \left[\frac{a}{1+g}\right](1+c)^y + \ldots + \left[\frac{a}{1+g}\right](1+c)^n$$

This series is in the same form as the series used to develop equations 6a and 6b in Appendix 1 Substituting $\left[\frac{a}{1+g}\right]$ for (a), and (d) for (i), except that the first term is "missing".[2] Therefore, using equation 6b from Appendix 1, the present value of a string of annual payments beginning next year is derived.

$$\sum_{y=1}^{n}\left[\frac{a}{1+g}\right](1+c)^y = \left[\frac{a}{1+g}\right]\left[\frac{(1+c)^{n+1} - 1}{c} - 1\right]$$

$$= a\left[\frac{(1+c)^{n+1} - 1 - c}{c\,(1+g)}\right]$$

$$= a\left[\frac{(1+c)\left[(1+c)^n - 1\right]}{c\,(1+g)}\right]$$

$$= a\left[\frac{\frac{1+g}{1+i}\left[(1+c)^n - 1\right]}{c\,(1+g)}\right]$$

(C) $$\sum_{y=1}^{n}\left[\frac{a}{1+g}\right](1+c)^y = a\left[\frac{(1+c)^n - 1}{c\,(1\ i)}\right]$$

$i < g$
DISCOUNTED ANNUAL PAYMENT FACTOR
(payment begins "next year")

1/ It is not mathematically necessary to define a separate composite compound rate (c). The composite discount rate (d) is sufficient if one is willing to work with negative d's in such cases. However, in practice it is more meaningful to work with positive compound and discount rates.

2/ The series developed here is a present value of an annual payment which "grows" by a higher rate than the discount rate. Therefore, the present value of each succesive year's payment increases. This results in a series of the same form as the series representing the compounded annual payment factor when g = 0.

The future value form of equation C is derived in the manner of equation B, using the same series, repeated below:

$$\sum_{y=0}^{n-1} a\ (1+g)^{n-y-1}\ (1+i)^{y} = a(1+g)^{n-1} + a(1+g)^{n-2}(1+i) + a(1+g)^{n-3}(1+i)^{2} + \ldots + a(1+g)^{n-y-1}(1+i)^{y} + \ldots + a(1+g)(1+i)^{n-2} + a(1+i)^{n-1}$$

$$\sum_{y=0}^{n-1} \left[\frac{a(1+g)^{n-1}\ (1+i)^{y}}{(1+g)^{y}}\right] = a\left[\frac{(1+c)^{n}-1}{c(1+i)}\right](1+i)^{n}$$

(D) $$\sum_{Y=0}^{n-1} \left[\frac{a(1+g)^{n-1}}{(1+c)^{y}}\right] = a\left[\frac{(1+i)^{n-1}\left[(1+c)^{n}-1\right]}{c}\right]$$

$i < g$
COMPOUNDED ANNUAL PAYMENT FACTOR
(payment begins "next year")

<u>Determination of present value factors and future value factors for annual payments beginning "this year"</u>

The usual case for a series representing the present value of annual payments beginning <u>"this year"</u> anc continuing for n <u>more</u> years is: 1/

$$\sum_{y=0}^{n} \frac{a}{(1+i)^{y}} = a + \frac{a}{(1+i)} + \frac{a}{(1+i)^{2}} + \ldots + \frac{a}{(1+i)^{y}} + \ldots + \frac{a}{(1+i)^{n}}$$

where,
- a = annual payment beginning <u>this year</u> and continuing for n <u>more</u> years
- n = number of years until last payment (there are n+1 payments)
- i = rate of interest (discount)
- y = any given year

However, again this only represents the special case when (a) remains constant. When (a) increases at a rate (g) each year, the following sequence results: 2/

$$a,\ a(1+g),\ a(1+g)^{2},\ \ldots,\ a(1+g)^{y},\ \ldots,\ a(1+g)^{n}$$

The present value of all (n+1) payments is represented by the following series:

$$a + a\left[\frac{(1+g)}{(1+i)}\right] + a\left[\frac{(1+g)^{2}}{(1+i)^{2}}\right] + \ldots + a\left[\frac{(1+g)^{y}}{(1+i)^{y}}\right] + \ldots + a\left[\frac{(1+g)^{n}}{(1+i)^{n}}\right]$$

As before, three possible relationships exist between (i) and (g).

1. If $i > g$, then a composite <u>discount</u> rate (d) can be defined.
2. If $i = g$, then a direct summing of terms is possible, and the present value of the payments becomes $a(n+1)$, while the future value of the payments becomes $a(n+1)(1+i)^{n}$.
3. If $i < g$, then a composite <u>compound</u> rate (c) can be defined.

1/ See Appendix 1, page 2.

2/ The rate (g) could also represent an annual decrease in (a).. In that case (g) would be negative. If two or more growth rates affected (a), (g) would be a composite of all growth rates.

Again, the composite discount rate (d) equals: $\left[\frac{(1+i)}{(1+g)} - 1\right]$.

Substituting $\left[\frac{1}{(1+d)}\right]$ into the above series for $\left[\frac{1+g}{1+i}\right]$, the following series results:

$$a + \frac{a}{(1+d)} + \frac{a}{(1+d)^2} + \ldots + \frac{a}{(1+d)^y} + \ldots + \frac{a}{(1+d)^n}$$

This series is in the same form as the series used to develop equation 5b in Appendix 1, except that (d') has replaced (i). Therefore, the following equation (analogous to equation 5b in Appendix 1) for the present value of a string of annual payments can be derived:

(E)
$$\sum_{y=0}^{n} \frac{a}{(1+d)^y} = a\left[\frac{(1+d)^{n+1} - 1}{d(1+d)^n}\right]$$

i>g
DISCOUNTED ANNUAL PAYMENT FACTOR
(payment starting "this year")

The future value equation is derived from equation E as follows:

$$\sum_{y=0}^{n} a(1+g)^{n-y}(1+i)^y = a(1+g)^n + a(1+g)^{n-1}(1+i) + a(1+g)^{n-2}(1+i)^2 + \ldots + a(1+g)^{n-y}(1+i)^y + \ldots + a(1+g)(1+i)^{n-1} + a(1+i)^n$$

$$\sum_{y=0}^{n} \left[\frac{a(1+g)^n (1+i)^y}{(1+g)^y}\right] = a\left[\frac{(1+d)^{n+1} - 1}{d\,(1+d)^n}\right](1+i)^n$$

$$\sum_{y=0}^{n} a(1+g)^n(1+d)^y = a\left[\frac{(1+i)^n\left[(1+d)^{n+1} - 1\right]}{d\,(1+d)^n}\right]$$

$$= a\left[\frac{(1+i)^n\left[(1+d)^{n+1} - 1\right]}{d\left[\frac{(1+i)^n}{(1+g)^n}\right]}\right]$$

(F)
$$\sum_{y=0}^{n} a(1+g)^n(1+d)^y = a\left[\frac{(1+g)^n\left[(1+d)^{n+1} - 1\right]}{d}\right]$$

i>g
COMPOUNDED ANNUAL PAYMENT FACTOR
(payment starting "this year"

The composite compound rate (c) again equals $\left[\frac{(1+g)}{(1+i)} - 1\right]$. 1/

Substituting (1+c) into the original series (p. 3 of this appendix) for $\left[\frac{1+g}{1+i}\right]$, the following series results:

$$a + a(1+c) + a(1+c)^2 + \ldots + a(1+c)^y + \ldots + a(1+c)^n$$

This series is in the same form as the series used to develop equations 6a and 6b in Appendix 1. (substituting "c" for "i"). 2/ Therefore, using equation 6b from Appendix 1, the present value of a string of annual payments beginning <u>this year</u> can be derived.

(G)
$$\sum_{y=0}^{n} a(1+c)^y = a\left[\frac{(1+c)^{n+1} - 1}{c}\right]$$

i<g
DISCOUNTED ANNUAL PAYMENT FACTOR
(payment begins "this year"

1/ See footnote 1, p2 of this appendix.
2/ See footnote 2, p2 of this appendix.

The future value form of equation G is derived in the manner of equation F, using the same series, repeated below:

$$\sum_{y=0}^{n} a(1+g)^{n-y}(1+i)^{y} = a(1+g)^{n} + a(1+g)^{n-1}(1+i) + a(1+g)^{n-2}(1+i)^{2} + \ldots + a(1+g)^{n-y}(1+i)^{y} + \ldots + a(1+g)(1+i)^{n-1} + a(1+i)^{n}$$

$$\sum_{y=0}^{n} \left[\frac{a(1+g)^{n}(1+i)^{y}}{(1+g)^{y}}\right] = a\left[\frac{(1+c)^{n+1} - 1}{c}\right](1+i)^{n}$$

$$\text{(H)} \quad \sum_{y=0}^{n} \frac{a(1+g)^{n}}{(1+c)^{y}} = a\left[\frac{(1+i)^{n}\left[(1+c)^{n+1} - 1\right]}{c}\right]$$

$i<g$
COMPOUNDED ANNUAL PAYMENT FACTOR
(payment begins "this year")

Annual capital recovery factors and sinking fund factors

Capital recovery factors and sinking fund factors generally do not consider growth rates. They are only affected by the interest rate (i), the number of payments (n), and the total amount of current or future principal to be repaid or accumulated. It is true that (i) might either be defined to be a real rate or a rate which includes an expectation of the average rate of inflation over the period. If (i) is defined as a real rate, the real value of each period's payment (a) would have to be multiplied by an inflation factor (CPI_y/CPI_o)[1/], to get the nominal value which would then equalize the real value of each payment. If (i) is defined to include inflation, each period's payment (a) would be equal in nominal terms, but each successive payment would decline in real value. In such a case, a real rate of discount could be used with a negative (g) used to represent the annual decline in the value of (a) (which would be defined in nominal - not real-terms). Then (g) could be used in the manner described in this appendix. Capital recovery factors and sinking fund factors which include (g) would be the reciprocals of the discounted annual payment factors and the compounded annual payment factors, respectively (applying the same conditions as do these equations, namely, the "i:g" relationship and the year in which payment begins). However, when one is not working in real value terms, using a negative (g) to represent the annual decline in real value caused by inflation is much more difficult than defining (i) to include inflation and using the simpler equations developed in Appendix 1. [2/]

Determination of present value factors and future value factors for periodic payments beginning "next period"

The usual case for a series representing the present value of periodic payments beginning in t years and continuing every t years for a total of n periods is: [3/]

$$\sum_{y=1}^{n} \frac{a}{(1+i)^{yt}} = \frac{a}{(1+i)^{t}} + \frac{a}{(1+i)^{2t}} + \ldots + \frac{a}{(1+i)^{yt}} + \ldots + \frac{a}{(1+i)^{nt}}.$$

where,
- a = periodic payment beginning in t years and continuing every t years for a total of n periods
- t = time interval between periods
- n = number of periodic payments (nt equals number of years until last payment)
- i = rate of annual interest (discount)
- y = any given period (yt equals any given year in which a payment occurs)

However, this only represents the special case when (a) remains constant. When (a) increases at a rate (g) each year the following sequence results:[4/]

1/ Consumer price index for year (y) divided by consumer price index for year zero.

2/ If: r=real rate of interest and f= inflation rate, $(1+r)(1+f)-1=i$, the composite rate of interest which includes inflation. Defining a negative (g), $g=-f/(1+f)$, is unnecessary and more complicated to apply.

3/ See Appendix 1, page 3.

4/ The rate (g) could also represent an annual decrease in (a). In that case (g) would be negative. If two or more growth rates affected (a), (g) would be a composite of all growth rates.

$a, a(1+g)^t, a(1+g)^{2t}, \ldots, a(1+g)^{(y-1)t}, \ldots, a(1+g)^{(n-1)t}$

The present value of all (n) payments is represented by the following series:

$$\frac{a}{(1+i)^t} + \frac{a(1+g)^t}{(1+i)^{2t}} + \frac{a(1+g)^{2t}}{(1+i)^{3t}} + \ldots + \frac{a(1+g)^{(y-1)t}}{(1+i)^{yt}} + \ldots + \frac{a(1+g)^{(n-1)t}}{(1+i)^{nt}}$$

Factoring out $\left[\frac{a}{(1+g)^t}\right]$, the series becomes:

$$\left[\frac{a}{(1+g)^t}\right]\left[\frac{1+g}{1+i}\right]^t + \left[\frac{a}{(1+g)^t}\right]\left[\frac{1+g}{1+i}\right]^{2t} + \left[\frac{a}{(1+g)^t}\right]\left[\frac{1+g}{1+i}\right]^{3t} + \ldots + \left[\frac{a}{(1+g)^t}\right]\left[\frac{1+g}{1+i}\right]^{yt} + \ldots + \left[\frac{a}{(1+g)^t}\right]\left[\frac{1+g}{1+i}\right]^{nt}$$

Three possible relationships exist between (i) and (g).

1. If $i > g$, then a composite <u>discount</u> rate (d) can be defined.
2. If $i = g$, then a direct summing of terms is possible, and the present value of the payments becomes: $n\left[\frac{a}{(1+g)^t}\right]$, while the future value of the payments becomes: $n\left[\frac{a}{(1+g)^t}\right](1+i)^{nt}$.
3. If $i < g$, then a composite <u>compound</u> rate (c) can be defined.

The composite <u>discount</u> rate (d) equals $\left[\frac{(1+i)}{(1+g)} - 1\right]$.
Substituting $\left[\frac{1}{(1+d)}\right]$ into the above series for $\left[\frac{1+g}{1+i}\right]$, the following series results:

$$\left[\frac{a}{(1+g)^t}\right]\left[\frac{1}{(1+d)^t}\right] + \left[\frac{a}{(1+g)^t}\right]\left[\frac{1}{(1+d)^{2t}}\right] + \left[\frac{a}{(1+g)^t}\right]\left[\frac{1}{(1+d)^{3t}}\right] + \ldots + \left[\frac{a}{(1+g)^t}\right]\cdot\left[\frac{1}{(1+d)^{yt}}\right] + \ldots + \left[\frac{a}{(1+g)^t}\right]\left[\frac{1}{(1+d)^{nt}}\right]$$

The composite discount rate (d) is in annual terms. To facilitate the simplification of the above series, it is useful to define a <u>periodic</u> composite discount rate, $p = (1+d)^t - 1$.

Then, substituting $(1+p) = (1+d)^t$, the following series results:

$$\left[\frac{a}{(1+g)^t}\right]\left[\frac{1}{(1+p)}\right] + \left[\frac{a}{(1+g)^t}\right]\left[\frac{1}{(1+p)^2}\right] + \left[\frac{a}{(1+g)^t}\right]\left[\frac{1}{(1+p)^3}\right] + \ldots + \left[\frac{a}{(1+g)^t}\right]\left[\frac{1}{(1+p)^y}\right] + \ldots + \left[\frac{a}{(1+g)^t}\right]\left[\frac{1}{(1+p)^n}\right]$$

This series is in the same form as the series used to develop equation A, except that $\left[\frac{a}{(1+g)^t}\right]$ has replaced $\left[\frac{a}{1+g}\right]$ and (p) has replaced (d). Therefore, the following equation (analagous to equation A) for the present value of a string of <u>periodic</u> payments can be derived:

$$\sum_{y=1}^{n}\left[\frac{\left[\frac{a}{(1+g)^t}\right]}{(1+p)^y}\right] = \left[\frac{a}{(1+g)^t}\right]\left[\frac{(1+p)^n - 1}{p(1+p)^n}\right]$$

Substituting, $p = (1+d)^t - 1$, and $(1+d)^t = (1+p)$

$$\sum_{y=1}^{n}\frac{\left[\frac{a}{(1+g)^t}\right]}{(1+d)^{yt}} = \left[\frac{a}{(1+g)^t}\right]\left[\frac{(1+d)^{nt} - 1}{(1+d)^{nt}\left[(1+d)^t - 1\right]}\right]$$

$$\text{(I)} \qquad \sum_{y=1}^{n}\frac{\frac{a}{(1+g)^t}}{(1+d)^{yt}} = a\left[\frac{(1+d)^{nt} - 1}{(1+g)^t(1+d)^{nt}\left[(1+d)^t - 1\right]}\right]$$

$i > g$
DISCOUNTED PERIODIC PAYMENT FACTOR
(payment starting in t years)

The future value (in year nt) of a "growing" periodic payment received in the next period (in t years) and continuing every (t) years for a total of (n) periods is derived from equation I as follows:

$$\sum_{y=0}^{n-1} a(1+g)^{(n-y-1)t}(1+i)^{yt} = a(1+g)^{(n-1)t} + a(1+g)^{(n-2)t}(1+i)^{t} + a(1+g)^{(n-3)t}(1+i)^{2t} + \ldots + a(1+g)^{(n-y-1)t}(1+i)^{yt} + \ldots + a(1+g)^{t}(1+i)^{(n-2)t} + a(1+i)^{(n-1)t}$$

$$\sum_{y=0}^{n-1} \left[\frac{a(1+g)^{(n-1)t}\ (1+i)^{yt}}{(1+g)^{yt}}\right] = a \left[\frac{(1+d)^{nt} - 1}{(1+g)^{t}\ (1+d)^{nt}\ \left[(1+d)^{t} - 1\right]}\right] (1+i)^{nt}$$

$$\sum_{y=0}^{n-1} a(1+g)^{(n-1)t}(1+d)^{yt} = a \left[\frac{(1+i)^{nt}\ \left[(1+d)^{nt} - 1\right]}{(1+g)^{t}\left[\frac{(1+i)^{nt}}{(1+g)^{nt}}\right]\ \left[(1+d)^{t} - 1\right]}\right]$$

(J) $$\sum_{y=0}^{n-1} a(1+g)^{(n-1)t}(1+d)^{yt} = a \left[\frac{(1+g)^{(n-1)t}\ \left[(1+d)^{nt} - 1\right]}{(1+d)^{t} - 1}\right]$$

<u>i g</u>

COMPOUNDED PERIODIC PAYMENT FACTOR
(payment starting in t years)

The composite <u>compound</u> rate (c) equals $\left[\frac{(1+g)}{(1+i)} - 1\right]$. 1/

Substituting (1+c) into the original series (p.6 of this appendix) for $\left[\frac{1+g}{1+i}\right]$, the following series results:

$$\left[\frac{a}{(1+g)^{t}}\right](1+c)^{t} + \left[\frac{a}{(1+g)^{t}}\right](1+c)^{2t} + \left[\frac{a}{(1+g)^{t}}\right](1+c)^{3t} + \ldots + \left[\frac{a}{(1+g)^{t}}\right](1+c)^{yt} + \ldots + \left[\frac{a}{(1+g)^{t}}\right](1+c)^{nt}$$

The composite compound rate (c) is in annual terms. To facilitate the simplification of the above series, it is useful to define a <u>periodic</u> composite compound rate, $p = (1+c)^{t} - 1$.

Then, substituting $(1+p) = (1+c)^{t}$, the following series results:

$$\left[\frac{a}{(1+g)^{t}}\right](1+p) + \left[\frac{a}{(1+g)^{t}}\right](1+p)^{2} + \left[\frac{a}{(1+g)^{t}}\right](1+p)^{3} + \ldots + \left[\frac{a}{(1+g)^{t}}\right](1+p)^{y} + \ldots + \left[\frac{a}{(1+g)^{t}}\right](1+p)^{n}$$

This series is in the same form as the series used to develop equation C, except that $\left[\frac{a}{(1+g)^{t}}\right]$ has replaced $\left[\frac{a}{1+g}\right]$ and (p) has replaced (c) Therefore, the following equation (analagous to equation C) for the future value of a string of <u>periodic</u> payments can be derived:

$$\sum_{y=1}^{n} \left[\frac{a}{(1+g)^{t}}\right](1+p)^{y} = \frac{a}{(1+g)^{t}} \left[\frac{(1+p)\ \left[(1+p)^{n} - 1\right]}{p}\right]$$

Substituting, $p = (1+c)^{t} - 1$, and $(1+c)^{t} = (1+p)$

$$\sum_{y=1}^{n} \left[\frac{a}{(1+g)^{t}}\right](1+c)^{yt} = a \left[\frac{(1+c)^{t}\ \left[(1+c)^{nt} - 1\right]}{(1+g)^{t}\ \left[(1+c)^{t} - 1\right]}\right]$$

$$= a \frac{\left[\frac{(1+g)^{t}}{(1+i)^{t}}\right]\ \left[(1+c)^{nt} - 1\right]}{(1+g)^{t}\ \left[(1+c)^{t} - 1\right]}$$

1/ See footnote 1, page 2 of this appendix.

$$(K)\quad \sum_{y=1}^{n} \left[\frac{a}{(1+g)^t}\right](1+c)^{yt} = a\left[\frac{(1+c)^{nt}-1}{(1+i)^t\left[(1+c)^t-1\right]}\right]$$

$i < g$

DISCOUNTED PERIODIC PAYMENT FACTOR
(payment starting in t years)

The future value form of equation K is derived in the manner of equation J, using the same series, repeated below:

$$\sum_{y=0}^{n-1} a(1+g)^{(n-y-1)t}(1+i)^{yt} = a(1+g)^{(n-1)t} + a(1+g)^{(n-2)t}(1+i)^t + a(1+g)^{(n-3)t}(1+i)^{2t} + \ldots + a(1+g)^{(n-y-1)t}(1+i)^{yt} + \ldots + a(1+g)^t(1+i)^{(n-2)t} + a(1+i)^{(n-1)t}$$

$$\sum_{y=0}^{n-1}\left[\frac{a(1+g)^{(n-1)t}(1+i)^{yt}}{(1+g)^{yt}}\right] = a\left[\frac{(1+c)^{nt}-1}{(1+i)^t\left[(1+c)^t-1\right]}\right](1+i)^{nt}$$

$$(L)\quad \sum_{y=0}^{n-1}\left[\frac{a(1+g)^{(n-1)t}}{(1+c)^{yt}}\right] = a\left[\frac{(1+i)^{(n-1)t}\left[(1+c)^{nt}-1\right]}{(1+c)^t-1}\right]$$

$i < g$

COMPOUNDED PERIODIC PAYMENT FACTOR
(payment starting in t years)

<u>Determination of present value factors and future value factors for periodic payments beginning "this year</u>

The usual case for a series representing the present value of <u>perfodic</u> payment beginning <u>this year</u> and continuing every (t) years for (n) <u>more</u> periods is:[1/]

$$\sum_{y=0}^{n} \frac{a}{(1+i)^{yt}} = a + \frac{a}{(1+i)^t} + \frac{a}{(1+i)^{2t}} + \ldots + \frac{a}{(1+i)^{yt}} + \ldots + \frac{a}{(1+i)^{nt}}$$

where, a = periodic payment beginning <u>this year</u> and continuing every t years for n <u>more</u> periods.

t = time interval between periods

n = number of periodic payments after first payment (there are n+1 payments and nt equals the number of years until the last payment)

i = rate of annual interest (discount)

y = any given period (yt equals any given year in which a payment occurs)

However, this again only represents the special case when (a) remains constant. When (a) increases at a rate (g) each <u>year</u>, the following sequence results:[2/]

$a,\ a(1+g)^t,\ a(1+g)^{2t},\ \ldots,\ a(1+g)^{yt},\ \ldots,\ a(1+g)^{nt}$

The present value of all (n+1) payments is represented by the following series:

$$a + a\left[\frac{1+g}{1+i}\right]^t + a\left[\frac{1+g}{1+i}\right]^{2t} + \ldots + a\left[\frac{1+g}{1+i}\right]^{yt} + \ldots + a\left[\frac{1+g}{1+i}\right]^{nt}$$

As before, three possible relationships exist between (i) and (g)

1. If $i > g$, then a composite <u>discount</u> rate (d) can be defined.
2. If $i = g$, then a direct summing of terms is possible, and the present value of the payments becomes $a(n+1)$, while the future value of the payments becomes $a(n+1)(1+i)^{nt}$.
3. If $i < g$, then a composite <u>compound</u> rate (c) can be defined.

1/ See Appendix 1, page 4.

2/ The rate (g) could also represent an annual decrease in (a). In that case (g) would be negative. If two or more growth rates affected (a), (g) would be a composite of all growth rates.

Again, the composite discount rate (d) equals: $\left[\frac{(1+i)}{(1+g)} - 1\right]$.

Substituting $\left[\frac{1}{(1+d)}\right]$ into the above series for $\left[\frac{1+g}{1+i}\right]$, the following series results:

$$a + \frac{a}{(1+d)^{t}} + \frac{a}{(1+d)^{2t}} + \ldots + \frac{a}{(1+d)^{yt}} + \ldots + \frac{a}{(1+d)^{nt}}$$

As before the composite discount rate (d) is in annual terms. To facilitate the simplification of the above series, it is useful to define a periodic composite discount rate, $p = (1+d)^{t} - 1$.

Then, substituting $(1+p) = (1+d)^{t}$, the following series results:

$$a + \frac{1}{(1+p)} + \frac{a}{(1+p)^{2}} + \ldots + \frac{a}{(1+p)^{y}} + \ldots + \frac{a}{(1+p)^{n}}$$

This series is in the same form as the series used to develop equation E, except that (p) has replaced (d). Therefore, the following equation (analagous to equation E) for the present value of a string of periodic payments can be derived:

$$\sum_{y=0}^{n} \frac{a}{(1+p)^{y}} = a\left[\frac{(1+p)^{n+1} - 1}{p(1+p)^{n}}\right]$$

Substituting, $p = (1+d)^{t} - 1$, and $(1+d)^{t} = (1+p)$

(M) $$\sum_{y=0}^{n} \frac{a}{(1+d)^{yt}} = a\left[\frac{(1+d)^{(n+1)t} - 1}{(1+d)^{nt}\left[(1+d)^{t} - 1\right]}\right]$$

$i > g$

DISCOUNTED PERIODIC PAYMENT FACTOR
(payment starting "this year")

The future value equation is derived from equation M as follows:

$$\sum_{y=0}^{n} a(1+g)^{(n-y)t}(1+i)^{yt} = a(1+g)^{nt} + a(1+g)^{(n-1)t}(1+i)^{t} + a(1+g)^{(n-2)t}(1+i)^{2t} + \ldots$$
$$+ a(1+g)^{(n-y)t}(1+i)^{yt} + \ldots + a(1+g)^{t}(1+i)^{(n-1)t} + a(1+i)^{nt}$$

$$\sum_{y=0}^{n} \left[\frac{a(1+g)^{nt}(1+i)^{yt}}{(1+g)^{yt}}\right] = a\left[\frac{(1+d)^{(n+1)t} - 1}{(1+d)^{nt}\left[(1+d)^{t} - 1\right]}\right](1+i)^{nt}$$

$$\sum_{y=0}^{n} a(1+g)^{nt}(1+d)^{yt} = a\left[\frac{(1+i)^{nt}\left[(1+d)^{(n+1)t} - 1\right]}{\left[\frac{(1+i)^{nt}}{(1+g)^{nt}}\right]\left[(1+d)^{t} - 1\right]}\right]$$

(N) $$\sum_{y=0}^{n} a(1+g)^{nt}(1+d)^{yt} = a\left[\frac{(1+g)^{nt}\left[(1+d)^{(n+1)t} - 1\right]}{(1+d)^{t} - 1}\right]$$

$i > g$

COMPOUNDED PERIODIC PAYMENT FACTOR
(payment starting "this year")

The composite compound rate (c) again equals $\left[\frac{(1+g)}{(1+i)} - 1\right]$.[1/]

1/ See footnote 1, p. 2 of this appendix.

Substituting (1+c) into the original series (p.8 of this appendix) for $\left[\frac{1+g}{1+i}\right]$, the following series results:

$$a + a(1+c)^{t} + a(1+c)^{2t} + \ldots + a(1+c)^{yt} + \ldots + a(1+c)^{nt}$$

As before, the composite compound rate (c) is in annual terms. To facilitate the simplification of the above series, it is useful to define a <u>periodic</u> composite compound rate, $p = (1+c)^{t} - 1$.

Then, substituting $(1+p) = (1+c)^{t}$, the following series results:

$$a + a(1+p) + a(1+p)^{2} + \ldots + a(1+p)^{y} + \ldots + a(1+p)^{n}$$

This series is in the same form as the series used to develop equation G, except that (p) has replaced (c). Therefore, the following equation (analagous to equation G) for the present value of a string of <u>periodic</u> payments beginning <u>this year</u> can be derived:

$$\sum_{y=0}^{n} a(1+p)^{y} = a\left[\frac{(1+p)^{n+1} - 1}{p}\right]$$

Substituting, $p = (1+c)^{t} - 1$, and $(1+c)^{t} = (1+p)$

(O)
$$\sum_{y=0}^{n} a(1+c)^{yt} = a\left[\frac{(1+c)^{(n+1)t} - 1}{(1+c)^{t} - 1}\right]$$

<u>$i < g$</u>
DISCOUNTED PERIODIC PAYMENT FACTOR
(payment starting "this year")

The future value form of equation O is derived in the manner of equation N, using the same series, repeated below:

$$\sum_{y=0}^{n} a(1+g)^{(n-y)t}(1+i)^{yt} = a(1+g)^{nt} + a(1+g)^{(n-1)t}(1+i)^{t} + a(1+g)^{(n-2)t}(1+i)^{2t} + \ldots + a(1+g)^{(n-y)t}(1+i)^{yt} \ldots + a(1+g)^{t}(1+i)^{(n-1)t} + a(1+i)^{nt}$$

$$\sum_{y=0}^{n} a(1+g)^{(n-y)t}(1+i)^{yt} = a(1+g)^{nt} + a(1+g)^{(n-1)t}(1+i)^{t} + a(1+g)^{(n-2)t}(1+i)^{2t} + \ldots + a(1+g)^{(n-y)t}(1+i)^{yt} + \ldots + a(1+g)^{t}(1+i)^{(n-1)t} + a(1+i)^{nt}$$

$$\sum_{y=0}^{n} \left[\frac{a(1+g)^{nt}(1+i)^{yt}}{(1+g)^{yt}}\right] = a\left[\frac{(1+c)^{(n+1)t} - 1}{(1+c)^{t} - 1}\right](1+i)^{nt}$$

(P)
$$\sum_{y=0}^{n} \left[\frac{a(1+g)^{nt}}{(1+c)^{yt}}\right] = a\left[\frac{(1+i)^{nt}\left[(1+c)^{(n+1)t} - 1\right]}{(1+c)^{t} - 1}\right]$$

<u>$i < g$</u>
COMPOUNDED PERIODIC PAYMENT FACTOR
(payment starting "this year")

<u>Periodic capital recovery factors and sinking fund factors</u>

Theoretically, periodic capital recovery factors and periodic sinking fund factors are possible (see Appendix 1, page 5). However, neither factor generally considers growth rates (see page 5 of this appendix). If some application of periodic capital recovery factors or sinking fund factors (which include an annual growth rate) exists, the present value and future value formulas for periodic payments (which include on annual growth rate) can be inverted and used in the manner described in Appendix 1.

<u>Determination of "infinite" factors</u>

When annual or periodic payments (which include an annual growth rate - g) continue for an infinite period (or for a sufficiently long finite period such that the "next" payment has no measureable present value) some of the equations for present values of such payments become much simplified. Only cases where $i > g$ can be simplified, since when $i < g$ the present values of successive payments do not decrease. Four equations described in this appendix can be simplified, equations A, E, I, and M.

(A) $$\sum_{y=1}^{n} \frac{\left[\frac{a}{1+g}\right]}{(1+d)^y} = a\left[\frac{(1+d)^n - 1}{d(1+d)^n\,(1+g)}\right] = a\left[\frac{1}{d(1+g)} - \frac{1}{d(1+d)^n\,(1+g)}\right]$$

$i > g$: FINITE
DISCOUNTED ANNUAL PAYMENT FACTOR
(payment begins "next year")

when $n \to \infty$, the second term $\left[\frac{1}{d(1+d)^n\,(1+g)}\right]$ approaches zero, leaving $\left[\frac{1}{d(1+g)}\right]$

Therefore:

(Q) $$\sum_{y=1}^{n=\infty} \frac{\left[\frac{a}{1+g}\right]}{(1+d)^y} = a\left[\frac{1}{d(1+g)}\right]$$

$i > g$: INFINITE
DISCOUNTED ANNUAL PAYMENT FACTOR
(payment begins "next year")

(E) $$\sum_{y=0}^{n} \frac{a}{(1+d)^y} = a\left[\frac{(1+d)^{n+1} - 1}{d\,(1+d)^n}\right] = a\left[\frac{1+d}{d} - \frac{1}{d(1+d)^n}\right]$$

$i > g$: FINITE
DISCOUNTED ANNUAL PAYMENT FACTOR
(payment begins "this year")

When $n \to \infty$, the second term $\left[\frac{1}{d(1+d)^n}\right]$ approaches zero, leaving $\left[\frac{1+d}{d}\right]$

Therefore:

(R) $$\sum_{y=0}^{n=\infty} \frac{a}{(1+d)^y} = a\left[\frac{1+d}{d}\right]$$

$i > g$: INFINITE
DISCOUNTED ANNUAL PAYMENT FACTOR
(payment begins "this year")

(I) $$\sum_{y=1}^{n} \frac{\left[\frac{a}{(1+g)^t}\right]}{(1+d)^{yt}} = a\left[\frac{(1+d)^{nt} - 1}{(1+g)^t(1+d)^{nt}\,\left[(1+d)^t - 1\right]}\right] = a\left[\frac{1}{(1+g)^t\,\left[(1+d)^t - 1\right]} - \frac{1}{(1+g)^t\,(1+d)^{nt}\,\left[(1+d)^t - 1\right]}\right]$$

$i > g$: FINITE
DISCOUNTED PERIODIC PAYMENT FACTOR
(payment starting in (t) years)

When $n \to \infty$, the second term $\left[\frac{1}{(1+g)^t\,(1+d)^{nt}\,\left[(1+d)^t - 1\right]}\right]$ approaches zero, leaving $\frac{1}{(1+g)^t\,\left[(1+d)^t - 1\right]}$

Therefore:

(S) $$\sum_{y=1}^{n=\infty} \frac{\left[\frac{a}{(1+g)^t}\right]}{(1+d)^{yt}} = a\left[\frac{1}{(1+g)^t\,\left[(1+d)^t - 1\right]}\right]$$

$i > g$: INFINITE
DISCOUNTED PERODIC PAYMENT FACTOR
(payment starting in t years)

(M) $$\sum_{y=0}^{n} \frac{a}{(1+d)^{yt}} = a\left[\frac{(1+d)^{(n+1)t} - 1}{(1+d)^{nt}\,\left[(1+d)^t - 1\right]}\right] = a\left[\frac{(1+d)^t}{(1+d)^t - 1} - \frac{1}{(1+d)^{nt}\,\left[(1+d)^t - 1\right]}\right]$$

$i > g$: FINITE
DISCOUNTED PERIODIC PAYMENT FACTOR
(payment starting "this year")

When $n \to \infty$, the second term $\left[\frac{1}{(1+d)^{nt}\,\left[(1+d)^t - 1\right]}\right]$ approaches zero, leaving $\left[\frac{(1+d)^t}{(1+d)^t - 1}\right]$

Therefore:

(T) $$\sum_{y=0}^{n=} \frac{a}{(1+d)^{yt}} = a\left[\frac{(1+d)^t}{(1+d)^t - 1}\right]$$

$i > g$: INFINITE
DISCOUNTED PERIODIC PAYMENT FACTOR
(payment starting "this year")

Appendix 3

Application of Electronic Calculators to Solutions of Discounting Problems

Several types of electronic calculators can automatically calculate the present value of a series of annual or periodic payments. Some electronic calculators also have the option of selecting either the end of year (column 1 of Table 2) factors or the beginning of year (column 3 of Table 3) factors.1/ For those calculators which only specify the end of year factors, one can calculate the beginning of year factors by adding in the value of the first payment. 2/ For example, the case illustrated in Table 4 applied the factor for the present value of an annual payment beginning immediately. The discounted value of net benefits for years 7-19 (W43.9 annually) was determined using a calculator to calculate the discounted value of a sequence of annual payments of W43.9 which began in year 8 and continued through year 19. The discounted value of these 12 annual payments (at a 10 percent discount rate) was automatically calculated to be W299.1. The value of net benefits for year seven was added (W299.1) + (W43.9) to give a total discounted value (in year 7) of W343.0.

To derive discounted and compounded values of periodic payments using electronic calculators, one first needs to calculate a periodic discount rate (p) equal to: $((1+i)^t-1)$, where (i) equals the annual discount rate and (t) equals the period between payments. Then the periodic payments can be treated as annual payments on the calculator, using (p) instead of (i). To calculate beginning of period and end of period payments, the same steps need to be followed as were described above for annual payments.

The use of a periodic discount rate can be illustrated by using the periodic net benefit (of W90.2) from Table 7. The periodic discount rate equals 21 percent $((1.10)^2-1)$, and 6 periodic payments occur after year 6. The discounted value of the 6 payments (using the 21 percent discount rate) equals W292.7. Adding the value (in year 6) of the first payment (W292.7 + W90.2), the total discounted value equals W382.9, as shown in Table 7.

A few of the annual payment formulas in Table 10 closely resemble their "non-growth" counterparts in Table 2. By substituting (d) or (c) for (i), and in some cases by further multiplying by an additional factor, electronic calculators with financial functions can be used to quickly calculate discounted or compounded values. However, for most of the formulas in Table 10, the adjustments necessary to "fit" the formulas into the electronic calculator's format are more time consuming than directly calculating the formula through multiplication of its componets.

1/ Table numbers refer to tables in the text.

2/ The "end of a given year" is equal to the "beginning of the following year" in this terminology, i.e., on a calendar year basis, 31 Dec., 11:59 p.m., 1976 equals 1 Jan., 00:01 a.m., 1977.

Appendix 4

Calculation of Net Present Worth (NPW) in Table 3

The calculation of the net present worth (NPW) for the Korea cash flow (Table 1) is demonstrated in Table 3.[1] Eight single years or annual series of years which have different net benefits (costs) are identified in the eight rows of Table 3. Column 1 of Table 3 lists each year or annual series of years, and column 2 shows the net benefits and costs from Table 1. These two columns contain all of the information found in row 8 of Table 1. Columns 3 and 4 of Table 3 identify the initial years of discount and the discounted values, respectively, for the two series of annual payments (rows 2 and 7). Since Table 3 illustrates the use of the "end of year" annual payment formula, each initial year of discount equals the year prior to the start of the annual series. The discounted values in column 4 are determined by using the "end of year" annual payment formula (from row 1, column 1 of Table 2) and a discount rate of 10 percent. Column 5 of Table 3 indicates the present value (year 0) of each single payment and series of annual payments. The present values of the 6 single payments in Table 3 (rows 1, 3, 4, 5, 6 and 8) equal the values in column 2 discounted at 10 percent for the number of years indicated in column 1. The present values of the 2 series of annual payments (rows 2 and 7) equal the values in column 4 discounted at 10 percent for the number of years indicated in column 3. The final step is to add all of the 8 present values in column 5 to get the net present worth (NPW), equal to W170.0.

---

[1] Tables referred to in this appendix are text tables.